AF366947

Tsunamis: 1992-1994

Their Generation, Dynamics, and Hazard

Edited by
Kenji Satake
Fumihiko Imamura

1995

Birkhäuser Verlag
Basel · Boston · Berlin

Reprint from Pure and Applied Geophysics
(PAGEOPH), Volume 144 (1995), No. 3/4

The Editor:

Dr. Fumihiko Imamura, Associate Professor
Disaster Control Research Center
Tohoku University
Aoba, Sendai 980-88, Japan
e-mail: imamura@tsunami2.civil.tohoku.ac.jp

Dr. Kenji Satake
Department of Geological Sciences
University of Michigan
Ann Arbor, MI 48109-1063
USA
e-mail: kenji@umich.edu

A CIP catalogue record for this book is available from the Library of Congress,
Washington D.C., USA

Deutsche Bibliothek Cataloging-in-Publication Data

Tsunamis: 1992–1994; their generation, dynamics, and hazard /
ed. by Kenji Satake; Fumihiko Imamura. – Basel ; Boston ;
Berlin : Birkhäuser, 1995
   ISBN-13: 978-3-7643-5102-1   e-ISBN-13: 978-3-0348-7279-9
   DOI: 10.1007/978-3-0348-7279-9
NE: Satake, Kenji [Hrsg.]

© 1995 Birkhäuser Verlag, P.O. Box 133, CH-4010 Basel, Switzerland
Printed on acid-free paper produced from chlorine-free pulp ∞

ISBN-13: 978-3-7643-5102-1

9 8 7 6 5 4 3 2 1

Contents

PAGEOPH, Vol. 145, Nos. 3/4 (1995)

Introduction to "Tsunamis: 1992–94"

KENJI SATAKE[1] and FUMIHIKO IMAMURA[2]

Background

Tsunamis have caused extensive human and property damage in the world. There had not been a tsunami disaster since 1983, when the Japan Sea earthquake tsunami caused 100 fatalities. In the three-year period (1992–94), however, several destructive tsunamis occurred in the world and caused more than 1,500 casualties. After each of these tsunami disasters, field survey teams consisting of scientists and engineers from various fields and countries were dispatched. They measured tsunami runup heights and estimated the velocity, direction and other local behavior. In addition to such hydrodynamic features, they searched for evidence of coseismic ground deformation and tsunami depositions and erosions. The first goal of this topical issue is to document data from these tsunamis for later use in various tsunami-related research.

The second goal is to present research results on various aspects of the recent tsunamis. During the "quiet" period of tsunami activity between 1983 and 1992, significant developments in tsunami-related research have occurred. In seismology, semi-real-time access and analysis of seismological data have become possible and tsunami data have been used to study earthquake sources. Geologists have found tsunami deposits from various geological events such as earthquakes, volcanic eruptions, submarine landslides and even a meteorite impact. On the observational side, deep ocean pressure gauges have been installed to record tsunami signals in the open ocean. Numerical modeling techniques of tsunami generation, propagation and runup processes have significantly improved, parallel to the development of computers. Three-dimensional runup processes have been modeled by theoretical, numerical, and experimental methods.

It is noteworthy that cooperation among various fields has advanced the research. For example, in order to model tsunami inundation on a particular coast, we have found that an offshore boundary condition of sinusoidal wave incidence is

---

[1] Department of Geological Sciences, University of Michigan, U.S.A. Now at: Seismotectonics Research Section, Geological Survey of Japan, Tsukuba, 305 Japan.

[2] School of Civil Engineering, Asian Institute of Technology, Bangkok, Thailand. Now at: Disaster Control Research Center, Tohoku University, Sendai, 980-77 Japan.

Table 1

Tsunamigenic Earthquakes in 1992–94

Year	mo	dy	hr	mn	sec	Latitude	Longitude	dep (km)	M_s	max. tsunami	Casualties	Region
1992	4	25	18	6	4.2	40.368°N	124.316°W	15	7.1	1 m	0	Cape Mendocino
1992	5	17	10	15	31.3	7.191°N	126.762°E	33	7.5	small	0	Mindanao
1992	5	27	5	13	38.8	11.122°S	165.239°E	19	7.0	small	0	Santa Cruz Is.
1992	7	18	8	36	58.7	39.419°N	143.330°E	29	6.9	0.5 m	0	Sanriku
1992	9	2	0	16	1.6	11.742°N	87.340°W	45	7.2	10 m	170	Nicaragua
1992	12	12	5	29	26.3	8.480°S	121.896°E	28	7.5	26 m	1713	Flores Is.
1993	2	7	13	27	42.0	37.634°N	137.245°E	11	6.2	0.2 m	0	Noto-oki
1993	6	8	13	3	36.4	51.218°N	157.829°E	71	7.3	0.1 m	0	Kamchatka
1993	7	12	13	17	11.9	42.851°N	139.197°E	17	7.6	30 m	239	Hokkaido
1993	8	8	8	34	24.9	12.982°N	144.801°E	59	8.0	1 m	0	Guam
1993	11	13	1	18	4.1	51.934°N	158.647°E	34	7.0	small	0	Kamchatka
1994	1	21	2	24	29.9	1.015°N	127.733°E	20	7.2	damaging	7	Halmahera
1994	4	8	1	10	40.8	40.608°N	143.683°E	13	6.3	0.2 m	0	Sanriku
1994	6	2	18	17	34.0	10.477°S	112.835°E	18	7.2	14 m	238	Java
1994	9	1	15	15	53.0	40.402°N	125.680°W	10	7.0	0.1 m	0	Cape Mendocino
1994	10	4	13	22	58.3	43.706°N	147.328°E	33	8.1	10 m	10	Kuril (Shikotan)
1994	10	8	21	44	09.1	1.222°S	127.992°E	31	6.8	damaging	1	Halmahera
1994	10	9	7	55	38.0	43.899°N	147.905°E	23	7.0	0.2 m	0	Kuril aftershock
1994	11	14	19	15	30.7	13.532°N	121.087°E	33	7.1	7 m	71	Mindoro
1994	12	28	12	19	23.6	40.451°N	143.491°E	33	7.5	1 m	0	Sanriku

According to Preliminary Determination of Epicenters (U.S.G.S). The casualties include those not due to tsunamis.

not appropriate, and tsunami generation and propagation from a geological fault model estimated from seismological analysis is necessary. The multidisciplinary nature of the tsunami phenomena has required geologists, seismologists, oceanographers and coastal engineers to interact.

Recent developments in computer networking have also promoted the interaction of researchers across disciplines. Immediately after the 1992 Nicaragua tsunami, the Tsunami Bulletin Board (originally called the Nicaragua Tsunami Bulletin Board) was established by the Pacific Marine Environmental Laboratory of N.O.A.A., U.S.A. This system (tsunami@pmel.noaa.gov) currently has about 130 subscribers worldwide and has been used to exchange information and data on tsunamis. It has been most useful after major tsunami events. Preliminary results of seismological analyses and numerical modeling were posted within a few days; field survey plans have been posted and discussed to form international survey teams; and the survey results were also posted.

Tsunamigenic Earthquakes in 1992–94

Figure 1 shows the epicenters of large ($M_s \geq 6$) earthquakes that occurred in the three-year period (1992–94), taken from the Preliminary Determination of Epicenters (PDE) catalog published by the U.S.G.S. Most of approximately 200 events in the world occurred around the rim of the Pacific Ocean, and 20 of these events generated tsunamis. The tsunamigenic earthquakes are shown as shaded circles in Figure 1 and listed in Table 1. About half the tsunamis were smaller than 1 m in maximum height and caused no damage. The maximum tsunamis heights for 6 events exceeded 5 m, and they resulted in substantial human loss and property damage. Chronologically, the 6 largest events were: Nicaragua (Sep. '92), Flores (Dec. '92), Hokkaido (July, '93), Java (June '94), Kuril (Oct. '94) and Mindoro (Nov. '94).

The 1992 Nicaraguan earthquake generated substantially larger tsunamis than expected on the basis of the surface wave magnitude. Several similar earthquakes have occurred in the world and have been called "tsunami earthquakes," but the Nicaragua event was probably the first event documented by extensive seismological and tsunami data. This was also the first tsunami event for which international survey teams were dispatched. The survey results have been reported in ABE *et al.* (1993), BAPTISTA *et al.* (1993) and SATAKE *et al.* (1993).

Fatalities from the Flores Island earthquake were 1,700; more than half due to tsunamis. Several interesting phenomena, such as very large runup in localized regions (up to 26 m) and the total devastation on a small circular-shaped island (Babi Island), have been reported (YEH *et al.*, 1993; and TSUJI *et al.*, this topical volume).

1992-94 Earthquakes (Ms>6.0) and Tsunamis

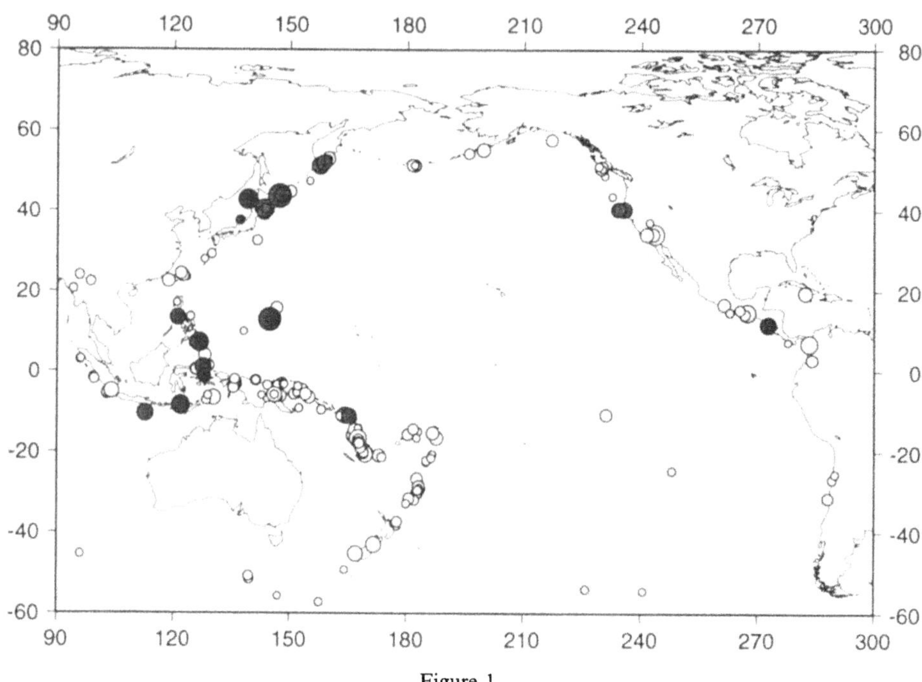

Figure 1

Epicenters of large shallow earthquakes (surface wave magnitude M_s equal or larger than 6.0) for the three-year period (1992–94) according to National Earthquake Information Center, the U.S.G.S. The size of the symbols is proportional to M_s. Tsunamigenic earthquakes are shown by shaded circles and listed in Table 1.

The Hokkaido earthquake of July 1993 produced a maximum runup height of 30 m on Okushiri Island, making it the largest tsunami in the three-year period. The fatalities which ensued from this earthquake and tsunami numbered about 230. This tsunami has been more extensively studied than any previous events. An initial report was made by the Hokkaido Tsunami Survey Group (1993). Eight follow-up papers appear in this volume.

In 1994, three hazardous tsunami events took place. The Java earthquake in June was another example of a "tsunami earthquake"; there was no damage due to ground shaking, while the maximum tsunami height was 14 m and over 200 people died. The largest earthquake in terms of surface wave magnitude (M_s 8.1) during the three-year period was the Kuril earthquake in October, 1994, although the tsunami caused little damage. The Mindoro, Philippines, earthquake in November was a strike-slip faulting, which is usually not effective in generating tsunamis. However, the runup height was as large as 7 m and about 70 people died.

Brief Summary of each Paper

This topical issue contains 26 papers treating various aspects of recent tsunamis (Table 2). They are arranged by earthquake in chronological order. SCHINDELE *et al.* report their seismological analysis of eight tsunamigenic earthquakes based on single station data at Papeete. They show that a "tsunami earthquake" can be detected in real time through their analysis. They further simulated the single station method for the eight earthquakes, and demonstrated the effectiveness of this method for real-time tsunami warning.

GONZÁLEZ *et al.* analyzed tsunami waveforms from the 1992 Cape Mendocino, California, earthquake. They demonstrated, through numerical and analytical modeling, that the maximum amplitude recorded about 3 hours after the initial onset of the tsunami was due to edge waves. THOMSON *et al.* report a rather unusual observation of the tsunami; they detected signals on the ocean cable connecting California and Hawaii and interpret them as tsunami-related signals induced by seawater movement.

Table 2

List of papers in this issue

Author(s)	Event(s)	Topic(s)
SCHINDELE *et al.*	8 tsunamis in '92–'94	Seismological Analysis
GONZÁLEZ *et al.*	'92/4 Cape Mendocino	Observation/Modeling
THOMSON *et al.*	'92/4 Cape Mendocino	Observation
KIKUCHI and KANAMORI	'92/9 Nicaragua	Seismological Analysis
SATAKE	'92/9 Nicaragua	Numerical Modeling
HATORI	central American tsunamis	Tsunami Magnitude Scale
TSUJI *et al.*	'92/12 Flores	Field Survey
SHI *et al.*	'92/12 Flores	Tsunami Deposits
HIDAYAT *et al.*	'92/12 Flores	Seismology/Modeling
IMAMURA *et al.*	'92/12 Flores	Numerical Modeling
BRIGGS *et al.*	circular island	Physical Modeling
TINTI and VANNINI	circular island	Theoretical Modeling
ABE and OKADA	'93/2 Noto-oki	Numerical Modeling
JOHNSON *et al.*	'93/6 & 11 Kamchatka	Seismological Analysis
SHUTO and MATSUTOMI	'93/7 Hokkaido	Field Survey
SHIMAMOTO *et al.*	'93/7 Hokkaido	Field Survey
SATO *et al.*	'93/7 Hokkaido	Tsunami Deposits
NISHIMURA and MIYAJI	'93/7 Hokkaido	Tsunami Deposits
ABE	'93/7 Hokkaido	Tsunami Magnitude Scale
TAKAHASHI *et al.*	'93/7 Hokkaido	Numerical Modeling
MYERS and BAPTISTA	'93/7 Hokkaido	Numerical Modeling
SATAKE and TANIOKA	'93/7 Hokkaido	Numerical Modeling
TANIOKA *et al.*	'93/8 Gaum	Seismological Analysis
TSUJI *et al.*	'94/6 Java	Field Survey
YEH *et al.*	'94/10 Shikotan	Field Survey
IMAMURA *et al.*	'94/11 Mindoro	Field Survey

The next three papers report the Nicaragua tsunami, as well as other tsunamis in central America. KIKUCHI and KANAMORI carried out detailed seismological analyses and showed that this tsunami earthquake was characterized by a slow and smooth rupture. SATAKE conducted numerical modeling of this tsunami and compared the results with the observations. He also examined the effects of various parameters on the computed tsunami waveforms and heights. HATORI examined the regional characteristics of the tsunami magnitude scale for the Nicaragua and other central American tsunamis.

The Flores earthquake generated the most fatalities in the three-year period. TSUJI et al. presented very detailed results of their field survey. SHI et al. analyzed the sand deposits from this tsunami. A detailed analysis of modern tsunami deposits is obviously very important and critical for paleotsunami studies. HIDAYAT et al. made seismological analyses and numerical computations of tsunamis to construct a source model of this earthquake. They illustrate that a fault model cannot explain the largest tsunami heights; an additional local source, presumably local slumping, is needed. IMAMURA et al. obtained similar results from their numerical modeling and also demonstrated that the extensive damage on Babi Island, a circular island, was due to reflected waves. More general research on tsunami behavior around a circular island is provided by BRIGGS et al., who measured tsunami runup heights around a circular island in a physical model and TINTI and VANNINI, who studied trapping of tsunamis around circular islands through theoretical computations.

Two smaller tsunamigenic earthquakes occurred in 1993, and small amplitude tsunamis were recorded on tide gauges. ABE and OKADA modeled the source of the 1993 Noto-Hanto-oki earthquake, the smallest earthquake included in this issue (M_s 6.3), through comparisons of computed waveforms with those recorded on tide gauges. JOHNSON et al. compared two earthquakes that occurred in June and November, 1993 off Kamchatka. Although the surface wave magnitudes are similar, they have different characteristics and only the June event generated observable tsunamis.

The tsunami from the 1993 Hokkaido earthquake has been the most extensively studied of all the recent tsunami events. SHUTO and MATSUTOMI report their field surveys of tsunami heights along Okushiri, Hokkaido and Tohoku coasts. SHIMAMOTO et al. provided a very detailed survey on Okushiri Island of tsunami behavior such as flow directions, velocity, and the wave heights above ground level. SATO et al. examined tsunami deposits from this and a previous earthquake, and discuss the difficulties of paleotsunami studies. NISHIMURA and MIYAJI also studied tsunami deposits from the Hokkaido event and compared features with those from 1640 tsunami generated by a volcanic eruption.

ABE used his tsunami magnitude scale, which was originally developed to estimate the earthquake size from tsunami heights, to predict tsunami runup heights for 1993 earthquake; he demonstrated that a quick prediction of tsunami heights is possible. Three papers present numerical modeling results of the 1993 tsunami.

TAKAHASHI *et al.* computed tsunamis from 24 different source models and compared the calculated heights with observed runup heights and the tide gauge records. MYERS and BAPTISTA made similar computations of tsunami generation and propagation and compared their results with the observations. They employed the finite-element method, while other modelers used finite-difference methods. SATAKE and TANIOKA combined tsunami waveforms recorded on tide gauges with seismic analysis and geodetic data to estimate the slip distribution on the fault. They also discussed the free oscillation of the Japan Sea excited by the earthquake.

TANIOKA *et al.* analyzed seismic and tsunami data from the 1993 Guam earthquake, the largest earthquake ever recorded in the Mariana trench. They point out that the tide gauge record would have been beneficial to study this unique seismic event if the sampling rate were smaller.

After the above papers were submitted to this topical issue, three large tsunami events occurred and survey teams were dispatched. We include field survey reports for these "late-breaking" tsunamis. TSUJI *et al.* report on the field survey of the June 1994 Java earthquake. YEH *et al.* report on their survey on the Kuril Island following the October 1994 earthquake. Finally, IMAMURA *et al.* report on their field survey conducted after the Mindoro Island, Philippines, earthquake in November 1994.

REFERENCES

ABE, KU., ABE, KA., TSUJI, Y., IMAMURA, F., KATAO, H., IIO, Y., SATAKE, K., BOURGEOIS, J., NOGUERA, E., and ESTRADA, F. (1993), *Field Survey of the Nicaragua Earthquake and Tsunami of 2 September 1992*, Bull. Earthq. Res. Inst., Univ. Tokyo *68*, 23–70 (in Japanese).

BAPTISTA, A. M., PRIEST, G. R., and MURTY, T. S., (1993), *Field Survey of the 1992 Nicaragua Tsunami*, Marine Geodesy *16*, 169–203.

HOKKAIDO TSUNAMI SURVEY GROUP (1993), *Tsunami Devastates Japanese Coastal Region*, EOS Trans. AGU *74*, 417, 432.

SATAKE, K., BOURGEOIS, J., ABE, KU., ABE, KA., TSUJI, Y., IMAMURA, F., IIO, Y., KATAO, H., NOGEURA, E., and ESTRADA, F. (1993), *Tsunami Field Survey of the 1992 Nicaragua Earthquake*, EOS Trans., Am. Geophys. Union *74*, 145, 156–157.

YEH, H., IMAMURA, F., SYNOLAKIS, C., TSUJI, Y., LIU, P., and SHI, S. (1993), *The Flores Island Tsunamis*, EOS Trans., Am. Geophys. Union *74*, 369, 371–373.

PAGEOPH, Vol. 144, Nos. 3/4 (1995)

Analysis and Automatic Processing in Near-field of Eight 1992–1994 Tsunamigenic Earthquakes: Improvements Towards Real-time Tsunami Warning

F. SCHINDELÉ,[1] D. REYMOND,[1] E. GAUCHER[1] and E. A. OKAL[2]

Abstract —We study eight tsunamigenic earthquakes of 1992–1994 with data from single near-field 3-component long-period stations. The analysis is made from the standpoint of tsunami warning by an automatic process which estimates the epicentral location and the seismic moment through the variable-period mantle magnitude M_m. Simulations of early warning based on the real-time computation of the seismic moment are also tested with this system, which would give a justified warning in each region of tsunami potentiality. By exploiting the dependence of moment rate release with frequency, the system has the capability of recognizing both "tsunami earthquakes" such as the 1992 Nicaragua and 1994 Java events, as well as instances of the opposite case of low-frequency deficiency, interpreted as indicating a deeper than normal source (1993 Guam event). We report both the results of delayed-time processing of the near-field stations, and the actual real-time warnings at PPT, which confirm the former.

Key words: Tsunami, tsunami earthquakes, seismic moment, mantle magnitude.

Introduction

During 1992–1994, eight earthquakes generated measurable tsunamis in the Pacific and Indian Oceans and their adjoining seas (Table 1 and Figure 1a). Six of them, destructive and deadly, claimed a total of more than 2500 lives: Nicaragua (Sept. 2, 1992); Flores Sea, Indonesia (Dec. 12, 1992); Hokkaido (July 12, 1993); Java, Indonesia (June 2, 1994); Kuriles (Oct. 4, 1994); and Mindoro, Philippines (Nov. 14, 1994). The other two (Guam, Aug. 8, 1993; and Halmahera, Indonesia, Jan. 21, 1994) caused no casualties or damage directly attributable to the tsunami. This sudden recrudescence in tsunami activity, although most probably fortuitous, provides a rare opportunity to apply tsunami warning algorithms to a variety of earthquakes featuring significantly distinct tsunamigenic characteristics.

The seismic moment of these events was estimated in real time at the French Polynesia Tsunami Warning Center (Centre Polynésien de Prévention des

---

[1] Laboratoire de Détection et Géophysique et Centre Polynésien de Prévention des Tsunamis, Commissariat à l'Energie Atomique, Boîte Postale 640, Papeete, Tahiti, French Polynesia.
[2] Department of Geological Sciences, Northwestern University, Evanston, Illinois 60208, U.S.A.

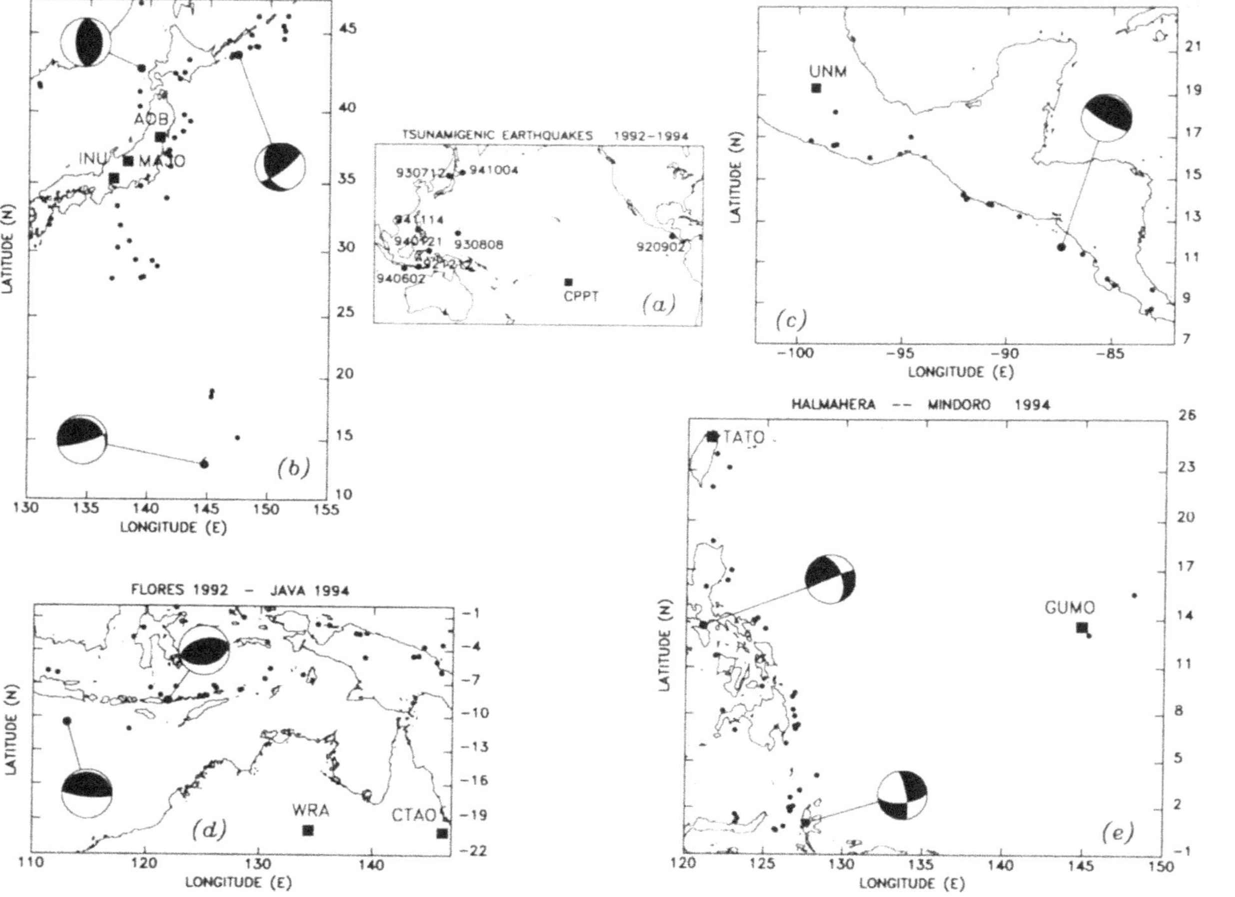

Figure 1

Maps of the eight tsunamigenic earthquakes studied in this paper (a): General map of the Pacific Ocean, showing the locations of the events (with date given as yymmdd), and of CPPT in Tahiti. (b–e): Close-up maps for each of the earthquakes under study. For each event, the CMT focal mechanisms are plotted with a solid line drawn to the exact epicenter. The station used in the near-field study is shown by a solid square. The background seismicity, extracted from the CMT catalogue (with $M_0 \geq 10^{19}$ N-m, and for the period starting in 1977 and ending before the earthquake studied) is shown as small solid circles. See text for details.

Table 1

Source parameters of the earthquakes studied

| | | Epicenter | | Depth | | | Moment |
Date	Region	°N	°E	(HRV; km)	m_b	M_s	10^{20} N-m
02 SEP 1992	Nicaragua	11.74	−87.34	10.0	5.3	7.2	3.4
12 DEC 1992	Flores Sea	−8.51	121.89	20.4	6.5	7.5	5.1
12 JUL 1993	Hokkaido	−42.84	139.25	16.5	6.7	7.6	4.7
08 AUG 1993	Guam	12.96	144.78	59.3	7.2	8.1	5.2
21 JAN 1994	Halmahera	1.01	127.73	15.0	6.2	7.2	0.32
02 JUN 1994	Java	−10.47	112.98	15.0	5.5	7.2	3.5
04 OCT 1994	Kuriles	43.71	147.33	59.0	7.4	8.1	37.
14 NOV 1994	Mindoro	13.57	121.08	15.0	6.0	7.1	0.59

Tsunamis, hereafter CPPT), and the estimates forwarded immediately to the National Earthquake Information Center and the Pacific Tsunami Warning Center. Ever since 1964, CPPT has been charged with the responsibility of issuing tsunami warnings (TALANDIER, 1993); the current procedure uses the automated algorithm TREMORS (Tsunami Risk Evaluation through seismic **MO**ment in a **R**eal time System). This system, implemented at CPPT in 1987 (REYMOND *et al.*, 1991; HYVERNAUD *et al.*, 1992), is based upon the real-time estimation of the seismic moment through the variable-period mantle magnitude M_m (OKAL and TALANDIER, 1989), following the automatic detection and location of seismic events using a single three-component long-period seismic station.

The goals of the present paper are several: we report the actual results from CPPT, as they were obtained in real time following each of the eight earthquakes studied. We then use near-field records acquired after the fact, and processed through the TREMORS algorithm, in order to simulate and assess the performance of TREMORS in the near field, with respect to reliable and early warnings. We also test the accuracy of the automatic phase detection and epicenter location algorithm in the near field. We illustrate the significant advantage of a regional center equipped with this automated system, in the vicinity of each region with substantial tsunami risk. Finally, we confirm that two of the four largest tsunamis were due to so-called "tsunami earthquakes", i.e., events which generated disproportionate tsunamis in relation to their standard magnitudes; these earthquakes are characterized by a slow seismic moment release, evidenced by a steady increase in the spectrum of their seismic moment rate release with period.

Background

TREMORS is an integrated system based on a single three-component long-period seismic station which delivers a rapid early warning, a first location with

12 JULY 1993 (HOKKAIDO)   Station: INU

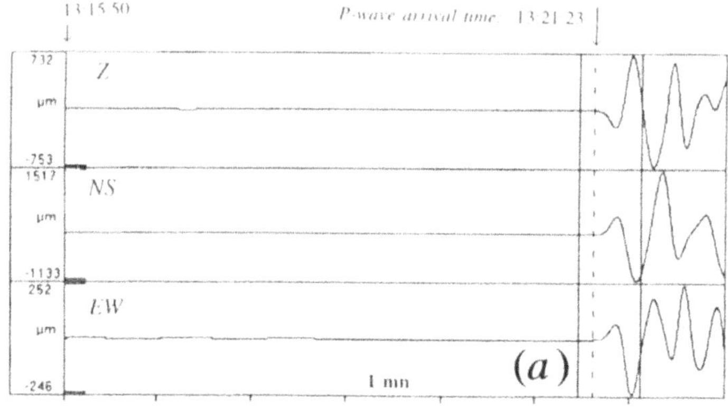

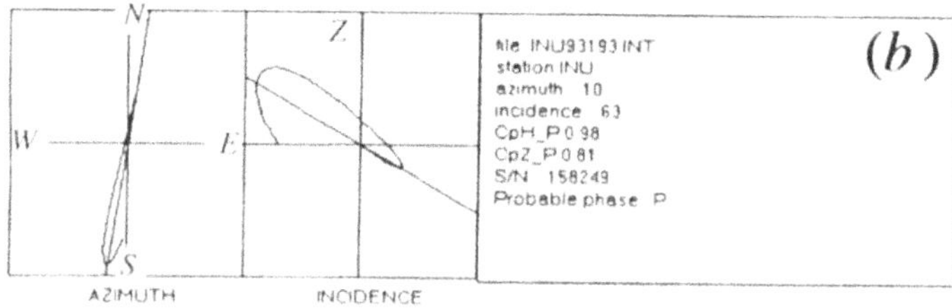

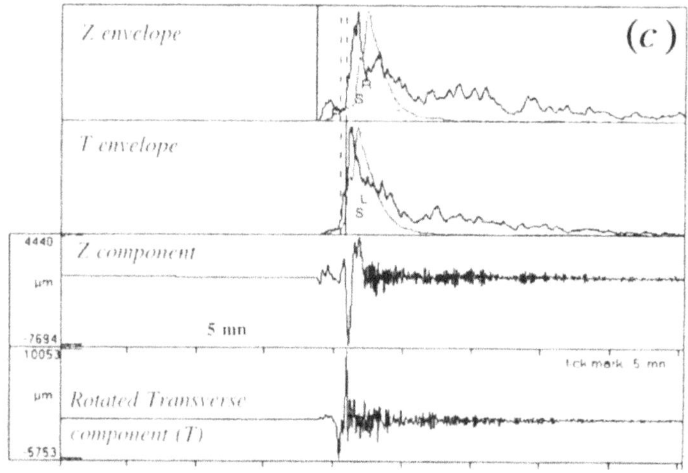

Back-azimuth: 10°
Distance from S – P: 7.7°   **Estimated Epicenter: 42.9°N; 138.9°E**

enough accuracy for immediate information and estimates of the seismic moment and tsunami risk. We give here a quick overview of the principal elements of the TREMORS algorithm, and refer the reader to REYMOND *et al.* (1991), for a more detailed description.

Location

The epicentral location is estimated from only two parameters: the $S - P$ time (or $R - P$ at short distances where the S and Rayleigh phases are often not well separated), and the back-azimuth best-fitting the polarization of P waves. This procedure is illustrated on Figure 2. These two parameters depend significantly on the signal-to-noise ratio, principally on the horizontal components, which can be noisy at long periods. Additionally, lateral heterogeneity in the earth on a global or regional scale can sway the ray outside of the expected great-circle plane. Our experience with teleseismic events recorded in Polynesia suggests an accuracy of 5° or better in azimuth when the horizontal signal-to-noise ratio (measured on energy) is 300 or more, and of 3° or better in epicentral distance when the signal-to-noise ratio on the horizontal instrument is at least 1000, these numbers relating to earthquakes at epicentral distances less than 107° (HYVERNAUD *et al.*, 1992).

Note that a *prelocation* can, in principle, be achieved immediately upon arrival of the P wave, by interpreting the angle of incidence of the P wave in terms of epicentral distance. However this procedure suffers from its sensitivity to crustal layering under the station, which in turn makes the result strongly frequency-dependent (see for example HASKELL (1962) for a discussion).

Estimation of the Seismic Moment

The seismic moment M_0 is estimated in realtime through the computation of the mantle magnitude M_m which was designed by OKAL and TALANDIER (1989) to obey the simple relation:

Figure 2

Automatic location procedure used by TREMORS, illustrated in the near field on the case of the Hokkaido earthquake recorded at Inuyama. (a) Close-up of the three-component seismogram showing the detection of the P wave. A 40-s fraction of time series (between the two vertical lines) is used in (b) to recover the back azimuth at the station (left) and the incidence angle of the arriving P wave (center). The latter could be used to compute a prelocation of the event, but note that the incidence angle is much less well constrained than the back azimuth. In (c), the horizontal records are then rotated using the back-azimuth information from (b), and the energy envelope of the transverse component used to detect the S_{ll} arrival, which in turn yields the epicentral distance.

$$\log_{10} M_0 = M_m + 13.0, \tag{1}$$

where M_0 is in N-m.

We refer the reader to OKAL and TALANDIER (1989, 1990) and OKAL (1990) for the details of the computation of M_m, which is defined in the frequency domain as:

$$M_m = \log_{10} X(\omega) + C_D(\omega, \Delta) + C_S(\omega) - 0.90, \tag{2}$$

where $X(\omega)$ is the spectral amplitude of ground motion in μm-s at the angular frequency ω, C_S a source correction and C_D a distance correction. This formalism can be applied both to Rayleigh and Love waves with obviously different expressions for the corrections. Note that the source correction is in principle depth-dependent; in practice, corrections are computed for 4 broad bins of hypocen-

12 JULY 1993 (HOKKAIDO)   **Station: INU**

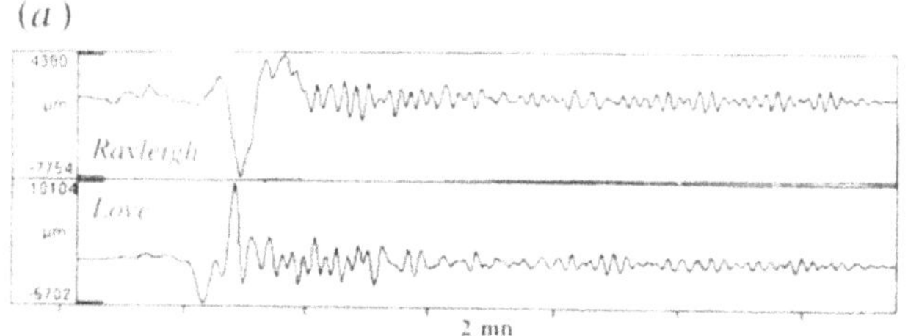
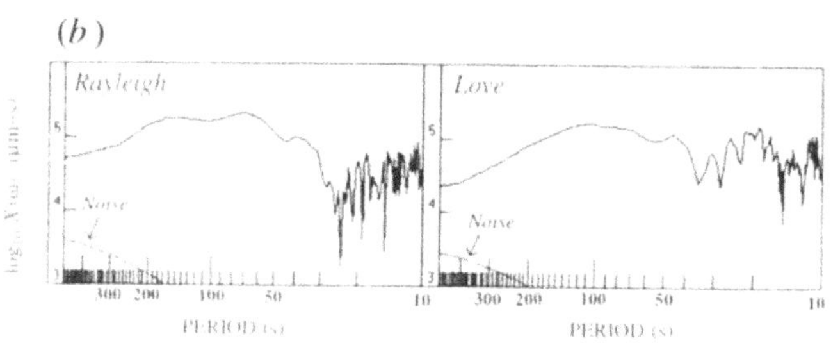

(c)

Period (s)	M_m	Moment (10^{20} N-m)
273.	7.82	6 61
256	7.81	6 46
213.	7.82	6 61
183.	7 81	6.46
160.	7.81	6.46
142.	7.85	7.08
128.	7.88	7 59
116	7 89	7 76
107	7.90	7 94
98	7.90	7.94
85.	7 91	8.13
75.	7 90	7.94
67	7 93	8 51
61	7 92	8 32
56	7.89	7 76
51	7 89	7.76

Figure 3

Automatic computation of mantle magnitude M_m and estimation of seismic moment M_0 by the TREMORS algorithm, illustrated in the near field on the case of the Hokkaido earthquake recorded at Inuyama. (a): Time series used in the final analysis (the Love wave trace has been offset for clarity due to difference in group velocities. (b): Raw spectra (in logarithmic scale) of the above records, before the distance and source corrections have been applied. The smaller trace in the bottom left corner represents the level of background noise, obtained by processing a window of similar duration occurring prior to the event. (c): Ouptut of the computed values of M_m and corresponding estimations of M_0 obtained from the Rayleigh spectrum.

tral depths (OKAL, 1990). In the case of Rayleigh waves at teleseismic distances ($\Delta \geq 10°$), M_m can also be computed directly in the time domain (OKAL, 1989).

In practice, the computation of the seismic moment can be initiated when a few minutes of surface waves have been recorded by the system (15 minutes or more in the absence of emergency). This procedure is illustrated on Figure 3. The computer systematically calculates M_m from Rayleigh waves for each category of focal depth (and from Love waves for a shallow source), and prints the results, with an operator on call 24 hours a day, having the final responsibility of deciding the most plausible depth. Further development is presently being undertaken to attempt to recognize automatically a deep focus (and eliminate it from the standpoint of tsunami danger), based on the combination of the incidence angle of the P wave, and of the total duration of the signal (which would discriminate against a teleseismic event with the same incidence angle).

Results compiled on a dataset of 474 earthquakes during the years 1987–91, and expressed through the residual

$$r = M_m(\text{measured}) - \log_{10} M_0(\text{published}) + 13 \tag{6}$$

feature an average $\bar{r} = 0.07$ units of magnitude, and a standard deviation $\sigma = 0.22$ units (HYVERNAUD et al., 1992). The quality of these results is comparable to that shown in previous studies from OKAL and TALANDIER (1989) and REYMOND et al. (1991). In the near field ($1.5° \leq \Delta \leq 15°$), an analysis made on a dataset of 115 records from shallow earthquakes yields $\bar{r} = 0.17$; $\sigma = 0.29$ units (TALANDIER and OKAL, 1992). These results express the excellent performance of the mantle magnitude M_m, as an estimator of the seismic moment M_0.

Tsunami Warning

● *P-wave warning:* Two warning systems are in effect at CPPT; a classical warning system is based on the amplitude of P waves, measured on a broad-band record, with a triggering threshold fixed at 100 μm. For a regular earthquake whose source time behavior approaches a step function, this corresponds approximately to a moment of 10^{20} N-m at typical teleseismic distances. However, this first warning can occasionally be deficient, since body waves are a poor descriptor of the long-period properties of the source: for example, in the case of the 1992 Nicaragua and 1994 Java earthquakes, the triggers were activated at PPT neither on the short-period nor on the long-period channels (also, false alarms could indeed be generated by deep earthquakes with strong P waves). Nevertheless, the warning based on P waves, when triggered, does retain some value, given that it is available as soon as the arrival time of P waves.

It would be inappropriate to lower the P wave warning threshold in order to trigger on slow, "tsunami", earthquakes, since this would result in the issuance of

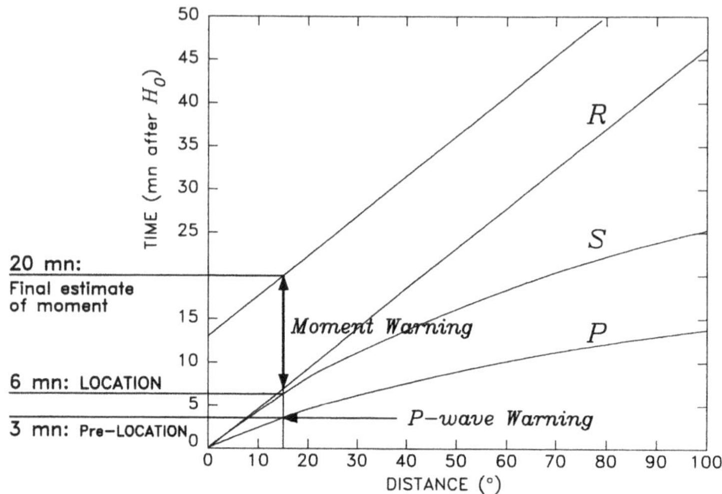

Figure 4

Time delay of P and M_0 warnings as a function of epicentral distance. The travel-time curves of P, S and the Airy phase of the Rayleigh wave (R) are plotted for a shallow earthquake. The fourth curve is obtained as ($R + 13$ mn). The case of $\Delta = 15°$ is highlighted: the P phase arrives 3 mn after origin time, triggering a P warning if the threshold is exceeded. The location is computed at the time of the S arrival (6 mn at 15°), and the seismic moment can exceed the assigned threshold anytime (generally a few minutes) after the R arrival. The final estimate of the moment is given 20 mn after H_0.

warnings for all earthquakes above approximately $m_b \geq 5.8$, most of which carry no tsunami danger. For this reason, the TREMORS warning is based on an estimate of the seismic moment. It is clearly superior as a tsunami warning, even though it only becomes available at a later time (see Figure 4).

● *TREMORS warning:* Once the seismic moment of the earthquake is computed, the tsunami warning proceeds from the observation that the amplitude of a tsunami on the high seas at teleseismic distance from the source is a linear function of the moment of the parent earthquake. This was both derived theoretically (OKAL, 1988), and verified experimentally on the basis of a dataset of 17 tsunamis at Papeete (TALANDIER and OKAL, 1992). In particular, we have argued in OKAL (1988) that the precise value of hypocentral depth (as long as $h \leq 70$ km), and the geometry of the focal mechanism have only minor influences on the amplitude of teleseismic tsunamis. In the near field, such effects are expected to become more important; however, the seismic moment M_0 remains the crucial seismological parameter controlling tsunami excitation.

The warning threshold at CPPT is fixed at 1.0×10^{20} N-m and six of the eight earthquakes under study generated a seismic moment warning, fifteen minutes after the Rayleigh arrival. As detailed in TALANDIER and OKAL (1989), various levels of warning (watch, alarm, etc.) have been defined as a function of the estimate of M_0, and of the particular epicentral region, in relation to Polynesia. It is clear that, in the context of a regional warning system, the warning threshold for M_0 would have

Table 2

PPT Real-time results

Event	USGS Location		TREMORS			
	Δ (°)	Back azimuth (°)	Δ (°)	Back azimuth (°)	M_m	Moment (10^{20} N-m)
Nicaragua	68.1	69.2	69	71	7.5	3.2
Flores	86.0	262.4	90	261	7.9	8.0
Hokkaido	88.8	315.9	93	305	8.0	10.0
Guam	71.6	290.6	73	283	7.95	9.0
Halmahera	83.3	273.2	96	272	6.9	0.87
Java	93.8	257.8	90	250	7.4	2.5
Kuriles	83.9	319.4	83	321	8.3	20.
Mindoro	93.4	283.0	94	275	7.3	2.0

to be adapted regionally, notably through the use of available runup and inundation maps for documented historical tsunamis.

Real-time Results at CPPT for the 1992–94 Tsunamis

Table 2 summarizes the parameters obtained in real time by the TREMORS algorithm running at CPPT. In all instances, these parameters were forwarded to the appropriate agencies (NEIC, PTWC) within one hour of the earthquake's origin time. In several cases, they are believed to have been the first estimates of the earthquake's seismic moment to be made available. However, in all cases, the large epicentral distances for PPT restricted the tsunami risk for Polynesia. In addition, in several cases, the tsunami occurred in seas with little if any opening into the main Pacific Ocean, actually alleviating the teleseismic risk totally.

The residuals in distance to PPT (true distance − distance evaluated by TREMORS) vary between −13° (in the case of the small Halmahera event; otherwise −4°) and +2°, and the residuals in azimuths between −2° and +10°. In a few instances, the accuracy of the TREMORS solution is insufficient to assess whether the epicenter is under the ocean, or under a continental mass or a large island. Yet, it is always sufficient to estimate the seismic moment in real time. A comparison between the moments obtained at PPT and later published by the Harvard group (DZIEWONSKI *et al.*, 1993a,b, 1994a,b; G. Ekström, pers. comm., 1994) shows that the TREMORS estimates are within a factor varying from 0.7 to 2.1 of the final CMT solutions, corresponding to an error on the mantle magnitude varying between −0.15 and +0.32 units (Figure 5). To put these numbers in perspective, it is useful to remember the general scatter of the estimates of

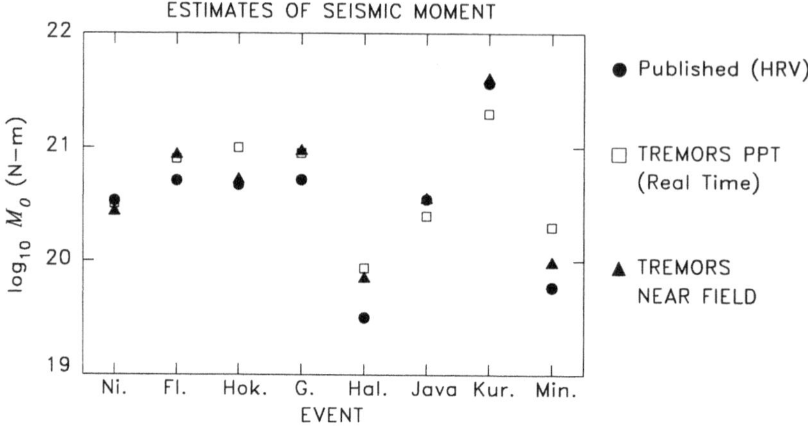

Figure 5

Comparison of the moment values obtained from CMT inversion by the Harvard group (solid circles) for each of the eight earthquakes studied, with estimates obtained in real time at PPT (open squares), and near-field estimates obtained with the TREMORS algorithm in the present study (solid triangles).

conventional magnitudes which are given in the hours immediately following a large earthquake.

Near-field Results

In a previous study, TALANDIER and OKAL (1992) demonstrated the validity of the mantle magnitude algorithm at regional distances down to 1.5° for the purpose of estimating the seismic moment of an event. However, their study required the knowledge of the exact epicenter, and as such was an after-the-fact exercise. In the context of TREMORS, it is necessary to investigate the performance of the method in providing in real time a reasonable estimate of the epicenter, from which the quantification of the source can proceed. We have recently adapted the TREMORS software to the particular case of near-field warning, for which the minimization of response times is crucially important. In order to avoid the pitfalls of classical warning systems which use a simple criterion based on P-wave amplitudes, TREMORS proceeds through a real-time computation of the seismic moment, which we will now briefly describe.

As the various phases composing the seismic signal are received at the station, a dataset of increasing duration is used for the determination of the seismic moment: this window is regularly increased in steps of 50 s, until mantle waves with the longest periods (300 s) have been included. This procedure is illustrated on Figure 6. The presumed location of the earthquake, which is necessary to estimate

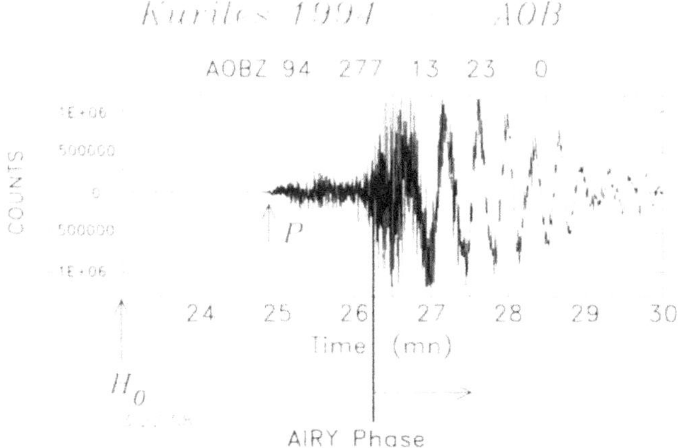

Start Analysis with growing window

Figure 6

Broadband record at Aobayama, Japan (AOB) of the great 1994 Kuriles earthquake. This figure illustrates the computation of M_m using a growing window: the computation starts upon arrival of the Airy phase (around 13:26:30 GMT); the beginning of the windows remains constant, but their durations are progressively increased in steps of 50 s.

the seismic moment, is calculated from its back azimuth (obtained from the polarization of P waves at trigger time), and distance (itself obtained from the identification of S in the growing window; at short distances, S and Rayleigh waves may not be differentiated). Once the location is estimated, the signal is Fourier-transformed, the magnitude M_m computed, and the seismic moment estimated, at each of the Fourier periods between 50 and 300 s. If the seismic moment is greater than the established threshold, a warning is sent automatically by the computer. This procedure is repeated regularly as more data become available, which in practice allows the study of the source at increasingly longer periods. However, the main advantage of the growing window is to minimize the response time of the algorithm, by allowing the issuance of a warning even before the longest periods have been received, should the characteristics of the early wavetrains warrant it. This point becomes crucial to an efficient warning system, as the epicentral distances are reduced, and will be particularly important in the case of the Kuriles event (see below).

For each of the eight tsunamis, we acquired very-broad-band data from stations located as close as possible to the relevant epicenters, and processed the datasets through the TREMORS automatic software (Table 3). Although conducted in delayed time, our experiment represents a test of the real-time performance of TREMORS, had the system been operational at a regional station. In practice, we use IRIS GEOSCOPE and POSEIDON Very Broad Band (VBB) channels. We refer to the IRIS annual reports, and to ROMANOWICZ et al. (1984), respectively,

Table 3

Near-field results

| Event | Station | USGS Location | | TREMORS | | | |
		Δ (°)	Back azimuth (°)	Δ (°)	Back azimuth (°)	M_m	Moment (10^{20} N-m)
Nicaragua	UNM	13.7	121.9	15	133	7.4	2.7
Fores	CTA	26.2	292.6	27	290	7.9	8.6
Hokkaido	INU	7.7	12.3	7.7	11	7.7	5.2
Guam	MAJO	24.2	164.2	23	167	8.0	9.3
Halmahera	GUMO	21.0	235.1	22	240	6.85	0.7
Java	WRA	22.6	291.4	24	289	7.55	3.5
Kuriles	AOB	7.33	39.9	8.1	45	8.6	40.
Mindoro	TATO	11.4	182.0	11.4	187	6.98	0.95

for a description of the networks, including the instrument responses at mantle
wave periods. Despite the occasionally low sampling rate of some channels (10 s),
the location algorithm performed well, as will be discussed in detail below.

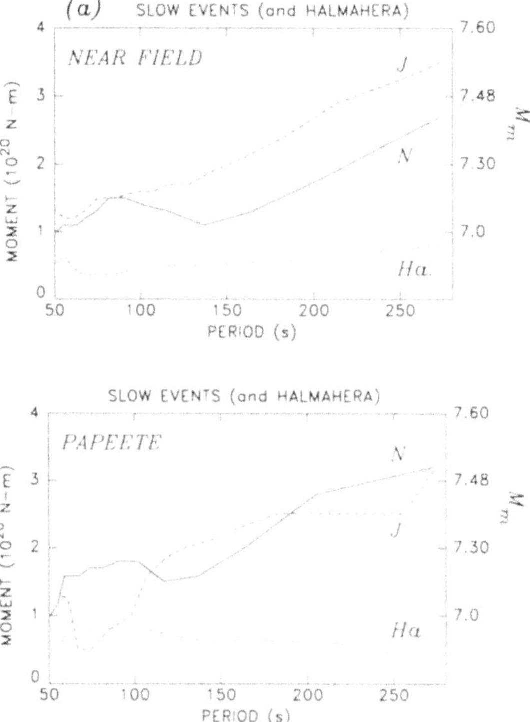

Figure 7(a).

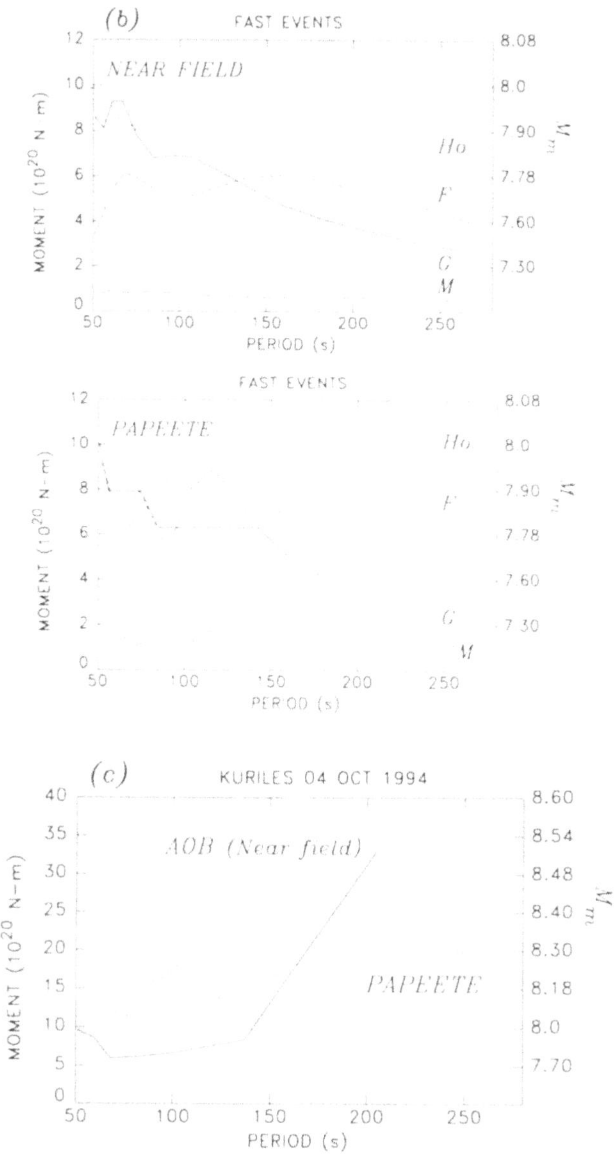

Figure 7

Value of the seismic moment M_0 obtained from TREMORS as a function of period. This quantity is equivalent to the spectral amplitude of the rate of moment release. These values are taken from the final window of analysis. The data has been split into (a) slow events ("tsunami earthquakes") and Halmahera; and (b) fast events. The Kuriles earthquake is plotted separately (c) due do its size. In (a) and (b), we plot separately the results at Papeete, and for the near-field station selected. In the case of Nicaragua (N) and Java (J), note the regular increase in seismic moment with period, indicative of a slow source. On the other hand, in the case of Hokkaido (Ho.) and Flores (F), the spectrum is fundamentally flat. Note that in the case of the Guam earthquake (G), the moment actually *decreases* with period.

We refer to the abundant existing literature, including the other papers in this volume, for complete descriptions of the events under study. For each of them, we will emphasize only the features most important from the standpoint of their tsunami generation, and proceed to discuss in detail the estimates of seismic moment and tsunami risk provided by our automated algorithm.

The earthquakes are presented in chronological order, except for the small Halmahera event, which is discussed last on account of its much smaller moment and tsunami.

● *Nicaragua, 1992*

The earthquake occurred on September 2, 1992 at 00:16 GMT, and was barely felt by residents along the coast of Nicaragua (Figure 1c). Its intensity was mostly II along the coast, and reached III only at a few places. Forty to seventy minutes after origin time, the tsunami reached the coast, with amplitudes of 4 m above the ambient sea level in most places along a stretch of 200 km of coastline, and a maximum runup height of 10.7 m (SATAKE *et al.*, 1993). The death toll, entirely due to the tsunami, was around 170.

The nearest broadband station from which we obtained data is the GEOSCOPE station UNM in Mexico City, at a distance of 13.7° from the epicenter. The TREMORS epicentral location is at {8.8°N, 88.1°W}, corresponding to errors of 1.3° in epicentral distance and 11.1° in azimuth. The latter is surprisingly large, and is probably due to a path refracted along the slab (based on numerous surface wave polarization studies, LASKE (1995) has shown that the orientation of the sensors at UNM is indeed very good, so that the effect of their misalignment can be dismissed). The TREMORS epicenter is situated in the Pacific continental margin, which is sufficient to warrant a warning for all the Pacific coasts of Nicaragua and Costa Rica. The seismic moment estimate (2.7×10^{20} N-m) is in good agreement with the Harvard CMT value (3.4×10^{20} N-m). The seismic moment rate spectrum shows a first peak (1.5×10^{20} N-m) at 70 s, followed by a slight decrease (to 1.0×10^{20} N-m) until 140 s (Figure 7a); beyond 140 s, the seismic moment increases rapidly, reaching 2.7×10^{20} N-m at 273 s. The large increase in the size of this event with period suggests a long source process, as independently demonstrated by many other studies (e.g., KANAMORI and KIKUCHI, 1993), and this event must be classified as a "tsunami earthquake," as confirmed by the large $m_b : M_s$ and $M_s : M_0$ discrepancies (Table 1).

● *Flores Sea, 1992*

On December 12, 1992 at 05:29 GMT, a very strong earthquake struck the eastern region of Flores Island (Figure 1d) and caused substantial damage and casualties. Because the rupture zone was very close to the coast, the tsunami attacked the coastal area within 5 minutes of the earthquake. KAWATA (1993) reveals that the residents had no information on tsunamis, so at the time of the earthquake, no one ran away from the shores. The tsunami had an average runup of 3 m, and a maximum height of 26 m was documented at the site of the village of Riangkrok, which was totally wiped out (YEH, 1993). This huge tsunami also attacked the southernmost part of Sulawezi Island, where it caused 22 deaths and damaged more than 400 buildings (SUNARJO, 1993); it reached Ambon-Baguala Bay on Ambon Island two hours after the earthquake.

The nearest station for which we obtained data is the IRIS station at Charter Towers, Australia (CTA), a distance of 26.2° from the epicenter. The TREMORS location is 150 km from the epicenter, with an error of 0.8° in distance and 2.6° in azimuth. This level of accuracy is insufficient to determine whether the earthquake is south of Flores Island, in the open Indian Ocean or North of the island, in the closed Flores Sea. The seismic moment is estimated at 8.6×10^{20} N-m, somewhat larger than the Harvard CMT value (5.1×10^{20} N-m). As shown on Figure 7b, the moment release remains very stable between 50 and 100 s, and then decreases slowly at periods larger than 150 s. Several factors did contribute to the exceptional amplitude of the tsunami waves, relative to the seismic moment: in addition to the general vicinity of Flores Island to the rupture zone, and to the shallow water depth in the epicentral area (only about 1000 m), it is probable that underwater landslides took place in the neighborhood of the highest observed runup, at the extreme northeast end of the island of Flores.

● *Hokkaido-Nansei-Oki, 1993*

This earthquake took place at 13:17 GMT on July 12, 1993 (Figure 1b), and was widely felt throughout Hokkaido, Northern Honshu and the neighboring islands. Two to five minutes later, the tsunami devasted Okushiri Island, then hit Hokkaido; the total loss of life was upward of 200. We refer to FURUMOTO (1993) and NAKANISHI et al. (1993) for a complete description of this earthquake and tsunami. Runup heights were an average of 5 m on Okushiri, reaching up to 30 m on the southwestern coast of the island (HOKKAIDO TSUNAMI SURVEY GROUP, 1993). Wave heights of more than 7 m were reported along Hokkaido's coasts; in 50 to 70 minutes, the tsunami crossed the Sea of Japan and reached the Russian and Korean coasts, with average runup heights of 2 m, reaching up to 4 m at some locations. The rupture zone of the earthquake reached within a few km of Okushiri Island, which explains both the short time interval between earthquake and tsunami, and the very large runup heights (TANIOKA et al., 1993).

The nearest station for which we obtained data is GEOSCOPE station INU at Inuyama, Japan, a distance of 7.7° from the epicenter. The TREMORS location based on this station is {42.9°N, 139.1°E}, 30 km from the epicenter, the distance being correct and the azimuth 1.3° off. The TREMORS epicenter is located very close to the actual epicenter and a warning for all coastal areas of the Sea of Japan (Hokkaido, Honshu, Russia, Korea) would be possible with this location.

Several factors contributed to the very large amplitude of the tsunami waves for this event, relative to its seismic moment: in addition to the extreme vicinity of the source from Okushiri Island, the reduced water depth (typically on the order of 1000 m in the epicentral area) enhanced significantly the tsunami excitation.

● *Guam, 1993*

On August 8, 1993, at 08:34 GMT, a powerful earthquake occurred near Guam (Figure 1b) causing extensive damage and injuring more than 50 people on the island of Guam. The event generated tsunami waves of 30 cm to 1 m amplitude recorded at tidal stations in Southern Japan.

The nearest station for which we obtained data is MAJO, the IRIS station at Matsushiro, Japan, a distance of 24.2° from the epicenter (power failure throughout

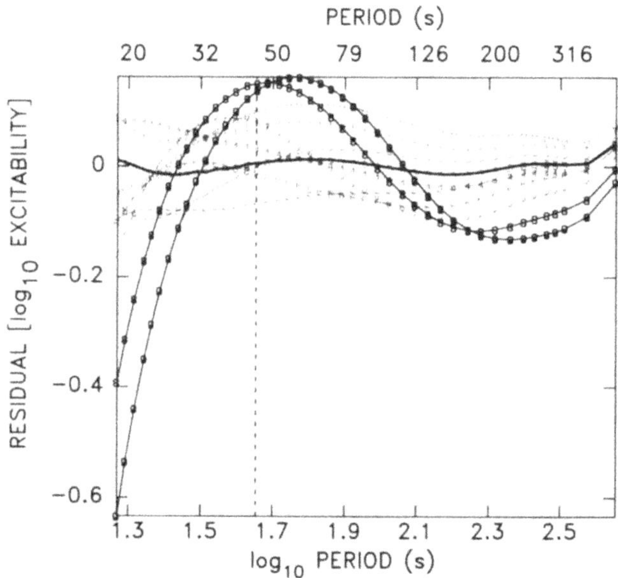

Figure 8

Residual of the logarithmic average excitability as a function of period for several depths of shallow earthquakes, after TALANDIER and OKAL (1989). For each depth (labeled 0 to 9 and extending from 0 to 75 km), the trace represents the average error introduced by using the correction C_S, instead of a proper excitability, in the computation of M_m. C_S is designed to model the excitability at 20 km (3; thick, nearly horizontal trace). Note that for depths of 60 (8) and 75 km (9), the excitability is larger than modeled at 50 s, but becomes smaller at the mantle periods (150–250 s). Accordingly, the magnitude of the event is overestimated (by about 0.15 units) at 50 s, and underestimated (by about 0.15) at 200 s. The decrease in magnitude (of about 0.3 units) between 50 and 200 s can therefore be an artifact of a deeper than usual source.

the island of Guam knocked down station GUMO at what should have been trigger time). The TREMORS location is $\{13.9°N, 143.3°E\}$, about 185 km from the epicenter, with an accuracy of $1.2°$ on the distance, and $2.8°$ on the azimuth. The TREMORS epicenter is situated in the Parece-Vela Basin, which is sufficient for a warning for the western coasts of the Marina Islands and the southwestern coasts of Japan.

The seismic moment estimated is 9.3×10^20 N-m, significantly larger than the Harvard CMT value (5.2×10^{20} N-m). It is remarkable that this estimate is obtained at a period of 67 s, and that the surface wave seismic moment rate spectrum is peaked around $60-70$ s, followed by a steady decrease to 2.5×10^{20} N-m at 273 s (Figure 7b). In order to compare more significantly our solution with the Harvard CMT value, we should use the estimate of the seismic moment at the period (135 s) used exclusively in the latter. The agreement (5.8×10^{20} as opposed to 5.2×10^{20} N-m) then becomes excellent.

The unexpected shape of the moment release rate spectrum for the Guam earthquake is due to its significant hypocentral depth (60 km, while most other sources are shallower than 20 km). As shown on Figure 8 (from OKAL and TALANDIER, 1989), which plots the logarithmic average excitability of Rayleigh waves for a combination of frequencies and source depths, hypocenters at 60 to 75 km (8 and 9 on the figure) exhibit maximum residual excitation (computed relative to $h = 20$ km) around 50 s, followed by a decrease of 0.3 orders of magnitude down to 200 s. The correction for the true hypocentral depth and focal mechanism of the event amounts to 0.4 orders of magnitude, and brings the moment in close agreement with the Harvard CMT value.

● *Java, 1994*

On June 2, 1994, at 18:18 GMT, a major earthquake took place in the Java trench about 200 km southeast of Malang (Figure 1d). Forty to sixty minutes after the earthquake, a tsunami hit the southern coast of Java, killing at least 250 people, and inflicting considerable damage. Tsunami runup heights averaged 4 m, and exceeded 11 m at several locations.

The nearest station at which we obtained data is WRA, the IRIS station at Warramunga Array, Australia, a distance of $22.6°$ from the epicenter. The TREMORS location ($10.6°S, 111.4°E$) is 125 km from the actual epicenter with an accuracy of $1.4°$ in distance and $2.4°$ in azimuth. The TREMORS epicenter is located in the Indonesian Trench, which is sufficient to warrant a warning for all the southern coast of Java and adjoining islands. The estimate of the seismic moment, 3.5×10^{20} N-m, is the same as the Harvard CMT value. The seismic moment rate spectrum shows a large and continuous increase in amplitude with period, from 1×10^{20} at 50 s to 3×10^{20} N-m at 273 s (Figure 7a). This large increase in amplitude suggests a long source time function, and this event must be classified as a "tsunami earthquake." Significant aerial landslides were observed in the eastern portion of Java, suggesting that underwater ones may also have taken place.

● *Kuriles, 1994*

This very large earthquake occurred on October 4, 1994 at 13:22 GMT, approximately 80 km from the islands of Shikotan and Iturup, in the Southwestern Kuriles. With a preliminary CMT moment of 3.7×10^{21} N-m, this event is the largest earthquake recorded worldwide since 1977, and possibly 1965 (KANAMORI, 1977). It generated a Pacific-wide tsunami, which prompted evacuation of certain coastal areas, notably in Hawaii. Local runup heights reached 9 m on Kunashir and Shikotan Islands. It is unclear if any of the 11 reported deaths were attributable to the tsunami.

The event was located by CPPT at $\{44°N, 150.5°E\}$, 250 km away from the true epicenter, an error of only $1.4°$ in distance, and $2°$ in azimuth. This location, and the moment (2×10^{21} dyn-cm) obtained at CPPT, predict a Pacific-wide tsunami, albeit one with only moderate amplitude; this explains that a warning, but no alarm, was issued for Polynesia. The amplitude at Papeete harbor was 23 cm.

The closest station for which we obtained data is the POSEIDON station at Aobayama, Japan (AOB), a distance of $7.3°$. TREMORS located the event about 110 km west of the true epicenter, and yields a moment of 4×10^{21} N-m, in excellent agreement with the preliminary Harvard CMT value. There is some suggestion of an increase of M_m with period, which may reflect the source duration (70 s according to the preliminary CMT, a figure in general agreement with the size of the event), and offset the effect of source depth (50 to 60 km).

● *Mindoro, Philippines, 1994*

This relatively small event took place on November 14, 1994 at 19:15 GMT in the immediate vicinity of the coastline of Mindoro (at the time of writing, it is still unclear if the epicenter was on land or at sea; C. E. SYNOLAKIS, pers. comm., 1994). It generated a local tsunami in the Bay of Marinduque, with runup heights of up to 6 m on Mindoro, resulting in approximately 70 deaths. Because of the combination of a large epicentral distance ($94°$), and of the relatively poor signal-to-noise ratio, the TREMORS location at CPPT is significantly offset ($8.2°$ in back azimuth) from the true epicenter.

The nearest station for which we obtained broadband data is TATO, the IRIS station at Taipei, Taiwan, a distance of $11.4°$ (Figure 1e). The solution obtained, $\{13.60°N; 120.07°E\}$, is about 250 km from the true epicenter, on the Palawan shelf. The seismic moment, 9.5×10^{19} N-m, is about 1.6 times the preliminary CMT solution. The spectrum of the moment release rate is essentially flat.

● *Halmahera, 1994*

Finally, and despite its much smaller size, we discuss the case of Halmahera event of January 21, 1994: it will allow us to get insight into a reasonable threshold for tsunami warning in the near field. At 02:04 GMT, a strong earthquake struck Halmahera Island (Figure 1e), killing 9 people and causing extensive damage to that Indonesian island. Minor tsunami waves, with a maximum runup of 2 m, were reported on its western coast. No deaths or significant damage were attributable to the tsunami.

The nearest station for which we obtained data is GUMO, the IRIS station at Agaña, Guam, a distance of 21.0° from the epicenter. The TREMORS epicenter (1.9°N, 126.0°E) is 210 km from the actual focus, with an accuracy of 1.0° in distance and 4.9° in azimuth. The seismic moment estimate is 7.3×10^{19} N-m, with the moment rate release spectrum basically flat from 50 to 250 s. The difference with the Harvard CMT value (3.2×10^{19} N-m) can be accounted for by the particular orientation of the station with respect to the focal mechanism. The low value of M_0, which is adequately recognized by TREMORS, explains the relatively benign character of the tsunami, and suggests that this range of moments constitutes the limit for the generation of destructive tsunamis in the near field.

Discussion

At this point, and based on our results for the eight earthquakes studied, we confirm TALANDIER and OKAL's (1992) result, namely that M_m performs flawlessly in the near field. In particular, a comparison between near-field results and the value computed for the same events at PPT confirms the validity of our estimates. In addition, we show that our detection algorithm is usually reliable at those short distances, indicating that the whole TREMORS software yields acceptable estimates of the seismic moment in real time (within ± 0.3 orders of magnitude for the present dataset), thus reaching a new milestone for efficient tsunami warning. Even in the case of a poor epicentral solution (Mindoro as located by TATO), the estimate of the seismic moment remains acceptable.

Beyond the mere estimation of seismic moment, it is particularly important to discuss the ability of TREMORS to recognize exceptional features of the earthquake source, and to do so in real-time. Among the eight earthquakes considered, the following characteristcs are of particular interest for tsunami warning:

* the character of the event as a "tsunami earthquake" (Nicaragua, Java), i.e., the existence of a slow source time function;
* the case of a truly large seismic moment (Kuriles); and
* the case of an earthquake of very large conventional magnitude generating only a benign tsunami (Guam), presumably due to its source depth.

We show here that TREMORS does recognize all these characteristics, and can do it in real time.

Anomalous Seismic Moment Rate Spectrum for Tsunami Earthquakes

As shown on Figure 7b, the moment spectra characteristics of the Flores, Hokkaido and Mindoro earthquakes correspond to a typical source time function, which cannot be distinguished from a step-function moment release at periods greater than 50 s. On the other hand, the Nicaragua and Java spectra (obtained both in near field and at PPT) feature an increase of about 0.4 orders of magnitude in the seismic moment rate spectrum from 50 to 273 s. This situation is particularly well expressed in the case of the Nicaraguan event, as analyzed at UNM with the growing time window: the first available estimate of the seismic moment is barely 10^{19} N-m, and this value grows by a factor of 20 over five minutes of elapsed time. In contrast, in the case of the Flores earthquake, the estimate of M_0 grows only by a factor 3. Figure 8 shows that this cannot be attributed to an artifact of source depth for shallow earthquakes; hence the Nicaragua and Java events are "tsunami earthquakes" with anomalous source properties leading to extended source durations. In terms of phenomena responsible for this behavior, leading contenders would be rupture velocities significantly slower than the commonly assumed 2.5 to 3.5 km/s, themselves a probable reflection of the influence of sedimentary structures in the immediate vicinity of the source (KANAMORI and KIKUCHI, 1993; TANIOKA and SATAKE, 1994). In turn, such structures may enhance tsunami excitation relative to seismic waves, and beyond its value expected on the basis of the zero-frequency value of the moment, as documented theoretically by OKAL (1988). Thus it is particularly important that TREMORS cannot only obtain the final static moment, but also document the slow character of the source.

The Case of a Truly Large Earthquake: Kuriles

In the case of the Kuriles event, the very first measurement obtained at station AOB, 4 minutes after origin time, is already in excess of 10^{21} N-m. While the final value, obtained a few minutes later grows to 4×10^{21} N-m (mainly due to the dimensions of the source), it becomes immediately clear that one deals with a major earthquake, which will generate a major tsunami, regardless of the values obtained later, and an alarm can be issued as early as 4 mn after origin time. This illustrates the power of our algorithm which begins computing estimates of M_0 immediately upon arrival of the Airy phase.

The Case of a Tsunami-deficient Event: Guam

Finally, in the case of the Guam earthquake, the strong decrease of moment rate spectra with period (Figure 7b) reflects the significant depth of the event (60 km);

as detailed above (and on Figure 5), the moment is overestimated (by about 0.4 units of magnitude) at the shorter periods, because of the particular behavior of the excitability of Rayleigh waves at 60 km. The low-frequency M_0 obtained by TREMORS is in excellent agreement with the CMT solution, and explains the relatively benign tsunami, despite a conventional surface wave magnitude $M_s = 8.1$. It is important to note that TREMORS can be used to recognize this property in real time by monitoring the evolution of M_0 with period. The only alternative explanation would be to seek a source time function deficient in low frequency with respect to a step function, which is improbable (it would require a significant rollback), and at any rate would not favor tsunami generation. We also note that a more stable rate of moment release is obtained if the earthquake is processed as having intermediate depth; however, the depth of the earthquake could not be determined independently in real time because of the difficulty of using automated algorithms for depths less than 150 km, due to the interference between pP and P on long-period records.

m_b vs. M_0: A Possible Identifier of Slow Earthquake Sources

In this section, we investigate the relationship between the body-wave magnitude m_b and the seismic moment M_0 as a possible means of identifying "tsunami

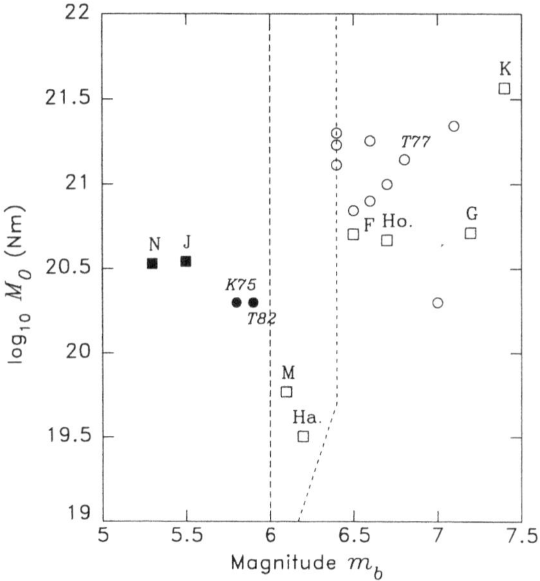

Figure 9

Comparison of m_b and M_0 values for a number of tsunamigenic earthquakes of the past 30 years (see Table 4). The eight events in the present study are shown as squares. "Tsunami earthquakes," exhibiting strong m_b:M_0 deficiencies are shown by solid symbols. The vertical long-dashed line is the $m_b = 6$ saturation level in the model of GELLER (1976). The segmented line with shorter dashes follows the model of OKAL and ROMANOWICZ (1994).

earthquakes" in real time. When first describing "tsunami earthquakes," KANA-MORI (1972) suggested that they result from an anomalously long source time function, which causes a large discrepancy between the seismic moment M_0 and its value expected from the conventional magnitude M_s, computed at higher frequency. This was later confirmed by a number of scientists who based most of their studies on a comparison between M_s and various seismic parameters (M_0, tsunami height, etc.).

In the present study, we elect to focus on the relationship between m_b and M_0 for a number of reasons. First and foremost, from an operational standpoint, m_b can be computed as early as the arrival time of S waves (the knowledge of $S - P$ is necessary to compute the distance correction), and does not require waiting for surface waves. Thus, and however incomplete and possibly misleading it may be, m_b remains the first available estimate of the size of the earthquake; on the other hand, as soon as the surface waves have arrived and M_s can be computed, the more significant M_m becomes available, and thus there is no advantage at this point in evaluating M_s for the purpose of quantifying the earthquake source.

It is well known (e.g., GELLER, 1976) that the magnitude scale m_b saturates around 6.0 to 6.5 (when properly measured at 1 s), but this saturation can actually

Table 4

Parameters of earthquakes used in Figure 9

Data	Region	m_b	M_s	M_0 (10^{20} N-m)	Symbol on Figure 9
17 OCT 1966	Peru	6.4	7.5	20	
11 AUG 1969	Kuriles	7.1	7.8	22	
17 JUN 1973	Japan	6.5	7.7	7	
03 OCT 1974	Peru	6.6	7.6	18	
10 JUN 1975	Kuriles	5.8	7.0	2	K75
22 JUN 1977	Tonga	6.8	7.2	14	T77
12 DEC 1979	Colombia	6.4	7.7	17	
01 SEP 1981	Samoa	7.0	7.7	2	
19 DEC 1982	Tonga	5.9	7.7	2	T82
03 MAR 1985	Chile	6.7	7.8	10	
19 SEP 1985	Mexico	6.8	8.1	11	
07 MAY 1986	Aleutian Is.	6.4	7.7	13	
20 OCT 1986	Kermadec	6.6	8.1	8	
		Events used in present study			
02 SEP 1992	Nicaragua	5.3	7.2	3.4	N
12 DEC 1992	Flores Sea	6.5	7.5	5.1	F
12 JUL 1993	Hokkaido	6.7	7.6	4.7	Ho.
08 AUG 1993	Guam	7.2	8.1	5.2	G
21 JAN 1994	Halmahera	6.2	7.2	0.3	Ha.
02 JUN 1994	Java	5.5	7.2	3.5	J
04 OCT 1994	Kuriles	7.4	8.1	37	K
14 NOV 1994	Mindoro	6.0	7.1	0.59	M

provide some insight in at least two situations: when m_b fails to reach its saturation value, and when m_b, correctly measured at 1 s, exceeds the expected level of saturation (e.g., reaches values of 7 and above).

Figure 9, which plots seismic moment as a function of body-wave magnitude m_b for the eight earthquakes studied here, shows a significant deficiency of about 1 unit of magnitude in m_b for "tsunami earthquakes", relative to other earthquakes of the same moment. Both the Nicaragua and Java events have $m_b \leq 5.6$. The anomalous $m_b : M_0$ relationship results of course from the source duration of the earthquakes, which is much longer than the period of 1 s used to determine m_b.

We also added on this figure 13 other tsunamigenic events including the last two "tsunami earthquakes" of record prior to 1992 (Kuriles, June 10, 1975 and Tonga, December 19, 1982), as well as the great Tonga earthquake of June 22, 1977 (Table 4). All "tsunami earthquakes" are well separated from the regular tsunamigenic earthquakes, and conversely. Note in particular that PELAYO and WIENS (1992) had argued that the 1977 Tonga earthquake could have been regarded as exhibiting as $M_s : M_0$ discrepancy, and thus classified as a "tsunami earthquake," even though its tsunami was typical of its seismic moment (1.4×10^{21} N-m). However, LUNDGREN and OKAL (1989) showed that the event extended significantly at depth, thus explaining the M_s deficiency. Furthermore, there is no evidence of the $m_b : M_0$ deficiency that would be required by a slow source; on the contrary, large values of m_b could be an artifact of an inadequate depth correction: while 20-s excitation decreases substantially with depth below 60 km, body-wave magnitude corrections, as given by RICHTER (1958) are practically constant in that range of depths, and actually decrease for near-field ranges (corresponding to an increase of excitation).

Thus, the real-time analysis of a possible $m_b : M_0$ deficiency, through an automatic computation of m_b incorporated into the location algorithm could be useful as additional evidence confirming the "tsunami earthquake" character of an event.

Near-field Tsunami Warning Using TREMORS

The present study was motivated fundamentally by the ultimate purpose of all tsunami warning systems: that of providing an accurate warning in a time frame fast enough to allow the evacuation of people from low-lying areas. In the case of the Flores and Hokkaido tsunamis, the nearmost communities were hit by the tsunami two to five minutes after the origin time (H_0); this interval drops to maybe one or two minutes for Mindoro. In such situations, it is probably illusory to envision an efficient tsunami warning system, and only the alertness of the residents can save them, by running for high ground as soon as the shaking of the earthquake has died down: Figure 10 shows that the seismic moment reaches the warning threshold 4 minutes after the Hokkaido earthquake, and 10 minutes after the Flores one. Even if the stations were at closer epicentral distances, it takes at

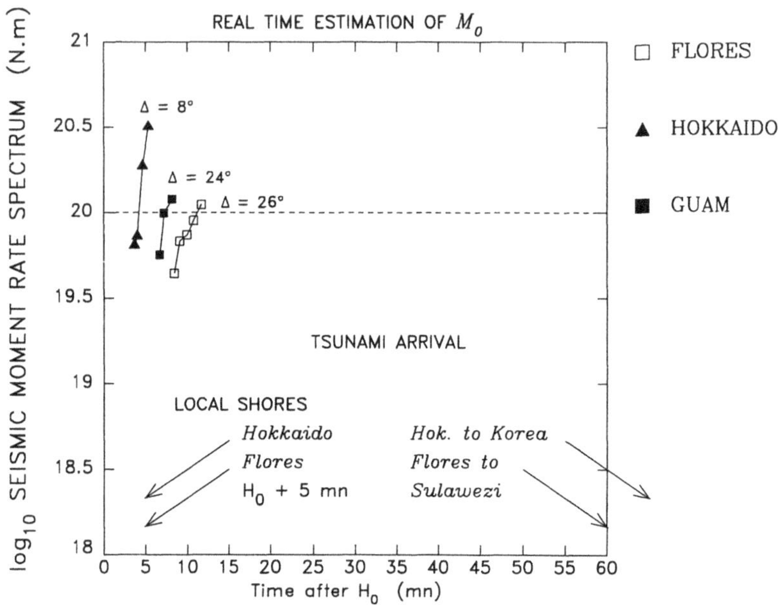

Figure 10

Real-time estimation of M_0 in the near field, in the case of the "fast" earthquakes (Flores and Hokkaido). On this figure (as on Figures 11 and 12), the various symbols show the estimates of the seismic moment rate spectrum obtained through the analysis of a growing window of seismic data, plotted as a function of time after H_0. The dashed line shows the moment threshold (10^{20} N-m) considered as representative as tsunami danger in the near field. A warning could have been issued as soon as M_0 reaches this threshold. Also indicated are the arrival times of the tsunami, both at the local shores, and at more distant locations. The Guam earthquake is also included for reference.

least 1 or 2 minutes to measure energy at frequencies lower than 20 mHz. With the rupture zones close to the coasts (possibly intersecting it in the case of Mindoro), early warning was impossible for the nearmost communities, but remains feasible for more distant locations, such as Honshu, Russia and Korea in the case of the Hokkaido event, and Sulawezi for the Flores one.

On the other hand, the Nicaragua and Java tsunamis, generated at the relatively distant trenches, hit the coast an average of 40 minutes after origin time. The Kuriles tsunami was generated closer than the trench, but in shallow seas, thereby slowing down its progression, and it reached Iturup and Kunashir 8 and 12 mn respectively after H_0.

The seismic moment of the Nicaraguan event exceeded 10^{20} N-m at the Mexican station UNM only 8 minutes after H_0. During the next 10 minutes, the crucial part of the Rayleigh waves (i.e., the long-period mantle waves) arrived, and it would have become possible to monitor the remarkable build-up of the seismic moment, and to detect the "tsunami earthquake" character of the event not more than 18 minutes after H_0 (Figure 11). Thus, a warning could have been issued for the

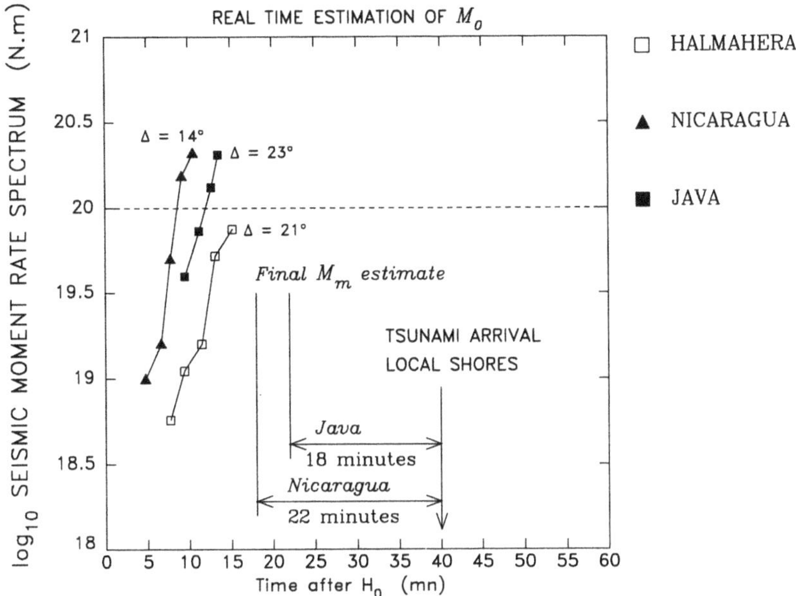

Figure 11

Same as Figure 10, for the "slow" events (or "tsunami earthquakes"). Note the spectacular increase (more than 1 order of magnitude) of the moment over about 5 mn in the case of Nicaragua, and a relatively similar situation for Java. The threshold is reached 7 and 11 mn respectively after H_0, with the final estimate of the mantle magnitude obtained about 20 mn after H_0, leaving adequate time before the arrival of the waves for these earthquakes, which occurred at the trench. In the case of Halmahera, there may be some evidence of a similar growth of the moment, but the final value fails to reach the threshold of danger.

Nicaraguan and Costa Rican coasts at least 22 minutes before the tsunami hit the coast. We recall that this earthquake was not even felt by residents of some sections of the coastline, who were to be swept away 40 minutes later.

In the case of the Java tsunami, the seismic moment exceeded 10^{20} N-m at station WRA 12 minutes after H_0. An efficient warning for Java residents could have been issued at least 18 minutes before the tsunami hit the coast.

In the case of the Kuriles tsunami, the very first estimate of the seismic moment, which can be obtained 4 minutes after H_0, is already 10 times the 10^{20} N-m threshold, and precedes the tsunami wave at Iturup by 4 minutes (Figure 12). While this lead time may be considered short, it increases to 8 minutes at Kunashir, and 20 minutes at Nemuro on the East coast of Hokkaido.

Conclusion

Our results show that TREMORS is indeed an efficient system for the rapid, automatic, estimation of the seismic moments of strong earthquakes both in the far

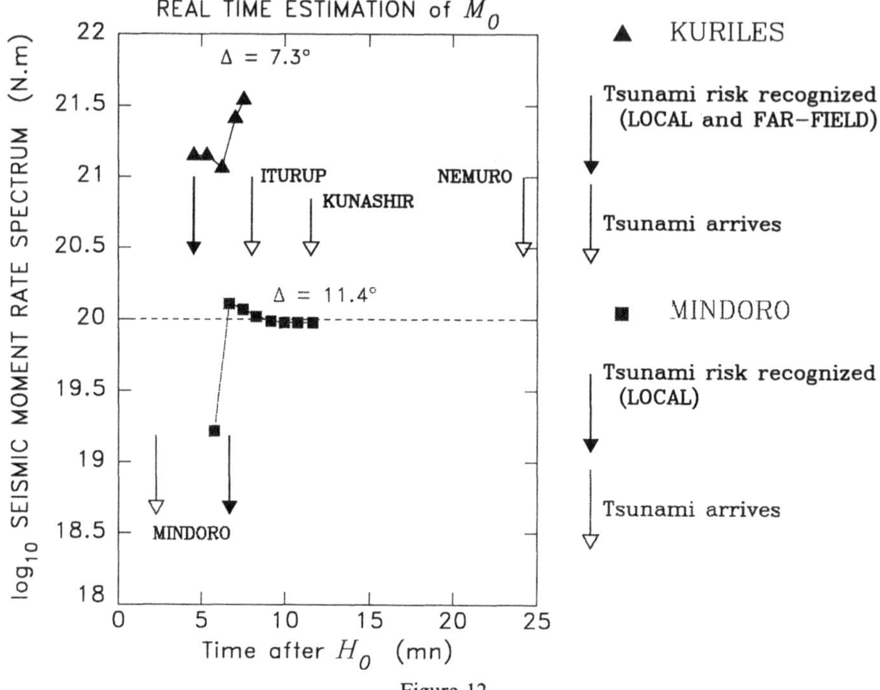

Figure 12

Same as Figure 10 for the more recent Kuriles and Mindoro events. In the case of Kuriles, note that the very first measurement is already considerably larger than the threshold of danger, which would warrant an immediate alarm, 4 mn after H_0. In the case of Mindoro, the alarm is also nearly immediate, but the station is clearly too distant to provide any efficient warning.

and near fields. As such, it can realistically be used as a reliable tsunami warning system. TREMORS can compute automatically an acceptable location, generate a warning in all the cases studied, except Halmahera, which is justified given both the smaller size of the earthquake, and the amplitude of the resulting tsunami, which incidentally did not cause any additional casualties or damage. Thus, our present experience suggests that a warning threshold of 10^{20} N-m seems an appropriate value in the near field for the generation of a dangerous tsunami. Our simulations obtained by computing seismic moments in windows of growing length, show that warnings would be issued within 4 to 11 minutes of H_0, depending on the epicentral distance, and that a final assessment of the seismic moment, including the possible recognition of "tsunami earthquake" character would be available within 18 to 20 minutes after H_0. In the case of a major earthquake, such as the Kuriles event, the warning is immediately available ($H_0 + 4$ minutes).

Unfortunately, at present, because of instrumental limitations, the traditional tsunami warning system continues to rely heavily on the 20-second M_s, a magnitude scale notoriously plagued by saturation effects in the precise range of earthquake sizes generating large tsunamis (GELLER, 1976; TALANDIER and OKAL, 1989). For slow "tsunami earthquakes" such as the Nicaragua or Java events, the classical magnitudes do not represent the true tsunami potential of the event. On the other hand, TREMORS has the capability of recognizing these events in real time, and of giving a lead time of about 20 minutes in the case of earthquakes generated at the subduction trenches. We want to stress that these figures were obtained using relatively distant broadband stations, situated about 15° from the epicenters. Obviously, better results could be obtained with closer stations, as demonstrated by the AOB/Kuriles combination. However, we think that a reliable global tsunami warning system could be built around several regional centers equipped with such an automated intelligent system. Each center would be in charge of a seismic zone with a 15° radius. This approach, whose cost and logistics are much lower than traditional dense seismic networks, should be a significant step towards effective tsunami warning and hazard reduction.

Acknowledgments

We thank the GEOSCOPE and IRIS data centers for providing the broadband data and Göran Ekström for the most recent 1994 CMT solutions. We also thank our colleagues in the tsunami community for many discussions. This research is supported by Commissariat à l'Energie Atomique (France).

REFERENCES

DZIEWONSKI, A. M., EKSTRÖM, G., and SALGANIK, M. P. (1993a), *Centroid-moment Tensor Solutions for July-September 1992*, Phys. Earth. Planet. Inter. *79*, 287–297.

DZIEWONSKI, A. M., EKSTRÖM, G., and SALGANIK, M. P. (1993b), *Centroid-moment Tensor Solutions for October-December 1992*, Phys. Earth. Planet. Inter. *80*, 89–103.

DZIEWONSKI, A. M., EKSTRÖM, G., and SALGANIK, M. P. (1994a), *Centroid-moment Tensor Solutions for July-September 1993*, Phys. Earth. Planet. Inter. *83*, 165–174.

DZIEWONSKI, A. M., EKSTRÖM, G., and SALGANIK, M. P. (1994b), *Centroid-moment Tensor Solutions for January-March 1994*, Phys. Earth. Planet. Inter. *86*, 253–261.

FURUMOTO, A. S. (1993), *Three Deadly Tsunamis in One Year*, Science of Tsunami Hazards, 111–121.

GELLER, R. J. (1976), *Scaling Relations for Earthquake Source Parameters and Magnitudes*, Bull. Seismol. Soc. Am. *66*, 1501–1523.

HASKELL, N. A. (1962), *Crustal Reflection of Plane P and SV Waves*, J. Geophys. Res. *67*, 4751–4767.

HOKKAIDO TSUNAMI SURVEY GROUP (1993), *Tsunami Devastates Japanese Coastal Region*, EOS, Trans. Am. Geophys. Un. *74*, 417–432.

HYVERNAUD, O., REYMOND, D., TALANDIER, J., and OKAL, E. A. (1992), *Four Years of Automated Measurements of Seismic Moments of Papeete using the Mantle Magnitude M_m: 1987–1991*, Tectonophysics *217*, 175–193.

KANAMORI, H. (1972), *Mechanisms of Tsunami Earthquakes*, Phys. Earth. Planet. Inter. *6*, 346–359.

KANAMORI, H. (1977), *The Energy Release in Great Earthquakes*, J. Geophys. Res. *82*, 2981–2987.

KANAMORI, H., and KIKUCHI, M. (1993), *The 1992 Nicaragua Earthquake: A Slow Tsunami Earthquake Associated with Subducted Sediment*, Nature *361*, 714–715.

KAWATA, Y. (1993), *Response of Residents at the Moment of Tsunamis.—The 1992 Flores Island Earthquake Tsunami, Indonesia.* Proceedings of the IUGC/IOC International Tsunami Symposium, Wakayama, Japan, Aug. 23–27, 1993, pp. 677–688.

LASKE, G. (1995), *Global Observation of Off-great Circle Propagation of Long-period Surface Waves,* J. Geophys. Res., in press.

LUNDGREN, P. R., and OKAL, E. A. (1988), *Slab Decoupling in the Tonga Arc: The June 22, 1977 Earthquake,* J. Geophys. Res. *93,* 13355–13366.

NAKANISHI, I., KODAIRA, S., KOBAYASHI, R., KASAHARA, M., and KIKUCHI, M. (1993), *Quake and Tsunami Devastate Small Town,* EOS, Trans. Am. Geophys. Un. *74* (34), 377–379.

OKAL, E. A. (1988), *Seismic Parameters Controlling Far-field Tsunami Amplitudes: A Review,* Natural Hazards *1,* 67–96.

OKAL, E. A. (1989), *A Theoretical Discussion of Time-domain Magnitudes: The Prague Formula for M_s and the Mantle Magnitude M_m,* J. Geophys. Res. *94,* 4194–4204.

OKAL, E. A. (1990), M_m: *A Mantle Wave Magnitude for Intermediate and Deep Earthquakes,* Pure and Appl. Geophys. *134,* 333–354.

OKAL, E. A., and ROMANOWICZ, B. A. (1994), *On the Variation of b-value with Earthquake Size,* Phys. Earth Planet. Inter. *87,* 55–76.

OKAL, E. A., and TALANDIER, J. (1989), M_m: *A Variable-period Mantle Magnitude,* J. Geophys. Res. *94,* 4169–4193.

OKAL, E. A., and TALANDIER, J. (1990), M_m: *Extension to Love Waves of the Concept of a Variable-period Mantle Magnitude,* Pure and Appl. Geophys. *134,* 355–384.

PELAYO, A. M., and WIENS, D. A. (1992), *Tsunami Earthquakes: Slow Thrust-faulting Events in the Accretionary Wedge,* J. Geophys. Res. *97,* 15312–15337.

REYMOND, D., HYVERNAUD, O., and TALANDIER, J. (1991), *Automatic Detection, Location and Quantification of Earthquakes: Application to Tsunami Warning,* Pure and Appl. Geophys *135,* 361–382.

REYMOND, D., HYVERNAUD, O., and TALANDIER, J. (1993), *An Integrated System for Real-time Estimation of Seismic Source Parameters and its Application to Tsunami Warning,* Tsunami in the World, 177–196.

RICHTER, C. F., *Elementary Seismology* (W. H. Freeman, San Francisco, 1958).

ROMANOWICZ, B., CARA, M., FELS, J.-F., and ROULAND, D. (1984) GEOSCOPE: *A French Initiative in Long-period Three-component Global Seismic Networks,* EOS, Trans. Am. Geophys. Un. *65,* 753–754.

SATAKE, K., BOURGEOIS, J., ABE, KU., ABE, KA., TSUJI, Y., IMAMURA, F., IIO, Y., KATAO, H., NOGUERA, E., and ESTRADA, F. (1993), *Tsunami Field Survey of the 1992 Nicaragua Earthquake,* EOS, Trans. Am. Geophys. Un. *74,* 145 and 156–157, 1993.

SUNARJO, (1993), *Experience in Handling the Flores Earthquake-tsunami of Dec. 12, 1992,* Proceedings of the IUGC/IOC International Tsunami Symposium, Wakayama, Japan, Aug. 23–27, 1993, pp. 861–869.

TALANDIER, J. (1993), *French Polynesia Tsunami Warning Center (CPPT),* Natural Hazards *7,* 237–256.

TALANDIER, J., and OKAL, E. A. (1992), *One Station Estimates of Seismic Moments from the Mantle Magnitude M_m: The Case of the Regional Field* ($1.5 \leq \Delta \leq 15°$), Pure and Appl. Geophys. *138,* 43–60.

TANIOKA, Y., RUFF, L. J., and SATAKE, K. (1993), *Unusual Rupture Process of the Japan Sea Earthquake,* EOS, Trans. Am. Geophys. Un. *74* (34), 377–380.

TANIOKA, Y., and SATAKE, K. (1994), *The 1994 Java Earthquake and Tsunami,* EOS, Trans. Am. Geophys. Un. *75* (44), 355–356 (abstract).

YEH, H. (1993), *Disaster on Flores Island,* Nature *361,* 686.

(Received September 1, 1994, revised January 21, 1995, accepted January 30, 1995)

PAGEOPH, Vol. 144, Nos. 3/4 (1995)

0033-4553/95/040409-18$1.50 + 0.20/0
© 1995 Birkhäuser Verlag, Basel

Edge Wave and Non-trapped Modes of the 25 April 1992 Cape Mendocino Tsunami

F. I. GONZÁLEZ,[1] K. SATAKE,[2] E. F. BOSS,[3] and H. O. MOFJELD[1]

Abstract—The 25 April 1992 Cape Mendocino earthquake generated a tsunami characterized by both coastal trapped edge wave and non-trapped tsunami modes that propagated north and south along the U.S. West Coast. Both observed and synthetic time series at Crescent City and North Spit are consistent with the zero-order edge wave mode solution for a semi-infinite sloping beach depth profile. Wave amplitudes at Crescent City were about twice that observed at North Spit, in spite of the fact that the source region was three times farther from Crescent City than North Spit. The largest observed amplitude was due to an edge wave which arrived almost three hours after the initial onset of the tsunami; since such waves are highly localized nearshore, this suggests that the enhanced responsiveness at Crescent City is at least partly due to local dynamic processes. Furthermore, the substantially delayed arrival of this wave, which was generated at the southern end of the Cascadia Subduction Zone, has significant implications for hazard mitigation efforts along the entire U.S. West Coast. Specifically, this study demonstrates that slow-moving but very energetic edge wave modes could be generated by future large tsunamigenic earthquakes in the CSZ, and that these might arrive unexpectedly at coastal communities several hours after the initial tsunami waves have subsided.

Key words: Tsunami, edge waves, Cape Mendocino, Cascadia Subduction Zone, hazard mitigation.

Introduction

On 25 April 1992, a tsunami was generated by a magnitude 7.1 earthquake at the southern end of the Cascadia Subduction Zone (CSZ) near Cape Mendocino, California (OPPENHEIMER *et al.*, 1993). The waves propagated north and south along the California and Oregon coasts and were recorded at nine tide gage stations maintained by the National Oceanic and Atmospheric Administration (NOAA). The measurements are presented in Figure 1. Although the waves in this event were small and not destructive, they have both intrinsic scientific interest and important

---

[1] Pacific Marine Environmental Laboratory/NOAA, 7600 Sand Point Way NE, Seattle, WA 98115, U.S.A.

[2] Dept. Geological Sciences, University of Michigan, 1006 CC Little Bldg., Ann Arbor, MI 48109, U.S.A.

[3] Joint Institute for Study of Atmosphere and Ocean, University of Washington, Seattle, WA 98195, U.S.A.

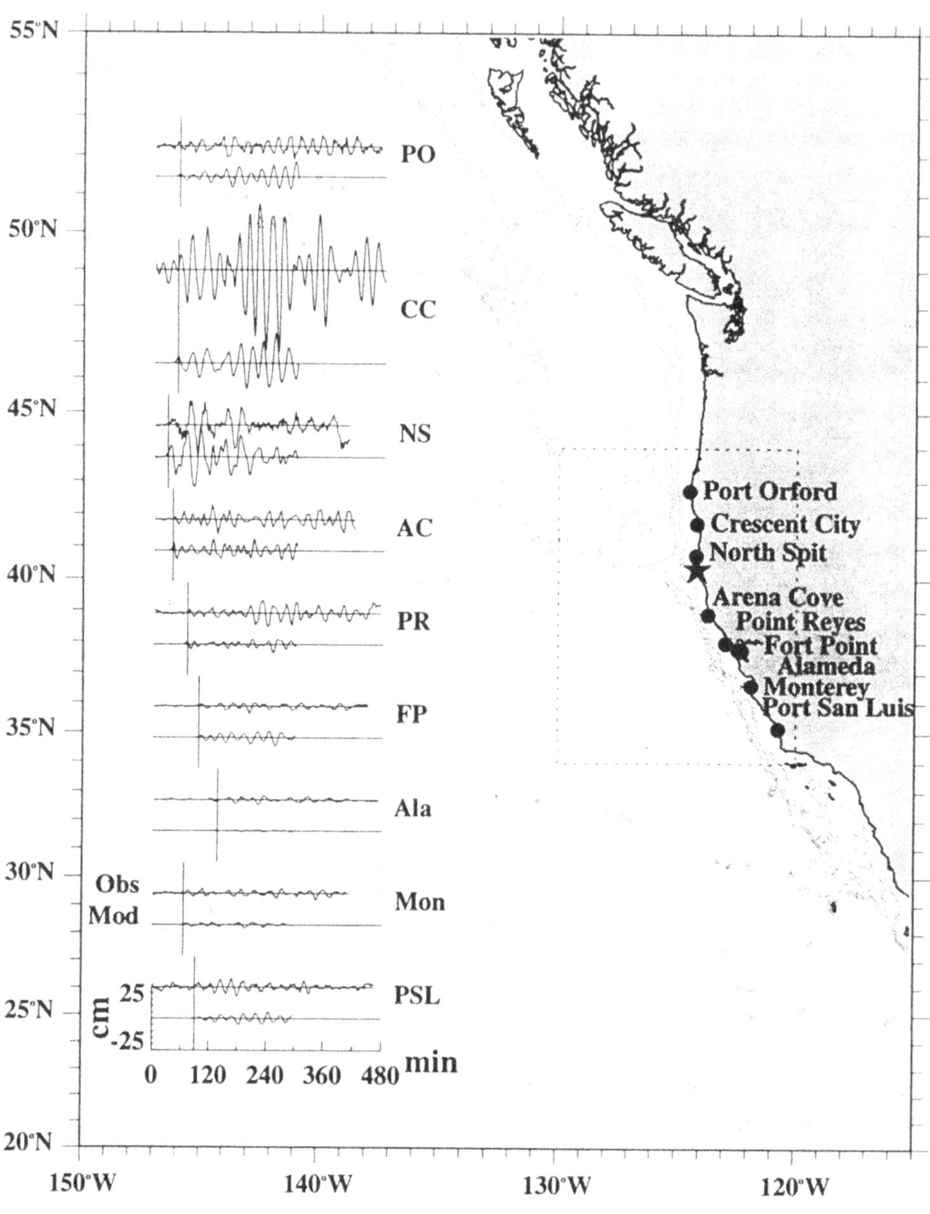

Figure 1

Summary chart presenting observed (top) and computed (bottom) time series for nine tide gage stations along the coasts of California and Oregon. Vertical axes are deviation from background sea level in cm; horizontal axes are minutes of elapsed time after the main shock, with a vertical line indicating the computed arrival time at each station listed in Table 1. Bathymetric contour interval is one kilometer. The dotted-line rectangle denotes the numerical model computational region. Cape Mendocino is just south of North Spit, and the solid star marks the earthquake epicenter.

implications for hazard assessment and mitigation along the California, Oregon and Washington coasts.

The practical implications of this event are related to the potential hazard posed by future large tsunamigenic earthquakes along the U.S. West Coast. Although the Cape Mendocino earthquake was large and inflicted substantial damage on land, the shallow dip angle of about 12 degrees induced vertical motion of the offshore ocean bottom of less than 1 m, producing correspondingly small ocean waves. However, there is increasing geological and seismological evidence that great ($M > 8$) earthquakes have previously occurred in the CSZ, and that these earthquakes have generated much larger, destructive tsunamis (ATWATER, 1987; HEATON and HARTZELL, 1987; WEAVER and SHEDLOCK, 1992). Furthermore, recent events in Japan (BERNARD et al., 1993), Indonesia (YEH et al., 1993), and Nicaragua (KANAMORI and KIKUCHI, 1993; SATAKE et al., 1993) have demonstrated that earthquakes as small as magnitude 7.1 can generate tsunamis capable of devasting communities in the near field, i.e., within a wavelength or so of the generation region (also see other articles in this issue on these tsunamis). Finally, the edge waves which accompanied the Cape Mendocino tsunami add a new dimension to the hazard along this coastline by virtue of two defining characteristics. First, such waves are coastal trapped, i.e., their amplitude is a maximum at the coastline and rapidly decays seaward; second, they propagate slowly along the coast and thus, depending on source distance and geometry, can arrive at coastal communities hours after the highest non-trapped tsunami wave incident from deeper water offshore. Near-source communities on the U.S. West Coast may thus be confronted by a hazard that has both a *rapid onset* and *long duration*; i.e., the non-trapped tsunami may strike from offshore in the first few minutes, then be followed hours later by slow-moving but energetic edge waves trapped along the coast.

Scientifically, edge wave research has a long and venerable history (STOKES, 1846; ECKART, 1951; URSELL, 1952; BALL, 1967), and their presence makes this event particularly interesting. They are apparently ubiquitous along the California coastline as a low level continuum of background noise; but more energetic modes can be excited by a number of mechanisms, including interactions with swell energy, atmospheric disturbances, and tsunamis incident on the coast from deep water (GREENSPAN, 1956; MUNK et al., 1956; FULLER and MYSAK, 1977). The present case is especially interesting because the tsunami edge waves were directly generated in the nearshore region by an impulsive vertical motion of the ocean bottom; they then propagated north and south from Cape Mendocino, trapped nearshore within the coastal waveguide for edge waves.

The purpose of this study is to present a simple but plausible analysis of the coastal tide gage time series that establishes the presence of both non-trapped and coastal trapped modes of tsunami wave energy. We will do this by comparing and interpreting the observations with both numerical model simulations and simple edge wave theory. Although both modes are tsunami waves, in conformance with common usage and for ease of exposition in what follows, we will sometimes refer to the non-trapped tsunami mode simply as the "tsunami," and to the coastal trapped tsunami modes as "edge waves."

Generation Mechanism: Impulsive, Nearshore Ocean Bottom Displacement

Figure 2 presents model results for the static vertical displacement of land and ocean bottom induced by crustal motion on a simple, rectangular fault plane (OKADA, 1985). We used values for the fault plane parameters of 21.5 km length, 16 km width, 6.3 km depth, 342° strike, 12° dip, 107° rake, 2.7 m slip, with a rigidity value of 3×10^{11} dyn/cm^2, with the plane situated such that the epicenter at (40° 18.08′N, 124° 11.80′W) was located in the southeast corner. This is the optimal fault plane model as described by OPPENHEIMER *et al.* (1993), and represents the

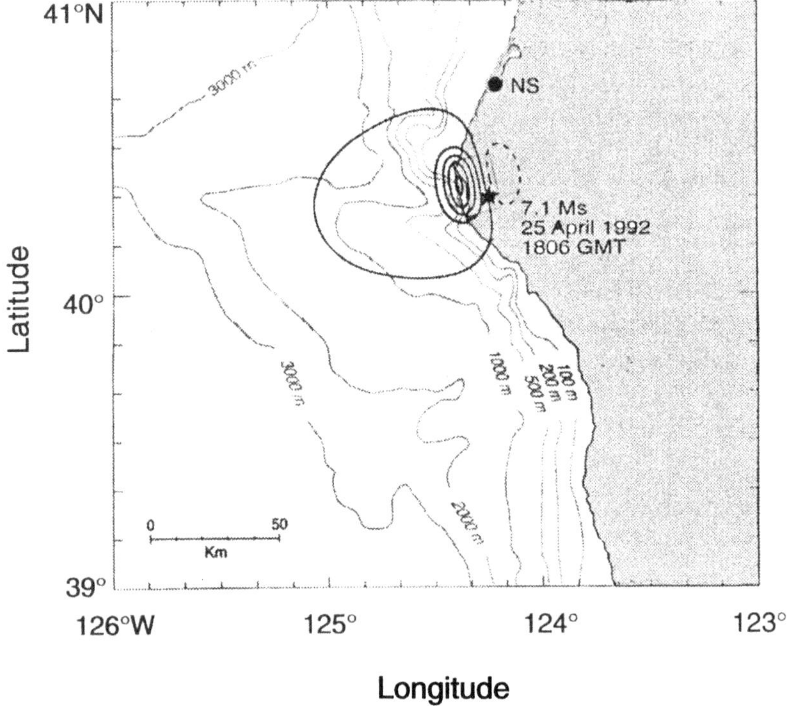

Figure 2

Computed vertical crustal deformation for a single fault plane model. Heavy, solid contours represent uplift, dashed subsidence; the first solid contour is 0.2 cm, and all other contours are at 20 cm intervals. The solid line rectangle delineates the surface projection of the fault plane (see *Generation Mechanism* section in text for fault plane parameters), and the location of the epicenter is indicated by a star.

best fit of vertical and horizontal model displacements to all available observations acquired from GPS and Geodolite measurements inland of the coast and from organism die-off horizons induced by uplift along the coast. A very reasonable fit to field observations is thereby obtained, although model variations are smoother and individual field measurements are somewhat higher. Good qualitative agreement is thus expected between reality and the model, although the real distribution of vertical displacement is probably not as smooth as the dipole pattern produced by this idealized model. As can be seen, the nearshore uplifted region delineated by the 20-cm level contour can be characterized as an ellipsoid with major and minor axes approximately 30 and 25 km long, respectively, with the major axis oriented in the local alongshore direction, and with the center located about 5 km offshore. The duration of this displacement was on the order of 10 s (Y. TANIOKA, personal communication), i.e., short enough to be considered an impulsive ocean wave generation mechanism (HAMMACK, 1973).

Coastal Sea-level Measurements

Analog tide gage records from each station were provided by NOAA's National Ocean Service. Unfortunately, only digital data with a 6-min sampling period was available for Arena Cove. The analog records were digitized at a sampling rate of 15 s, then subjected to a 2–90 min band-pass filter to remove the tides. To aid in the identification of tsunami energy in the tide gage records, we also performed travel time computations for selected West Coast stations, using an iterative grid refinement technique (BRADDOCK, 1969). The computations assume that the deep water tsunami waves travel at the long wave group velocity, $(gh)^{1/2}$, where g is the

Table 1

Wave ray travel time computations, assuming propagation speed $c_g = (gh)^{1/2}$

Station	Travel Time (min)
Port Orford	50
Crescent City	47
North Spit	26
Arena Cove	37
Point Reyes	69
Fort Point	94
Alameda	135
Monterey	64
Port San Luis	97

acceleration due to gravity and h the water depth. The results are summarized in Table 1, and are also presented graphically in Figure 1 in the form of vertical lines that intersect the time axes of the individual time series plots.

At the tide gage station nearest the epicenter, North Spit (in Humboldt Bay), a very small positive wave is observed to arrive after about 25 min. Here, the clear positive sense of the first wave is consistent with the uplift assumed in the generating region (Figure 1). Two distinct wave packets are evident, with the second arriving about 135 min after the main shock. The first wave packet is more energetic than the second, with maximum positive amplitudes of 20 and 15 cm, respectively, and the arrival time of the first is consistent with the computed estimate of 26 min (Table 1, Figure 1), corresponding to a wave ray path (not shown) that traverses deeper offshore water.

However, the largest amplitudes (about twice that at North Spit) were recorded at Crescent City, even though this station is three times more distant from the source region than North Spit (Figure 1). Again, two well-defined packets of wave energy are apparent in the first five hours of the record. In contrast to North Spit, however, the second packet is more energetic than the first, with maximum positive amplitudes of 53 and 35 cm, respectively. Neither the precise arrival time nor the polarity (positive or negative sense) of the first wave are clear in this record, due to the presence of background noise. However, the arrival of the first packet of tsunami wave energy is again consistent with the travel time estimate of 47 min (Table 1, Figure 1), corresponding to the non-trapped mode. The second wave packet arrives about 155 min after the main shock; the analysis presented below suggests that these are coastal trapped waves, i.e., edge waves, characterized by much slower propagation speeds and by amplitudes which decrease rapidly with distance offshore.

As previously noted, the Arena Cove record suffers from a longer sea-level sampling interval (6 min) than that of the other stations (15 s), so that waveforms may be somewhat distorted. Nonetheless, it is again apparent that two packets of tsunami energy are present, the first arriving about 35 min and the second about 210 min after the main shock. At Point Reyes, a very small amplitude, low frequency wave packet arrives after 65 min, followed by a larger amplitude packet which arrives 180 min after the main shock. Much weaker signals are discernible at Fort Point and Alameda in San Francisco Bay, starting at about 90 and 110 min, respectively; it is possible to identify and match individual features in these two records, as the wave propagates into the bay with relatively little distortion. At Monterey, a weak low frequency signal is also seen to arrive at about the predicted time of 64 min.

Taken together, the last four stations demonstrate the importance of two local bathymetric features—Monterey Canyon, just offshore of Monterey Bay, and the broad shelf region that extends north from the canyon to Arena Cove. Thus, the tsunami arrives at Point Reyes and Monterey at about the same time, even though

Monterey is 200 km farther south; this is because the speed of the tsunami is significantly reduced in the relatively shallow approach to Point Reyes, but increased in the deeper waters offshore of Monterey. The significantly later arrival times at Fort Point and Alameda, inside San Francisco Bay, also reflect the influence of the broad, shallow shelf outside the bay, the relatively narrow entrance to the bay, and the shallow depth of the bay itself.

Both the extreme southern and northern West Coast records, Port San Luis and Port Orford (Figure 1), are difficult to interpret. The first two hours of each are somewhat noisy, but coherent low frequency oscillations appear to begin at about the predicted arrival times. It is unclear whether more than one wave packet is present in either record. At Point Arguello and Point Concepcion, just south of Port San Luis, there is an abrupt change in the trend of the coastline (Figure 1). Evidently, no significant wave energy propagated around and to the southeast of this feature, since records we examined for stations along this coast displayed no tsunami signal.

Numerical Model Results

Tsunami propagation was simulated as an initial value problem, by integrating the nonlinear long wave equations with friction by means of a leap-frog finite difference scheme with upwind advection terms (SATAKE, 1995). Boundary conditions required zero normal flow at the coastline, and a constant sea-surface slope was maintained at open ocean boundaries through application of a radiation condition (HWANG et al., 1972). The integration time step was 5 s, the resolution of the computational grid was 1 arc-minute, and the computational region extended over the 10 degree by 10 degree rectangle delineated by the dotted line in Figure 1.

Initial sea-level values were specified by assuming that the ocean surface instantaneously conformed to the ocean bottom displacement while horizontal water velocities were set to zero. We note that, as a consequence of the position and orientation of the modeled crustal displacement (Figure 2), the initial sea-surface displacement was characterized as a mound of water with a maximum value of 80 cm situated less than 5 km offshore in 50–100 m of water, an alongshore length scale of about 30 km, and an offshore length scale of about 8–10 km extending from the coast to about the 500 m bathymetry line.

The observed and computed records at the eight coastal tide gage stations are compared in Figure 1. The agreement in period and phase of individual waves is quite good at all stations and, although amplitudes do not agree as well, individual wave packets with 2–5 cycles per group are reproduced at several stations. First wave arrivals of both model and observations at each station agree well with those predicted by wave ray tracing (Table 1, Figure 1). These initial waves are thus the non-trapped tsunami waves traveling with group velocity $(gh)^{1/2}$ and have mini-

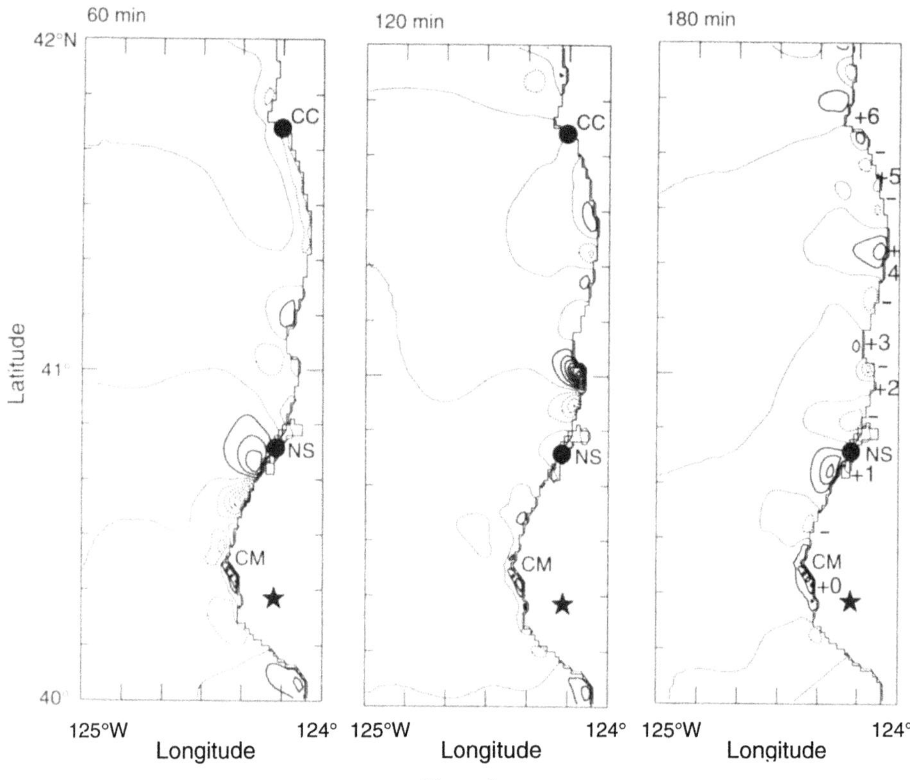

Figure 3

Sea-level distributions computed by the tsunami numerical model at 60, 120, and 180 min after generation; contour interval is 5 cm, and solid and dotted lines indicate values above and below mean sea level, respectively. At 180 min, six cycles of edge wave extrema along the coast are marked by numbered plus signs and unlabeled minus signs. (The extrema are actually at the coastal boundary, not offshore; see text for explanation.) The star marks the position of the earthquake epicenter, and the locations of Cape Mendocino (CM), North Spit (NS) and Crescent City (CC) are also indicated.

mized their travel time by propagating along paths traversing the deeper offshore water associated with higher group speeds.

In the next section, we show that the subsequent wave packets are edge waves, i.e., waves characterized by rapidly decreasing amplitude offshore (trapped), and very slow speeds of alongshore propagation. This is clearly seen in Figure 3, a series of snapshots of computed sea level, in which individual, coastal trapped waves can be seen propagating alongshore away from the source region. (Note that in this figure, maximum and minimum values are actually at the coast, not offshore; the apparent offshore locations of these extrema are false artifacts of the contouring algorithm, which interpolates and inserts contour lines between zero-value land cells and adjacent nonzero sea-level values.)

Edge Wave Analysis

The simplest bathymetry that gives rise to edge waves is a one-dimensional semi-infinite sloping beach for which depth, h, increases linearly with the offshore coordinate, x, as

$$h(x) = sx, \tag{1}$$

where s is a constant and represents the slope of the beach. In this case, solutions exist which decay exponentially offshore and are sinusoidal in the alongshore coordinate, y. They are of the form (e.g., LeBlond and Mysak, 1978; Mei, 1983)

$$\eta_n(x, y) = A \, e^{i(k_n y - \omega_n t)} \, e^{-k_n x} L_n(x). \tag{2}$$

Here, n is the mode number, η_n is the wave amplitude, k_n is the alongshore wave number, ω_n is the angular frequency, and the $L_n(x)$ are the n-th order Laguerre polynomials, of which the first three are

$$L_0(x) = 1$$

$$L_1(x) = (1 - 2k_1 x) \tag{3}$$

$$L_2(x) = (1 - 4k_2 x + 2k_2^2 x^2).$$

The dispersion relation for each mode is then given by

$$\omega_n^2 = (2n + 1)sgk_n. \tag{4}$$

The phase and group velocities can then be written, respectively, as

$$c = \omega_n/k = [(2n + 1)sg/k]^{1/2}, \tag{5}$$

$$c_g = d\omega_n/dk = 1/2[(2n + 1)sg/k]^{1/2} = c/2, \tag{6}$$

where the mode number subscript on the variables c, c_g and k are not explicit, but understood in context.

We note that these phase speeds are equivalent to that of a deep water gravity wave multiplied by the beach slope and a factor dependent on the mode number. The phase speed thus increases with mode number, but low modes on beaches of small slope propagate considerably slower than either deep water gravity waves at the same frequency or long, shallow water gravity waves such as tsunamis. These edge waves are also highly dispersive with the group speed only half that of the phase speed.

If these dispersive edge waves were impulsively generated as a mound of water similar to the displacement pattern shown in Figure 2, then the initial mound should evolve into a wave group traveling at celerity c_g, with individual wave crests propagating through the group at the higher phase speed, c. In particular, we expect

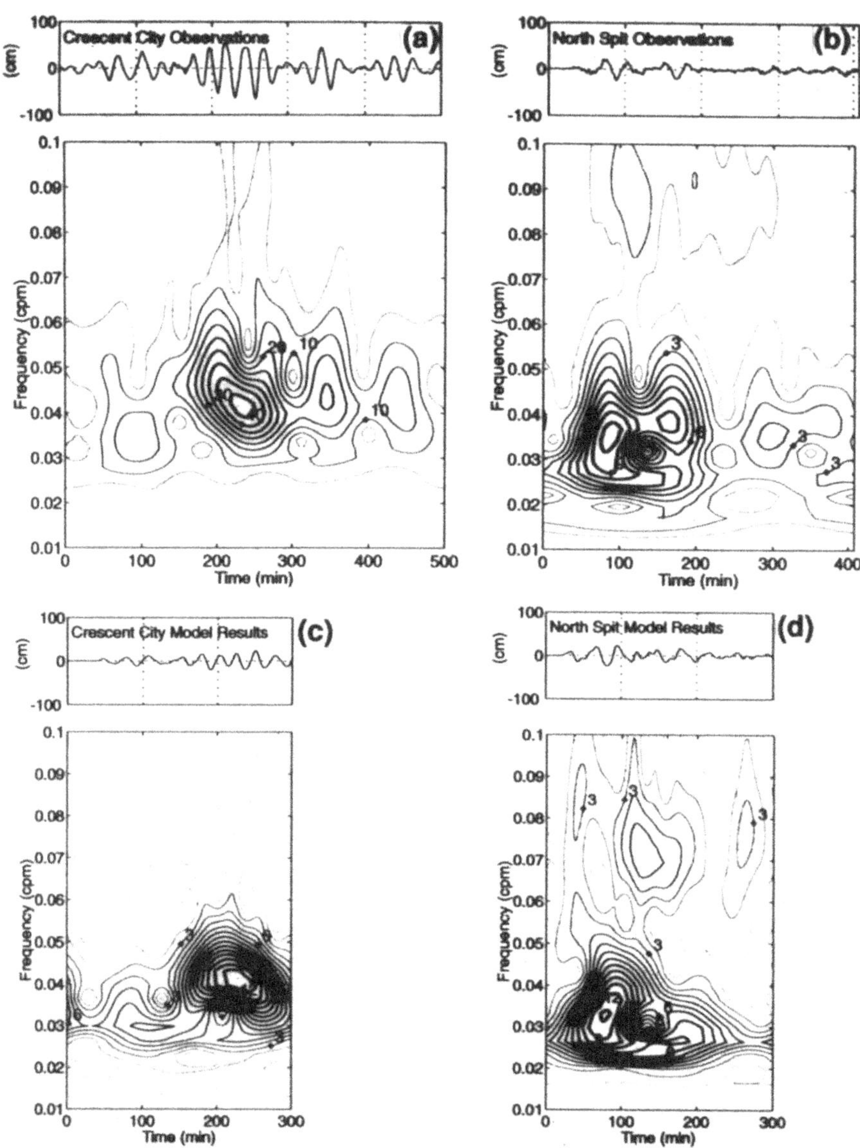

Figure 4

Observed and modeled time series (top of each panel) and temporal evolution of the amplitude spectra (bottom of each panel) for Crescent City and North Spit. In the spectral plots, note that the contour interval of the Crescent City observations [4(a)] is 5 cm, while contour intervals of the other three spectra are 1 cm; contour lines have been broadened to facilitate differentiation of local maxima (relatively broad) and local minima (relatively narrow).

the maximum of the mound to travel at the group velocity, so that arrival of the group maximum can be used to estimate the group velocity as

$$c_g = \Delta r / \Delta t, \qquad (7)$$

where Δr is the distance traveled and Δt is the elapsed time from generation.

The nearshore bathymetry to the north of Cape Mendocino is simpler and more reasonably modeled as a semi-infinite beach (Eq. (1)) than that to the south. However, the quality of our numerical model bathymetric grid also degrades somewhat north of Crescent City; this is because high quality, 15 arc-second resolution NOAA bathymetry was not available north of 42°. Luckily, the wave amplitudes are highest and the signal-to-noise ratio is most favorable at North Spit and Crescent City, stations that are located nearest the generation region and in the area of our most reliable bathymetric data (Figure 1). For these reasons, we focus our analyses on these two stations.

The bathymetric data provide estimates of Δr, the propagation distance along the coast from the generation zone to the observation point, and estimates of the mean bottom slope near the coast, s. We computed Δr as the distance along the 50 m depth contour from the generation region to Crescent City and North Spit and found them to be 59 km and 174 km, respectively. We next computed the mean depth profile for the offshore region extending from the generation area to Crescent City; successive linear least square fits to the first 10, 20, and 25 km of this profile yielded estimates of $s = 0.008$, 0.010 and 0.011, respectively, so that $s \approx 0.01$ appears to be a robust estimate over the first 25 km.

The tide gage observations provide estimates of ω and Δt. To aid in the estimates of these two parameters, both observed and computed time series at North Spit and Crescent City were subjected to an analysis of the temporal evolution of their energy spectrum (DZIEWONSKI, 1969); the results are presented in Figure 4. Table 2 summarizes these estimates, and Table 3 provides a comparison of observed group velocities with the first three theoretical edge wave modes.

At Crescent City, it appears that the $n = 0$ edge wave mode explains the observed group velocity of the second wave packet (Table 3). Figure 5 graphically

Table 2

Estimates of wave group parameters at Crescent City and North Spit (refer to Figure 4)

Station	Wave Group	$f = \omega/2\pi$ (cpm)	$T = 1/f$ (min)	Δt (min)	Δr (km)	c_g (km/min)
Crescent City	Second Observed	0.0412	24.3	234	174	0.74
Crescent City	*Second Modeled*	*0.0407*	*24.5*	*230*	*174*	*0.75*
North Spit	First Observed	0.0348	28.7	89	59	0.66
North Spit	*First Modeled*	*0.0326*	*30.7*	*77*	*59*	*0.76*

Table 3

Comparison of edge wave theory and observations

Station	n	k (km^{-1})	$L = 2\pi/k$ (km)	Computed c_g (km/min)	Observed c_g (km/min)
Crescent City	0	0.1895	33.2	0.68	0.74
Crescent City	1	0.0632	99.5	2.05	0.74
Crescent City	2	0.0379	165.8	3.41	0.74
North Spit	0	0.1358	46.3	0.81	0.66
North Spit	1	0.0453	138.8	2.42	0.66
North Spit	2	0.0271	231.3	4.03	0.66

summarizes this simple model, presenting the mean profile data, the linear least square fit to these data, and the offshore (exponential) dependence of the zero-mode edge wave solution with unit amplitude at the coast, i.e., for Equations (2) and (3) with $n = 0$ and $A = 1$. As a consistency check on the offshore length scales involved, we note that for $n = 0$ and $k = 0.1895$ km^{-1} (Table 3), the amplitude is reduced to 10% of the maximum value at an offshore distance of 12.2 km, comfortably within the 20–25 km offshore limit we have assumed for our simple model.

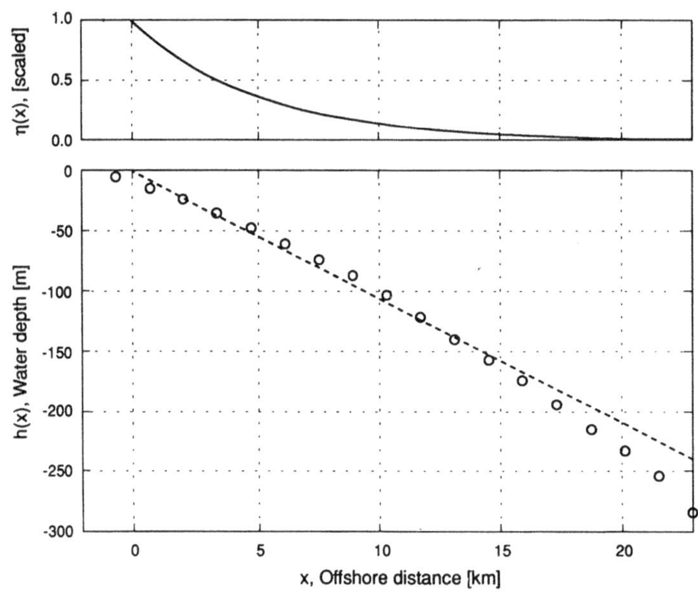

Figure 5

Lower panel presents the mean depth profile of the coastal region from Cape Mendocino to Crescent City (open circles), and the linear least square fit for the first 20 km offshore (dashed line). The offshore distance indicated by the horizontal axis corresponds to the linear least square fit model. The upper panel presents the offshore profile (scaled to unit amplitude at the coastline) of the zero-mode edge wave solution.

This agreement of Crescent City observations with simple edge wave theory allows us to recognize and interpret certain features of the numerical model results in terms of the underlying edge wave dynamics. In the sea-level distribution computed by the model at 180 min (Figure 3), six wave cycles are discernible over the distance of 174 km along the 50 m bathymetric contour from the generation region to Crescent City. Although the separation between extrema varies due to bathymetric variations and coastline irregularities (FULLER and MYSAK, 1977), the average wavelength of these six cycles is 29.0 km, which compares well with the zero-mode edge wave result of the simple theoretical model, $L = 2\pi/k = 33.2$ km (Table 3). We thus believe that this feature of the numerical model reflects the dynamics approximated by the simple analytic theory for edge waves presented above.

At North Spit, however, the interpretation is less straightforward. The measured waveform is delayed relative to theory, with an apparent group velocity that is 19% smaller than predicted (Table 3). This discrepancy is primarily due to two factors. First, the theory applies to the open coast while the observations were made inside of Humboldt Bay, about 1.5 km from the interior end of the narrow entrance channel that is approximately 0.6 km wide and 2 km long. Consequently, wave propagation through the channel and into the bay should result in arrival times at the gage that are delayed relative to arrival at the coast. Second, both non-trapped and edge wave mode tsunami energy are present in the first wave group of the North Spit record. The relative amplitude and phasing of these two modes creates constructive and destructive interference that, in general, results in a hybrid wave group with different arrival time for the group maximum. At Crescent City, these effects are also present but much less important because this station is substantially farther from the source. As a consequence, delays in arrival at the gage relative to the coast are a smaller fraction of the total travel time, and superposition of the non-trapped and edge wave modes is minimized because their respective arrivals are well separated in time.

The first factor, the effect of a narrow channel entrance on tsunami propagation, has been simulated for Humboldt Bay by WHITMORE (1993), who utilized a fine-scale computational grid (12 × 12 arc-seconds) and found that the waveform near the North Spit station was both attenuated in amplitude and delayed relative to arrival at the open coast (P. Whitmore, private communication). In contrast, our numerical model utilizes a relatively coarser computational grid with individual cells which are about three times the approximate width of the entrance channel. However, we did not vary the frictional coefficient or the channel depth to delay the wave and improve agreement; as a consequence, our numerical model waveforms lead the observed by a few minutes at both Crescent City and North Spit (Figures 1 and 4).

The second complicating factor in the North Spit record, superposition of the tsunami and edge waves, can be illustrated by examining the period and expected arrival times of these two modes. For the tsunami, note that the estimated arrival time at North Spit is 26 min (Table 1) and that the period at Crescent City is about 30 min (Figure 4a); thus, since the period should be invariant, the second tsunami wave cycle would arrive at North Spit $26 + 30 = 56$ min after generation. For the edge wave, the best estimates of period and group velocity are again those observed at Crescent City (Table 2), from which we can estimate an arrival time at North Spit of $\Delta t = \Delta r/c_g = 59$ km$/0.74$ km/min $= 80$ min for the center of a zero-mode group with a dominant period of 24 min. Thus, the cycle preceding the center of the group would arrive at $80 - 24 = 56$ min, i.e., about the arrival time estimated for the second tsunami cycle. We conclude that the first wave packet observed in the North Spit time series (Figure 4b) is at least partly composed of the sum of both the non-trapped and the zero-mode edge wave tsunami modes.

Finally, we note that the expected arrival time of the first mode ($n = 1$) at North Spit is $\Delta t = \Delta r/c_g = 59$ km$/2.05$ km/min $= 29$ min, and that of the second mode ($n = 2$) is $\Delta t = 59$ km$/4.03$ km/min $= 15$ min (Tables 2 and 3). Since Figure (4b) displays no significant energy during the first 50 min of the record, we conclude that the first and second edge wave mode energy is absent or negligible in the North Spit record. Because they must traverse North Spit to reach Crescent City, we also conclude that these modes are absent in the Crescent City record.

Discussion

The edge wave model we have used to interpret the observations is the simplest possible, i.e., a semi-infinite sloping beach, even though the actual physical setting deviates significantly from this ideal. For example, although the mean depth profile presented in Figure 5 is reasonably well fit by a linear function, some curvature in the data is apparent, so that a quadratic or exponential fit (not shown) provides a better representation. Furthermore, some alongshore variations in slope do occur, and coastal irregularities are not accounted for. Nonetheless, the correspondence of observations, numerical model, and this very simple edge wave theory are quite convincing. We believe that the essential features of the observed time series can be explained by the presence of both deep water tsunami waves and shallow water edge waves trapped along the coast.

A more detailed comparison and discussion of the observed and modeled time series is provided by the spectral evolution analyses presented in Figure 4. The North Spit records in Figures 4(b) and 4(d) display excellent agreement in the most important parameters. The main energy lobes are similar in both the amplitude and in the details of its structure and evolution in the frequency-time plane; this close match in the frequency domain reflects the remarkably close correspondence of individual features in the original time series. There is, however, an interesting difference in a detail of the main lobes. Both have a high frequency ridge, but that of observations is more pronounced and oriented more vertically than that of the model. Inspection reveals a vee-shaped trough in the observed time series after about 90 min have elapsed, while the corresponding trough in the model time series is more sinusoidal; in contrast, the peak that follows is more vee-shaped in the model than in the observations. Higher frequency Fourier components are required to resolve such peaks and troughs, so the difference in the two ridges may be due to the difference in the location and severity of these sharp, nonsinusoidal features in the time series. A more important disagreement in the two North Spit records is most clearly revealed as a difference in the location of the secondary lobe, which appears earlier and at a lower frequency in the observations. This difference appears to be due to late arrival of edge wave energy in the simulation; small differences in the phasing of the deep water tsunami and edge waves could have produced the differences apparent in both the waveform and the spectral evolution diagram.

At Crescent City, there is also qualitative agreement between observed and modeled evolution of the spectra displayed in Figures 4(a) and 4(c). Two wave groups are apparent which arrive at nearly the same time, and in each time series the second is more energetic than the first. But important differences are also apparent. The most striking is the disparity in amplitude; both the non-trapped and edge wave tsunami modes are observed to be two to three times larger than the model results. The source of this discrepancy is not presently understood, but a number of factors could be responsible, e.g., errors in bathymetry or initial conditions. The simulated non-trapped tsunami is also somewhat lower frequency than observed, and the observed edge waves are characterized by sharp, vee-shaped troughs that create a high frequency ridge absent in the modeled results.

Why were the largest waves observed at Crescent City rather than North Spit (Figures 1 and 4), even though Crescent City is located three times farther from the source than the North Spit station (Table 2)? This is an important and interesting issue. It is possible that the lower amplitudes observed at North Spit are due to destructive interference of tsunami and edge wave energy. But the Crescent City community has a well-known history of especially severe tsunami damage, and a number of investigators have suggested possible mechanisms that may make this area especially responsive to incident long wave energy. Various studies of individual tsunami events have cited source directionality (WIEGEL, 1976; HATORI, 1993), focusing of wave energy by offshore bathymetry (ROBERTS and CHIEN, 1964), and resonant shelf oscillations (WILSON and TORUM, 1968) to explain high tsunami amplitudes at this location. HOUSTON and GARCIA (1978) performed finite difference simulations of the 1964 Alaskan tsunami. Their model results indicate that the high amplitudes on the West Coast were generally due to source directionality but that the Crescent City area in particular was also characterized by quite localized resonant response, extending over length scales only 2 to 4 miles (4 to 7 km) alongshore. Both the coastal tide gage observations and our numerical model results are consistent with a mechanism involving local effects since the coastal trapped edge wave amplitudes are larger than the deep water tsunami waves in both the computed and observed Crescent City record.

Both the Crescent City and North Spit observations are characterized by additional wave groups that appear after the edge wave. They are more pronounced at Crescent City, where they occur at approximately the same frequency as the first and largest edge wave, and are regularly spaced at intervals of about 100 min (Figure 4a). The nature and origin of these waves is not clear, but we speculate that these observed characteristics may indicate a mechanism involving reflection of the zero-mode edge waves.

Summary and Conclusions

A small tsunami was generated at the southern end of the Cascadia Subduction Zone on 25 April 1992 by a magnitude 7.1 earthquake near Cape Mendocino, California. Wave energy propagated north and south along the West Coast and was detected at NOAA coastal tide gage stations in California and Oregon. The records were characterized by several distinct wave groups arriving over the course of 8–10 hours. We have processed these data to remove the tidal signal and isolate the energy in the tsunami frequency band and have estimated the temporal evolution of the spectra for each time series. We have compared these records with arrival times estimated by wave ray computations, nonlinear long wave numerical model simulations and, for the records at North Spit and Crescent City, simple edge wave theory based on a linear fit to the mean depth profile. We obtain consistent results between observations, model results and theory, and our analyses lead us to conclude

(1) that both non-trapped and coastal trapped edge wave tsunami modes were generated, and

(2) that the edge wave mode was the gravest, or zero-mode with alongshore wavelengths of about 30 km and a maximum amplitude at the coast which decayed exponentially offshore with a length scale of 5–10 km.

Furthermore, since this event occurred at the southern end of the Cascadia Subduction Zone, it is particularly significant to issues of tsunami hazard mitigation along the adjacent California, Washington and Oregon coastlines. At Crescent City, the edge wave packet was characterized by 50% higher amplitude, and therefore twice the energy of the non-trapped tsunami mode, and arrived several hours after tsunami waves were incident from offshore.

Considerably more energetic edge wave generation by future large earthquakes in the CSZ must therefore be considered a distinctly hazardous possibility. In such a scenario a community might be subjected to a damaging tsunami within minutes of an earthquake, assumed from the subsidence of these waves that the danger had passed, then be struck unexpectedly a number of hours later by even larger waves which had propagated slowly along the coast as trapped edge wave tsunami modes.

We thus also conclude

(3) that energetic edge waves generated by future tsunamigenic earthquakes in the CSZ pose an additional and particularly insidious hazard to U.S. West Coast communities, since the edge wave energy may appear unexpectedly, hours after the non-trapped tsunami waves have arrived and dissipated and

(4) that local dynamic processes may make Crescent City particularly responsive, and therefore more vulnerable, to such coastal trapped long wave energy.

Acknowledgments

Coastal tide gage records were provided by NOAA's National Ocean Service. This study is contribution no. 1592 of the Tsunami Research Project in the Ocean Environment Research Division of NOAA's Pacific Marine Environmental Laboratory, and contribution no. 299 of the Joint Institute for the Study of the Atmosphere and Ocean. Partial funding for this work was provided by NOAA's Coastal Ocean Program.

REFERENCES

ATWATER, B. F. (1987), *Evidence for Great Holocene Earthquakes Along the Outer Coast of Washington State*, Science *236*, 942–944.

BALL, F. K. (1967), *Edge Waves in an Ocean of Finite Depth*, Deep-Sea Res. *14*, 79–88.

BERNARD, E. N., GONZÁLEZ, F. I., SIGRIST, D., TSURUYA, H., KATO, K., SHUTO, N., MATSUTOMI, H., TSUJI, Y., ITO, H., YAMAMOTO, K., IWASAKI, Y., SASAKI K., HATTORI, A., SUZUKI, Y., TAKAHASHI, S., GOTO, C., HASHIMOTO, N., HOSOYAMADA, T., NAGAO, T., MIZUNO, Y., YANO, K., NAGAOKA, O., HONDA, S., and TATEHATA, H. (1993), *Tsunami Devastates Japanese Coastal Region*, EOS, Trans. Am. Geophys. Union *74*, pp. 417 and 432.

BRADDOCK, R. D. (1969), *On Tsunami Propagation*, J. Geophys. Res. *74*, 1952–1957.

DZIEWONSKI, A. S. (1969), *A Technique for the Analysis of Transient Seismic Signals*, Bull. Seismol. Soc. Am. *59*, 427–444.

ECKART, C., *Surface Waves in Water of Variable Depth*, Scripps Institution of Oceanography Wave Report 100 (1951).

FULLER, J. D., and MYSAK, L. A. (1977), *Edge Waves in the Presence of an Irregular Coastline*, J. Phys. Ocean *7*, 846–855.

GREENSPAN, H. P. (1956), *The Generation of Edge Waves by Moving Pressure Distributions*, J. Fluid Mech. *1*, 574–592.

HAMMACK, J. L. (1973), *A Note on Tsunamis: Their Generation and Propagation in an Ocean of Uniform Depth*, J. Fluid Mech. *60*, 769–799.

HATORI, H., *Distribution of tsunami energy on the circum-Pacific zone*, In *Proceedings, IUGG/IOC International Tsunami Symposium* (Wakayama, Japan, 1993).

HEATON, T. H., and HARTZELL, S. H. (1987), *Earthquake Hazards on the Cascadia Subduction Zone*, Science *236*, 162–168.

HOUSTON, J. R., and GARCIA, A. W. (1978), *Type 16 Flood Insurance Study: Tsunami Predictions for the West Coast of the Continental United States*, U.S. Army Engineer Waterways Experiment Station Technical Report H-78-26.

HWANG, L.-S., BUTLER, H. L., and DIVOKY, D. J. (1972), *Tsunami Model: Generation and Open-sea Characteristics*, Bull. Seismol. Soc. Am. *62*, 1579–1596.

KANAMORI, J., and KIKUCHI, M. (1993), *The 1992 Nicaragua Earthquake: A Slow Tsunami Earthquake Associated with Subducted Sediments*, Nature *361*, 714–716.

LEBLOND, P. H., and MYSAK, L. A., Waves in the Ocean (Elsevier Scientific Publishing Company, 1978).

MEI, C. C., *The Applied Dynamics of Ocean Surface Waves* (John Wiley and Sons, 1983).

MUNK, W. H., SNODGRASS, F. E., and CARRIER, G. (1956), *Edge Waves on the Continental Shelf*, Science *123*, 127–132.

OKADA, Y. (1985), *Surface Deformation due to Shear and Tensile Faults in a Half-space*, Bull. Seismol. Soc. Am. *75*, 1135–1154.

OPPENHEIMER, D., BEROZA, G., CARVER, G., DENGLER, L., EATON, L., GEE, L., GONZÁLEZ, F., JAYKO, A., LI, W. H., LISOWSKI, M., MAGEE, M., MARSHALL, G., MURRAY, M., McPHERSON, R., ROMANOWICZ, B., SATAKE, K., SIMPSON, R., SOMERVILLE, P., STEIN, R., and VALENTINE, D. (1993), *The Cape Mendocino, California, Earthquake of April 1992: Subduction at the Triple Junction*, Science *261*, 433–438.

ROBERTS, J. A., and CHIEN, C.-W. (1964), *The Effects of Bottom Topography on the Refraction of the Tsunami of 27–28 March 1964: The Crescent City Case*, Meteorology Research, Inc. Report.

SATAKE, K. (1995), *Linear and Nonlinear Computations of the 1992 Nicaragua Earthquake Tsunamis*, Pure and Appl. Geophys., this volume.

SATAKE, K., BOURGEOIS, J., ABE, K., TSUJI, Y., IMAMURA, F., IIO, Y., KATAO, H., NOGUERA, E., and ESTRADA, F. (1993), *Tsunami Field Survey of the 1992 Nicaragua Earthquake*, EOS, Trans. Am. Geophys. Union *74*, pp. 74, 145 and 156–147.

STOKES, G. G. (1846), *Report on Recent Researches in Hydrodynamics*, 16th Meeting Brit. Assoc. Adv. Sci., 1–20.

URSELL, F. (1952), *Edge Waves on a Sloping Beach*, Proc. R. Soc. Lond. *A214*, 79–97.

WEAVER, C. S., and SHEDLOCK, K. M. (1992), *Estimates of Seismic Source Regions from Consideration of the Earthquake Distribution and Regional Tectonics in the Pacific Northwest*, U.S. Geological Survey, Open-File Report 91–441.

WHITMORE, P. M. (1993), *Expected Tsunami Amplitudes and Currents along the North American Coast for Cascadia Subduction Zone Earthquakes*, Natural Hazards *8*, 59–73.

WIEGEL, R. L. (1976), *Tsunamis*. In *Seismic Risk and Engineering Decisions* (eds. Lomnitz, C., and Rosenblueth, E.) (Elsevier Scientific Publishing Company, Amsterdam, 1976) pp. 225–286.

WILSON, B. W., and TORUM, A. (1968), *The Tsunami of the Alaskan Earthquake, 1964*, U.S. Army Corps of Engineers Coastal Engineering Research Center Technical Memorandum *25*.

YEH, H., IMAMURA, F., SYNOLAKIS, C., TSUJI, Y., LIU, P., and SHI, S. (1993), *The Flores Island Tsunamis*, EOS, Trans. Am. Geophys. Union *74*, pp. 369 and 371–372.

(Received October 17, 1994, revised February 3, 1995, accepted February 9, 1995)

PAGEOPH, Vol. 144, Nos. 3/4 (1995)

0033–4553/95/040427–14$1.50 + 0.20/0
© 1995 Birkhäuser Verlag, Basel

Ocean Cable Measurements of the Tsunami Signal from the 1992 Cape Mendocino Earthquake

D. J. THOMSON,[1] L. J. LANZEROTTI,[1] C. G. MACLENNAN,[1] and L. V. MEDFORD[1]

Abstract — The movement of the seawater across the earth's magnetic field produces a large-scale motional electric field. Using the Point Arena, California, to Hanauma Bay, Hawaii, unpowered HAW-1 cable, we have studied the geopotential across this distance to look for possible tsunami-induced fields that might have been produced following the April 1992 Cape Mendocino earthquake. We have used a ten-day interval prior to and including the earthquake as a reference for geopotential signals and for geomagnetic activity. We have also used geomagnetic data from Point Arena, Honolulu and Boulder as reference data. The results of the analyses show that there are tsunami-related effects in the cable geopotential data. These are (a) larger voltage prediction errors (residuals) for the interval following the main shock; (b) enhanced (compared to the 10d reference interval) geopotential spectral power following the main shock: two enhancements are larger than geomagnetically-induced spectral power enhancements in the same time interval; and (c) strong evidence for an ∼30 min "echo" in the cable geopotential signal following the main shock.

Key words: Tsunami, geopotential, geomagnetism.

Introduction

The 25 April 1992 Cape Mendocino, California, earthquake (1806 UT; magnitude 7.1) and its two principal aftershocks on 26 April (0741 UT; 1141 UT) occurred ∼140 km north of the AT&T Pacific telecommunications cable station that is located on Manchester beach (some 10 km north of the town of Point Arena, from which the cable station derives its name). The tsunami generated by the earth movement in the earthquake was clearly seen in tide gauge measurements along the California coast north and south of the epicenter, as well as in records from Hawaii and other Pacific islands (OPPENHEIMER *et al.*, 1993; GONZÁLEZ *et al.*, 1995). In particular, a significant tide gauge signal (average variation of order ± 10 cm; OPPENHEIMER *et al.*, 1993) was seen at Arena Cove, Point Arena.

The cessation in the late 1980s of commercial usage of the first Pacific telecommunications cable, HAW-1, provided the opportunity for the reuse of the unpow-

---

[1] AT&T Bell Laboratories, Murray Hill, NJ 07974, U.S.A.

ered cable for research purposes. Since 1989 we have been using this cable to monitor the geopotential between California and Hawaii in order to investigate topics in such areas as solid earth physics (e.g., LANZEROTTI *et al.*, 1986), ionospheric physics (e.g., LANZEROTTI *et al.*, 1992, 1993), and ocean physics (CHAVE *et al.*, 1992). Thus, the occurrence of the earthquake and resultant tsunami close to the cable route encouraged us to analyze our data in order to ascertain the nature of the signals that might be recorded from these natural events.

The measurement of the movement of seawater by its motional electromagnetic induction has been investigated since the time of Faraday's initial attempts in studies of the tides in the Thames (FARADAY, 1832). The tidal signal can be large, and numerous studies of tides have been carried out using short cables such as across the Irish Sea (e.g., PRANDLE and HARRISON, 1975) and on trans-oceanic cables (e.g., THOMSON *et al.*, 1986). Studies, using cables, of the flow of ocean current systems such as the Gulf Stream have been the subject of considerable attention (e.g., WERTHEIM, 1954; LARSEN and SANFORD, 1985; LARSEN, 1992). Most of the studies of motional EM induction in seawater are accomplished with sea-bed electric and magnetic sensors (e.g., CHAVE and FILLOUX, 1985; FILLOUX, 1987; CHAVE *et al.* 1989; SEGAWA and TOH, 1992; LILLEY *et al.*, 1993). This paper uses large-scale geopotential data as acquired with a long oceanic cable to investigate the response of the geopotential to motionally-induced electromagnetic fields from the Cape Mendocino tsunami. Since little, if any, work has previously been reported related to this topic, this paper concentrates on detailed analysis rather than modeling.

The location of the earthquake epicenter and the AT&T cable station is shown in Figure 1. The HAW-1 cable consists of two parallel coaxial cables separated by about 100 km throughout their ocean crossing. The south cable broke near the Hawaii coast in 1991, and the center conductor is now at sea ground. The north cable is grounded at Hanauma Bay (HB), Hawaii, using the standard iron rod telecommunications ground array that is located in the bay. At Point Arena (PA), the potentials between the cable center conductors and the local ground (a single standard 8 m long, 2.5 cm diameter telecommunications iron rod surrounded by 13-cm diameter coke that is sunk to sea level) are monitored by an AT&T computer-based data acquisition system. The cable length between PA and HB is ~4050 km. There are 58 (57) unpowered repeaters in the south (north) cable, with a total impedance of 8,150 (7,470) ohms.

A three axis fluxgate magnetometer is also installed at the PA cable station. The three magnetic axes are oriented in the geomagnetic south-north (H component), west-east (D component), and vertical (Z, positive increasing downward; a right-handed system) directions. The analog outputs of the three magnetometer channels as well as the cable potential are multiplexed and sampled at 2 s intervals before being written on removable hard disk cassettes. Engineering parameters of the system are monitored and written to disk at the beginning of every 20-m data file. In the analyses that follow, geomagnetic observatory data from Honolulu, Hawaii,

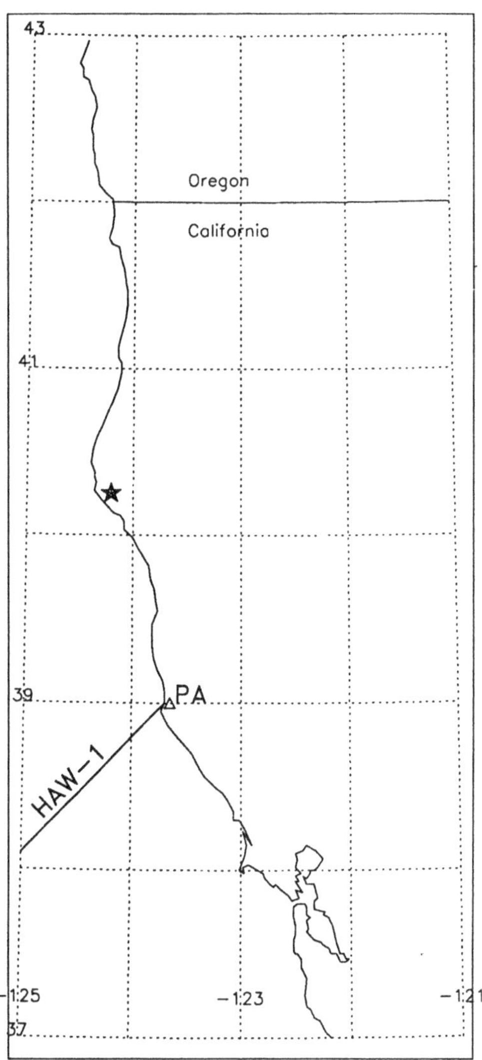

Figure 1

Sketch of the northern California coast showing the location of the 25 April 1992 earthquake epicenter (★), the AT&T cable station near Point Arena (PA), and the two HAW-1 cables to Hawaii.

and from Boulder, Colorado, are also used (provided on CDROMs courtesy of the U.S. Geological Survey).

Analyses

Plotted in Figure 2 are the three magnetometer components, the two cable potentials, and the difference between the two cable potentials for the time interval

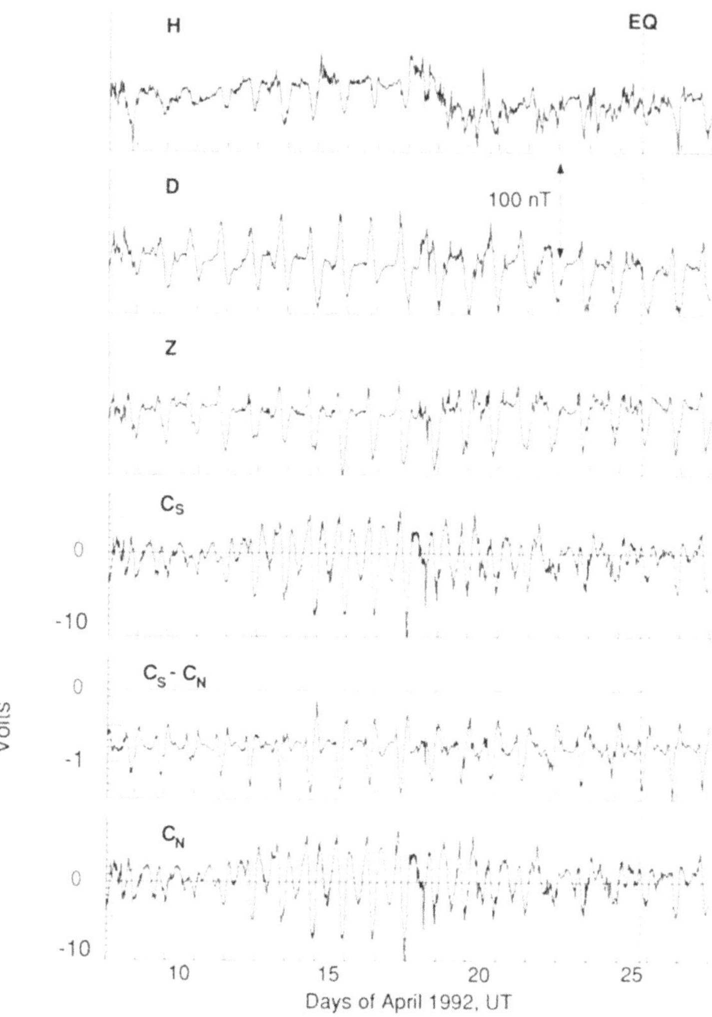

Figure 2

Data for April 8–27 (inclusive), 1992. The three top panels show the three magnetic field components as measured at PA; the traces labeled C_S and C_N are the induced voltages in the south and north cables; $C_S - C_N$ is the difference between the two cable potentials. EQ marks the time of the 25 April earthquake.

0700 UT 8 April to 0700 UT 28 April. The time of the main earthquake shock is indicated by the vertical line through the data set. The daily Sq variations in the D and Z components of the magnetometer are particularly noticeable. The H component shows evidence of magnetic activity beginning on 18 April. The principal large variations in the cable potentials arise from ocean tide (12.42-h period) and ionospheric Sq (12-h period) effects. The 14.9-day beating between these two periods produces the reduced amplitudes of the variations that are seen on 8–12 April and during the interval including the earthquake. The tides tend to be large scale, so that the potential difference between the cables is dominated by the small spatial scale Sq signal, as is evident visually in the trace of $C_S - C_N$ (second trace from bottom in the figure).

In order to study any tsunami-induced geopotential changes from the cable data, it was first necessary to remove the geomagnetic-induced effects. Magnetic field data were obtained from the PA magnetometer as well as from the geomagnetic observatories at Honolulu and Boulder. Prediction of the voltage requires that the transfer functions between the various magnetic field components and the cable be known, at least approximately. Estimation of transfer functions is a classical time-series problem; see e.g., GOODMAN (1957), AKAIKE and YAMANOUCHI (1962), MUNK and CARTWRIGHT (1966), or for recent synthesis, SHUMWAY (1988). In addition, geomagnetic induction has been extensively studied in its own right and the problem of estimating magnetotelluric transfer functions is still a subject of active research; see LANZEROTTI et al. (1986), EGBERT and BOOKER (1989), EGBERT (1989), CHAVE and THOMSON (1989), THOMSON and CHAVE (1990).

Here, a slightly unconventional method was used to predict the cable voltages from the magnetic fields. Recall that magnetotelluric transfer functions are typically estimated on a logarithmic frequency scale while time-domain modeling is usually done on a uniform grid. In this analysis, we made a time-domain estimate of the impulse responses using approximately exponentially-spaced time steps. This represents a compromise between simplicity and numerical stability; in principle, estimating the impulse response on a uniform grid would be preferable (and equivalent to the Fourier transform of the usual MT transfer function) but, as many more parameters must be estimated from the finite data records available, the estimate tends to become numerically unstable. Thus, for this analysis, the fit was made by minimizing the weighted square prediction residuals

$$\sum_{t=T_0}^{T_1} w(t)[\delta V(t)]^2 \tag{1}$$

where T_0 and T_1 represent the start and end time of the fitting interval and $w(t)$ is a weight described below. The residual, $\delta V(t)$ is just the difference between the observed and predicted voltages,

$$\delta V(t) = V(t) - \hat{V}(t) . \tag{2}$$

The prediction, $\hat{V}(t)$, is the sum of two components:

$$\hat{V}(t) = \sum_{k=1}^{P} a_k V(t - \tau_k) + \sum_{m=1}^{M} \sum_{j=1}^{J} \beta_{m,j} B_m(t - \xi_j) \tag{3}$$

an autoregressive term (first term) on previous values of the voltage, and induction from the magnetic fields (second term).

In the autoregressive term the a_k's are the coefficients used to predict the cable voltage from its value τ_k samples earlier. In the induction terms, M is the number of magnetic field components used, and $B_m(t)$ is a magnetic field record. M and B_m are used to denote both station and orientations. Finally, $\beta_{m,j}$ represents the regression coefficient from field component B_m at time $t - \xi_j$ onto the voltage at time t. Typically $P = 4$ autoregressive delays were used with $\tau_k = \{8, 15, 30, 60\}$ minutes while $J = 11$ coefficients at delays of $\xi = \{-60, -30, -8, -2, 0, 2, 4, 8, 15, 30, 60\}$ minutes were used with each of the magnetic field components. Defining

$$A(\omega) = 1 - \sum_{k=1}^{P} a_k e^{-i\omega\tau_k} \tag{4}$$

the overall transfer function from field component B_m to V is given by

$$H_m(\omega) = \frac{\sum_{j=1}^{J} \beta_{m,j} e^{-i\omega\xi_j}}{A(\omega)}. \tag{5}$$

Following the work on robust estimation of power spectra (THOMSON, 1977; KLEINER *et al.*, 1979), it is now generally recognized (see CHAVE *et al.*, 1987;

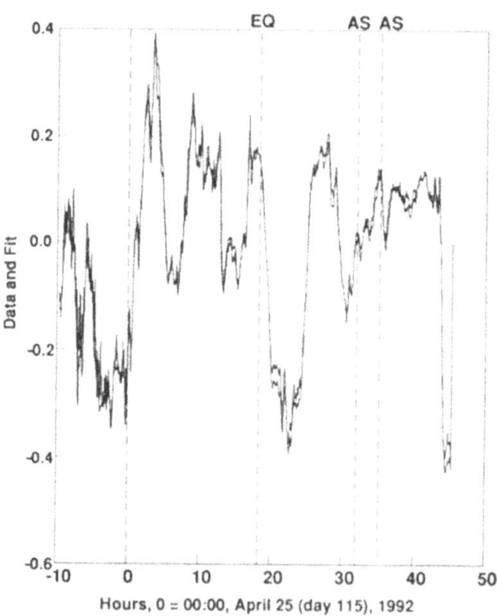

Figure 3
Potentials measured on C_N for 60 h on 24–27 April. The superimposed dotted line is a fit derived from the 10-day interval 17–26 April. The time of the main shock (EQ) and two principal aftershocks (AS) are shown.

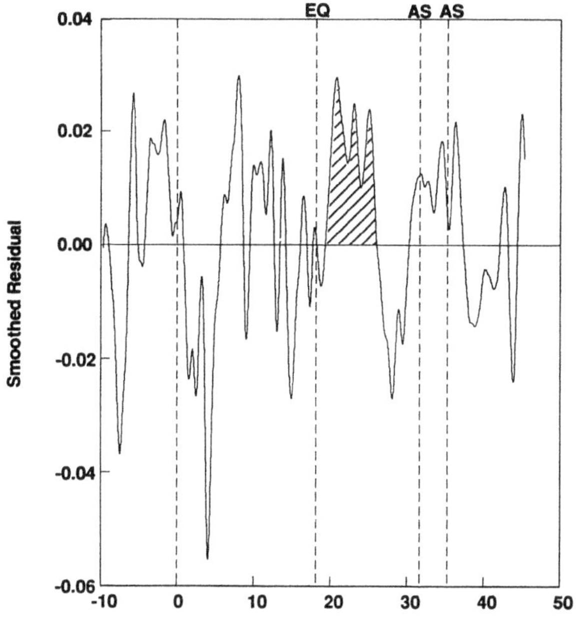

Figure 4

Smoothed residuals from the data and fit shown in Figure 3. The time of the main shock (EQ) and two aftershocks (AS) are shown. Note that following the EQ, the residuals remain appreciably of one sign for an extended interval (marked by shading).

CHAVE and THOMSON, 1989; LARSON, 1989; SUTARNO and VOZOFF, 1989) that magnetotelluric transfer functions must be estimated robustly. Examining the prediction errors from this present analysis in the fitting region showed several short intervals, in total about 14 percent of the data, where the average residual power was 5.8 times larger than for the remainder and where the worst residuals had 50 times the power expected from a stationary Gaussian process. Taken at face value, the data in these short intervals dominate the transfer function estimates and typically have been found to result in poorer overall estimates than are obtained where these data are discounted. We note, however, that the superficial quality of the direct (non-robust) least-squares fit is better than that of the robust fit. Overall using the time-domain robust estimate, the variance of the residual cable voltage is found to be a factor of 200 smaller than the variance of the raw cable voltage, and the range of the corresponding power spectrum is reduced by a factor of 500. We emphasize, however, that these are simple fits and, while adequate for our purposes here, we do not recommend the above method for estimating MT transfer functions.

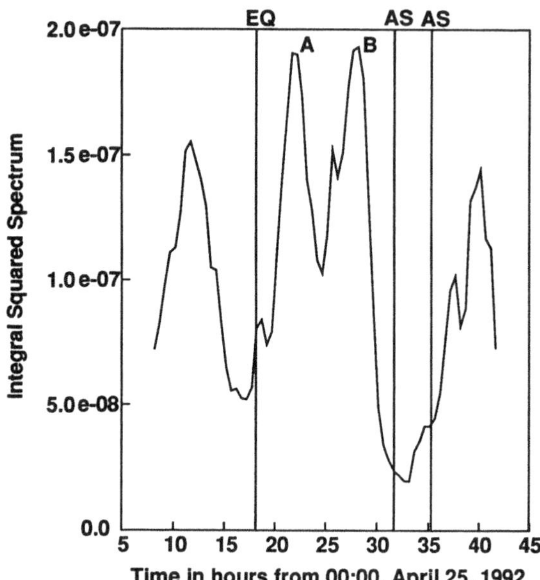

Figure 5

Integral squared power spectrum of the voltage calculated from individual 6-h spectra as a function of time from 00:00 UT 25 April 1992.

Plotted in Figure 3 are the north cable potentials for 1400 UT 24 April to 0200 UT 27 April (a total of 60 hours). The fit to the data as derived from Eq. (3) from the entire ten-day interval is also shown and is often essentially indistinguishable from the data in this plot.

Figure 4 plots the smoothed residuals between the observed cable potentials and the fit for the same 60-hour interval as in Figure 3. The smoothing was done with a parabolic tapered window with a two-hour time span. The times of the main shock and the two aftershocks are indicated. There is a considerable fluctuation of the residuals about the null line. A principal feature is the residual behavior following the earthquake and prior to the first aftershock (marked by shading). The residuals remain appreciably of one sign for an extended interval and their mean value for the interval is considerably enhanced compared to the mean value for other intervals of comparable length in the period shown. The residuals spanning the two aftershocks also remain of one sign for an extended interval, but their mean level is considerably less than the residuals in the interval following the main shock.

Figure 5 is a plot vs. time of the integral of the squared power spectra.

$$V_V(t) = \int S^2(f, t)\, df \tag{6}$$

for the residuals for the interval 0500 UT 25 April to 1800 UT 26 April. The individual 6-h spectra are calculated from 1 min residual values using a multiwindow estimate (THOMSON, 1982; THOMSON and CHAVE, 1990) with a time-bandwidth product of 4.0 and with 5 windows on each block. The time window for each 6-h spectrum is slid by 30 min before calculating the subsequent spectrum.

The integral of the squared spectra in Figure 5 is a measure of the stability of the power estimates. Simple functionals of a spectrum often have simple physical interpretations; for example, the average over frequency is the process variance, the geometric mean is the one-step prediction variance, and the harmonic mean is the single-sample interpolation variance. V_V in Eq. (6) is the asymptotic variance of a sample variance; that is, it is a measure of how reliable a power estimate is. As can be seen in Figure 5, this measure peaked significantly at two distinct times following

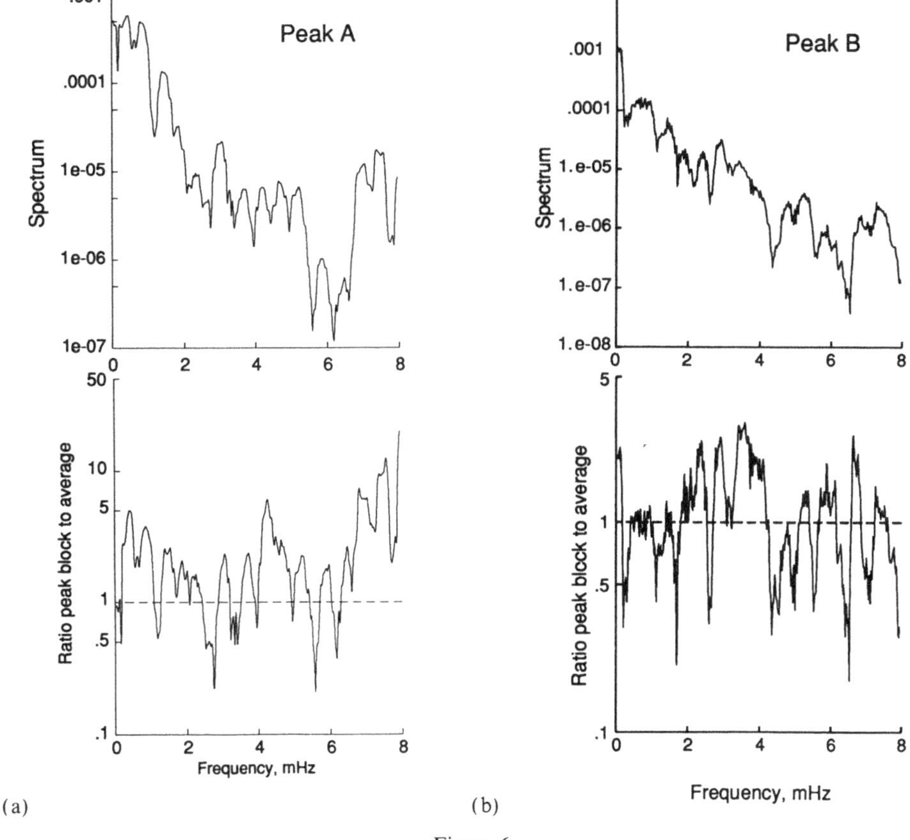

Figure 6

Top panel: Multiwindow spectra during the 6-h intervals containing Peaks A and B. Lower panel: Ratio of the Peak spectra to the average spectrum. Multiwindow spectra during the 6-h intervals containing Peak A (left) and Peak B (right). Lower panels: Ratio of the peak spectra to the average spectrum.

the main shock and prior to the time of the first aftershock. As will be seen below, examination of these intervals showed that both contained strong "echoes" at reasonably long time-delays.

In addition to the two largest peaks in the integrated squared spectrum in Figure 5, there are two other enhancements as well. The first, between hours 11 and 14 UT (02–05 LT at the middle of the cable route) corresponds to an interval of enhanced geomagnetic activity that began with a Pi2 hydromagnetic wave event at ~0820 UT (not shown here; near magnetic midnight for the cable route). The second smaller peak, on the day following the earthquake (near hour 40 in the figure), also corresponds to enhanced geomagnetic activity.

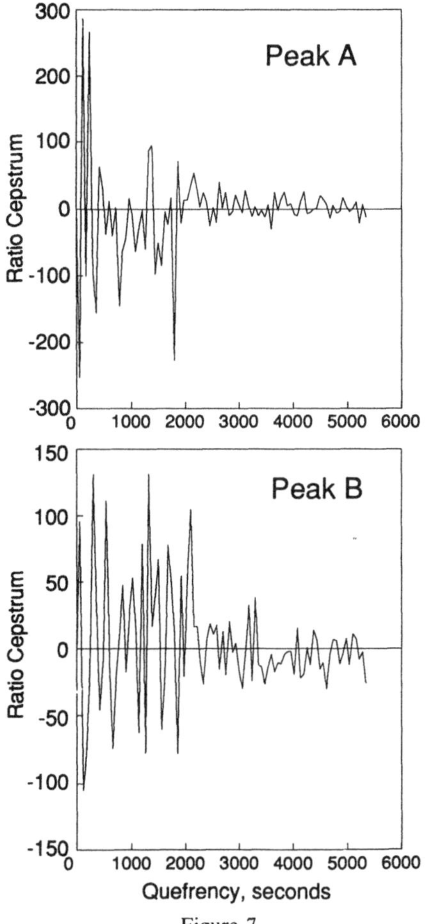

Figure 7

Cepstra calculated for the ratios shown in the lower panels of Figure 6. The upper panel cepstrum is for the Peak A period shows a strong peak at 30 min; the lower panel (Peak B period) is more complicated with a number of smaller peaks.

As noted above, the two largest peaks of integrated residual power occur in the interval following the main shock of the earthquake and prior to the first after-shock. In order to examine further the nature of these power enhancements in the residuals, a multiwindow power spectrum was calculated for those residuals (Figure 4) that occur in each interval: 1840–2440 UT 25 April and 0410–1010 UT 26 April. These spectra are shown in the top panels of Figures 6a and 6b.

Each of these spectra was then ratioed as a function of frequency to the average residual power spectrum for the entire interval 0500 UT 25 April to 1800 UT 26 April. The ratios of the residual spectral values from the interval of Peak A to the average spectrum are shown in the lower panel of Figure 6a for the frequency range 0–8 mHz. A quite noticeable periodic variation is seen in the ratio values. In a similar manner, the ratioed residual spectral values during Peak B are shown in the lower panel of Figure 6b. The visual impression reflects a higher frequency content in these (Figure 6b) ratios than in the ratios in Figure 6a.

The cepstrum, the Fourier transform of the logarithm of the ratios shown in Figures 6a,b, was calculated for the 0–8 mHz band. The cepstrum for the ratios from the Peak A interval is shown in the top panel of Figure 7. In addition to the two large cepstrum values at very low frequencies ($\lesssim 250$ s) that correspond to the overall shape of the ratio plot of Figure 6a, there is another, even larger, value of the cepstrum at a period of nearly 30 min.

The larger high frequency content in the spectrum from the Peak B interval is confirmed by the cepstrum of the ratios shown in the lower panel of Figure 7. In addition to two low frequency peaks (of opposite sign), again corresponding to the overall shape of the spectral ratio plot in the range 0–8 mHz, there are a number of peaks in the period interval spanning about 20–35 min. These cepstrum peaks are all smaller in value than the large, ~ 30 min peak in the top panel.

Discussion

Potential variations on long oceanic cables are produced by several different geophysical sources (e.g., MELONI *et al.*, 1983). Geomagnetic field fluctuations with periods on the order of minutes to hours are responsible for the most evident of these variations. Removing the geomagnetic field fluctuations from the potential measurements that were made on the unpowered HAW-1 cable has enabled us to search for signals that can be attributed to the tsunami associated with the Cape Mendocino earthquake. The geomagnetically-induced potential signals on the cable were removed by use of magnetometer measurements made at PA as well as at geomagnetic observatories in Honolulu and in Boulder.

Three principal effects were found in the analyses of the cable data that appear to have time associations with the earthquake-produced tsunami. The first of these involves the predictability of the cable voltages. The cable voltage predictions were determined from predictive coefficients that were derived from ten days of data, eight days preceding and two days including the main and two aftershocks. An examination of the residuals between the actual measured cable voltage data and the predicted voltages shows higher than usual residuals in the 5–6-hour interval following the main shock than in comparable intervals in the two-day interval around the time of the main shock (Figure 4). There was also an interval of enhanced residuals around the time of the two aftershocks, but the overall mean level was significantly smaller than in the interval following the main shock. This is entirely consistent with the absence of a tsunami signal associated with the aftershocks (GONZÁLEZ *et al.*, 1995).

The second effect found in the analysis of the cable voltage data was an enhancement in the integral of the squared spectral power of the residuals in the 12 h or so following the main shock (Figure 5). The enhanced power occurred in two separate ~6-h intervals, and was larger in value in the two days around the main shock than was the enhanced power from the two intervals of geomagnetic activity. Enhanced spectral power was not found in the interval around the time of occurrence of the two aftershocks, although the residuals were enhanced for an extended interval at the time (Figure 4). A spectrum of the residual data (top of Figure 6a) during the interval of the first peak in the integral squared residual spectrum (Peak A) shows some enhanced power at about 1 mHz or less. This is probably evidence of the oscillatory tsunami signal that was evident in the tide gauge data along the California coast (OPPENHEIMER *et al.*, 1993), and that has been recently modeled by GONZÁLEZ *et al.* (1995).

GONZÁLEZ *et al.* (1995) have modeled the tsunami ray travel times assuming a propagation speed proportional to the square root of the product of the water depth and the acceleration of gravity. They show that the travel times to Arena Cove (near the cable landing point), south of the epicenter, and to Crescent City, north of the epicenter, are similar, 37 min and 47 min, respectively. The modeling shows that the peak response of the signal at about 0.67 mHz at Crescent City occurs about 4 h or so following the main shock. This is similar to the delay from the main shock to the time of the maximum of the integrated squared spectrum of Peak A in Figure 5. The model broad-band frequency response at about 0.67 mHz corresponds to the low frequency peak in Figure 6a.

The third effect found in the data is the strong cepstrum value that occurred at a period of ~30 min in the first ~6 h following the main shock (i.e., during Peak A). This strong cepstral peak can be interpreted as an "echo" in the signal recorded on the cable. That is, a signal seen by the cable was matched, with opposite sign (and therefore out-of-phase) with a signal that had been seen in the interval some 30 min earlier. The interpretation of this echo in terms of the tsunami wave process is not easy without additional modeling. The modeling results of GONZÁLEZ *et al.* (1995) suggest that there may be a "series" of wave intensity enhancements and rarefactions during the 8–10-h interval following the main shock. The spacing of these events, about 1.5 h at Crescent City and about 1 h at North Spit (also north of the epicenter), is longer than the cepstrum value of 0.5 h. However, a precise match of these numbers might not be expected, especially since the cable is measuring an integrated water flow effect while the modeling is reproducing the tide gauge data at specific shore locations. Nevertheless, the order-of-magnitude match is suggestive that such wave effects are related to those reflected in the cable data.

The second interval of enhanced residual spectral power (Peak B) had a much less well-defined "echo" time, with a range of cepstrum peaks in the interval $\sim 20-35$ m, all with smaller values than in the case of the Peak A interval. It is clear in the case of this second peak (B) that there were some delayed cable voltage responses, but they are difficult to interpret.

Acknowledgments

We thank the AT&T Point Arena cable station personnel for their interest, support, and assistance with the instrumentation located in their building.

REFERENCES

AKAIKE, H., and YAMANOUCHI, Y. (1962), *On the Statistical Estimation of Frequency Response Function*, Ann. Inst. Math. *14*, 23–56.

CHAVE, A. D., LUTHER, D. S., LANZEROTTI, L. J., and MEDFORD L. V. (1992), *Geoelectric Field Measurements on a Planetary Scale: Oceanographic and Geophysical Applications*, Geophys. Res. Lett., *19*, 1411.

CHAVE, A. D., and FILLOUX, J. H. (1985), *Observation and Interpretation of the Seafloor Vertical Electric Field in the Eastern North Pacific*, J. Geophys. Res. *12*, 793.

CHAVE, A. D., FILLOUX, J. H., LUTHER, D. S., LAW, L. K., and WHITE, A. (1989), *Observations of Motional Electromagnetic Field During EMSLAB*, J. Geophys. Res. *94*, 14153.

CHAVE, A. D., THOMSON, D. J., and ANDER, M. E. (1987), *On the Robust Estimation of Power Spectra, Coherences, and Transfer Functions*, J. Geophys. Res. *92*, 638–48.

CHAVE, A. D., and THOMSON, D. J. (1989), *Some Comments on Magnetotelluric Response Function Estimation*, J. Geophys. Res. *94*, 14,215–14,225.

EGBERT, G. D., and BOOKER, J. R. (1989), *Multivariate Analysis of Geomagnetic Array Data. 1. The Response Space*, J. Geophys. Res. *94*, 14,227–14,247.

EGBERT, G. D. (1989), *Multivariate Analysis of Geomagnetic Array Data. 2. Random Source Models*, J. Geophys. Res. *94*, 14,249–14,265.

FARADAY, M. (1932), *Experimental Researches in Electricity*, Phil Trans. R. Soc. *163*.

FILLOUX, J. H., *Instrumentation and experimental methods for oceanic studies*. In *Geomagnetism*, 1 (ed. J. A. Jacobs) (Academic Press 1987), 143 pp.

GONZÁLEZ, F. I., SATAKE, K., BOSS, E. G., and MOFJELD, H. O. (1995), *Offshore and Edgewave Tsunami Modes Generated by the 25 April 1992 Cape Mendocino Earthquake*, Pure and Appl. Geophys., this issue.

GOODMAN, N. R. (1957), *On the Joint Estimation of the Spectrum, Cospectrum, and Quadrature Spectrum of Two-dimensional Stationary Gaussian Processes*, Scientific Paper 10, Engineering Statistical Laboratory, New York Univ. (Also AD 134919, Defense Technical Information Center).

KLEINER, B., MARTIN, R. D., and THOMSON, D. J. (1979), *Robust Estimates of Spectra (with discussion)*, J. Royal Statist. Soc. *B41*, 313–351.

LANZEROTTI, L. J., THOMSON, D. J., MELONI, A., MEDFORD, L. V., and MACLENNAN, C. G. (1986), *Electromagnetic Study of the Atlantic Continental Margin Using a Section of a Transatlantic Cable*, J. Geophys. Res. *B91*, 7417–7427.

LANZEROTTI, L. J., MEDFORD, L. V., KRAUS, J.S., MACLENNAN, C. G., and HUNSUCKER, R. D. (1992), *Measurements of Small Amplitude TIDs Using Parallel, Unpowered Telecommunications Cables*, Geophys. Res. Lett. *19*, 253.

LANZEROTTI, L. J., SAYRES, C. H., MEDFORD, L. V., KRAUS, J. S., MACLENNAN, C. G., and THOMSON, D. J. (1993), *Statistical Study of Induced Voltages Across Oceanic Telecommunications Cables*, Proc. 1992 Solar-Terrestrial Pred. Conf. *1*, 224.

LARSEN, J. C. (1989), *Transfer Functions: Smooth Robust Estimates by Least-squares and Remote Reference Methods*, Geophys. J. *99*, 645.

LARSEN, J. C., and SANFORD, T. B. (1985), *Florida Current Volume Transports from Voltage Measurements*, Science *227*, 302.

LARSEN, J. C. (1992), *Transport of the Florida Current at 27°N Derived from Cross-stream Voltages and Profiling Data: Theory and Observations*, Phil. Trans. R. Soc. Lond. *A338*, 169.

LILLEY, F. E. M., FILLOUX, J. H., MULHEARN, P. J., and FERGUSON, L. J. (1993), *Magnetic Signals from an Ocean Eddy*, J. Geomagn. Geolectr. *45*, 403.

MELONI, A., LANZEROTTI, L. J., and GREGORI, G. P. (1983), *Induction of Currents in Long Submarine Cables by Natural Phenomena*, Rev. Geophys. *21*, 795.

MUNK, W. H., and CARTWRIGHT, D. E. (1966), *Tidal Spectroscopy and Prediction*, Phil. Trans. R. Soc. Lond. *A259*, 533–581.

OPPENHEIMER, D., BEROZA, G., CARVER, G., DENGLER, L., EATON J., GEE, L., GONZÁLEZ, F., JAYKO, A., LI, W. H., LISOWSKI, M., MAGEE, M., MARSHALL, G., MURRAY, M., McPHERSON, R., ROMANOWICZ, B., SATAKE, K., SIMPSON, R., SOMERVILLE, P., STEIN, R., and VALENTINE, D. (1993), *Cape Mendocino, California, Earthquakes of April 1992: Subduction at the Triple Point*, Science *261*, 433.

PRANDLE, D., and HARRISON, A. J. (1975), *Recordings of the Potential Difference across the Port Patrick-Donaghedee Submarine Cable*, Rep. Inst. Oceanog. Sci. Bidston Obs. *21*.

SEGAWA, J., and TOH, H. (1992), *Detecting Fluid Circulation by Electric Field Variations at the Nankai Trough*, Earth Planet. Sci. Letts. *109*, 469.

SHUMWAY, R. H., *Applied Statistical Time Series Analysis* (Prentice Hall, Englewood Cliffs, NJ 1988).

SUTARNO, D., and VOZOFF, K. (1989), *Robust M-estimation of Magnetotelluric Impedance Tensors*, Exploration Geophys. *20*, 383–398.

THOMSON, D. J. (1977), *Spectrum Estimation Techniques for Characterization and Development of WT4 Waveguide*, Bell System Tech. J. *56*, Part I, 1769–1815, Part II, 1983–2005.

THOMSON, D. J. (1982), *Spectrum Estimation and Harmonic Analysis*, Proc. IEEE, *70*, 1055–96.

THOMSON, D. J., and CHAVE, A. D., *Jackknifed error estimates for spectra, coherences, and transfer functions*, Ch. 2. In *Advances in Spectrun Analysis* (ed. S. Haykin) (Prentice-Hall 1990).

THOMSON, D. J., LANZEROTTI, L. J., MEDFORD, L. V., MACLENNAN, C. G., MELONI, A., and GREGORI, G. P. (1986), *Study of Tidal Periodicities Using a Trans-Atlantic Telecommunications Cable*, Geophys. Res. Lett. *13*, 525.

WERTHEIM, G. K. (1954), *Studies of Electrical Potential Between Key West, Florida, and Havana, Cuba*, Trans. AGU *35*, 872.

(Received August 22, 1994, revised January 13, 1995, accepted January 28, 1995)

PAGEOPH, Vol. 144, Nos. 3/4 (1995)

0033–4553/95/040441–13$1.50 + 0.20/0

Source Characteristics of the 1992 Nicaragua Tsunami Earthquake Inferred from Teleseismic Body Waves

MASAYUKI KIKUCHI[1] and HIROO KANAMORI[2]

Abstract—We analyzed the broadband body waves of the 1992 Nicaragua earthquake to determine the nature of rupture. The rupture propagation was represented by the distribution of point sources with moment-rate functions at 9 grid points with uniform spacing of 20 km along the fault strike. The moment-rate functions were then parameterized, and the parameters were determined with the least squares method with some constraints. The centroid times of the individual moment-rate functions indicate slow and smooth rupture propagation at a velocity of 1.5 km/s toward NW and 1.0 km/s toward SE. Including a small initial break which precedes the main rupture by about 10 s, we obtained a total source duration of 110 s. The total seismic moment is $M_o = 3.4 \times 10^{20}$ Nm, which is consistent with the value determined from long-period surface waves, $M_o = 3.7 \times 10^{20}$ Nm. The average rise time of dislocation is determined to be $\tau \approx 10$ s. The major moment release occurred along a fault length of 160 km. With the assumption of a fault width $W = 50$ km, we obtained the dislocation $D = 1.3$ m. From τ and D the dislocation velocity is $\dot{D} = D/\tau \approx 0.1$ m/s, significantly smaller than the typical value for ordinary earthquakes. The stress drop $\Delta\sigma = 1.1$ MPa is also less than the typical value for subduction zone earthquakes by a factor of 2–3. On the other hand, the apparent stress defined by $2\mu E_s/M_o$, where μ and E_s are respectively the rigidity and the seismic wave energy, is 0.037 MPa, more than an order of magnitude smaller than $\Delta\sigma$. The Nicaragua tsunami earthquake is characterized by the following three properties: 1) slow rupture propagation; 2) smooth rupture; 3) slow dislocation motion.

Key words: Tsunami earthquake, Nicaragua earthquake, body wave inversion, source process.

Introduction

In 1992–1993, three large earthquakes that caused great tsunami damage occurred: the Nicaragua earthquake of September 2, 1992; the Flores Sea earthquake of December 12, 1992; the Hokkaido-Nansei-Oki earthquake of July 12, 1993. Of these earthquakes the Nicaragua event was a tsunami earthquake (KANAMORI, 1972) that caused large tsunamis despite its moderate surface wave magnitude, $M_s = 7.2$. Several earthquakes have to date been identified as tsunami earthquakes (e.g., FUKAO, 1979; PELAYO and WIENS, 1992), however, the 1992 Nicaragua earthquake is the first tsunami earthquake since the global network of

---

[1] Department of Physics, Yokohama City University, Seto, Kanazawa-ku, Yokohama 236, Japan.
[2] Seismological Laboratory, California Institute of Technology, Pasadena, California 91125, U.S.A.

broadband seismographs has been deployed. The broadband body waves are especially useful for the study of tsunami earthquakes because they contain information pertaining to both the complex rupture propagation and the slow dislocation motion.

In this paper we analyze broadband records of teleseismic body waves to clarify the characteristics of the Nicaragua tsunami earthquake. This is an extension of KANAMORI and KIKUCHI (1993) with additional broadband data.

Preliminary Inspection of Broadband Records

Figure 1 shows the epicentral distribution of the main shock and aftershocks. The hypocentral parameters of the main shock given by the National Earthquake Information Service (NEIS) are: epicenter = (11.74°N, 87.34°W), depth = 45 km, origin time = 00:16:01.6 UT, $M_s = 7.2$. The aftershock area is about 200×100 km^2,

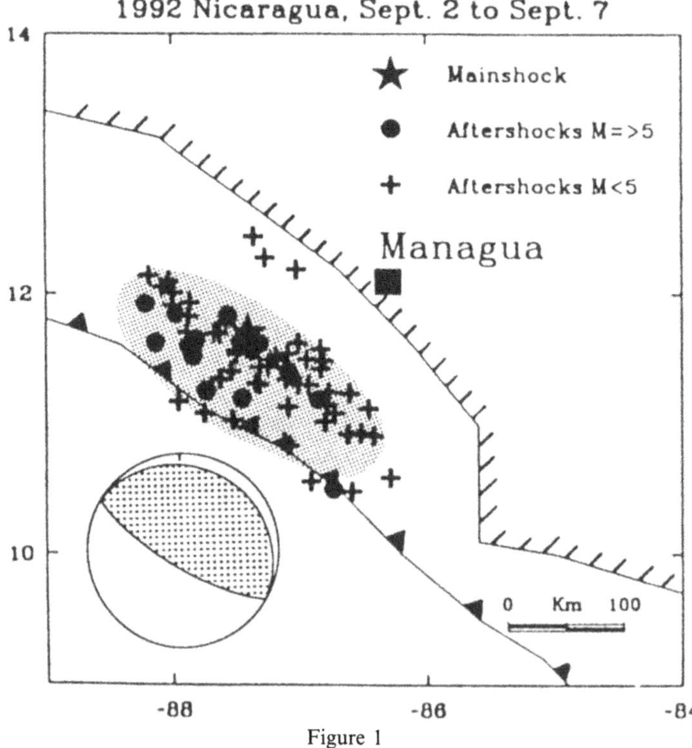

Figure 1

Epicentral distribution of the main shock and aftershocks. The mechanism solution obtained from long-period surface waves is shown with an equal area projection of the lower hemisphere.

Figure 2

IRIS and GEOSCOPE broadband seismograph stations. Inner and outer circles indicate the epicentral distances of 30° and 90°, respectively.

unusually large for the above surface wave magnitude. The mechanism solution determined from long-period surface waves is a nearly pure dip-slip given by: (strike, dip, slip) = (303°, 15°, 91°) (KANAMORI and KIKUCHI, 1993). The mechanism diagram is shown by an equal area projection of the lower hemisphere ir Figure 1.

Figure 2 shows the IRIS and GEOSCOPE broadband seismograph stations which recorded the Nicaragua earthquake at the epicentral distances between 30° (inner circle) and 90° (outer circle). The azimuthal coverage of stations is not very uniform, but the station coverage in EW direction is sufficient to resolve the rupture pattern in the strike direction. The direction cosine, γ, of the ray with respect to the fault strike ranges from -0.50 (for ZOBO) to 0.50 (for PAS). Thus an epicentral distance of $l = 20$ km results in a difference of P-wave arrival times at PAS and ZOBO of

$$\delta t = l\delta(\gamma)/\alpha = 20(0.50 - (-0.50))/6.2 = 3.2 \text{ s} \tag{1}$$

where $\alpha = 6.2$ km/s is used for the P-wave velocity. The above time difference can be resolved in the observed broadband records.

Figure 3 displays P and SH waves which are converted to the ground-motion displacement and bandpassed between 2 mHz and 0.5 Hz. A small phase is seen at all stations preceding the main pulse by about 10 s. Using the O-C values of P-wave arrival times, we relocated the initial break relative to the NEIS's location as

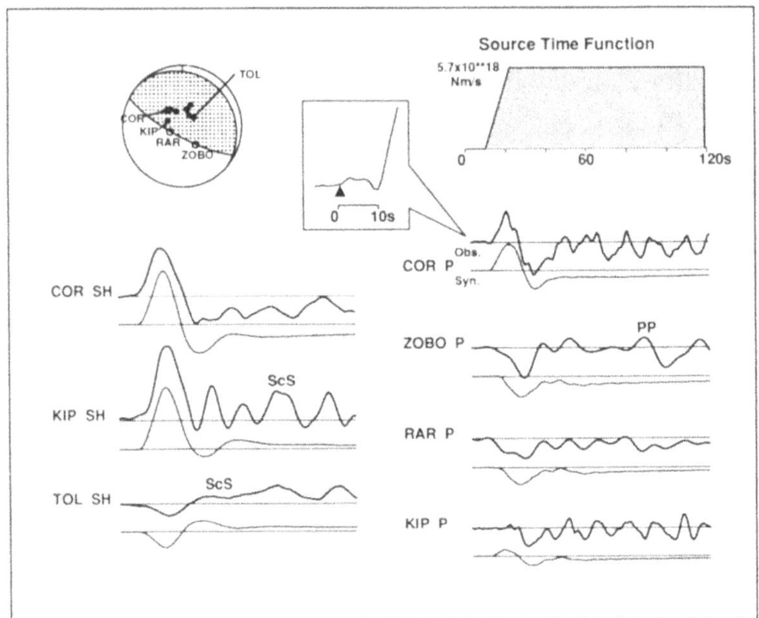

Figure 3

Upper traces show observed body waves. Small initial phases at time $t = 0$ are visible before the main pulses. Lower traces show synthetic waves generated by a point dislocation source with a ramp time history. The best-fit mechanism solution is plotted at top left.

(North, East) $= (5$ km, 22 km), and the relative origin time as $t = -5$ s, where we fixed the source depth at 20 km. With this relocation, the standard deviation of the O-C values was reduced from 0.94 s to 0.54 s. The mechanism and the seismic moment of this initial break are estimated in the waveform inversion as (strike, dip, slip) $= (140°, 75°, 50°)$ and $M_o = 2.5 \times 10^{17}$ Nm $(M_w = 5.5)$, respectively. The moment value is less than 0.1% of the total moment of the main rupture. In the following analysis, we use this initial break as a reference point in space and time: epicenter $= (11.78°N, 87.14°W)$ and origin time $= 00:15:57$ UT.

The most anomalous feature observed in body waves is that the wave train after the main pulse remains on one side above or below the zero-line for a long duration, about 100 s. This feature strongly suggests that the source time function is given by a long and smooth function. It should be noted here that the low-cut frequency of the bandpass filter is important in detecting this feature. Conventionally, the low-cut frequency is chosen at around 10 mHz (IDE *et al.*, 1993). With this cut-off frequency, the baseline offset could not be observed. The synthetic waves caused by a point dislocation are shown in Figure 3 under the observed records, where we used a ramp time history with a rise time of 12 s. The arrivals of *PP* and *ScS* phases, which are not considered in the synthetics, are indicated in the observed records.

Table 1

Structures

α(km/s)	β(km/s)	ρ(kg/m³)	d(km)
		Source region	
1.50	0.00	1.00×10^3	3.0
5.10	2.94	2.60	8.0
6.20	3.58	2.70	13.0
6.60	3.81	2.87	22.0
8.10	4.68	3.30	–
		Receiver region	
5.57	3.36	2.65	15.0
6.50	3.74	2.87	18.0
8.10	4.68	3.30	–

$\alpha,\beta = P$-, S-wave velocities; ρ = density; d = thickness

The layered structures used in the synthetics are given in Table 1. For a near-source structure we adopted the result obtained by MATSUMOTO *et al.* (1977) with an additional layer of 3-km thick water. The one-sided wave trains are seen to be reproduced in the synthetic waves. Although some long-period energy in this time window could be due to the W phase (KANAMORI, 1993) which is not included in the synthetics, W phase is usually more pronounced after the PP phase which is, except for ZOBO, outside of the time window used in Figure 3.

The mechanism solution obtained for a point dislocation depends on the source depth, h, slightly. The residual error (normalized variance), Δ, is minimized at

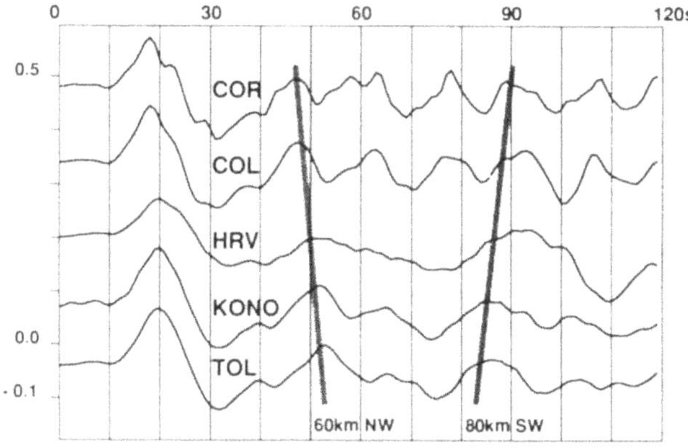

Figure 4

P waves arranged in the order of γ (the direction cosine of the ray with respect to the fault strike: N300°E). Gray lines denote the P-wave arrival times from the two sources located along the fault strike on opposite sides of the epicenter.

$h = 20$ km, and then $\Delta = 0.27$. If we take an allowable range of the residual error as $\Delta < 0.30$, then the estimate of the source depth may vary from 10 to 30 km. For this depth range, the mechanism of point dislocation is obtained as (strike, dip, slip) $= (315 \pm 5°, 16 \pm 2°, 103 \pm 5°)$, and shown in Figure 3. This mechanism is very similar to that obtained from long-period surface waves as shown in Figure 1. The height of the ramp source time function, $\dot{M}_o$, derived in the least-squares method, depends considerably on the source depth. For example, $\dot{M}_o = 7.5,\ 5.7,\ 5.0 \times 10^{18}$ Nm/s for $h = 10, 20, 30$ km. This suggests that there may be a large uncertainty in the moment estimate.

Directivity is not very clear in the amplitude azimuthal variation, but there are some noticeable wave packets that exhibit azimuthal dependence of the arrival times. In Figure 4, P waves observed at the stations are arranged in the order of the direction cosine of the ray with respect to the fault strike

$$\gamma = \cos(\phi_a - \phi_s)\sin i_h \tag{2}$$

where ϕ_a, ϕ_s and i_h are respectively the station azimuth, the fault strike and the take-off angle of the ray. We used $\phi_s = 300°$ as inferred from the aftershock area. The records are lined up such that the small initial phases are set at time $t = 0$ s. On the (t, γ) plane, the arrival time t of the phase generated from a point source with the origin time t_0 and the distance l_0 from the epicenter is given by

$$t = t_0 - (l_0/\alpha)\gamma. \tag{3}$$

Two sources, although not distinct, are inferred as shown in Figure 4 by gray lines: one is a source at $t_0 \approx 50$ s and $l_0 \approx 60$ km in the NW direction from the epicenter, the other is at $t_0 \approx 85$ s and $l_0 \approx 80$ km in the SE direction. Considering that the main rupture starts at time $t = 10$ s, we estimate the interval velocity of rupture propagation of these two sources as $v \approx 60/40 = 1.5$ km/s toward NW and $v \approx 80/75 \approx 1$ km/s toward SE.

Waveform Inversion

The above inspection of body waves suggests that the Nicaragua earthquake can be described mainly by a "smooth" and "slow" rupture propagation rather than a series of distinct subevents. In order to determine this feature in more detail and more quantitatively, we use here the method of KIKUCHI and KANAMORI (1991) with a slight modification.

Method

We assume a line source at a fixed depth and take 9 grid points along the fault strike at distances from the epicenter: $l_k = 0, \pm 20, \ldots, \pm 80$ km. We fix the mechanism as obtained above. Let $w_j(t, l_k)$ denote the Green's function at j-th station for a unit source at k-th grid point and time $t = 0$ s. We discretize the time axis for the moment rate function with an interval of Δt ($= 5$ s) and represent the moment rate at k-th grid point as

$$s_k(t) = \sum_i m_{ik} u(t - i\Delta t) \tag{4}$$

where $u(t)$ is a triangle time function with unit moment and the base-width $2\Delta t$ ($= 10$ s). The total moment-rate function is given by

$$s(t) = \sum_k s_k(t). \tag{5}$$

The synthetic wave at j-th station is

$$y_j(t) = \sum_i \sum_k m_{ik} w_j(t - i\Delta t, l_k). \tag{6}$$

Let $x_j(t)$ denote the observed record at j-th station, consequently the least-squares criterion is given by

$$\Delta = \sum_j \int [x_j(t) - y_j(t)]^2 \, dt = \text{minimum}. \tag{7}$$

Without any constraint, this is a linear inversion problem with respect to the seismic moments, m_{ik}. However, because it is ill-conditioned, the inversion often becomes unstable, producing negative moments and noncausal sources. To reduce the number of unknown parameters (model parameters) and stabilize the inversion, we impose some constraints on m_{ik}. First we assume that a rupture front monotonically propagates along the fault strike, i.e., the source at grid points farther from the epicenter rupture later. Also the duration of $s_k(t)$ is bound within 35 s (6 triangle sources for each k where $k = 0, \pm 1, \ldots, \pm 4$). Under these constraints we performed an inversion with a starting model (consisting of $6 \times 9 = 54$ model parameters) in which a rupture front propagates bilaterally at constant speeds of v_1 and v_2 toward NW and SE directions, respectively. Considering possible subevents with rupture velocity higher than 1.5 km/s as inferred from the preliminary inspection, we took several combinations of v_1 and v_2 which are less than or equal to 2.0 km/s, and applied the least-squares method. We then obtained the best waveform match for $v_1 = 2.0$ km/s and $v_2 = 1.0$ km/s. Some of the derived moments, m_{ik}, are positive and others are negative. This situation is illustrated in Figure 5a.

We then imposed an additional constraint of nonnegative moment, i.e.,

$$m_{ik} \geq 0. \tag{8}$$

This was done in a trial-and-error manner as follows. We assume that a negative source resulted from the poor spatial resolution of inversion and is not significant.

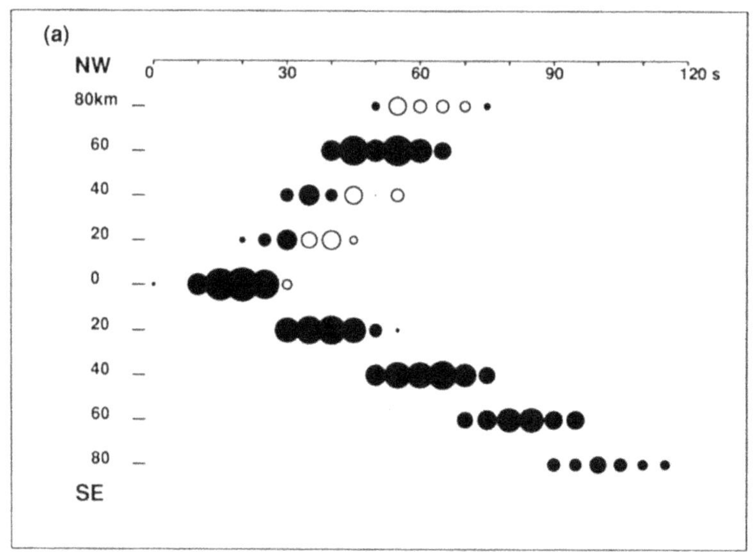

(a)
NW
0
30
60
90
120 s
80km
60
40
20
0
20
40
60
80
SE

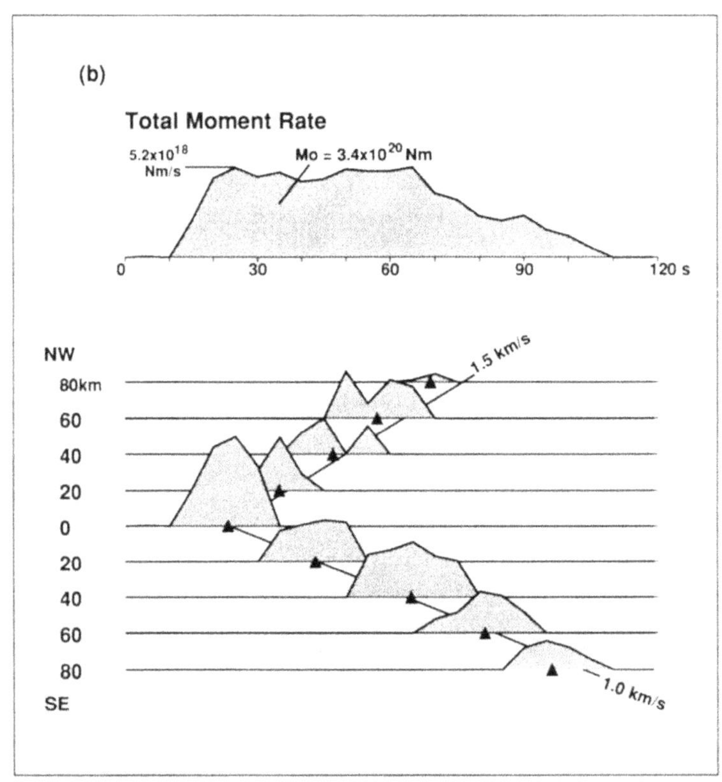

(b)
Total Moment Rate
5.2x10^18 Nm/s
Mo = 3.4x10^20 Nm
0
30
60
90
120 s
NW
80km
60
40
20
0
20
40
60
80
SE
1.5 km/s
1.0 km/s

Hence, when m_{ik} was found negative, we performed three independent inversions with the source at the grid (i, k), $(i, k - 1)$, and $(i, k + 1)$ removed, and chose the solution with the best fit if the resultant moments at the remaining grids are positive. This procedure was repeated until all m_{ik}'s became positive.

Results

We used the record sections of body waves with a duration of 120 s. The final result of inversion is shown in Figure 5b where the moment-rate functions at individual grids, $s_k(t)$, as well as the total moment rate, $s(t)$, are displayed. Figure 6 compares the synthetic waveforms with the observed ones. Approximate arrivals of *PP* and *ScS* phases, which are not considered in the present synthetics, are shown by their symbols on observed records. The waveform match is generally good at all stations except KIP. The poor match of *P* wave at KIP is perhaps due to the structure beneath this island station and partly due to its location being close to the *P*-wave nodal plane. However, considering the severe constraints made on the synthetics such as fixed depth and fixed mechanism, the waveform match is satisfactory.

In Figure 5b the centroid times of the individual moment-rate functions are indicated by triangles. This figure suggests that the main rupture propagates bilaterally in an asymmetric manner. The average velocity of rupture propagation is about 1.5 km/s toward NW while it is about 1.0 km/s toward SE. In the NW fault segment, the velocity of a rupture front may exceed higher than that of centroid by about 2.0 km/s. Such an asymmetric rupture pattern is identical to that inferred from the direct inspection of *P* waves in Figure 4. The onset time, the duration and the seismic moment at each grid point is given in Table 2. The total seismic moment is $M_o = 3.4 \times 10^{20}$ Nm. This agrees well with the seismic moment determined from long-period surface waves: $M_o = 3.7 \times 10^{20}$ Nm (KANAMORI and KIKUCHI, 1993). The total duration of the main rupture is 100 s. If we add the duration of the initial rupture to this, we have a duration of 110 s for the entire rupture process.

On the other hand, the average of the duration of the subevents is 26.1 s. This value includes rupture propagation time for a fault segment of 20 km. Since the rupture velocity is about 1.0–1.5 km/s, the rupture time is about 13–20 s. Therefore, the rise time of dislocation motion is estimated to be about 6 to 13 s, the average being about 10 s.

Figure 5

(a) Seismic moments estimated for the grid scheme with constant rupture front speeds: 2.0 km/s to NW and 1.0 km/s to SE. Closed and open circles denote positive and negative moments, respectively. The radius is proportional to the seismic moment. (b) Final solutions for the moment-rate functions at individual grid points as well as the total moment-rate function obtained with the body wave inversion. Closed triangles show the centroid times of the moment-rate function for each subevent.

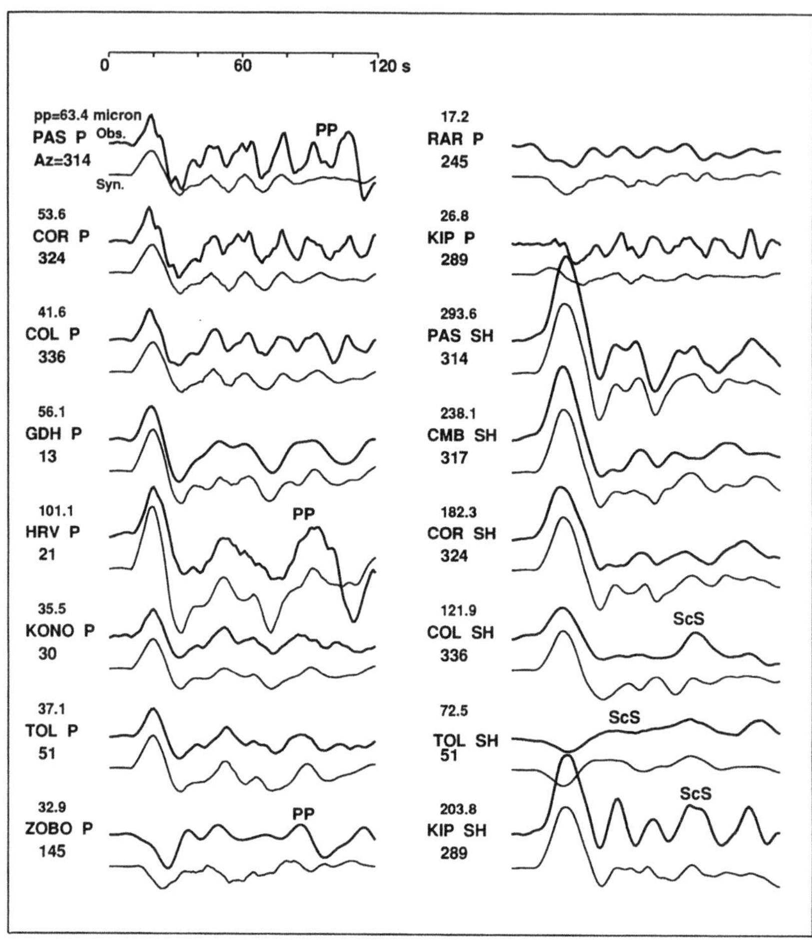

Figure 6
Comparison between observed (upper trace) and synthetic (lower trace) waves. Amplitude scale is common to all the traces for each of P and SH components. The peak-to-peak value (micrometer) and the station azimuth (degree) are given above and below the station code, respectively.

The fault length inferred from the present analysis is about 160 km: 70 km toward NW and 90 km toward SE. Furthermore, if we assume a fault width of $W = 50$ km, the fault area is $S = 8.0 \times 10^9 \, \text{m}^2$, and the average dislocation D is

$$D = M_o/\mu S = 1.4 \, \text{m} \tag{9}$$

where we used $\mu = 30$ GPa for the shear modulus. The above fault width is considerably narrower than that inferred from the aftershock area, only one half or so. SATAKE (1994) suggests that such a narrow width, as well as a shallow depth, is necessary to make the seismological fault model compatible with the large tsunami height. For the rise time $\tau = 10$ s, the dislocation velocity is estimated to be

Table 2

Source parameters

Distance(km)	Onset(s)	Duration(s)	Moment(Nm)
80	60	15	0.3×10^{19}
60	45	25	3.8
40	35	25	2.5
20	25	20	2.7
0	10	25	7.6
-20	30	25	4.4
-40	50	30	6.5
-60	65	30	3.9
-80	85	25	2.3
Ave. and Total		26.1	33.9×10^{19}

References: location = (11.7°N, 87.14°W); time = 00:15:57 UT

$\dot{D} = D/\tau \approx 0.14$ m/s, significantly slower than the typical value for subduction zone earthquakes.

For a mode-II shear crack as in the present case, the stress drop is estimated as

$$\Delta\sigma = \mu D/(1 - v)W = 1.1 \text{ MPa} \tag{10}$$

where we used $v = 1/4$ for the Poisson's ratio. The above value is significantly smaller than the typical stress drop of 3 MPa for subduction zone earthquakes. It should be noted, however, that the discrepancy is not very large, only about a factor of 2–3.

Once the spatio-temporal distribtution of moment release is obtained, we put this source model in a homogeneous space to calculate the seismic wave energy, E_s, numerically (KIKUCHI and FUKAO, 1988). The value thus obtained is $E_s = 2.1 \times 10^{14}$ J. The energy-moment ratio is then $E_s/M_o = 6.2 \times 10^{-7}$. This ratio is more than one order of magnitude less than the typical value for ordinary earthquakes. Consequently, the apparent stress defined by $\sigma_a = 2\mu E_s/M_o$ is much smaller than the stress drop

$$\sigma_a = 0.037 \text{ MPa} \ll \Delta\sigma = 1.1 \text{ MPa}. \tag{11}$$

VELASCO *et al.* (1994) also obtained a very low estimate of 0.01–0.07 MPa for the Orowan stress drop which is essentially the same as the apparent stress defined above. The low value of σ_a seems to be one of the most important characteristics of tsunami earthquakes.

Discussion

The Nicaragua earthquake can be modeled by slow asymmetric bilateral rupture. Similar results have been obtained by other investigators. For example,

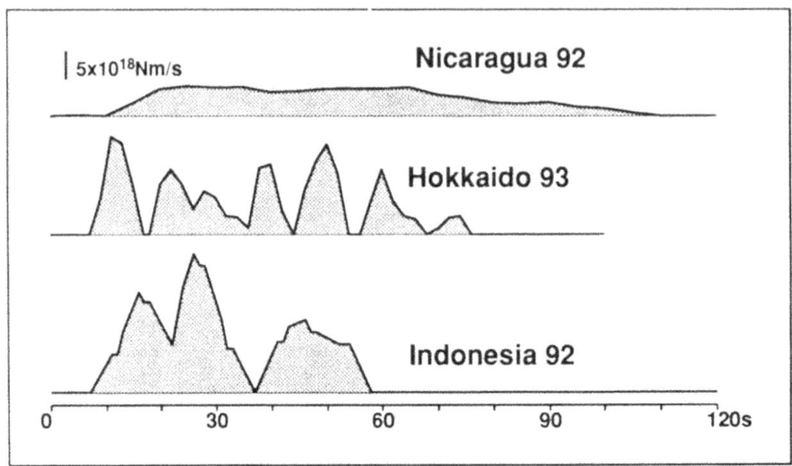

Figure 7

Comparison of the moment-rate functions for three large earthquakes, one tsunami earthquake (Nicaragua) and two tsunamigenic earthquakes (Hokkaido-Nansei-Oki and Indonesia). Vertical axis is common to all the traces.

when IDE *et al.* (1993) analyzed teleseismic body waves, they obtained three subevents, and determined the interval velocities of 2.2 km/s toward NW and 1.5 km/s toward SE. On the other hand, VELASCO *et al.* analyzed body and surface waves using empirical Green's functions and inverse Radon transform methods to obtain asymmetric bilateral rupture with a slow velocity ranging from 0.9 to 2.3 km/s.

Our results support these general conclusions, but the quantitative details of the rupture extent and velocity are slightly different. Our results also reveal that the stress drop is a factor of 2–3 smaller than the typical value for subduction zone earthquakes while the apparent stress is more than one order of magnitude smaller than the stress drop. In summary we can characterize the Nicaragua tsunami earthquake with the following three properties:

1) slow rupture propagation with a velocity of 1–1.5 km/s,
2) smooth rupture with no prominent subevents,
3) slow dislocation motion with a velocity of about 0.1 m/s.

The second property is a key to distinguish this event from some ordinary multiple shocks with a relatively long duration. The latter are associated with many distinct multiple subevents which rupture in sequence with some pauses in between, thereby resulting in a relative long total duration time. In this respect it is interesting to compare the moment-rate functions for both types of earthquakes. Figure 7 compares the Nicaragua earthquake with two "ordinary" multiple shocks: the Flores Island earthquake of 12 December, 1992 (IMAMURA and KIKUCHI, 1995); the Hokkaido-Nansei-Oki earthquake of 12 July, 1993 (KUGE *et al.*, 1994). The smoothness of rupture propagation of the Nicaragua earthquake is remarkable.

The slow and smooth source process of the Nicaragua earthquake cannot be explained in terms of the simple brittle failure model in which kinetic friction is essentially constant during failure. It probably requires time-dependent friction which holds the slip at slow speed while allowing large total displacement. KANAMORI and KIKUCHI (1993) suggested that large amounts of unconsolidated sediments on the Nicaraguan subduction boundary are responsible for the slow Nicaragua earthquake. Sediments are likely to have the rheological property which allows such time-dependent friction.

Acknowledgement

This research was partially supported by the Grant-in-Aid for Scientific Research No. 04452067 from the Ministry of Education, Japan, the National Science Foundation Grant EAR-9303804, the U.S.G.S. grant 1434-93-G-2305. Contribution No. 5448, Division of Geological and Planetary Sciences, California Institute of Technology, Pasadena, California 91125.

REFERENCES

FUKAO, Y. (1979), *Tsunami Earthquake and Subduction Processes near Deep Sea Trenches*, J. Geophys. Res. *84*, 2303–2314.

IDE, S., IMAMURA, F., YOSHIDA, Y., and ABE, K. (1993), *Source Characteristics of the Nicaragua Tsunami Earthquake of September 2, 1992*, Geophys. Res. Lett., 863–866.

IMAMURA, F., and KIKUCHI, M. (1995), *Moment Release of the 1992 Flores Island Earthquake Inferred from Tsunami and Teleseismic Data*, Sci. Tsunami Hazards (in print).

KANAMORI, H. (1972), *Mechanism of Tsunami Earthquakes*, Phys. Earth Planet. Inter. *6*, 346–359.

KANAMORI, H. (1993), *W Phase*, Geophys. Res. Lett. *20*, 1691–1694.

KANAMORI, H., and KIKUCHI, M. (1993), *The 1992 Nicaragua Earthquake: A Slow Tsunami Earthquake Associated with Subducted Sediments*, Nature *361*, 714–716.

KIKUCHI, M., and FUKAO, Y. (1988), *Seismic Wave Energy Inferred from Long-period Body Wave Inversion*, Bull. Seismol. Soc. Am. *78*, 1707–1724.

KIKUCHI, M., and KANAMORI, H. (1991), *Inversion of Complex Body Waves-III*, Bull. Seismol. Soc. Am. *81*, 2335–2350.

KUGE, K., KIKUCHI, M., and ZHANG, J. (1994), *Complex Source Process of the July 12, 1993 Hokkaido, Japan, Earthquake Inferred from Teleseismic Body Waves and Surface Waves*, Kaiyo Monthly, Special Issue No. 7, 21–28 (in Japanese).

MATSUMOTO, T., OHTAKE, M., LATHAN, G., and UMANA, J. (1977), *Crustal Structure in Southern Central America*, Bull. Seismol. Soc. Am. *67*, 121–134.

PELAYO, A. M., and WIENS, D. A. (1992), *Tsunami Earthquakes: Slow Thrust-faulting Events in the Accretionary Wedge*, J. Geophys. Res. *97*, 15321–15337.

SATAKE, K. (1994), *Mechanism of the 1992 Nicaragua Tsunami Earthquake*, Geophys. Res. Lett. *21*, 2519–2522.

VELASCO, A. A., AMMON, C. J., LAY, T., and ZHANG, J. (1994), *Imaging a Slow Bilateral Rupture with Broadband Seismic Waves: The September 2, 1992 Nicaragua Tsunami Earthquake*, Geophys. Res. Lett. *21*, 2629–2632.

(Received September 12, 1994, revised February 27, 1995, accepted March 3, 1995)

PAGEOPH, Vol. 144, Nos. 3/4 (1995)

0033-4553/95/040455-16$1.50 + 0.20/0
© 1995 Birkhäuser Verlag, Basel

Linear and Nonlinear Computations of the 1992 Nicaragua Earthquake Tsunami

KENJI SATAKE[1]

Abstract—Numerical computations of tsunamis are made for the 1992 Nicaragua earthquake using different governing equations, bottom frictional values and bathymetry data. The results are compared with each other as well as with the observations, both tide gauge records and runup heights. Comparison of the observed and computed tsunami waveforms indicates that the use of detailed bathymetry data with a small grid size is more effective than to include nonlinear terms in tsunami computation. Linear computation overestimates the amplitude for the later phase than the first arrival, particularly when the amplitude becomes large. The computed amplitudes along the coast from nonlinear computation are much smaller than the observed tsunami runup heights; the average ratio, or the amplification factor, is estimated to be 3 in the present case when the grid size of 1 minute is used. The factor however may depend on the grid size for the computation.

Key words: Tsunami, numerical computation, finite-difference method, Nicaragua earthquake.

1. Introduction

Numerical computation has become a powerful and popular tool to study tsunamis. In this topical issue, a number of papers are found on modeling recent tsunamis for various purposes such as to reproduce the observed tsunamis, to estimate unobserved offshore tsunami heights or the effects on coastal structures, or to study earthquake source processes. SHUTO (1991), in a review of tsunami numerical computations, mentioned that numerical computations can predict runup heights with errors smaller than 15%. Tsunami runup heights of the recent tsunamis such as the 1992 Nicaragua tsunami, however, seem to be much larger than those predicted using numerical computations from seismological fault models.

In this paper, I describe a numerical computation method of tsunamis with the 1992 Nicaragua earthquake tsunami as an example. I discuss various factors that affect the computations: the governing equations, linear and nonlinear shallow water equations; bottom frictional values; bathymetric data and the grid size. The

---

[1] Department of Geological Sciences, University of Michigan, Ann Arbor, MI 48109-1063, U.S.A.

computations are compared with the observed data, both tide gauge records and runup heights.

2. The 1992 Nicaragua Earthquake and Tsunamis

2.1 Seismolgical Analyses

The 1992 Nicaragua earthquake occurred at 00 h 15 m 57.5 s on September 2 (GMT). The seismic waves from the Nicaragua earthquake were recorded on global digital seismic networks and the data were available through computer network in almost real time. The aftershock epicenters located by the National Earthquake Information Service, USGS, were also available by computer network. Seismological analyses of these data (IDE *et al.*, 1993; KANAMORI and KIKUCHI, 1993; SATAKE *et al.*, 1993; VELASCO *et al.*, 1994; KIKUCHI and KANAMORI, 1995) showed that the focal mechanism exhibits a thrust-type fault, with a plane dipping shallowly toward the northeast and the strike parallel to the Middle America trench. This is consistent with a subduction of the Cocos plate beneath the Caribbean plate. The seismic moment M_0 estimated from long-period seismic waves ranges 3–4×10^{20} Nm ($M_W = 7.6$–7.7). The duration of the rupture process is about 100 s, unusually long for its size. Aftershocks occurred in an area about 200 km along the strike and 100 km downdip of the trench (Figure 1).

2.2 Observed Tsunami Data

The tsunami runup heights along the Nicaraguan coast were measured by field surveys soon after the earthquake and are reported elsewhere (ABE *et al.*, 1993; BAPTISTA *et al.*, 1993). The maximum runup height was 9.9 m above mean sea level (MSL), but the runup heights are mostly between 3 and 8 m along the Nicaraguan coast. The tsunamis were also recorded on tide gauges at Corinto and Puerto Sandino as shown in Figure 2. The Corinto tide gauge record shows an impulsive tsunami with its maximum at 61 min after the earthquake origin time. There is a small fall (about 10 cm) of sea level before the first rise. The trough-to-peak amplitude is 49 cm. The Puerto Sandino record also shows a sea-level fall of about 10 cm, followed by a very abrupt sea level rise which made the gauge go off-scale at 65 min after the origin time. The trough-to-peak amplitude is at least 117 cm. TADEPALLI and SYNOLAKIS (1994) called such a waveform, a small trough followed by a larger peak, a leading depression N wave.

2.3 Fault Models

The fault parameters of this earthquake have been estimated from seismological analysis (IDE *et al.*, 1993) and from tsunami modeling (IMAMURA *et al.*, 1993;

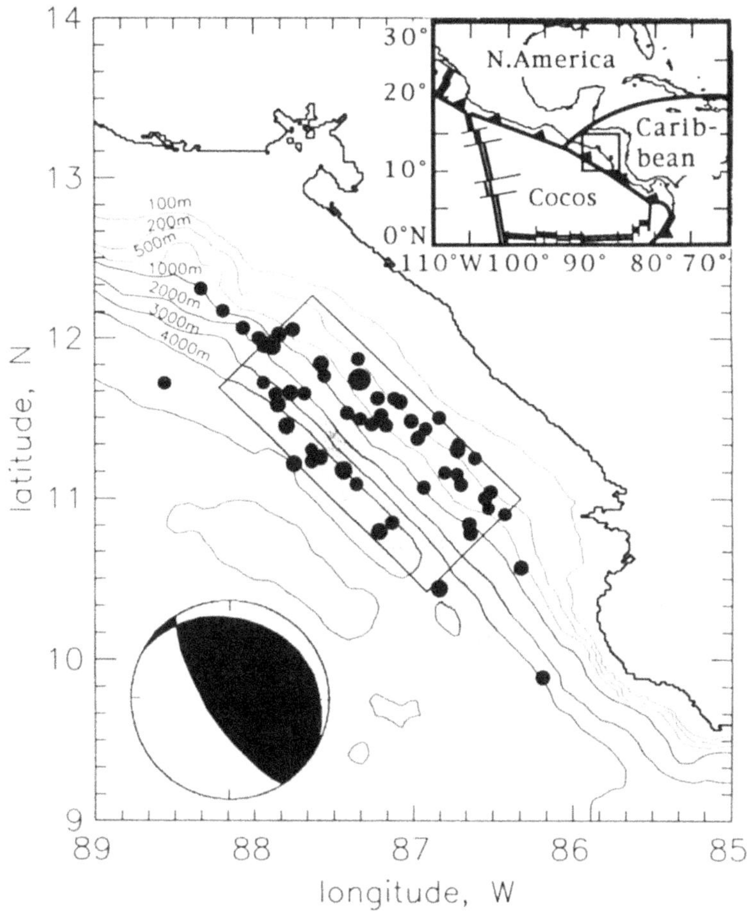

Figure 1

The source region of the 1992 Nicaragua earthquake. Solid circles shows the aftershocks within 1 day of the main shock (from Preliminary Determination of Epicenters). Open frame shows the horizontal projection of the seismological fault model, whereas the shaded frame is for the tsunami fault model. The focal mechanism of the main shock (SATAKE *et al.*, 1993) is also shown. The inset shows the tectonic framework of central America.

TITOV and SYNOLAKIS, 1993). IDE *et al.* (1993) estimated that the fault size is 200 km × 100 km and the average slip is 0.5 m. IMAMURA *et al.* (1993) made linear tsunami computations from a similar model and showed that the computed tsunamis are too small, a factor of 5.6 to 10, compared to the observed runup heights. They then suggested that the fault slip is larger by the factor than the seismological model; their seismic moment is 3×10^{21} Nm, an order of magnitude larger than the seismological results. TITOV and SYNOLAKIS (1993) roughly reproduced the

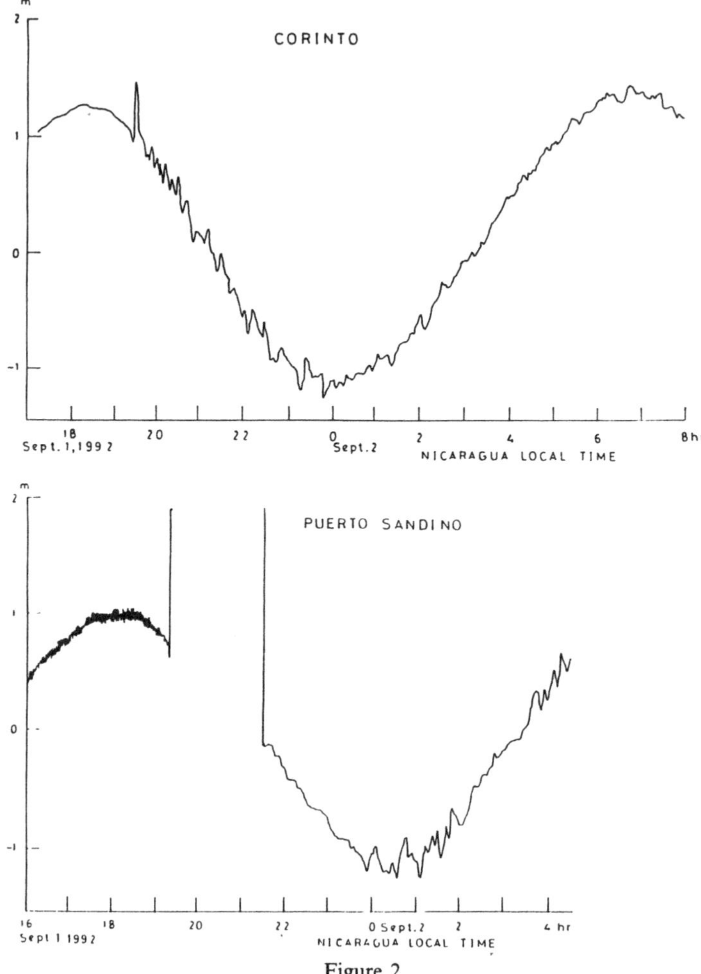

Figure 2

Tide gauge records at Corinto and Puerto Sandino showing the tsunamis (ABE *et al.*, 1993). The Puerto Sandino record went off-scale soon after the first tsunami arrival. Time scale is in the Nicaragua local time in which the earthquake origin time was 18:16.

observed runup heights from this seismic moment, using a combination of a two-dimensional nonlinear equation with a one-dimensional runup equation.

Recently, SATAKE (1994) estimated the fault parameters that are consistent with both seismological and tsunami data. Tsunami waveforms recorded on tide gauges required the fault width to be 40 km, much narrower than the aftershock area, and it extends only into the upper 10 km of the ocean bottom. Slip amount on the fault is estimated to be 3 m from the comparison of tsunami amplitudes on tide gauges. The fault length is estimated to be 250 km, slightly longer than the aftershock area,

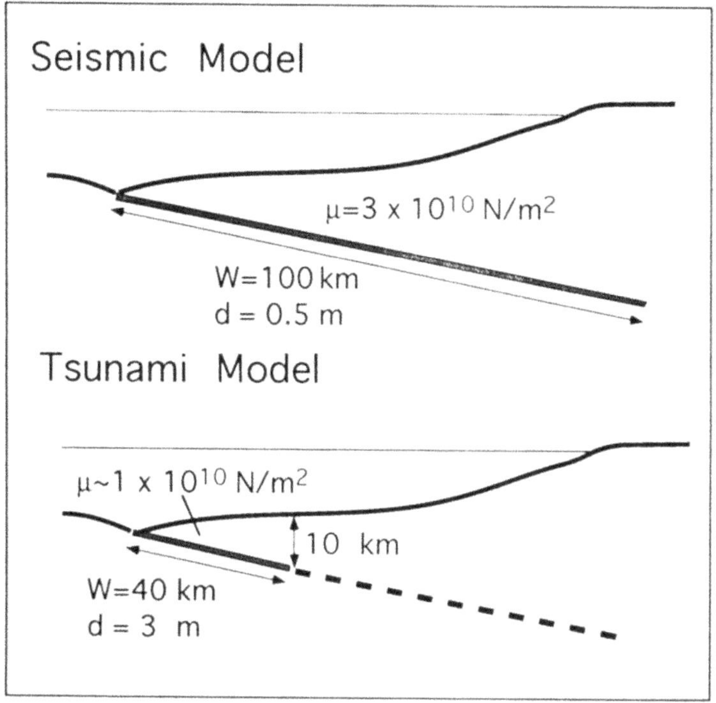

Figure 3

Schematic cross-sectional view of the seismic fault model (*top*: IDE *et al.*, 1993) and tsunami fault model (*bottom*: SATAKE, 1994). Both give the same seismic moment, but very different tsunamis.

from comparison of the tsunami height distribution. If the faulting occurred at such a shallow depth, the rigidity around the fault may be small. If we use $1 \times 10^{10}\ \text{N/m}^2$ as the average rigidity for the top 10 km of crust, compared to a standard value of $3\text{–}4 \times 10^{10}\ \text{N/m}^2$ for subduction-zone earthquakes, then the seismic moment becomes $3 \times 10^{20}\ \text{Nm}$. This value is consistent with the seismic observations, $3\text{–}4 \times 10^{20}\ \text{Nm}$. In the following, I use this fault model as the tsunami source model (Figure 3). SATAKE (1994) showed that the seismic and tsunami models produce very different tsunamis, while they are indistinguishable from seismic observations.

3. Shallow Water Theory

3.1 Linear and Nonlinear Equations

Tsunamis are usually treated as shallow water waves, which is also referred to as long waves. Shallow water theory is valid when the wavelength is much larger than the water depth, which applies to the source region of most earthquake tsunamis. The vertical acceleration is ignored and the horizontal velocity is assumed to be uniform. In other words, water from bottom to surface moves uniformly. The equation of motion, or conservation of momentum, can be written as (e.g., MADER, 1988; KOWALIK and MURTY, 1993)

$$\frac{\partial V}{\partial t} + (V \cdot \nabla)V = -g\nabla h - C_f \frac{V|V|}{d + h} \tag{1}$$

where V is the depth-averaged horizontal velocity vector, h is the water elevation or tsunami amplitude, d is the water depth and g is the gravitational acceleration. The first term on the left-hand side represents local acceleration, the second term is nonlinear advection term. The first term on the right-hand side represents pressure gradient, or restoring force due to gravity, and the second term is bottom friction where C_f is the nondimensional friction coefficient.

When the tsunami amplitude h is very small compared to water depth d, we can use "small amplitude" approximation, $h \ll d$. Further, if we can ignore the bottom friction, then the above equation can be linearized. The linear long wave equation is

$$\frac{\partial V}{\partial t} = -g\nabla h. \tag{2}$$

Linear theory is valid in deep ocean or small amplitude tsunamis. The advantage of linear theory is its "linearity." Since the slip amount on the fault and the crustal deformation are linearly related, the tsunami amplitude is also linearly related to the slip amount. Therefore, once we compute tsunamis for a certain amount of slip on the fault, tsunami amplitude for different slip amount can be easily estimated by multiplication of the appropriate factor. This principle has been used for inversion of tsunami waveforms to estimate the slip distribution on an earthquake fault (e.g., SATAKE, 1989).

The equation of continuity, or conservation of mass, is written as

$$\frac{\partial(d + h)}{\partial t} = -\nabla \cdot \{(d + h)V\}. \tag{3}$$

Note that water depth, d, may be a function of time because of the coseismic bottom movement.

3.2 Bottom Friction

The bottom friction of tsunamis has not been well-known and various forms and values adopted in other areas have been used. In engineering hydraulics, two frictional coefficients are often used: the De Chézy coefficient C and the Manning's roughness coefficient n. These have different dimensions. The nondimensional frictional coefficient C_f in equation (1) is related to these as

$$C_f^{\,2} = \frac{g}{C^2}$$

and

$$C_f = \frac{gn^2}{(d+h)^{1/3}}. \tag{4}$$

The Manning's roughness coefficient n is used for a uniform turbulent flow on a rough surface. It indicates that the bottom friction varies with water depth.

A typical value of C_f for tidal flow is $2\text{--}5 \times 10^{-3}$ in rivers and $2.4\text{--}2.8 \times 10^{-3}$ in shallow sea (e.g., DRONKERS, 1964). For runup of solitary waves, KAJIURA (1984) estimated that $C_f \sim 1 \times 10^{-2}$ from comparisons of theory and experiments. SYNOLAKIS and SKJELBREIA (1993) compiled post-breaking solitary wave data that range $C_f = 4$ to 9×10^{-2} for beach slopes of $1:166$ to $1:50$. They commented that the value for nonbreaking waves or waves before breaking would be smaller. KOWALIK and WHITMORE (1991) used the value of $C_f = 3.3 \times 10^{-3}$ for tsunamis. AIDA et al. (1988) used different values for their numerical computation of tsunami runup; $C_f = 5 \times 10^{-3}$ for coastal water and 1×10^{-2} on land without obstacles.

A typical value of n for coastal water is $0.03 \ \mathrm{m}^{-1/3}\,\mathrm{s}$ (e.g., BAPTISTA et al., 1989). If we use this value, the frictional coefficient C_f becomes 2.3×10^{-3} and 1×10^{-2} for a total depth $(d+h)$ of 50 m and 0.6 m, respectively, and they agree well with the above observational values of tidal flow and runup of solitary waves.

In this paper, I use the formulation (4) with $n = 0.03 \ \mathrm{m}^{-1/3}\,\mathrm{s}$. For comparison, I also calculate for two depth-independent frictional values, $C_f = 3 \times 10^{-3}$ and 1×10^{-2}, that may correspond to tidal flow and tsunami runups.

4. Numerical Computations of Tsunamis

4.1 Finite-difference Method

We take the spherical coordinate system (r, θ, ϕ) with the origin at the earth's center, but r is constant and equal to the earth's radius R. Note that θ is colatitude and measured southward from the North Pole and ϕ corresponds to longitude measured eastward from the Greenwich meridian. If we write the east and south components of the depth-averaged horizontal velocity V as u and v, respectively, then the equation of motion (1) can be written in each component as

$$\begin{aligned}
\frac{\partial u}{\partial t} + \frac{u}{R \sin \theta} \frac{\partial u}{\partial \phi} + \frac{v}{R} \frac{\partial u}{\partial \theta} &= -\frac{g}{R \sin \theta} \frac{\partial h}{\partial \phi} - C_f \frac{u \sqrt{u^2 + v^2}}{d + h} \\
\frac{\partial v}{\partial t} + \frac{u}{R \sin \theta} \frac{\partial v}{\partial \phi} + \frac{v}{R} \frac{\partial v}{\partial \theta} &= -\frac{g}{R} \frac{\partial h}{\partial \theta} - C_f \frac{v \sqrt{u^2 + v^2}}{d + h}
\end{aligned} \tag{5}$$

and the equation of continuity (3) becomes

$$\frac{\partial(d+h)}{\partial t} = -\frac{1}{R \sin \theta}\left[\frac{\partial\{u(d+h)\}}{\partial \phi} + \frac{\partial\{v \sin \theta(d+h)\}}{\partial \theta}\right]. \tag{6}$$

The equations (5) and (6) are solved by finite-difference method. In this paper, the staggered leap-frog method is used. For the advection term, upwind difference scheme is used (e.g., PRESS *et al.*, 1992). On the ocean boundary, radiation condition, in which the tsunami wave is assumed to go out without changing its shape, is assumed. On land boundary (coast), total reflection is assumed. The coastal boundary is fixed; i.e., no runup on land is considered.

The time step of computation is determined to satisfy the stability condition (Courant condition) of the finite-difference computation. It is set to 5 s in the present case. Numerical computations of tsunamis are made for three hours of tsunami propagation. The tsunami waveforms are output on coastal points including the two tide gauge locations. The computations were made using both linear and nonlinear shallow water equations.

4.2 Bathymetry Data

The tsunami computational area extends from 9°N to 14°N and 89°W to 85°W. The ETOPO5 data, now on the Global Relief CD-ROM and available from the National Geophysical Data Center, NOAA, for the corresponding region is shown in Figure 4 (*left*). The ETOPO5 bathymetry data are reasonably accurate in the deep (> 1000 m) ocean, but known to be inaccurate in shallow water. As can be seen in the figure, the ETOPO5 data shows very shallow (about 10 m depth) elongated bank running parallel to the coast between 11° and 12°N, but such a bank does not exist in any nautical charts. In addition, the coastal shape cannot be accurately represented by 5 minute grids.

I compiled the bathymetry data with a grid size of 1 minute (about 1.6 km) as shown in Figure 4 (*right*). The number of grid points is 72,000 (240 × 300). I used nautical charts (Defense Map Agency Nos. 21520, 21540 and 21550) and manually updated the bathymetry data for the coastal region (depth < 1000 m). The depths for deeper grids are interpolated from the ETOPO5 data. The shallowest depth along the coast is about 10 m.

Around the two tide gauge stations (Corinto and Puerto Sandino), I also prepared bathymetry data with finer (12 seconds) grid size. They are shown in Figure 5. The higher resolution (up to 1 : 12,500) nautical charts (DMA 21524 and 21542) are used to make the gridded data. The shallowest depth is as small as 1 m, although the depth at the Corinto tide gauge station is 10 m and Puerto Sandino is 4 m.

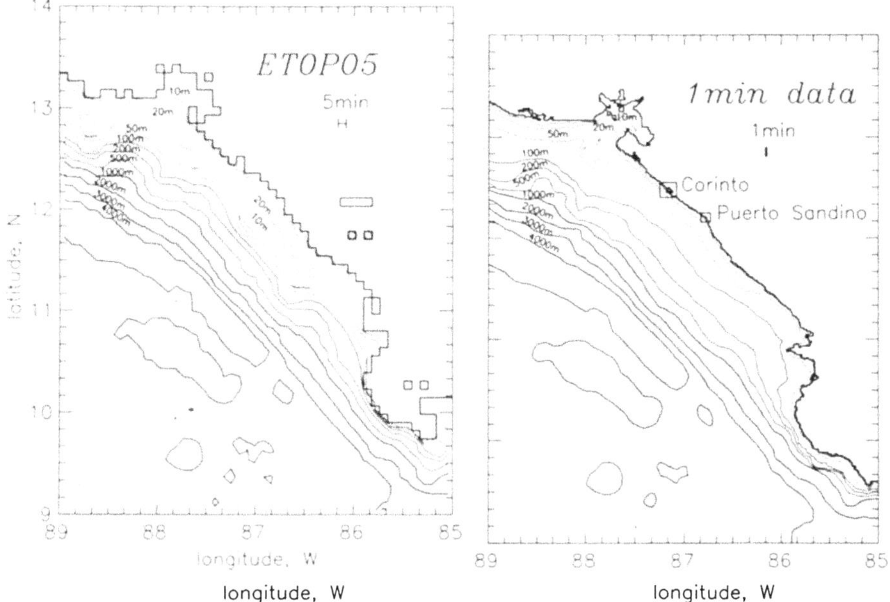

Figure 4

Comparison of the ETOPO5 data (*left*) and the 1 minute gridded data used in this study (*right*). Around the Corinto and Puerto Sandino tide gauge stations (shown by squares), more detailed bathymetry data (see Figure 5) are used.

4.3 Initial Condition

The initial bottom displacement is computed from the fault parameters estimated by SATAKE (1994). They are: the fault length 250 km, the fault width 40 km, the fault extends from ocean bottom to 10 km depth with a dip angle of 15°. The

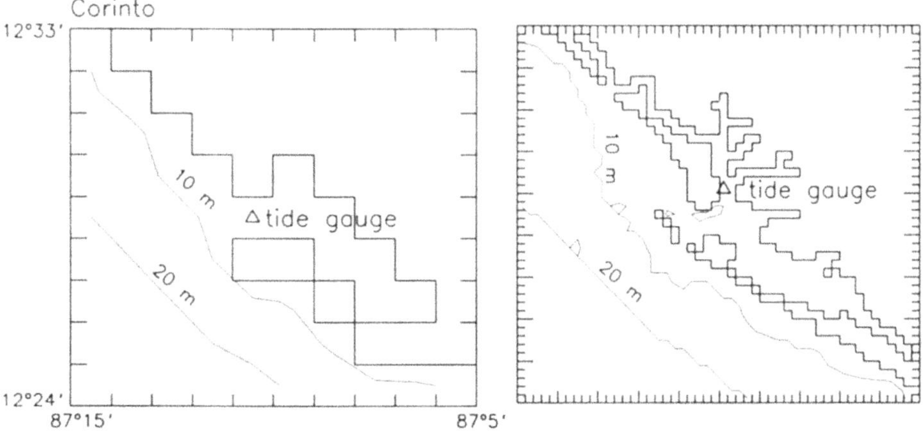

Figure 5(a)

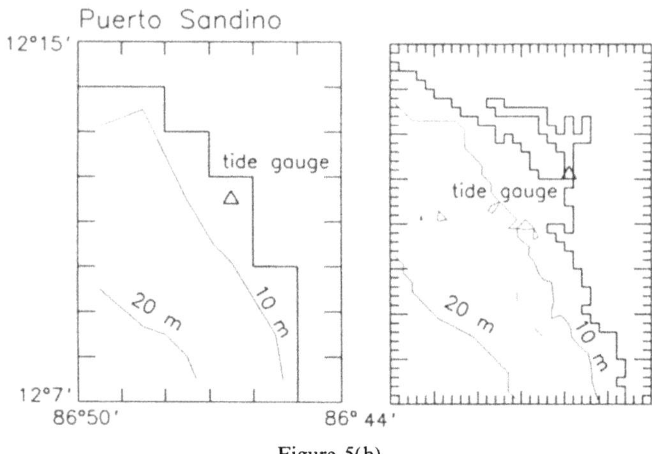

Figure 5(b)

Figure 5

Gridded bathymetry around Corinto (a) and Puerto Sandino (b) tide gauge stations. At each station, the left panel shows 1 minute grid and the right is 1/5 minute (12 seconds) grid.

slip on the fault is a pure dip slip (slip angle 90°) and the amount is 3 m. The fault strike is 315° and the location is shown in Figure 6. The vertical crustal deformation is calculated from these parameters (e.g., OKADA, 1985) and also shown in Figure 6. The maximum uplift is about 133 cm and occurs at the top edge of the fault, which is located at the trench axis. The maximum subsidence of 56 cm occurs above the deeper end of the fault, which is closer to the coast. The tsunami waveforms on the Nicaraguan coast are therefore expected be leading depression N waves, as actually observed. TADEPALLI and SYNOLAKIS (1994) showed that the runup heights from such a wave are larger than solitary waves or leading elevation N waves.

5. Computation of Tide Gauge Records

5.1 Effect of Grid Size

The tsunami waveforms are first computed at the two tide gauge locations, Corinto and Puerto Sandino, both on 1 minute and 12-second grid systems, and they are compared with the observation (Figure 7). As can be seen in Figure 7, the computed waveforms are different on the 1 minute and 12-second grid systems. The latter is closer to the observed waveforms. In particular, the sharp rise of water at Puerto Sandino could be reproduced only by using the 12-second grids. The

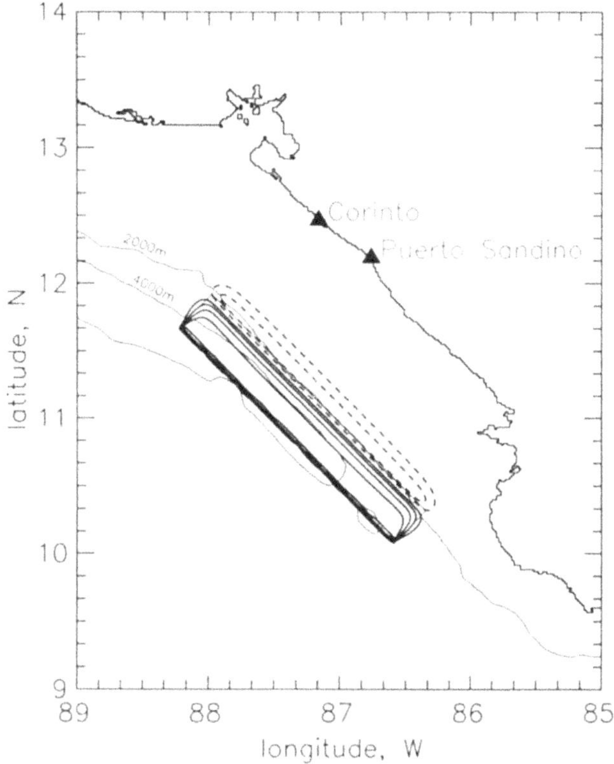

Figure 6

The initial bottom deformation overlaid on bathymetry. The solid lines are for uplift and the dashed lines are for subsidence. The contour interval is 20 cm. The location of two tide gauge stations are also shown.

computation on the 1-minute grid system predicted earlier arrival of tsunami than both that on the 12-second grids and the observed. This is not surprising if we look at Figures 5. The 1 minute grid system cannot represent the coastal topography around the stations very well.

As far as the same bathymetric data are used, the computed waveforms are similar for the linear and nonlinear equations. The amplitude at Puerto Sandino on 12-second grids are different, but the waveforms are still similar. On 1-minute grid system, both amplitude and waveforms are very similar, because of the much smaller amplitude than the water depth. In other words, the tsunamis can be well approximated as linear shallow water waves.

The above comparison shows that it is more important to use detailed and accurate bathymetry data than to include nonlinear terms for accurate tsunami computation, as far as the tide gauge records are concerned.

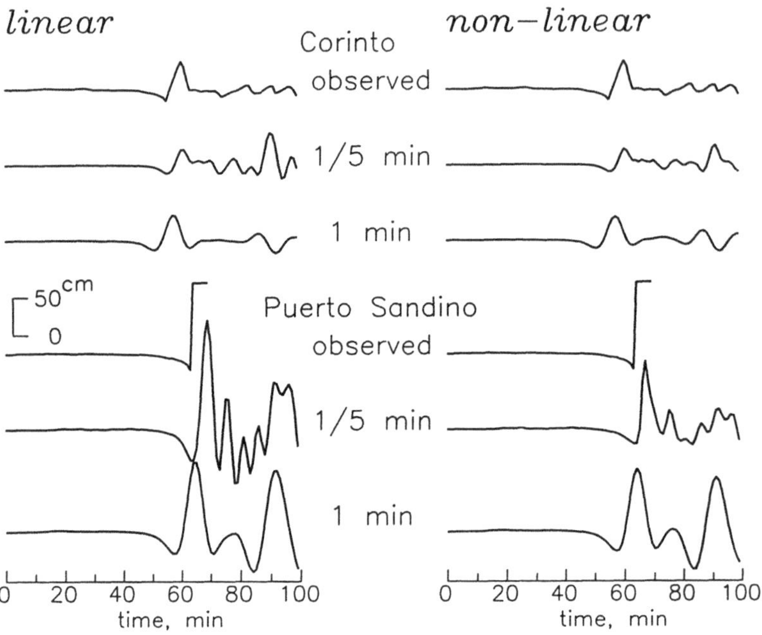

Figure 7

Tsunami waveforms calculated at the tide gauge locations, both 1 minute and 1/5 minute (12 seconds) grid systems, are shown for both linear (*left*) and nonlinear (*right*) computations. The observed tsunami waveforms, after the tidal components are removed, are also shown.

5.2 Effect of Bottom Friction

The effect of the bottom friction is examined next. Figure 8 shows the computed waveforms at the tide gauge stations on the 12-second grid system for different values of the frictional coefficients: the Manning's coefficients $n = 0.03 \text{ m}^{-1/3}$ s, nondimensional frictional coefficients $C_f = 3 \times 10^{-3}$ and $C_f = 1 \times 10^{-2}$. The first case assumes the depth-dependent bottom friction while the latter two are depth-independent friction. The waveforms calculated from the linear equation without bottom friction are also shown. The figure shows that the amplitudes become smaller as the friction coefficient becomes larger. This is more evident for later phase, which often registers the largest amplitude. The effect of bottom friction is more evident at Puerto Sandino, where the tsunami amplitude is larger than Corinto.

The above comparison shows that the nonlinear effect, including the bottom friction, plays more important roles for later phases than the first arrival and when tsunami amplitude becomes larger.

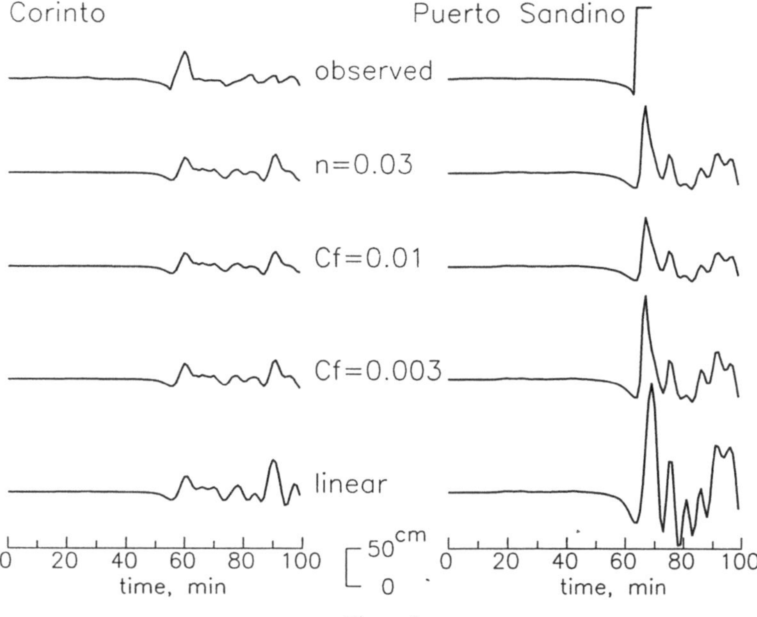

Figure 8

Tsunami waveforms calculated at the tide gauge locations, on 1/5 minute (12 seconds) grid system, are shown for different bottom frictional values. The observed tsunami waveforms, after the tidal components are removed, are also shown.

6. Comparison with Runup Heights

The computed tsunami heights are then compared with the maximum runup heights measured along the entire Nicaraguan coast. For the computed heights, the maximum amplitudes within 3 hours of the origin time are used and plotted in Figure 9 for different computations, linear and nonlinear with different bottom friction. The figure shows that the linear computation produces the largest amplitudes, and the case with the largest frictional coefficient $C_f = 1 \times 10^{-2}$ produces the smallest tsunami heights. The maximum amplitudes for two cases $n = 0.03$ m$^{-1/3}$ s and $C_f = 3 \times 10^{-3}$ are almost identical. The difference between the two extreme cases, linear and $C_f = 1 \times 10^{-2}$, is as large as 0.8 m, or 70% of the amplitude.

The figure also shows the observed runup heights above mean sea level (MSL) compiled by ABE et al. (1993) and BAPTISTA et al. (1993). The tsunami arrived about an hour after the high tide when the sea level was about 50 to 80 cm above MSL, depending on the location and the actual arrival time of the maximum tsunamis. Even when we consider this correction, the observed runup heights are much larger than the computed heights, by about a factor of 3. In other words, the average amplification factor is 3.

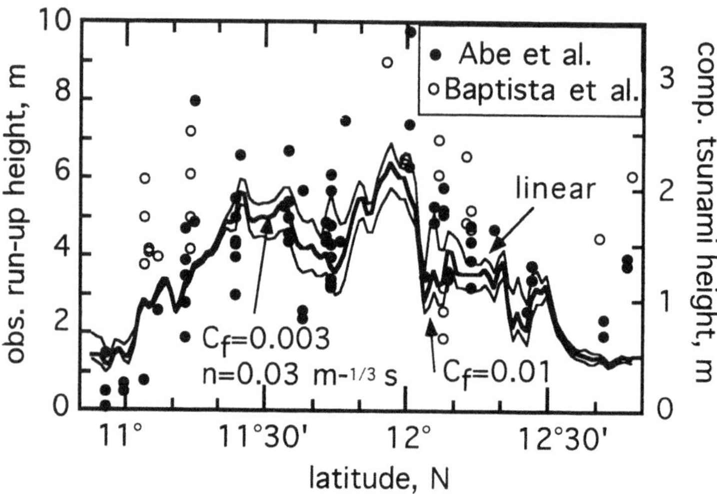

Figure 9
Calculated maximum amplitudes from different bottom frictional values along the Nicaraguan coast.
The observed runup heights (ABE *et al.*, 1993 and BAPTISTA *et al.*, 1993) are also shown with a different
scale.

The tsunami runup heights were measured in the vicinity of the tide gauge
stations at Corinto and Puerto Sandino (ABE *et al.*, 1993). At Corinto, the runup
height was 2.7 m above MSL, or about 2 m above the sea level at the tsunami
arrival. This is about 5 times larger than the amplitude (40 cm) on the tide gauge
record. At Puerto Sandino, the observed runup height was 3.7 m above MSL or
2.9 m above the sea level at tsunami arrival. The tide gauge went off-scale, which
means that the amplitude was at least 1 m. Hence the amplification factor at this
site is less than 3. The average factor 3, estimated in our computation, lies between
these observed values.

There is no simple theoretical work to compare the predicted tsunami heights
and the runup heights. IMAMURA *et al.* (1993) assumed that the factor is 2, based
on experimental data on a beach with slope of 1/10. TITOV and SYNOLAKIS (1993)
combined a two-dimensional linear computation with a one-dimensional nonlinear
runup computation and successfully reproduced the tsunami runup heights from a
source model of IMAMURA *et al.* (1993), indicating that the amplification factor
becomes one if runup calculation is made on a smaller grid size. Hence the
amplification factor estimated above is not to be taken as universal, but may be a
function of grid size of computation. SATAKE and TANIOKA (1995) indeed show for
the 1993 Hokkaido tsunami that the computed maximum tsunami heights are larger
when smaller grids are used.

7. Conclusions

Both linear and nonlinear computations of tsunami propagation are made for the Nicaragua earthquake as a test case. The comparisons show that

(1) In order to reproduce tsunami waveforms on tide gauges, the use of detailed bathymetry data with a small grid size is more effective than to include nonlinear terms in the tsunami computation. This is probably because of the small amplitude of tsunamis relative to the water depth at the tide gauges.

(2) For maximum amplitude of tsunamis, the linear computation produces larger amplitudes than the nonlinear computations. In the present case, the depth-independent bottom frictional value of $C_f = 3 \times 10^{-3}$ and Manning's coefficient $n = 0.03\ \mathrm{m}^{-1/3}$ s gave very similar maximum amplitudes.

(3) The observed tsunami runup heights are still much larger than the computed maximum amplitudes using nonlinear equation. The amplification factor, in the present case with a grid size of 1 minute, is 3 on the average; the factor however may depend on the grid size used for the computation.

In order to make tsunami computation comparable to the observed runup heights, we must use runup calculations. This involves an approximation of water flux on the moving boundary and requires very fine grids of topography as well as bathymetry (e.g., SHUTO, 1991; TITOV and SYNOLAKIS, 1993). While such a computation has become possible, an alternative way is to apply an appropriate amplification factor to the computed tsunami amplitude on a coarse grid system. More comparisons of the observed runup heights with computed tsunami amplitudes on various types of coast are needed to estimate the appropriate amplification factor.

Acknowledgments

I thank Antonio Baptista, Fumihiko Imamura, Costas Synolakis and Yuichiro Tanioka for helpful discussion and/or comments on the manuscript. This work was supported by National Science Foundation (EAR-9117800 and EAR-9405767).

REFERENCES

ABE, Ku., ABE, Ka., TSUJI, Y. IMAMURA, F., KATAO, H., IIO Y., SATAKE, K., BOURGEOIS, J. NOGUERA, E., and ESTRADA, F. (1993), *Field Survey of the Nicaragua Earthquake and Tsunami of 2 September 1992*, Bull. Earthq. Res. Inst., Univ. Tokyo *68*, 23–70 (in Japanese).

AIDA, I., TSUBOKAWA, H., and KAWAGUCHI, M. (1988), *Numerical Experiments on Behavior of Tsunamis Exceeding the Design Height of a Sea Wall: Case Studies for Matsuzaki, Shizuoka Prefecture and Taro, Iwate Prefecture*, Zisin, J. Seismol. Soc. Japan *41*, 343–350 (in Japanese).

BAPTISTA, A. M., PRIEST, G. R., and MURTY, T. S. (1993), *Field Survey of the 1992 Nicaragua Tsunami*, Marine Geodesy *16*, 169–203.

BAPTISTA, A. M., WESTERINK, J. J., and TURNER, P. J. (1989), *Tides in the English Channel and Southern North Sea. A Frequency Domain Analysis Using Model TEA-NL*, Adv. Water Resources *12*, 166–183.

DRONKERS, J. J., *Tidal Computations in Rivers and Coastal Waters* (North-Holland Publishing Company, 1964).

IDE, S., IMAMURA, F., YOSHIDA, Y., and ABE K. (1993), *Source Characteristics of the Nicaragua Tsunami Earthquake of September 2, 1992*, Geophys. Res. Lett. *20*, 863–866.

IMAMURA, F., SHUTO, N., IDE S., YOSHIDA, Y., and ABE, K. (1993), *Estimate of the Tsunami Source of the 1992 Nicaragua Earthquake from Tsunami Data*, Geophys. Res. Lett. *20*, 1515–1518.

KAJIURA, K. (1984), *On Runup of Solitary Waves*. Tsunami Engin. Tech. Rep., Tohoku Univ. *1*, 49–62 (in Japanese).

KANAMORI, H., and KIKUCHI, M. (1993), *The 1992 Nicaragua Earthquake: A Slow Tsunami Earthquake Associated with Subducted Sediments*, Nature *361*, 714–716.

KIKUCHI, M., and KANAMORI, H. (1995), *Source Characteristics of the 1992 Nicaragua Tsunami Earthquake Inferred from Teleseismic Body Waves*, Pure and Appl. Geophys, this issue.

KOWALIK, Z., and MURTY, T. S., *Numerical Modeling of Ocean Dynamics* (World Scientific, 1993).

KOWALIK, Z., and WHITMORE, P. M. (1991), *An Investigation of Two Tsunamis Recorded at Adak, Alaska*, Sci. Tsunami Hazards *9*, 67–83.

MADER, C. L., *Numerical Modeling of Water Waves* (Univ. California Press, 1988).

OKADA, Y. (1985), *Surface Deformation due to Shear and Tensile Faults in a Half-space*, Bull Seismol. Soc. Am. *75*, 1135–1154.

PRESS, W. H., TEUKOLSKY, S. A., VETTERLING, W. T., and FLANNERY, B. P., *Numerical Recipes in FORTRAN: The Art of Scientific Computing* (Cambridge Univ. Press, 1992).

SATAKE, K. (1989), *Inversion of Tsunami Waveforms for the Estimation of Heterogeneous Fault Motion of Large Submarine Earthquakes: The 1968 Tokachi-oki and the 1983 Japan Sea Earthquakes*, J. Geophys. Res. *94*, 5627–5636.

SATAKE, K. (1994), *Mechanism of the 1992 Nicaragua Tsunami Earthquake*, Geophys. Res. Lett. *21*, 2519–2522.

SATAKE, K., BOURGEOIS, J., ABE, Ku., ABE, Ka., TSUJI, Y., IMAMURA, F., IIO, Y., KATAO, H., NOGUERA, E., and ESTRADA, F. (1993), *Tsunami Field Survey of the 1992 Nicaragua Earthquake*, *EOS*, Trans. Am. Geophys. Union *74*, 156–157.

SATAKE, K., and TANIOKA, Y., (1995), *Generation and Propagation Characteristics of the 1993 Hokkaido Nansei-oki Earthquake Tsunamis*, Pure and Appl. Geophys., this issue.

SHUTO, N. (1991), *Numerical Simulation of Tsunamis — Its Present and Near Future*, Natural Hazards *4*, 171–191.

SYNOLAKIS, C. E., and SKJELBREIA, J. E., (1993), *Evolution of Maximum Amplitude of Solitary Waves on Plane Beaches*, J. Waterway, Port, Coastal and Ocean Engin. *119*, 323–342.

TADEPALLI, S., and SYNOLAKIS, C. E. (1994), *The Runup of N Waves on Sloping Beaches*, Proc. R. Soc. Lond. A. *445*, 99–112.

TITOV, V. V., and SYNOLAKIS, C. E., *A numerical study of wave runup of the September 2, 1992 Nicaraguan tsunami*, Proc. IUGG/IOC Inter. Tsunami Symposium, (Wakayama, Japan, 1993) pp. 627–635.

VELASCO, A. A., AMMON, C. J., LAY, T., and ZHANG, J. (1994), *Imaging a Slow Bilateral Rupture with Broadband Seismic Waves: The September 2, 1992 Nicaraguan Tsunami Earthquake*, Geophys. Res. Lett. *21*, 2629–2632.

(Received August 24, 1994, revised February 24, 1995, accepted April 25, 1995)

PAGEOPH, Vol. 144, Nos. 3/4 (1995)

0033-4553/95/040471-09$1.50 + 0.20/0
© 1995 Birkhäuser Verlag, Basel

Magnitude Scale for the Central American Tsunamis

TOKUTARO HATORI[1]

Abstract — Based on the tsunami data in the Central American region, the regional characteristic of tsunami magnitude scales is discussed in relation to earthquake magnitudes during the period from 1900 to 1993. Tsunami magnitudes on the Imamura-Iida scale of the 1985 Mexico and 1992 Nicaragua tsunamis are determined to be $m = 2.5$, judging from the tsunami height-distance diagram. The magnitude values of the Central American tsunamis are relatively small compared to earthquakes with similar size in other regions. However, there are a few large tsunamis generated by low-frequency earthquakes such as the 1992 Nicaragua earthquake. Inundation heights of these unusual tsunamis are about 10 times higher than those of normal tsunamis for the same earthquake magnitude ($M_s = 6.9$-7.2). The Central American tsunamis having magnitude $m > 1$ have been observed by the Japanese tide stations, but the effect of directivity toward Japan is very small compared to that of the South American tsunamis.

Key words: Tsunami source, magnitudes of earthquake and tsunami, Central American region.

1. Introduction

The description of tsunamis generated in Central America since 1700 were compiled by IIDA *et al.* (1967), and SOLOVIEV and GO (1985). Tsunami magnitudes on the Imamura-Iida scale were listed in the tsunami catalogs. The epicenters and magnitudes of the earthquakes since 1900 were determined from the instrumental data (Ka. ABE, 1981; UTSU, 1990).

In this paper, the magnitudes of the 1985 Mexico and 1992 Nicaragua tsunamis and other tsunamis are examined on the basis of the tsunami height-distance diagrams (HATORI, 1986). By comparing with the statistical relation of earthquake and tsunami magnitudes on the data of Japan and some great earthquakes in the world, the deviation of magnitude values for the Central American tsunamis is investigated. Abnormal tsunamis caused by the low-frequency earthquakes are examined from the relation between earthquake magnitudes and tsunami heights. The radiating energy of tsunamis toward Japan is discussed, based on the tide-gauge records.

---

[1] Suehiro 2-3-13, Kawaguchi, Saitama 332, Japan.

2. Distribution of Tsunami Sources

The tsunamigenic earthquakes which occurred in the Central American region (Mexico ~ Panama) during the period 1900–1993 are listed in Table 1. The surface-wave magnitudes of earthquakes, M_s, were mainly determined by Ka. ABE (1981). Tsunami magnitudes on the Imamura-Iida scale, m, are quoted from SOLOVIEV and GO (1985), and those of some tsunamis are examined in the present study.

Figure 1 shows the distribution of the epicenters of tsunamigenic earthquakes from 1900 to 1993. Generating years, earthquake magnitudes, M_s, and tsunami

Table 1

List of tsunamigenic earthquakes in Central America, 1900–1993

Date			Location	Lat. N	Long. W	M_s	m	Remarks
1902	II	26	El Salvador	13.0°	89.0°	–	2	
1905	I	20	Costa Rica	–	–	–	–	
1907	IV	15	Mexico	17.0	100.0	7.7	1	Acapulco 185 cm
1909	VII	30	Mexico	17.00	100.5	7.4	0	Acapulco inundated
1915	IX	7	Guatemala	14.0	92.0	7.9	0.5	
1916	V	25	El Salvador	12.0	90.0	7.5	–	
1925	XI	16	Mexico	18.5	107.0	7.0	2*	Sihuatanejo 10 m
1928	VI	17	Mexico	16.2	98.0	7.8	1	Hilo 40 cm
°1932	VI	3	Mexico	19.5	104.3	8.2	2	Manzanillo swept
°1982	VI	18	Mexico	19.5	104.3	7.8	1*	Manzanillo inundated
1932	VI	22	Mexico	19.0	104.5	6.9	2*	Cuyutlan 6 m
1934	VII	18	Panama	8.0	82.5	7.6	1.5	
1941	XII	5	Costa Rica	8.5	83.0	7.6	– 1*	Puntarenas 22 cm
1941	XII	6	Costa Rica	8.5	84.0	7.0	– 2	Puntarenas 8 cm
1948	XII	4	Mexico	22.0	105.5	6.9	1.5	
1950	X	5	Nicaragua	11.0	85.0	7.7	– 1*	Puntarenas 10 cm
1950	X	23	Guatemala	14.5	91.5	7.2	– 1*	San Jose 20 cm
1950	XII	14	Mexico	16.0	97.5	7.1	– 1.5	Acapulco 30 cm
1957	VII	28	Mexico	16.7	99.2	7.5	1*	Acapulco 255 cm
1962	V	11	Mexico	17.0	99.6	7.0	– 1*	Acapulco 75 cm
1973	I	30	Mexico	18.5	102.9	7.3	0*	near source 1 m
1978	XI	29	Mexico	16.1	96.6	7.6	–	
1985	IX	19	Mexico	18.3	102.5	8.1	2.5*	Acapulco 141 cm
1985	IX	21	Mexico	17.8	101.7	7.5	–	Acapulco 123 cm
°1992	IX	2	Nicaragua	11.8	87.4	7.2	2.5*	El Transito 9.9 m

M_s: Surface-wave magnitude of earthquakes is quoted from Ka. ABE (1981).
m: Tsunami magnitude of Imamura-Iida scale.
* Values estimated in the present study.
° Observed in Japan.

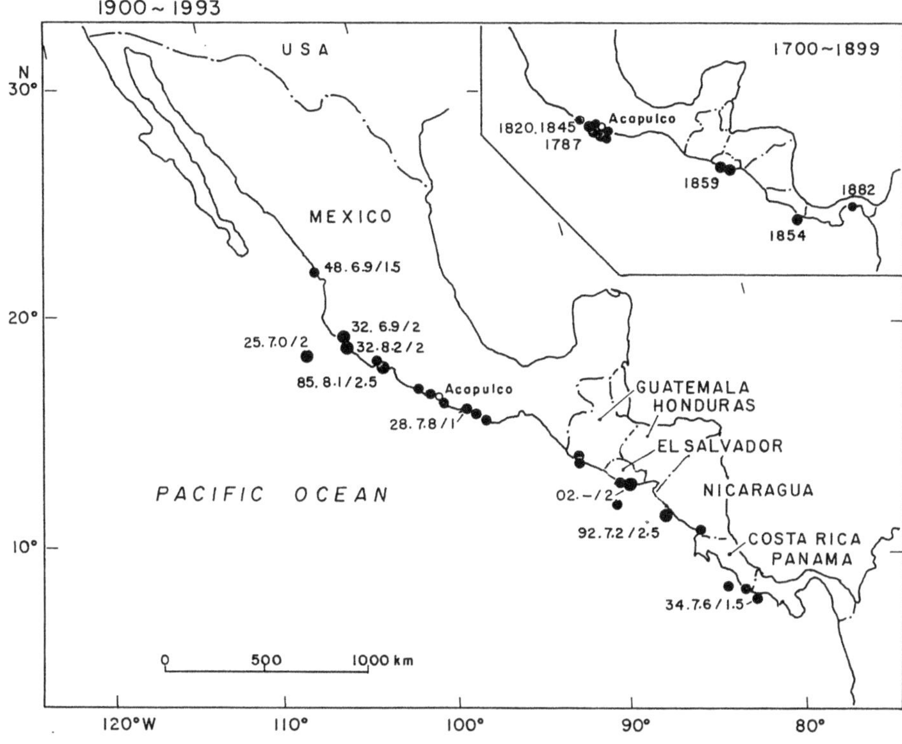

Figure 1

Distribution of the epicenters of tsunamigenic earthquakes during 1900–1993. Generating years, earthquake magnitudes, M_s, and tsunami magnitudes, m, are indicated. Inset shows the distribution of principal historical tsunamis during 1700–1899.

magnitudes, m, are indicated. The activity areas of the principal historical tsunamis (1700–1899) are similar to those of recent tsunamis.

3. Determination of Tsunami Magnitudes

The examples of magnitude determination are shown by using the author's method (HATORI, 1986) and the data of inundation heights near the source areas and the tide-gauge records at distant stations.

The Mexico tsunami of Sept. 19, 1985 hit a 500 km segment from Manzanillo to Acapulco, and the maximum inundation height reached 4–5 m at Lazaro Cardenas (Ka. ABE et al., 1986). Figure 2 shows the distribution of tsunami heights projected along the coast. The segment, L, is the distance measured from the epicenter. The tsunami magnitude scale is classified by tsunami height of 2.24 intervals: Tsunami

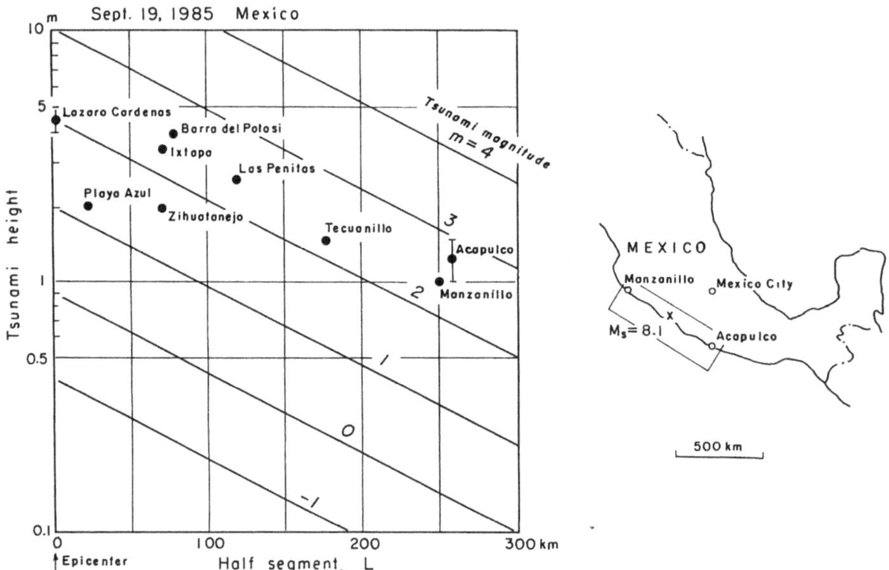

Figure 2
Distribution of inundation heights of the 1985 Mexico tsunami projected in parallel with the coast, where the origin is the epicenter. Straight lines indicate classification of the tsunami magnitude scale.

energy is reduced by one-fifth for the decreasing magnitude by a unit. The relation of tsunami magnitude, m, and inundation height, H, is empirically expressed as follows:

$$m = 0.008L + 2.7 \log H + 0.31 \quad (L; \text{km}, H; \text{m}).$$

Thus, the tsunami magnitude is determined to be $m = 2.5$ from Figure 2.

It was noted that the Nicaragua earthquake of Sept. 2, 1992 is one of the typical low-frequency earthquakes (Ku. ABE *et al.*, 1993). Although the seismic intensity near the source area was only 1 to 2 (JMA scale), the tsunami struck about a 300 km segment of the Nicaraguan coast. The maximum inundation height reached 9.9 m at El Transito. Judging from Figure 3, the tsunami magnitude is determined to be $m = 2.5$.

The above-mentioned tsunamis were observed at Pacific tide stations. Figure 4 displays another method for magnitude determination using the tide-gauge records at distant stations. The tsunami magnitude scale is classified by the attenuation of wave-height (semi-amplitude) with distance from the epicenter, $\Delta^{-1/2}$. Although the data of the 1992 Nicaragua tsunami are widely scattered, the magnitude of the 1985 Mexico tsunami is obtained with the same value as $m = 2.5$. For tsunamis with a few data, the magnitudes are mainly determined from wave heights at the near field. The re-examined magnitudes are shown in Table 1.

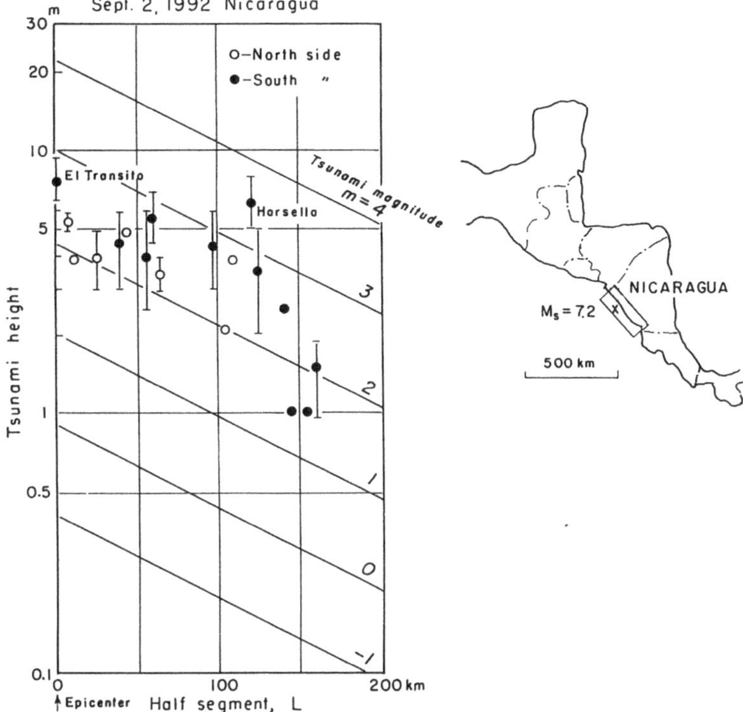

Figure 3

Distribution of inundation heights of the 1992 Nicaragua tsunami projected in parallel with the coast, where the origin is the epicenter. Straight lines indicate classification of the tsunami magnitude scale.

4. Relation of Magnitudes of Earthquake and Tsunami

Tsunami magnitudes strongly depend on earthquake magnitude and the focal depth, and tsunami waves carry information concerning the source mechanism. Figure 5 shows the relation between the surface-wave magnitude of an earthquake, M_s, and the tsunami magnitude on the Imamura-Iida scale, m. The symbol A indicates the empirical equation obtained from the data near Japan and some great global earthquakes (KOYAMA and KOSUGA, 1985), as follows:

$$m = 3.9M_s - 28.6.$$

The line shown by symbol B fits the tsunami data in the Philippine and Indonesia regions (HATORI, 1994). For most of the Central American tsunamis, the magnitude values are relatively small compared with other regions. However, the magnitude values of a few tsunamis generated by the low-frequency earthquakes are

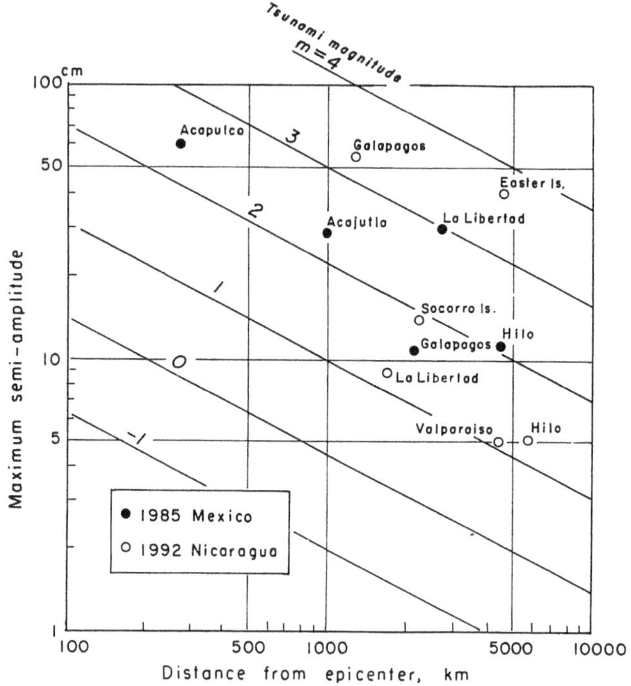

Figure 4

Relation between the maximum semi-amplitude and distance from the epicenter. Tsunami magnitude scale is classified by the attenuation of wave-height with distance, $\Delta^{-1/2}$.

three magnitude units (about 10 times for wave-height) higher than normal tsunamis due to the same earthquake magnitude.

According to the data of tsunamis in the vicinity of Japan, the relation between tsunami magnitude and the maximum runup heights, H_{max}, near the source are expressed as follows:

$$\log H_{max} = 0.35m - 0.05 \quad (H_{max}, \text{meter}).$$

As shown in Figure 6, the relation fits the data of principal tsunamis in the world. However, the maximum tsunami heights caused by the low-frequency earthquakes exceed two times that of the same tsunami magnitude.

5. Tide-gauge Records in Japan

A few tsunamis generated in the Central American region have been observed at the Japanese tide stations. According to the tide-gauge record of the 1992

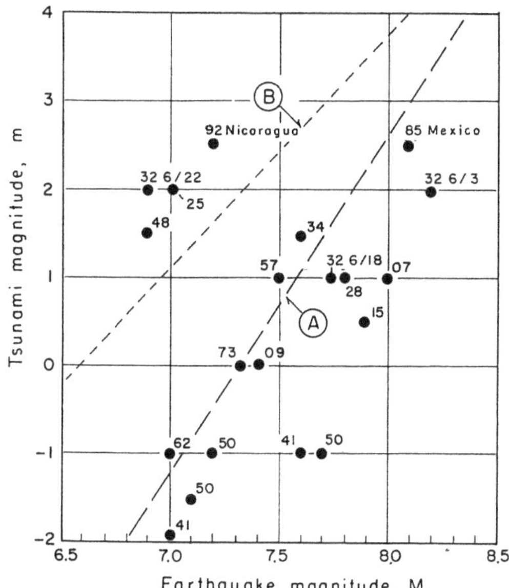

Figure 5

Relation between earthquake magnitude and tsunami magnitude in the Central American region. Relation A: Earthquake near Japan and predominate earthquakes in the Pacific zone (KOYAMA and KOSUGA, 1985). Relation B: Earthquakes in the Philippine and Indonesia regions (HATORI, 1994).

Nicaragua tsunami at Kesennuma, northeastern Japan, the initial wave was observed at about 16 hr 50 min after the main shock. The maximum double amplitude was 13 cm (Ku. ABE *et al.*, 1993).

Figure 7 shows examples of tide-gauge records from the Mexico tsunami on June 3 and 18, 1932 observed at Mera (JMA station). Travel times are 14 hr 43 min and 15 hr 03 min. The maximum double amplitudes are 26 cm and 10 cm, respectively.

The 1960 Chile tsunami ($m = 4.5$) impacted the entire east coast of Japan and killed 142 persons. Eleven tsunamis generated in the South American region have been observed at the Japanese tide stations since 1900 (HATORI, 1989). On the other hand, the Central America tsunamis have been observed in Japan with small amplitudes. The radiating energy toward Japan is strikingly weak because of a directivity effect.

In the Central American region, the amount of energy generated from distant tsunamis is relatively small due to the geographical condition. For the total energy radiating from the circum-Pacific zone, the percentage of the received energy was only 6.4% in Central America during 1900–1992 (HATORI, 1993).

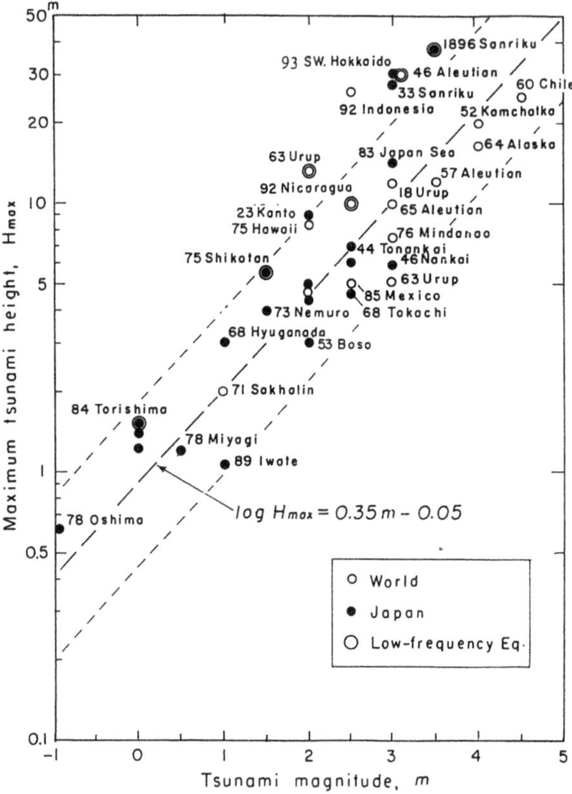

Figure 6

Relation between tsunami magnitude and the maximum runup height.

6. Conclusion

For the tsunamis which were generated in the Central American region during 1900–1993, the tsunami magnitudes are discussed in relation to earthquake magnitude. The magnitude values of tsunamis are relatively small compared with other regions. However, the inundation heights of a few tsunamis caused by the low-frequency earthquakes ($M_s = 6.9–7.2$) exceed about 10 times the normal tsunamis. A second tsunami has been generated by the aftershocks of large earthquakes ($M_s > 7.7$). The tsunami magnitudes are under control of the focal depth and the dip angle of fault. For the tsunami hazard, it is indispensable to strengthening the warning system.

REFERENCES

ABE, Ka. (1981), *Magnitudes of Large Shallow Earthquakes from 1904 to 1980*. Phys. Earth Plan. Inter. 27, 72–92.

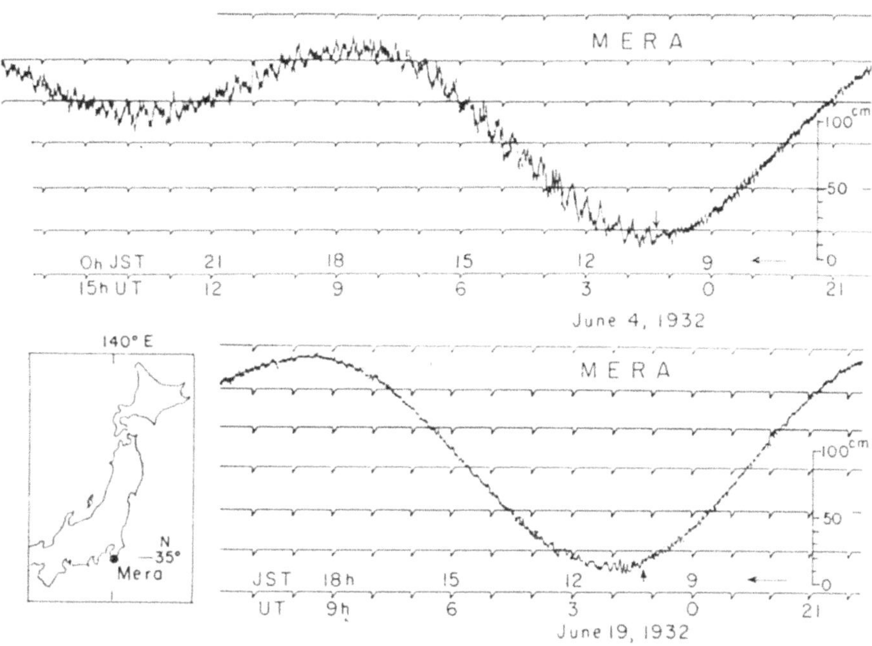

Figure 7
Tide-gauge records of the Mexico tsunamis on June 3 and 18, 1932 observed at Mera, East Japan.

ABE, Ka., HAKUNO, M., TAKEUCHI, M., and KATADA, T. (1986), *Survey Report on the Tsunami of the Michoacan, Mexico Earthquake of Sept. 19, 1985*, Bull. Earthq. Res. Inst., Univ. Tokyo *61*, 475–481.

ABE, Ku., ABE, Ka., TSUJI, Y., IMAMURA, F., KATAO, H., IIO, Y., SATAKE, K., BOURGEOIS, J., NOGUERA, E., and ESTRADA, F. (1993), *Field Survey of the Nicaragua Earthquake and Tsunami of Sept. 2, 1992*, Bull. Earthq. Res. Inst., Univ. Tokyo *68*, 23–70.

HATORI, T. (1986), *Classification of Tsunami Magnitude Scale*, Bull. Earthq. Res. Inst., Univ. Tokyo *61*, 503–515.

HATORI, T. (1989), *Distribution of Wave Energy Received from Distant Tsunamis along the Coast of Japan*, Zisin *42*, 467–473.

HATORI, T. (1993), *Distribution of Tsunami Energy on the Circum-Pacific Zone*, Proc. IUGG/IOC Intern. Tsunami Symposium, Wakayama, Japan, pp. 165–173.

HATORI, T. (1994), *Tsunami Magnitudes in Taiwan, Philippines, and Indonesia*, Zisin *47*, 155–162.

IIDA, K., COX, D. C., and PARARAS-CARAYANNIS, G. (1967), *Preliminary Catalog of Tsunamis Occurring in the Pacific Ocean*, Hawaii Inst. Geophy., Hawaii Univ. Data Report *5*, HIG–67–10.

KOYAMA, J., and KOSUGA, M. (1985), *Tsunami Magnitude and Fault Parameters*, Zisin *38*, 610–613.

SOLOVIEV, S. L., and GO, Ch. N. (1985), *Catalog of Tsunamis on the Western Shore of the Pacific Ocean* (in Russian). Translated by Canada Institute for Scientific and Technical Information National Research Council, Ottawa, Canada K1A 0S2, 285 pp.

UTSU, T. (1990), *Catalog of Large Earthquakes in the World from Ancient through 1989*, Memorial Press of Prof. Tokuji Utsu, 243 pp. (in Japanese).

(Received July 23, 1994, revised November 14, 1994, accepted November 16, 1994)

PAGEOPH, Vol. 144, Nos. 3/4 (1995)

0033–4553/95/040481–44$1.50 + 0.20/0
© 1995 Birkhäuser Verlag, Basel

Damage to Coastal Villages due to the 1992 Flores Island Earthquake Tsunami

YOSHINOBU TSUJI,[1] HIDEO MATSUTOMI,[2] FUMIHIKO IMAMURA,[3]
MINORU TAKEO,[1] YOSHIAKI KAWATA,[4] MASAFUMI MATSUYAMA,[5]
TOMOYUKI TAKAHASHI,[6] SUNARJO[7] and PRIH HARJADI[7]

Abstract — A field survey of the 1992 Flores Island earthquake tsunami was conducted during December 29, 1992 to January 5, 1993 along the north coast of the eastern part of Flores Island. We visited over 40 villages, measured tsunami heights, and interviewed the inhabitants. It was clarified that the first wave attacked the coast within five minutes at most of the surveyed villages. The crust was uplifted west of the Cape of Batumanuk, and subsided east of it. In the residential area of Wuring, which is located on a sand spit with ground height of 2 meters, most wooden houses built on stilts collapsed and 87 people were killed even though the tsunami height reached only 3.2 meters. In the two villages on Babi Island, the tsunami swept away all wooden houses and killed 263 of 1,093 inhabitants. Tsunami height at Riang-Kroko village on the northeastern end of Flores Island reached 26.2 meters and 137 of the 406 inhabitants were killed by the tsumani. Evidence of landslides was detected at a few points on the coast of Hading Bay, and the huge tsunami was probably formed by earthquake-induced landslides. The relationship between tsunami height and mortality was checked for seven villages. The efficiencies of trees arranged in front of coastal villages, and coral reefs in dissipating the tsunami energy are discussed.

Key words: Coastal damage due to tsunamis, coeismic crustal motion, aftershock area, secondary tsunamis by induced landslides, short arrival time of tsunami, liquefaction, sand blow, relationship between tsunami height and ratio of mortality.

1. Introduction

A large earthquake with magnitude M_w 7.8 occurred on the north coastal area of the eastern part of Flores Island, Indonesia, at 5 h 29 m GMT (13 h 29 min Flores local time) on December 12, 1992. Most buildings were damaged in Maumere City and its vicinity, where seismic intensity was estimated at 9 to 10 on the Modified

---

[1] Earthquake Research Institute, University of Tokyo, Japan.
[2] Faculty of Mine Engineering, Akita University, Japan.
[3] Asian Institute of Technology, Bangkok, Thailand.
[4] Disaster Prevention Research Institute, Kyoto University, Japan.
[5] Central Research Institute of Electric Power Industry, Abiko City, Japan.
[6] Tohoku University, Sendai City, Japan.
[7] Meteorological and Geophysical Agency, Jakarta, Indonesia.

Mercalli scale or at 6 on the JMA intensity scale. In the city area, evidence of liquefaction, sand blows, and cracks was observed at many places. Additionally, landslides occurred in the mountainous regions.

Several minutes after the main shock, a tsunami attacked the villages on the north coast of the eastern part of Flores Island. The residential area of Wuring village, which is located about 2 km northwest of Maumere on a small sand spit 700 meters long, was hit by the tsunami resulting in the collapse of 80 percent of the wooden houses and the death of 87 persons of the 1,400 inhabitants. The height of the tsunami was only 2.5–3.2 m above mean sea level.

Babi Island, with a diameter of 2.5 km, is located offshore some 40 kilometers northeast of Maumere City. Two villages are situated on the south coast of the island; Muslim in the western part and Christian in the eastern, totalling a population of 1,093. About 3 minutes after the main shock, the first wave attacked both villages, and all houses were washed away; leaving no trace of buildings. The tsunami took the lives of 263 persons.

On the coast near Cape Bunga, the most northeastern point of Flores Island, an extraordinarily large wave attacked the coastal villages. Sea water ran up along a slope and reached a height of 26.2 m above sea level at Riang-Kroko village. Houses were entirely swept away, and nothing remained to provide evidence of human lives there. In this village 137 persons perished due to the tsunami.

We conducted a survey along the coast of the eastern part of Flores Island from December 29, 1992 to January 6, 1993 and visited about 40 coastal villages including Wuring, Babi Island, and Riang-Kroko. We also visited the refugees' tents at Nangahale village, about 30 kilometers east of Meumere, where the survivors from Babi Island stayed temporarily. We measured the tsunami inundation height by detecting traces of sea-water submergence and by questioning eyewitness accounts. We also researched building damage in each village. Additionally, we copied the Indonesian newspaper, "Kompas", dated from December 13 to 30, 1992, from which we could obtain statistical information of human and building damage with the total number of houses and inhabitants of several villages.

In the present study, the characteristics of the earthquake and tsunami are mentioned briefly. Tsunami heights and inundation at coastal villages are described in detail with the statistics of human casualties and building damage. Furthermore, we compare this damage with tsunamis in Japan.

2. Historical Background of Tsunamis in Indonesia

2.1 Distribution of Earthquakes and Tsunamis in the Central and Eastern Parts of Indonesia

Virtually all Indonesian territory consists of island arcs, accompanied by parallel ocean trenches. Naturally, the seismicity of most of the territory is generally active.

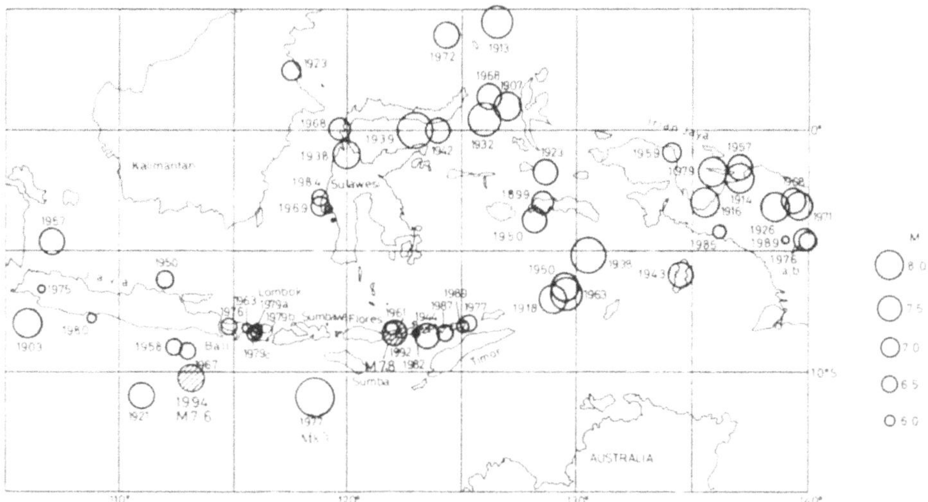

Figure 1

Distribution of the epicenters of earthquakes with damage, which occurred in the central and eastern parts of Indonesia since the beginning of this century. Shadings represent the 1992 Flores and the 1994 East Java earthquakes.

A tsunami catalog of the western Pacific was published by SOLOVIEV and GO (1974), and they also published an additional catalog for the entire Pacific coast (SOLOVIEV and GO, 1987). In addition, SUNARJO and TAJAN (1993) created a table of tsunamis in Indonesia. UTSU (1991) edited a catalog of all destructive earthquakes in the world, in which tsunamigenic earthquakes in Indonesia are listed. A short article in the newspaper KOMPAS, dated December 14, 1992, introduced destructive earthquakes and tsunamis in the region of Flores-Alor Archipelago and its vicinity. On the basis of those materials, we made the map of the distribution of damaging earthquakes in the central and eastern parts of Indonesia (Fig. 1). Figure 1 shows the locations of the epicenters.

A group of earthquakes occurred since 1964 in the Flores-Alor region, and seismic activity has increased since 1977. In 1977, a massive earthquake of normal fault type with magnitude 8.3 occurred in the sea south of Sumba Island, and 189 deaths were attributed mainly to the tsunami (KATO *et al.*, 1993). Another group of earthquakes struck the region of Bali and Lombok Islands, where seismic activity has increased since 1963. Earthquakes causing damage occur rarely in the region from Sumbawa Island to the western part of Flores Island, and a seismic gap seems to form in the area between the two seismically active areas.

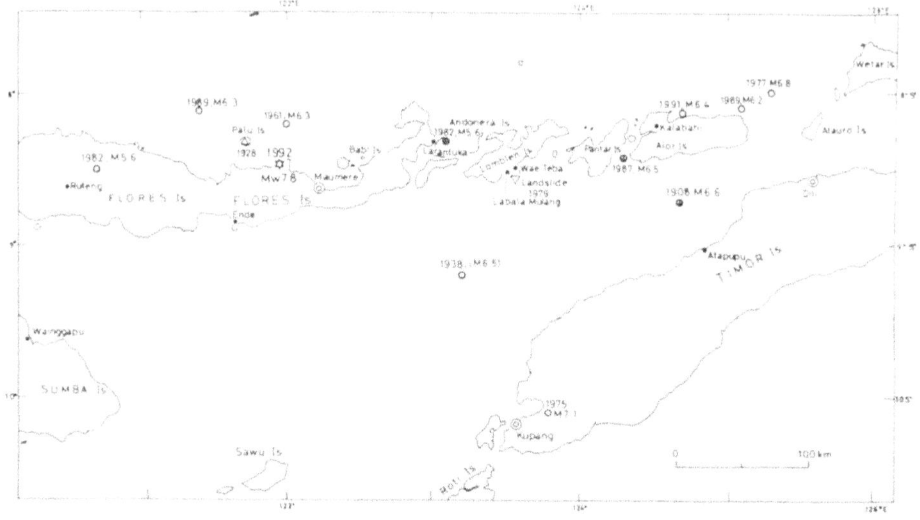

Figure 2

Distribution of earthquakes and tsunamis with damage, which occurred in the Flores-Alor region since the beginning of this century. Open circles show earthquakes with damage, circles with crosses show earthquakes with tsunamis. A triangle shows a tsunami caused by a volcanic eruption. An inverted down triangle shows a tsunami caused by a landslide.

2.2 Distribution of Earthquakes and Tsunamis in the Region of the Flores-Alor Island Arc

Figure 2 shows the distribution of earthquakes and tsunamis in the Flores-Alor region during this century. The volcanic Island of Palu erupted August 4 to 5, 1928, and a tsunami was induced. Both the eruption and tsunami killed 226 persons in total, 128 of which were victims of the tsunami. On July 18, 1979, a large-scale landslide occurred on the southeast coast of Lomblen Island and generated a large tsunami. The tsunami hit two neighboring villages, Wae-Teba and Labala-Mulang; 539 people were killed and about 700 were missing due to the tsunami (MIYOSHI, 1993). This tsunami was not associated with any earthquake and there were no irregular meteorological conditions on that day. The crust beneath the islands of the Flores-Alor region is basically formed of multiple thick basalt layers, and there are many steep cliffs on the coast in this region. This suggests that landslides can be easily induced by earthquakes or heavy rainfall. In this region, damaging earthquakes have had a tendency to occur more frequently since 1977, the time of the Sumbawa tsunamigenic earthquake of August 19, 1977 with magnitude M_w 8.3.

2.3 Earthquakes and Tsunamis in the Past Ten Years in the Region of Flores Island and its Vicinity

In 1982, two damaging earthquakes occurred on Flores Island. One occurred on August 6, near Retung, an inland city in the central part of the island. The magnitude of the earthquake was 5.6, causing slight building damage. The other occurred in the city of Larantuka, on the easternmost coast of the island. The magnitude of the earthquake was 5.6, and it caused damage in the area extending from the eastern part of Flores Island to the Adonara and Solor Islands. Submarine landslides were induced in the channel between the islands of Flores and Adonara and a small tsunami was generated. Thirteen people were killed, 17 people were seriously injured, 400 people were slightly injured, and 1,875 houses totally collapsed due to the event.

On July 15, 1989, an earthquake with magnitude 6.2 struck the northeastern tip of Alor Island; 7 people were killed and 95 houses caved in. Sixteen days after the previous event, on July 31, 1989, another sizeable earthquake with magnitude 6.3 occurred offshore north of the central part of Flores Island, and a few persons were killed in Maumere City. The damaged area of this event overlapped that of the 1992 Flores Earthquake. Seismic intensity at Maumere was 5 on the MM scale.

One year before the present event, July 4, 1991, an earthquake with magnitude 6.2 (M_s 6.4) occurred along the north coast of Alor Island, near Kalabahi City; 23 persons perished, 181 persons were injured, and 1,150 houses were destroyed. The records of broadband seismographs show that two events occurred sequentially, separated by 2.5 seconds.

2.4 Characteristics of Historical Earthquakes and Tsunamis in the Region of Flores Island and its Vicinity

We can judge that the present event took place in the zone of increased seismic activity in the Flores-Alor region since 1977 (Fig. 3). In particular, we note that

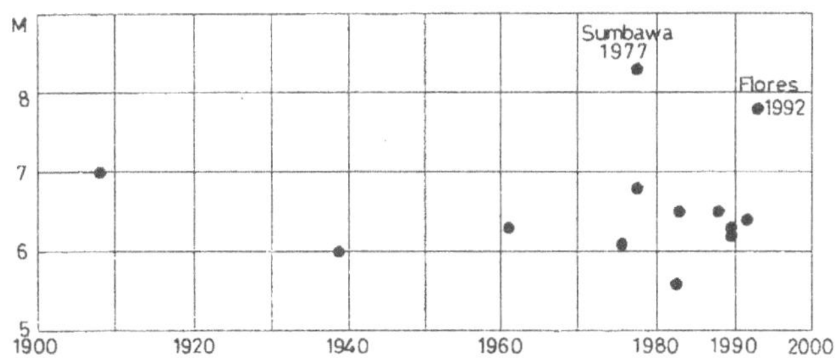

Figure 3

Chronological diagram of earthquakes with damage in the Flores-Alor region.

three earthquakes with magnitudes over 6.0 occurred in the Flores-Alor region within the three years prior to the 1992 Flores earthquake and may be considered as precursors of that event. We also notice that landslides can be easily induced by earthquakes in this district, due to the many steep cliffs of basalt layers and to poor vegetation conditions on the slopes caused by negligible precipitation on land. We should also note that tsunamis were often caused by landslides, which may have been induced both by some seismic or meteorological conditions.

3. The 1992 Flores Earthquake

3.1 Tsunami Source Area

The epicenter was located on the north coastal region of the eastern part of Flores Island, near the Cape of Batumanuk, 35 kilometers northwest of Maumere City. The CMT solution by Harvard University demonstrates that the event had a thrust mechanism with a dip angle of 32° on the southward dipping plane.

Figure 4 displays the distribution of aftershocks until 30 days after the main shock (until January 11, 1993), as located by the USGS. It suggests that the fault plane lies between the epicenter near the Cape of Batumanuk and the Cape of Bunga, on the northeastern tip of Flores Island. The length of the fault is about

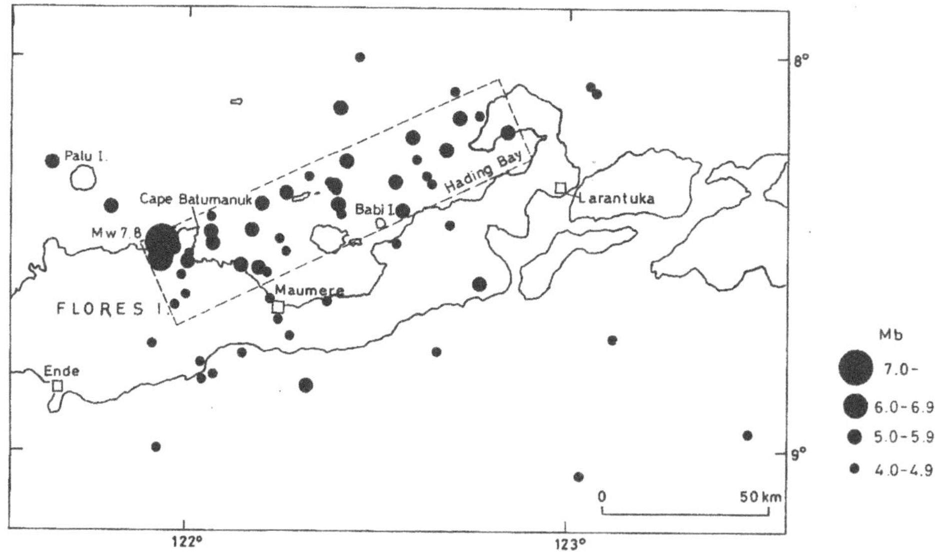

Figure 4
Distribution of aftershocks until January 11, 1993, that is, thirty days after the main shock from USGS. Dashed rectangle shows the area of the estimated fault.

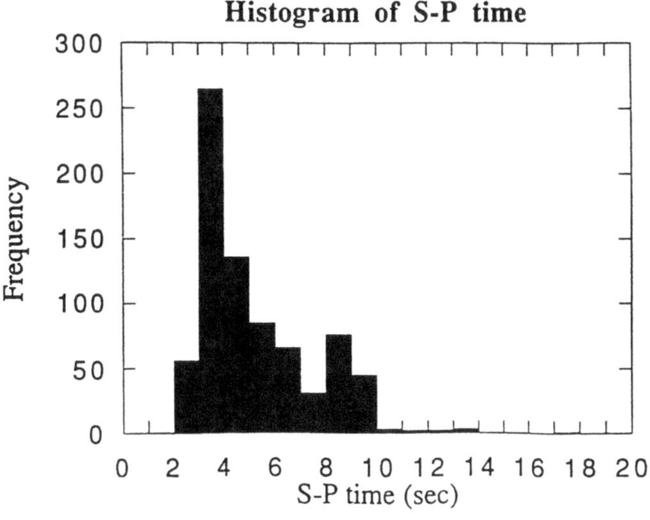

Figure 5

S-P time distribution of aftershocks from noon, December 30, 1992 to evening January 5, 1993 observed at the top of Broadcasting Tower Hill behind the residential area of Maumere.

110 km, and the horizontal width is 35 km. The fault plane area partially underlies the coastal regions near Cape Batumanuk and Cape Bunga, at the northeast end of Flores Island. Babi, Besar and Pomana Islands are estimated to be located above the fault plane area.

We temporarily observed aftershocks by setting short-period seismographs on a rock hill of the local TV station behind the residential area of Maumere City, from the afternoon of December 30, 1994 to the evening of January 5, 1993. During this period, about 1,000 aftershocks were recorded and S-P times were estimated for 773 events. Most of the S-P times of the events are distributed between 2 and 10 seconds (Fig. 5), indicating the occurrence of most aftershocks within 80 km of Maumere City. As it is well-known that the tsunami source area coincides generally with the aftershock area, we can judge that the tsunami source area also extended close to the north coast of Flores Island.

Coseismic crustal deformation was observed by inhabitants at several points in the coastal villages, and we generated a map of the distribution of vertical displacement shown in Figure 6. The shore line was uplifted on the coast west of Cape Batumanuk by 0.5 to 1.1 meters, and subsided at many places east of the Cape. The amount of subsidence reached 1.6 m at Kolisia village, 25 km northwest of Maumere. Evidence of subsidence can also be seen in the port area of Maumere. However, here liquefaction took place, making it difficult to distinguish the amount of subsidence purely due to coseismic crustal deformation. The subsidence on the shore of Babi Island was also observed by an eyewitness to be 75 centimeters.

Figure 6
Distribution of vertical crustal deformation in meters. Positive values show uplift, while negative ones
show subsidence.

Based on eyewitness accounts, the tsunami arrival times after the main shock at most of the coastal villages were distributed between 2 and 5 minutes. We back-projected the tsunami propagation from the observation points, and the result is shown in Figure 7, which also suggests that the tsunami source area was located close to the north coast of the eastern part of Flores Island. We also notice that the tsunami arrival time was observed as only 2 minutes, even in Hading Bay.

3.2 *Distribution of Seismic Intensity and Damage*

The seismic intensity on the Flores Island region was presented by the Meteorological and Geophysical Agency of Indonesia as shown in Figure 8. Seismic intensities on the MM scale were estimated as 9 to 10 at Maumere and 8 to 9 at Ende on the southern coast of Flores Island. Most of the brick houses were damaged in both cities. Small and moderate size landslides and rock falls appeared throughout the mountainous regions of those districts. Traffic on the 147-km long highway between Ende and Maumere was blocked for more than one month.

The earthquake was felt at Ujung Pandang (MM intensity 4) on Sulawesi Island, Kupang (3) on Timor Islands, Waingapu (4) on Sumba Island, and at Bima (3) on Sumbawa Island. It was also felt at Denpasar (2) on Bali Island, 700 km west of the epicenter.

The Military Commander of Maumere identified the areas of most severe damage due to the earthquake and initiated relief efforts. Hatched zones in Figure 9

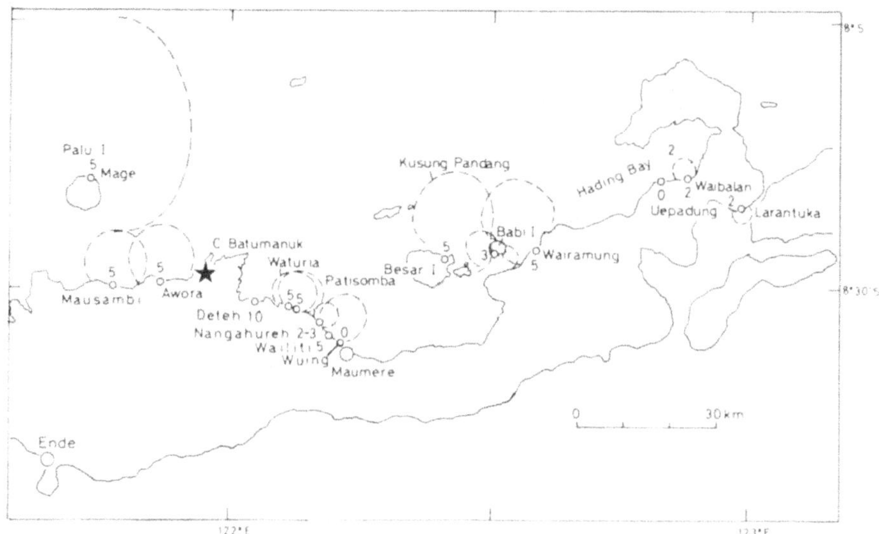

Figure 7

Tsunami arrival time (minutes) after the earthquake in minutes. Broken line circles show the inversely drawn refraction lines of tsunami propagation. Dotted area shows the location of the estimated fault by the aftershocks in Figure 4.

show the damaged areas for both the earthquake and the tsunami. As tsunami damage occurred only on the coast of the Flores Sea, the damaged areas on land and on the south coast are purely due to the earthquake. Arrows on the north coast of Flores Island indicate places with only slight damage due to the tsunami. Tsunami damaged coast extended as far as Reo Port on the western part of Flores Island.

Damage to public constructions and buildings was reported in detail by HAKUNO (1993) and SHIBUYA (1993).

4. Tsunami Heights and Damage in Coastal Villages

4.1 Field Survey of the Tsunami

We conducted a field survey of the tsunami damaged coast on the eastern part of Flores Island from December 29, 1992 to January 5, 1993. We divided our members into four teams, who visited about forty coastal villages to determine seismic intensity, crustal movement, tsunami arrival time, and inundation area. In addition, we also asked whether they had correct knowledge of tsunamis. We prepared questionnaires written in Indonesian, and asked Indonesian translators to conduct interviews at each coastal village.

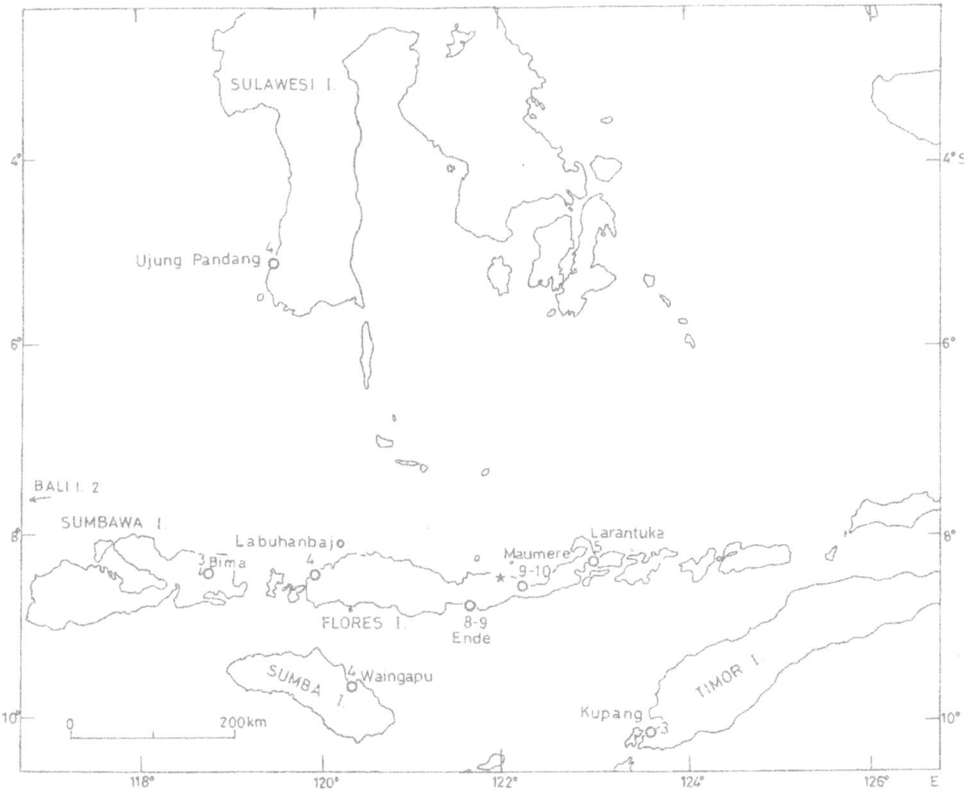

Figure 8
Distribution of seismic intensities on the Modified Mercalli scale.

On January 3 and 5, we were given the opportunity to ride in helicopters of the Indonesian Air Force, and we visited Ende, Palu Volcanic Island, and the coast of the northeastern tip area of Flores Island.

As there are no benchmarks on the coast, we measured tsunami height above sea level at the time of the survey. The astronomical tide components at tsunami damaged villages were corrected, resulting in values of tsunami heights above the mean sea level. In order to check how the numerically predicted astronomical tide agrees with the actual one, we made observations of sea-level change on New Year's Day, 1993 at the Sari Beach Hotel for calibration, and confirmed that they agree well. Astronomical tide change on the day of the tsunami is shown in Figure 10. At that time, sea level was higher than mean sea level by 24–36 centimeters.

Figure 11 shows the locations of surveyed points with evident damage. We surveyed approximately 40 points, but little damage occurred on the coasts between Maumere and Talibura, 30 kilometers to the east, thus only the measured heights will be described at these locations.

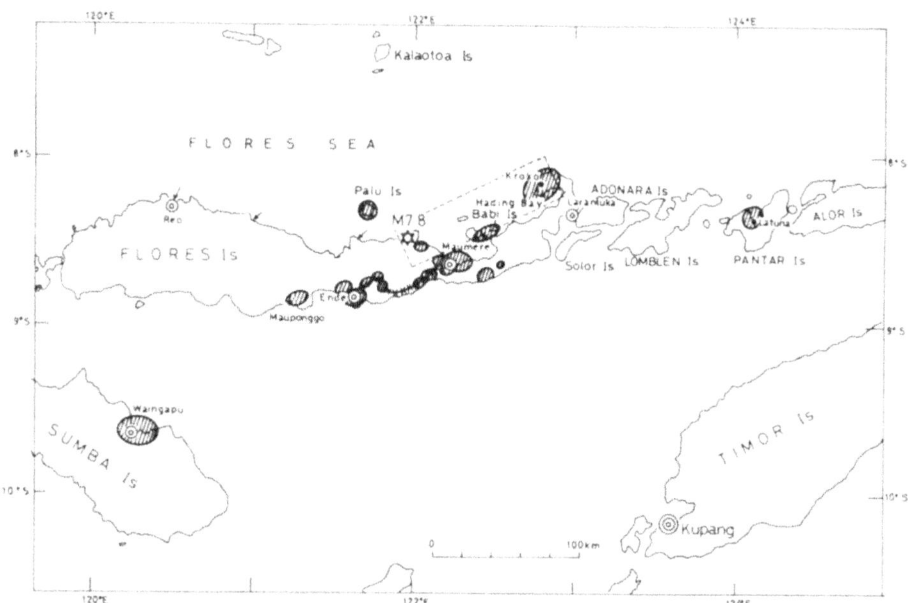

Figure 9

Damaged areas (hatched zone) identified by the Military Commander of Maumere.

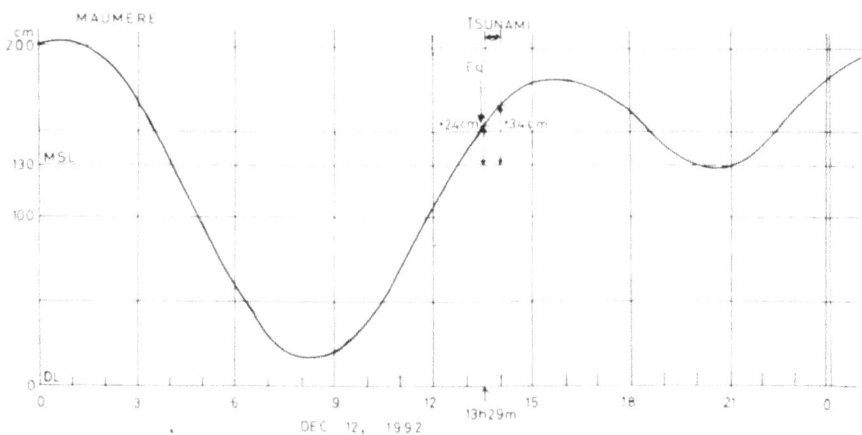

Figure 10

Astronomical tide at Maumere on the day of the 1992 Flores earthquake. The first tsunami wave arrived at the coast in the source region at tide stage of 24–36 cm above mean sea level.

4.2 Tsunami Damage at Maumere Port

Evidence of severe liquefaction was observed in the Maumere Port area. A line of large cracks could be seen on the ground in front of the market yard (Fig. 12).

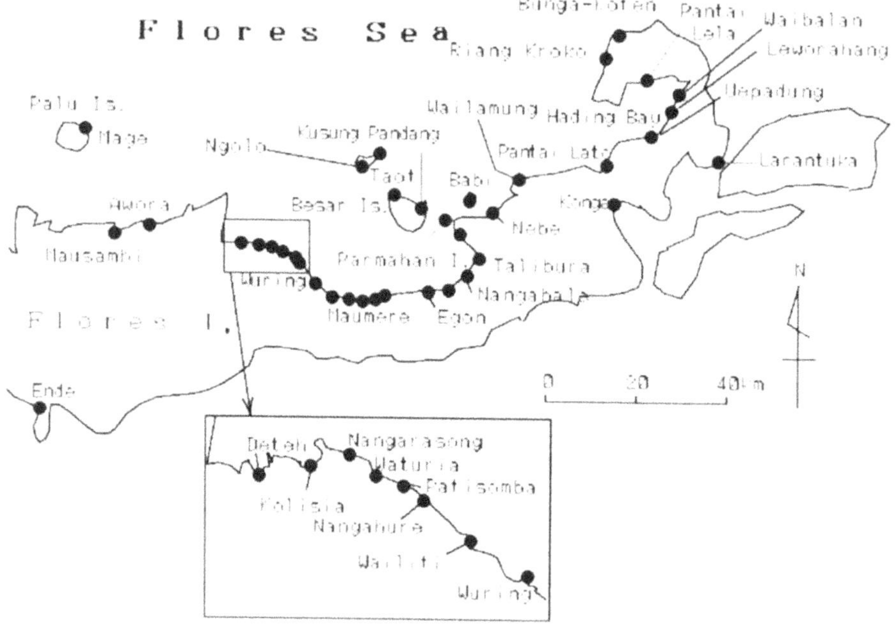

Figure 11
Locations of surveyed villages.

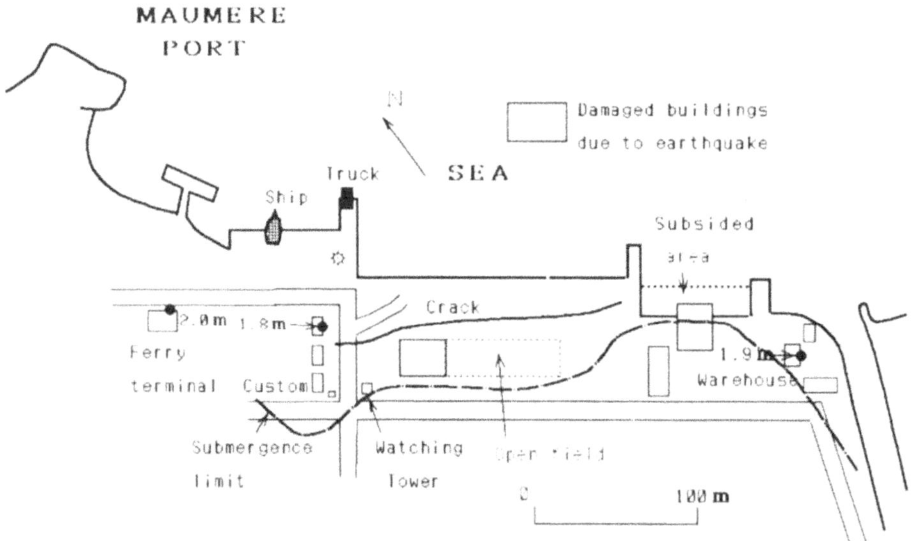

Figure 12
Inundated area of Maumere Port. The broken line shows the limit of sea-water inundation. Numbers show the tsunami height above mean sea level (m). The locations of a displaced truck and a boat are noted, as well as the portion of the port which was observed to have subsided.

The ground evidently subsided by 0.5–1 m in this area, however we could not distinguish whether this was caused by crustal deformation or by liquefaction of the soil.

Sea water invaded the end of the front row of blocks and a clear high water mark could be observed. We measured the height of the water mark on an outside wall of the port office building as 1.8 m (after compensations for the astronomical tide component at the time of the measurement). The water mark was 0.6 m above the ground. Although buildings were submerged, tsunami damage was minor at Maumere. A truck in a parking lot was carried by the wave onto a pier, and a ship was washed onto a street along the waterfront (Fig. 12). A resident stated that the depth of the port area changed from 9–15 m at a point 20 m away from the sea wall (see Fig. 12).

4.3 Tsunami Heights and Damage on the West Coast of Maumere

(a) Wuring Village

The residential area of Wuring is on a spit about 3 km northwest of Maumere. The length of the spit is about 650 m. The main road runs on the axis of the spit, and the ground height is only 1.3–2.1 m above sea level.

Tsunami waves attacked the area four times, and the first wave came just after the main shock. Prior to the first wave, sea level dropped slightly. The first wave originated from the north and east directions. The second wave was the strongest and was accompanied by a loud noise. The third wave was the highest.

On this sand spit about 1,400 people had been living in houses built on stilts. Most of the wooden houses collapsed, and 87 persons were fatalities of the tsunami. The distribution of inundation height, and sea-water thickness above the ground are shown in Figure 13. The tsunami height was only 1.8–3.6 m above the mean sea level, and water height above ground was as much as 1.7 m maximum.

Most houses in Wuring were destroyed. The majority of the destroyed houses leaned towards the southeast, indicating flow from the northwest direction. The mosque, a concrete building, was only submerged and not destroyed. Traces of three water levels were visible on the wall at heights of 0.91, 1.91, and 1.56 meters, respectively. MATSUTOMI (1993) analyzed these traces in detail, and estimated the fluid speed as 2.7 to 3.6 m/s from west to east, with the resistance force required of the building calculated as 8.0 to 10.8 tons. He also estimated the force applied to a wooden pillar 15 centimeter square and one meter in length as 57 to 100 kilograms. Many wall supports of houses on stilts could not resist the horizontal force of the flow of sea water. Diagonal bracing of walls and supports is recommended for this type of house.

The coral reef is wider at the tip of the spit, and narrow at the landward side, consequently house damage was more severe at the landward side than that at the tip. At the time of the tsunami, many fishing boats drifted into the residential area,

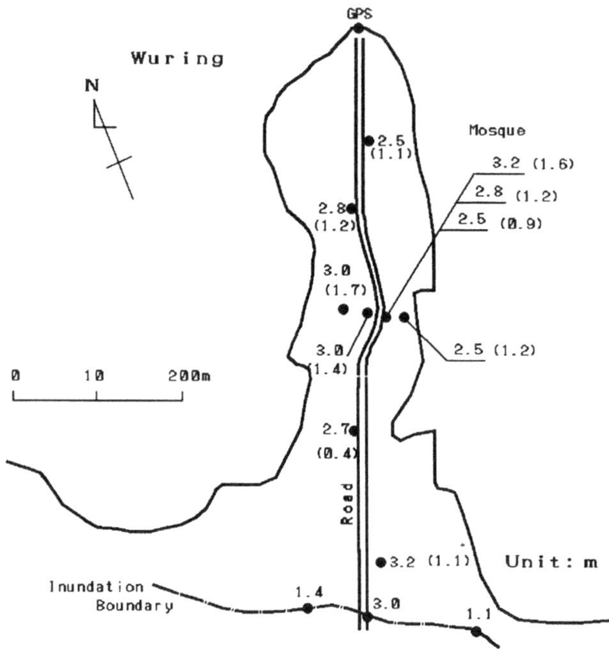

Figure 13

Distribution of tsunami heights (m) above mean sea level in the residential area of Wuring. Numbers with parentheses show sea-water thickness on the ground during the tsunami inundation.

which intensified the damage. The following day, the people made a bonfire of the rubbish. They did not carefully enough dispose of the embers, which led to a fire (the KOMPAS, Dec. 15) increasing the loss of property in Wuring.

(b) Wailiti, GPS 8° 35′ 16.2″S, 122° 11′ 02.8″E

Wailiti village is 5 km northwest of Maumere and the area inundated by the tsunami is shown in Figure 14. The first wave arrived 5 minutes after the earthquake. Six fishing boats were thrown on shore. The front wall of a factory was partially destroyed by the tsunami and a clear trace of inundated sea water was detected on the side wall of the building of 2.1 m height. One person was killed. The north coast of the river eroded and a new surface of sand step with a height of one meter appeared.

(c) Nangahureh, GPS 8° 34′ 18.3″S, 122° 10′ 14.3″E

Two or three minutes after the earthquake the first withdrawal of sea water was seen horizontally about 50 m. Sea water (not always being the first wave) broke the bank of a lagoon, whose fresh water became salty after the tsunami (Fig. 15). Sea water invaded a corn field reaching a height of 1.9 m. A crack line appeared on the field and white sand gushed into the cornfield near the crack.

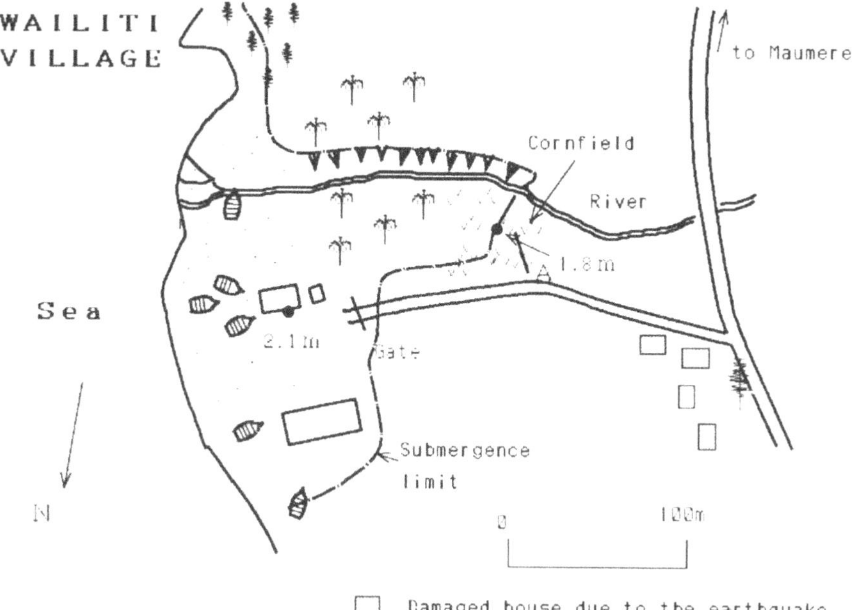

Figure 14

Inundation area in Wailiti village. A large crack appeared in the ground in the cornfield (A). Six boats were displaced on the sandy shore.

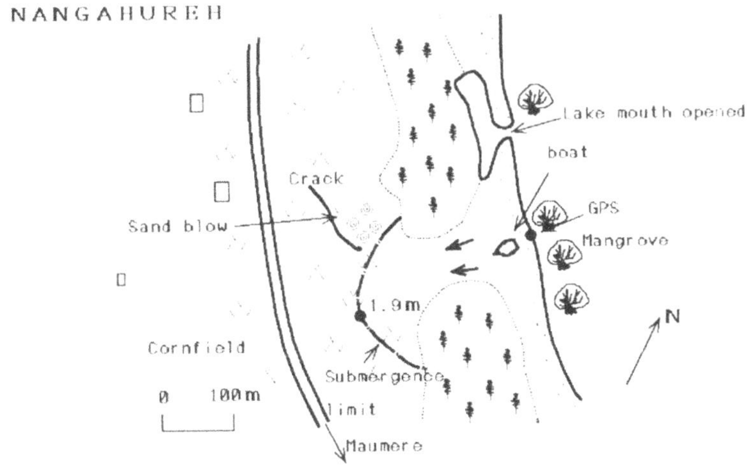

Figure 15

Inundation area in Nangahureh village. Sea water eroded the beach so that the lagoon became open to the sea. An eminent crack appeared on the ground of a cornfield, where sand blowing was observed. A boat was displaced on the sandy shore.

(d) Patisomba, GSP 8° 33′ 01.8″S, 122° 08′ 38.5″E

Strong shocks were felt twice within an interval of 15 min, and a long crack which measured 1 km appeared on the ground. The tsunami arrived 5 min after the shocks without lowering the sea level before the arrival of the initial wave. The wave front looked like a wall. Waves arose repeatedly with the first one incoming from the east. The waves inundated the area to 3.3 m, sweeping away four houses (Fig. 16).

(e) Waturia, GPS 8° 32′ 48.8″S, 122° 07′ 57.2″

The tsunami wave hit 3 times, the second one being the largest. Before the first wave, the shore line regressed by 200 m. The first wave arrived 5 min after the earthquake from the direction of Besar Island (ENE). Three houses collapsed in the village (Fig. 17). Due to the coseismic crustal deformation, sea-water level rose and the shore line shifted horizontally by 5 m inland. The amount of subsidence was measured as 30 cm vertically.

Water in the well at point A (Fig. 17) flowed better after the earthquake, while that at Waliki (the next village to the east) flowed insufficiently.

(f) Nagarasong, GPS 8° 32′ 24.8″S, 122° 07′ 14.9″E

The tsunami waves struck 3 times, and before the first wave sea water drew down and the sea bottom appeared. Subsequent inundation by sea water reached the foot of a mountain slope, where many rocks fell. Cracks with widths reaching 30 cm and gushing sand appeared on the ground (Fig. 18). Residents were jeopardized by both the tsunami and rocks falling from the steep mountain slopes.

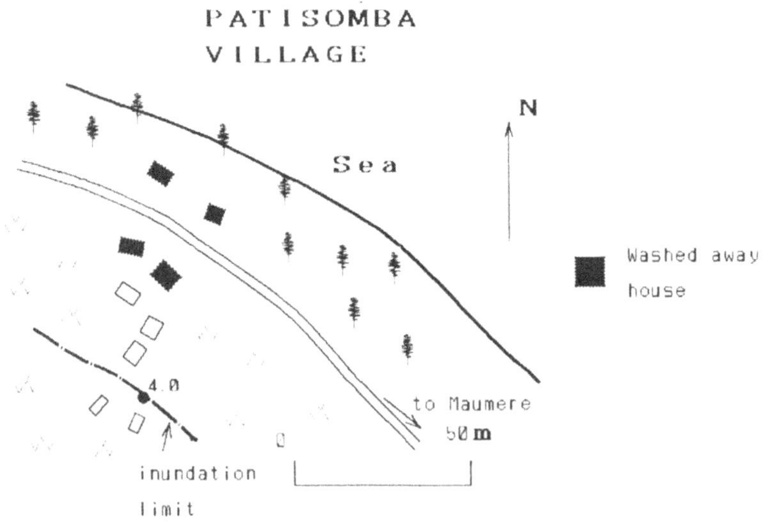

Figure 16
Inundation area in Patisomba village.

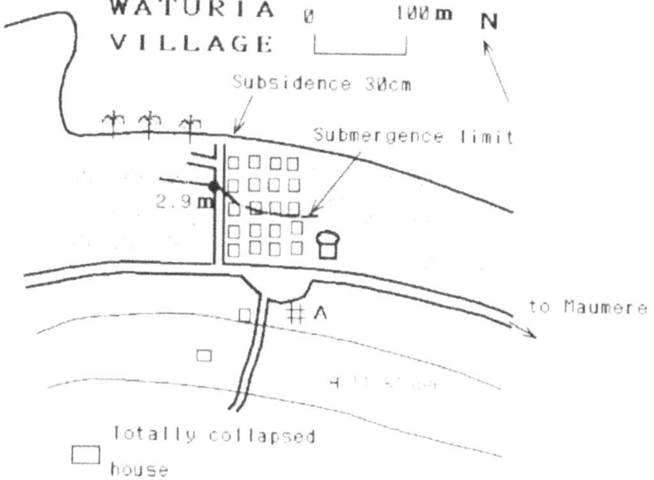

Figure 17
Inundation area in Waturia village.

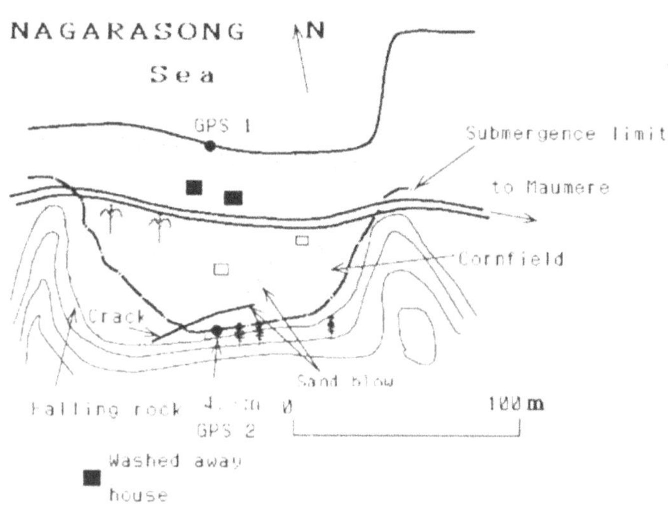

Figure 18
Inundation area in Nagarasong village. Rocks fell along the slope of the hill behind the village.

(g) Kolisia, GPS 8° 32′ 33.5″S, 122° 05′ 17.5″E

Sea water crossed the highway and flooded a rice field as much as 400 m from the shoreline. Maximum runup was 5.2 m. Earthquake damage was serious and 8 people were killed. The ground subsided 1.6 m and the beach became narrower (Fig. 19). A section of highway (1 km length) between the villages of Kolisia and

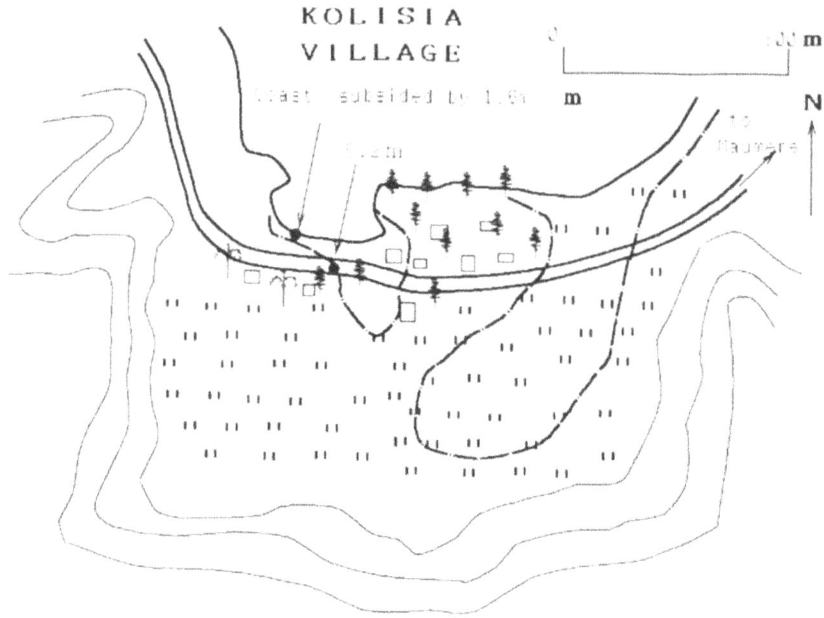

Figure 19
Inundation area in Kolisia village, where rice fields were flooded by sea water.

Deteh was seriously damaged due to the falling rocks and collapsed shoulders of the road.

(h) Deteh in Magepandang village, GPS 8° 31′ 49.1″S, 122° 02′ 10.6″E

Five to ten minutes after the earthquake, the first of three waves struck. The third was the highest, and the second one was the smallest. Maximum runup was 2.3 m. Sea water submerged most of the residential area of the village and a few houses were destroyed (Fig. 20). This inundation claimed two lives.

The shore line moved landward by 4 m and the width of sandy shore narrowed to only one or two meters due to the subsidence of the ground by about one meter.

The highway across the Batumanuk Cape area leads to Arowa and Mausambi villages. However, we could not reach those villages by car because traffic was interrupted by landslides and rock falls in this mountainous region. We were forced to navigate the Batumanuk Cape by boat on January 2, 1993.

(i) Awora, GPS 8° 29′ 36.2″S, 121°51′ 08.5″E

Awora village is located 20 km westward of Deteh beyond Cape Batumanuk. There is a small cape some 500 m west of the village, and a rock there appeared above the sea. People in the village said that the rock emerged above sea level after the earthquake. The ground seemed to be uplifted about 50 cm by eye measurement of the rock. The split of the coral reef on the east end of the village became wider, caused by the uplifting of the ground (Fig. 21). The tsunami arrived here 5 min

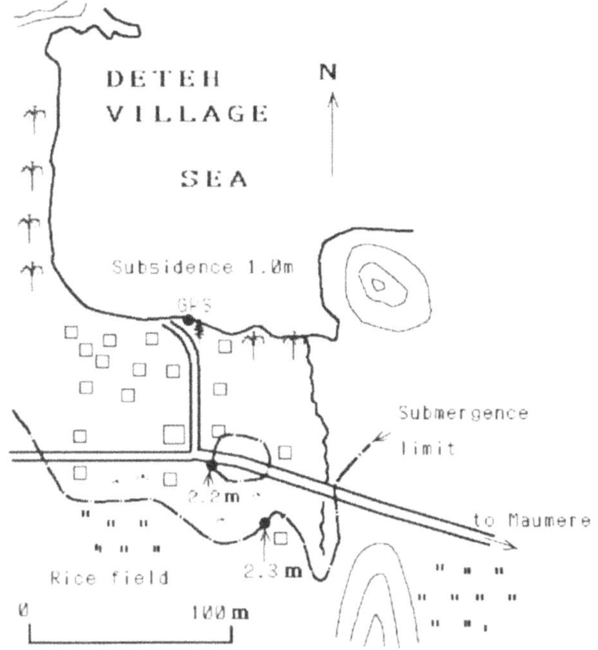

Figure 20
Inundation part of the residential area of Deteh village in the Magepanda region.

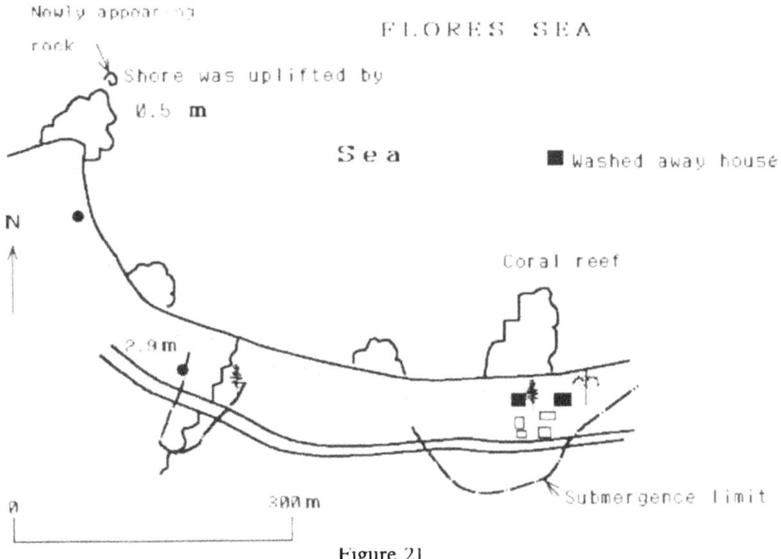

Figure 21
Tsunami damage in Awora village.

after the earthquake and runup was 2.9 m. Several houses on the coast were swept away and one person perished.

(j) Mausambi, GPS 8° 30' 34.1"S, 121° 47' 10.1"E

The tsunami arrived 5 min after the earthquake. Before the first wave arrived, the sea withdrew. Three waves arrived, the second being the largest. Sea water crossed the main road and invaded a rice field behind the residential area. In the residential area sea water rose up to the height of an adult's waist, 80 cm above the ground, that is 3.4 m above mean sea level. Two people were killed due to the tsunami.

Inhabitants said that another tsunami which besieged the village in 1973 exceeded the height of the present event. We could find no corresponding descriptions in any of the authorized tsunami catalogs mentioned in Chapter 2.

The ground was uplifted by 1.1 m on the shoreline.

(k) Mage (Palu Island)

Mage village lies on the northeast coast of the volcanic island of Palu. After the earthquake, sea level gradually subsided and five minutes later the first wave forming a wall of water attacked the village from the north. The initial wave regressed about 20 min before the arrival of several other waves. The first one was the largest. Sea-water inundation reached a stone wall surrounding the field of the elementary school, where runup was as large as 2.8 m.

4.4 Tsunami Heights and Damage on the Shores of Islands North of Flores Island

(a) Ngolo village on Pomana Besar Island

There are more than 50 houses in Ngolo village on Pomana Besar (Big) Island. Within the residential area evidence of liquefaction was observed on the ground. During the earthquake, numerous cracks and water spouts appeared, and many houses were destroyed by shaking. Soon after the earthquake a wall-like tsunami wave attacked the village. There were five waves of which the first one was the highest. Water withdrew by 20 m between waves.

The entire residential area of the village was submerged and tsunami height was measured at 2.7 and 3.2 m (Fig. 24). The ground height was 2.0 m in front of the Mosque, located in the central part, thus the covering thickness of the sea water was about 1 m in the residential area.

Prior to the earthquake, fish which usually live in the deep sea died and floated on the sea surface. One week before the earthquake inhabitants easily caught many deep sea fish in the shallow sea region.

Twenty-five persons lost their lives on Pomana Besar Island, including Buton.

(b) Buton on Pomana Besar Island

A clear trace of sea water remained on the wall of a house, 74 cm above the ground from which we established the tsunami height as 1.5 m.

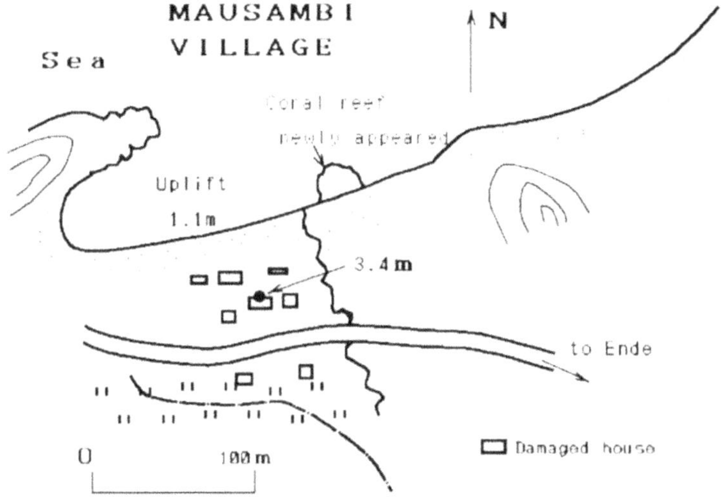

Figure 22
Inundation area in Mausambi village.

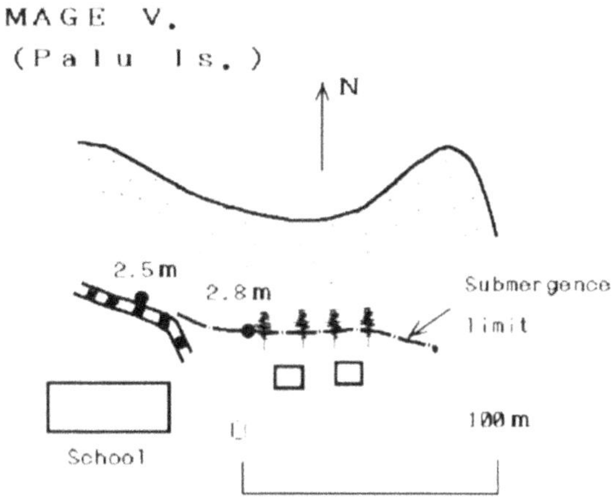

Figure 23
Inundation area in Mage village on Palu volcanic island.

(c) Taot on Besar Island, GPS 8° 25″ 30.7″S, 122° 20′ 09.3″E

Taot village is located on the northwest coast of Besar (Big) Island where 5 families totalling 66 persons reside (Fig. 25). The tsunami height reached 2.8 m causing minor wave damage. The Mosque collapsed due to shaking.

Subsidence is indicated by the submergence of the bases of trees below high-tide level.

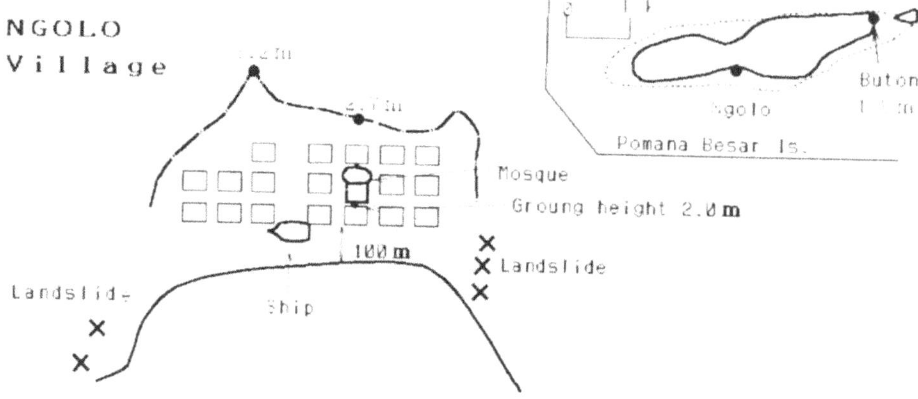

Figure 24
Inundation area of Ngolo village on Pomana Besar Island.

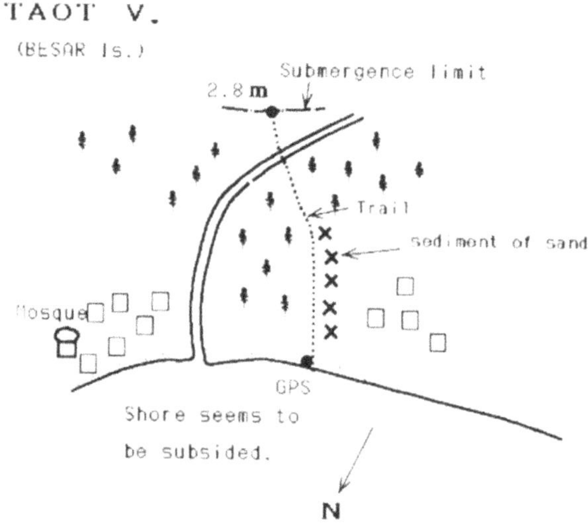

Figure 25
Inundation area of Taot village on the west coast of Besar Island.

(d) Kusung Pandang on Besar Island, GPS 8° 26′ 48.4″S, 122° 24′ 32.1″E

Kusung Pandang village is located on a small bay on the northeast coast of
Besar Island (Fig. 26). In this village 55 people comprising 11 families lived, but no
one was killed. Several people were carried away by the tsunami waves, but all were
rescued. The tsunami wave arrived at the coast as a wall of water. Three waves
arrived, with the initial one appearing five minutes after the earthquake. Sea water
ran up along the gentle slope and reached a height of 4.1 m, 180 m inland from the

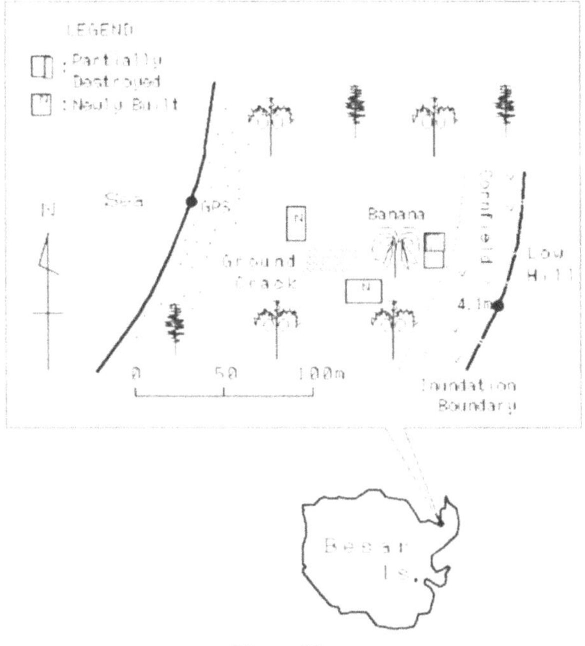

Figure 26
Inundation area in Kusung Pandang village on Besar Island.

shore line. The ground was cracked and fissured locally. Most houses on stilts were damaged severely or slightly due to both the earthquake and the tsunami waves.

(e) Pangabatang Island

There are two islands between Besar Island and the mainland of Flores Island; the western one is Damhilah Island and the eastern one is Pangabatang Island. The latter is a flat small island with a length of about one kilometer, part of which was "lost" as a result of the tsunami. The portion which disappeared had been a sandy beach.

4.5 Tsunami Damage on Babi Island

Babi Island is located approximately 40 km northeast of Maumere city. It is a round island with a diameter of 2.5 km, and is surrounded by a coral reef. A mountain of 351 m elevation was in the center of the island. There is no flat land on the coast except for the south coast facing the coast of Nebe (Nanga Merah) on the mainland of Flores Island. The width of the strait between Babi and Flores islands is about 5 km. Two villages are situated on the flat land of the southern coast: a Muslim village (Kampungbaru) on the western side and a Christian village

(Pagaraman) on the eastern part. The coral reef is considerably narrower on the south coast, where the two villages are situated, than it is around other parts of the island. The population of the island was 1,093 and the ratio of populations of Muslim to Christian villages is 5 to 1.

The number of dead and lost persons on Babi Island was announced by journalists as 600 to 700 within a few days after the catastrophe, but after the refugee camp was constructed at Nangahale on December 17, several hundred temporarily 'missing' persons were recovered in the camp, and the exact number of victims finally given on January 4, 1993 as 263 persons.

On Babi Island the number of women (175) killed was more than twice that of men (88). This tendency can be also found in other tsunami events. For example, TSUJI and HINO (1993) pointed out the same tendency in coastal villages on the Kumamoto Prefecture, Japan, in the case of the Ariake Bay Tsunami of 1792. As a result of our interviews in the field, we found that men have a tendency to take refuge for themselves, but women want to protect small children and the elderly during an emergency. This difference in motivation may affect the number of tsunami victims.

(a) Muslim village (Kampungbaru) on Babi Island

The residential area of the Muslim village, Kampungbaru, faces the sea and a row of palm trees were located along the shore fronting the village. These provided virtually no protection for the houses behind. All the houses were uprooted from their original locations and the materials of the houses, furniture and other goods were carried into the palm tree forest behind the village. The Mosque was completely washed away, leaving only the roof on the ground. The ground in most of the residential area was covered by coral sand, which was brought in by the tsunami waves.

We measured the inundation height by the traces on tree bark. Tsunami height was 3.6 m at the eastern point of the residential area of the Muslim village. The JAPAN DISASTER RELIEF TEAM (1993) also measured tsunami heights at 5 points in the residential area, and determined the heights between 3.28–3.68 m above MSL. YEH *et al.* (1993) reported tsunami inundation slightly higher (4.6 m) than our result. The ground height was only 1.2–2.0 m above mean sea level. Thus, the tsunami with a water thickness of only 2 m entirely washed away the houses. The speed of the sea water is estimated to have exceeded 5 meters/seconds or more.

We also measured the tsunami height at 7.2 m at two points on the foot of the steep mountain slope behind the palm tree forest, west of the residential area (Fig. 27). No houses stood in front of the measured points. We assume that the actual height of the tsunami surface at the time of inundation of the residential area was about 4 m. The height of 7.2 m can be explained as a doubling of the height due to a reflected wave. The inundation height of 3.6 m can be explained as the height of a progressive wave after passing the high friction area.

As a result of interviews of survivors at the refuge camp at Nangahale, we learned that the tsunami arrived 3 min after the earthquake, and that the shore

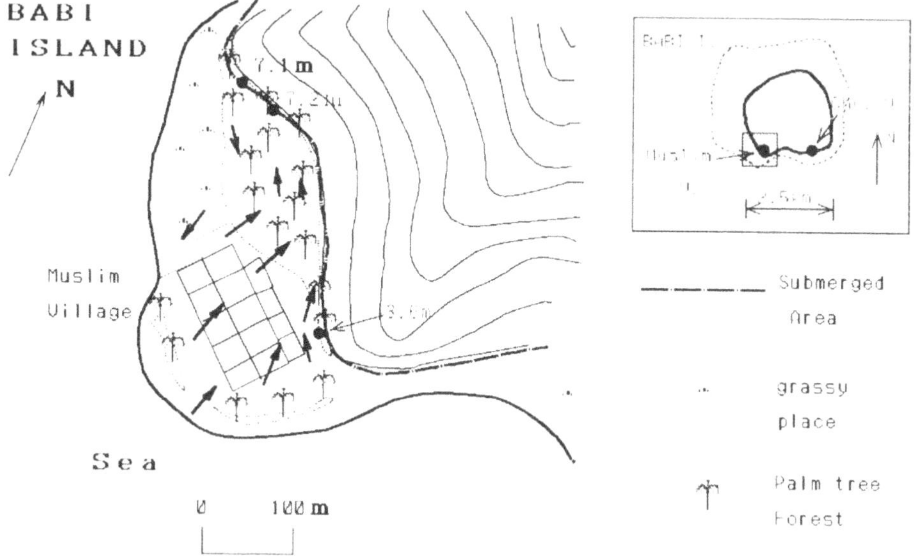

Figure 27

Inundation in the Muslim village (Kampungbaru) on Babi Island. Arrows show the direction of sea-water movement. Ground elevation of the residential area was about 2 m. Approximately 900 persons lived in the village. All the houses were swept away.

receded vertically by about 75 cm. Some witnesses indicated that the waves originated from the south, which would suggest that the waves which reflected from the mainland of Flores Island effected Babi Island more seriously than the direct waves from the source. On the slope of the mountain behind the village, rocks fell to the foot as a result of the earthquake, preventing inhabitants from evacuating from the tsunami.

(b) Christian village (Pagaraman) of Babi Island

The Christian village, Pagaraman is located on the southeast coast of Babi Island, directly facing Nebe village on the mainland of Flores Island. All houses in this village were swept away, and only foundations remained. Most of the residential area was surrounded by a palm tree forest, and the materials of the houses were carried into the forest (Fig. 28). Two lines of large cracks appeared on the ground of the grassy plain behind the forest (XX' and YY' in the figure). We easily detected the direction of tsunami flow by checking prostrated grass on the plain. Behind the grassy plain a sand dune 5 to 6 m high runs in an east-west direction, and sea water flowed over it. We could clearly trace the inundation height on the surface of palm trees and measured it as 5.6 m. Sea water surged across the dune and flowed westward along the valley behind. The ground height of the residential area was about 2 m, thus the thickness of sea water can be estimated as about 3 m.

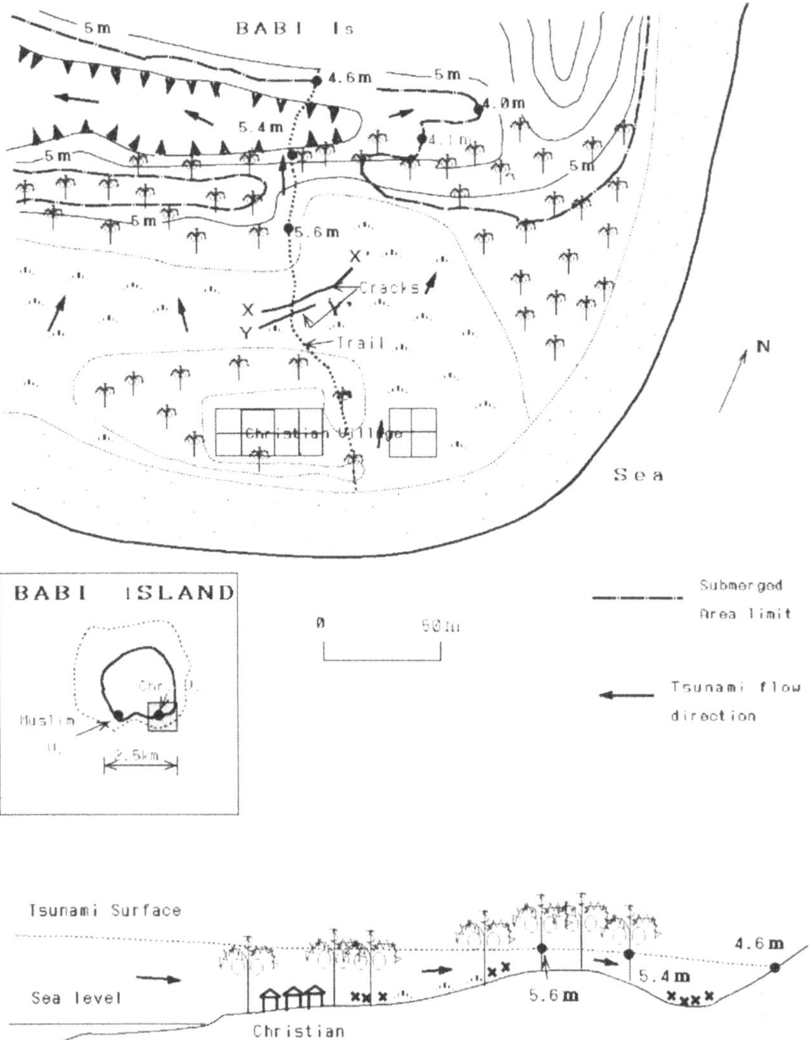

Figure 28

Inundation in the Christian village (Pagaraman) on Babi Island. Some 200 persons lived in the village. The ground height of the residential area is 1.5–2 m. All the houses were swept away. Crosses in the lower figure show sedimentation of coastal sand.

4.6 Tsunami Height and Damage from Nebe to Hading Bay

(a) Nebe (Nanga Merah) GPS 8°27′ 50.0″S, 122° 32′ 24.0″E

The coast of Nebe village is located on Flores Island facing Babi Island, separated by a channel 3 km wide. The residential area of Nebe is located within a palm tree forest and the ground elevation is 1.7–2.6 m (Fig. 29), and was inhabited

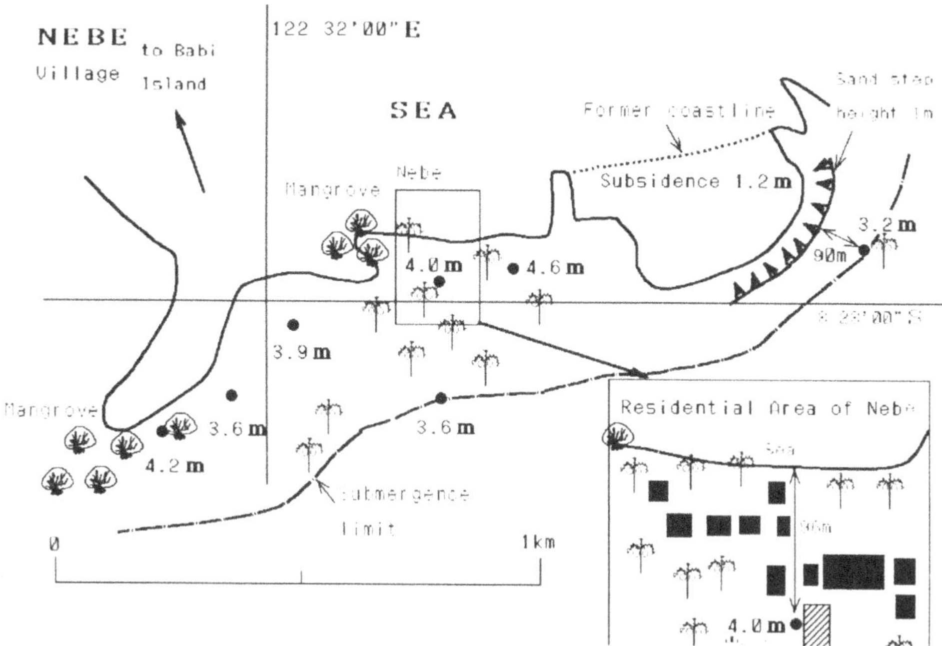

Figure 29
Inundation area of Nebe village and its vicinity.

by 35 families of which two elderly persons were killed by the tsunami. The first
wave came 5 min after the earthquake (2 min according to another witness), and
began with a regression of the shoreline by 200–300 m. Three waves came of which
the first was the largest. One person noted that the shoreline shifted landward by
15 m compared to that before the earthquake. The crust. also subsided there. Most
of the houses were constructed of brick, and totally collapsed from the impact of
the earthquake and the tsunami. The tsunami inundation height in the residential
area was measured by the trace on the bark of trees to be 4.0 and 4.6 m. Sea water
reached a point 320 m inland from the shoreline, where the height was 3.2 m.
Cracks and gushings of sand appeared on the ground in the tsunami submerged
area. Some traces of gushings were not disturbed by the flow of sea water, and they
seemed to be formed by aftershocks.

We should note that only two persons of about 150 inhabitants were killed,
despite the fact that most houses were totally destroyed. This suggests that the
palm tree forest was effective in dissipating the tsunami energy, mitigating the
human toll.

(b) Wailamung, GPS 8° 25′ 29.4″S, 122° 35′ 22.0″E

The entire residential area was surrounded by a palm tree forest inhabited by
about 600 persons; 6 lives were claimed by the tsunami. The first wave arrived 5 min

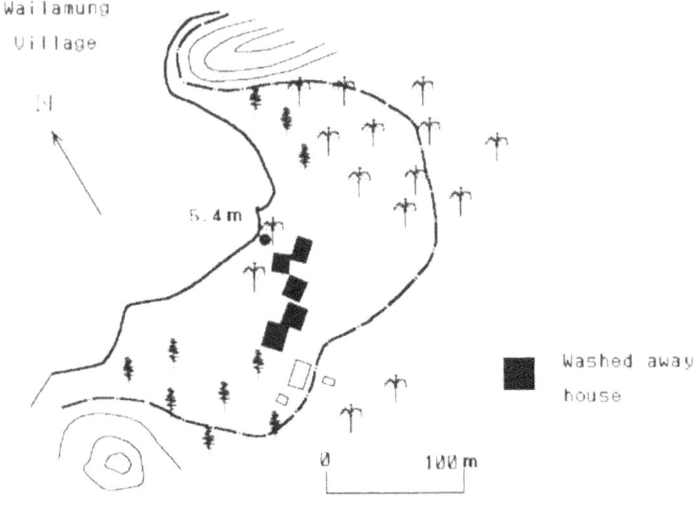

Figure 30
Inundation area in Wailamung village.

after the earthquake. The shore line subsided and shifted 10 m inland. The amount of vertical subsidence was measured as 1.0 m. The sandy shore was narrowed after the earthquake. The tsunami inundation height was 5.5 m, measured from the trace, where sea weeds were carried on a palm tree near the shore (Fig. 30).

(c) Pantai Lato, GPS 8° 21′ 20.0″S, 122° 46′ 05.5″E

Pantai Lato (= Lato Coast) lies on the southern coast of the mouth of Hading Bay. The submerged area is shown in Figure 31. The paved coastal road was obstructed by a crack at the southwest end of the village. A large-scale landslide occurred at the northeast coast of the village, burying the road there. Sea water did not reach the main road (further inland), and the submerged area covered about half of the sea-side residential area (Fig. 31), where tsunami height was measured at 3.5 to 3.8 m. Damage to houses was caused mainly by the earthquake. The numbers of completely and partially collapsed and inclined houses in the sea-side area were 12, 8, and 3, respectively. Only three houses were undamaged.

About 2–3 km southwest of the village the width of the submergence zone due to the tsunami reached 140 m from the shoreline. We measured a tsunami height of 6.9 m at this point GPS 8° 21′ 30.9″S, 122° 45′ 57.1″E (Fig. 32).

(d) Uepadung, GPS 8° 17′ 42.6″S, 122°49′ 56.8″E

Uepadung is located on the south coast of Hading Bay. The coastal road was lost due to the regression of the coast line. The top of a palm tree was seen on the surface of sea, which established that a large-scale landslide of the coastal region occurred here; the land moved into the sea and the palm tree remained standing. Sea water flowed along a river to the level of 11.0 m above mean sea level. One person testified that the wave arrived immediately after the earthquake.

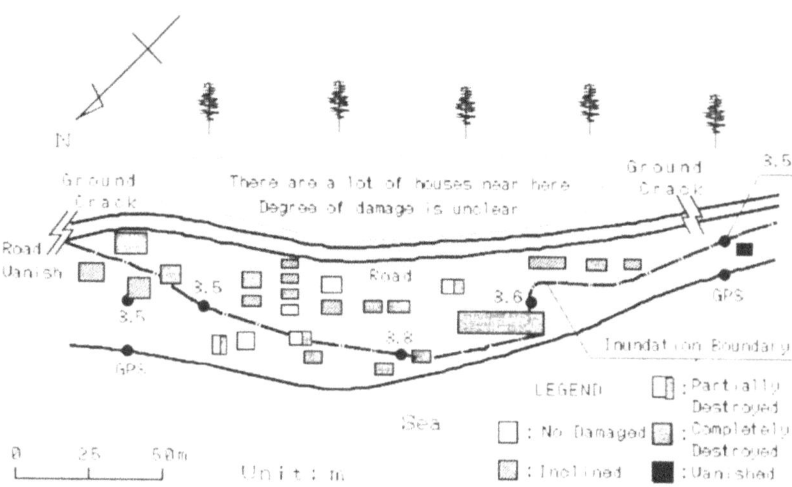

Figure 31

Tsunami submerged area and house damage due to the earthquake in the residential area of Pantai Lato.

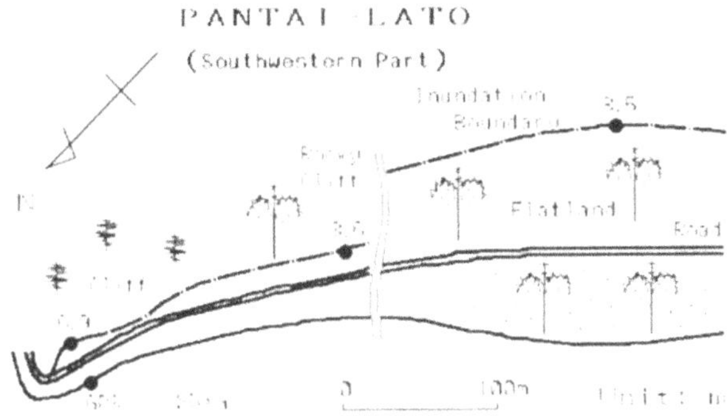

Figure 32

Southeastern part of Pantai Lato, about 2 km away from the main residential area shown in Figure 31.

(e) Leworahang, 8° 17′ 23.1″, 122° 52′ 39.2″E

Leworahang is located some 5 km east of Uepadung. The tsunami height was estimated by eye to be 10–14 m (Fig. 34). A palm tree moved into the sea area and stood there about 10 m apart from the shoreline with only its top appearing above the sea surface. The height of the tree was estimated at 12 m. It would appear that this was also the result of a landslide on the coast. YEH *et al.* (1993) reported that 12 houses were swept away and 24 people killed, and that near Lewobele, about 6 km west of Leworahang, at least two more subacueous slumps, about 1 km long each, were found.

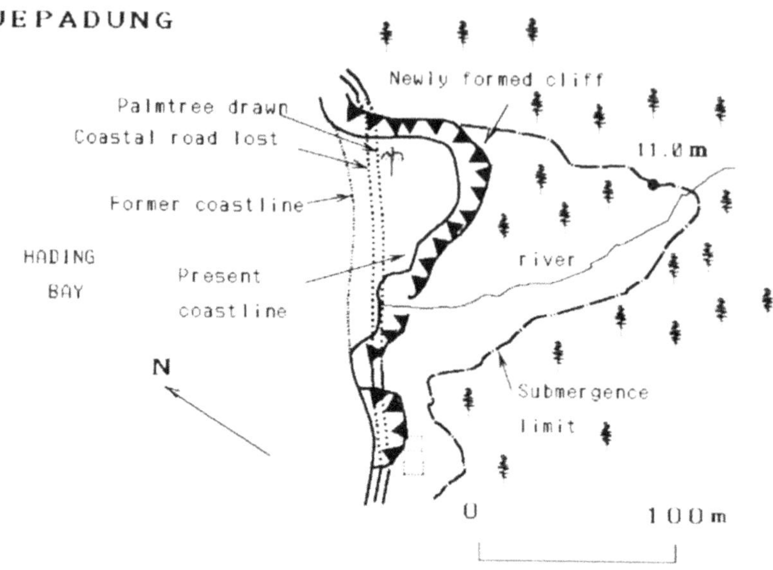

Figure 33
Shoreline regression and the inundated area of the coast of Uepadung village.

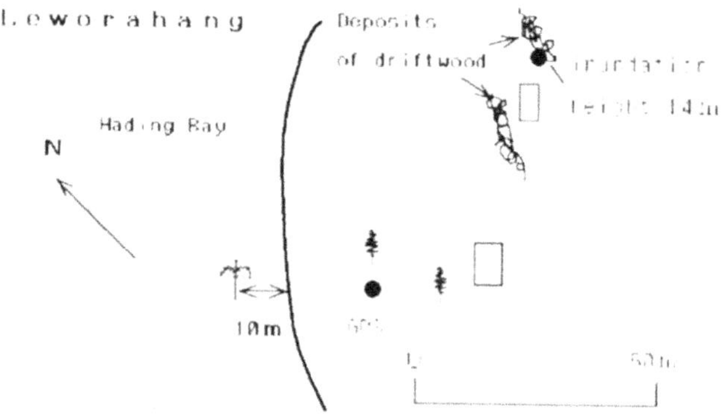

Figure 34
Tsunami inundation of Leworahang village.

(f) Waibalan, GPS 8° 16′ 51.9″, 122° 53′ 13.3″E

Waibalan lies on the coast close to the innermost point of Hading Bay. Tsunami waves arrived from the north three times. The first wave was the largest, arriving 2 min after the earthquake. The first wave looked like a white (or dark yellow) wall of water. The sea surface rose from the beginning without an initial draw. We measured two points on the submergence limits whose heights were 7.9 m (121 m from the shoreline) and 10.6 m (108 m from shore), respectively.

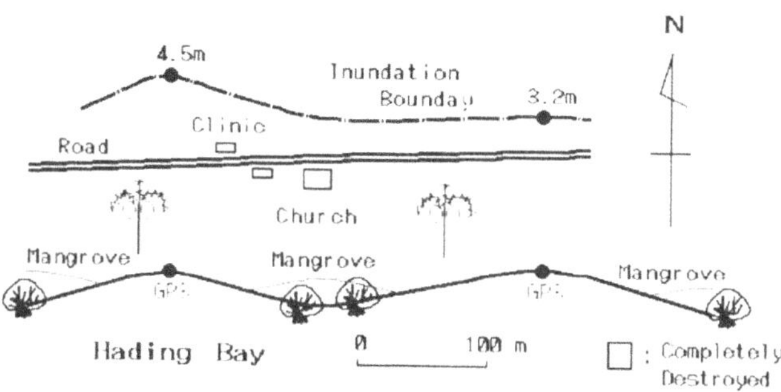

Figure 35
Inundation area of Pantai Lela village.

It is worthy of observation that the tsunami arrival time was only 2 min after the earthquake. If this is true, and taken together with the account of the witness at Uepadan that the tsunami arrived just after the earthquake, the origin of the initial wave should be within Hading Bay. This could have been caused by landslide(s) in the coastal area. That the water color of the first wave was white or dark yellow also indicates that some sea-water disturbance arose near the south coast of Hading Bay.

(g) Pantai Lela, GPS 8° 11′ 08.2″S, 122° 50′ 14.7″E

Pantai Lela (Lela coast) is located on a gentle slope on the north coast of Hading Bay. Tsunami waves arrived three times and the first was the largest, as a result of which the church building caved in (Fig. 35). In addition, two brick buildings were destroyed. The inundation height and length were 4.5 m and 140 m, respectively. The shoreline was shifted inland by 10 m which suggests subsidence of the crust.

4.7 Tsunami Heights on the Coast of the Cape of Watupajung

(a) Riang-Kroko (in Turubeang village), GPS 8° 09′ 01.4″S, 122° 47′ 0.03″E

An extraordinarily large tsunami hit the north and west coasts of the Cape of Watupajung. At Riang-Kroko village, sea water surged along the slope to a level of 26.2 m. Figure 36 displays the inundation area with tsunami heights at four points. YEH (1993) reported the average height of these values as 19.8 m. All vegetation was washed away in this area. In the residential area, with a ground elevation of 3.6 m, 69 families totalling 406 persons had lived in more than 200 houses, but no evidence of human life remained. Even foundations of houses were washed away. Here 137 persons lost their lives. Even large trees were uprooted. We could easily distinguish the inundation boundary due to the sharp color contrast of the ground; light brown soil for the submerged area, and green chlorophyll for the unsubmerged

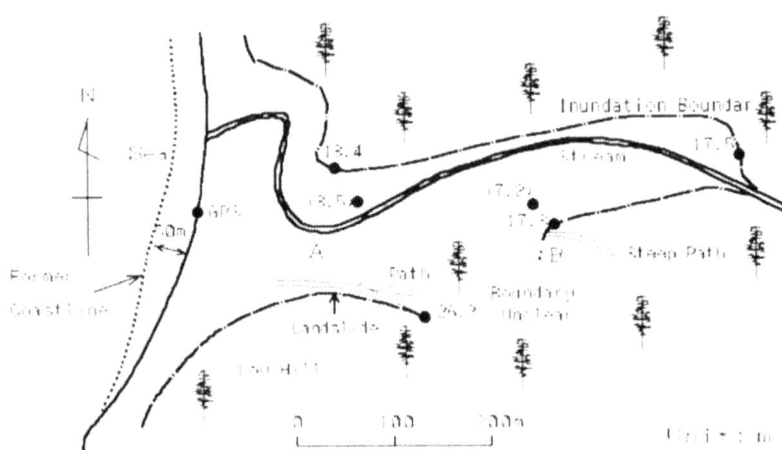

Figure 36

Inundated area in Riang-Kroko village, where 139 persons were killed due to the huge tsunami. Houses were entirely swept away and even foundations disappeared. No evidence of human life remained. No vegetation survived below the limit of the submerged area.

area. The tsunami waves struck five times and the first one was the largest. The coastline was shifted by 30 m inland. It is difficult to distinguish whether the shifting of the coastline was due to erosion by the tsunami or resulted from a landslide on the coast.

(b) Bunga-Koten (Bou Tanabeten), GPS 8° 06′ 40.6″S, 122° 47′ 44.3″E

We watched the coast between Riang-Kroko to Bunga-Koten on January 3, 1993 from two helicopters. The trace of inundation of the tsunami was clearly distinguished from the air because of the sharp contrast in ground color.

Bunga-Koten village is located 6 km northeast of Riang-Kroko. Here all 100 houses, one church, and the elementary school building were entirely swept away. The residential area here is called the colony (RT) of Bou Tanabeten (Fig. 37). On the ground, traces of the foundations of several houses could scarcely be detected. The tsunami height on the east slope was measured at 12.3 m. In contrast to such severe building damage, only three persons lost their lives here (KOMPAS, December 18).

4.8 Tsunami Height on the South Coast of Flores Island

Since the epicenter was located in the north coastal region of Flores Island, on the south coast of the island only minor evidence of the tsunami was observed by the native people.

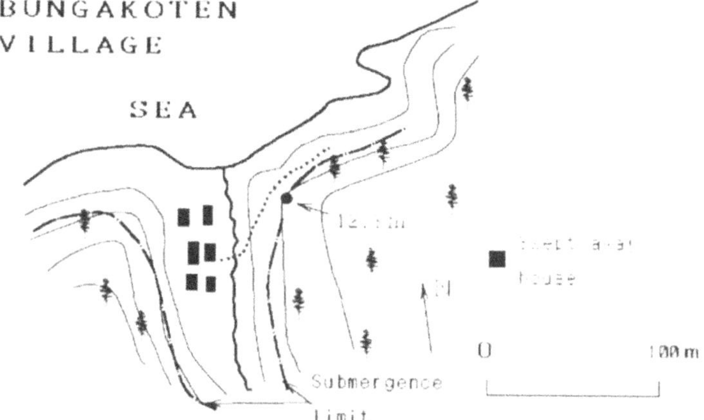

Figure 37

Inundated area of Bunga-Koten village (Bon Tanabeten). All one hundred houses were entirely swept away.

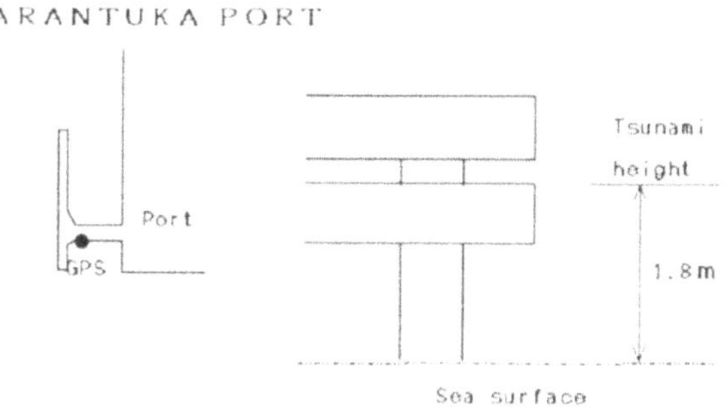

Figure 38

Sea water rose to 1.8 m above mean sea level at Larantuka Port.

(a) Larantuka Port

The earthquake was felt with an intensity of 5 on the MM scale at Larantuka Port, on the easternmost coast facing Adonara Island. Sea water rose to the upper edge of a concrete beam on the pier in the port, and it was measured 1.8 m high (Fig. 38). The tsunami was reported to have arrived only two minutes after the earthquake, and eight waves appeared within 5 min. The first wave was the largest, and sea surface gradually rose. It is difficult to explain such an early arrival and the short period of the waves. The possibility exists that the tsunami in this area was generated by a local landslide and was independent of the tsunami observed on the north coast.

No tsunami information was given at Waibalun Port (8° 20′ 42.5″S, 122° 56′ 58.9″E).

Konga

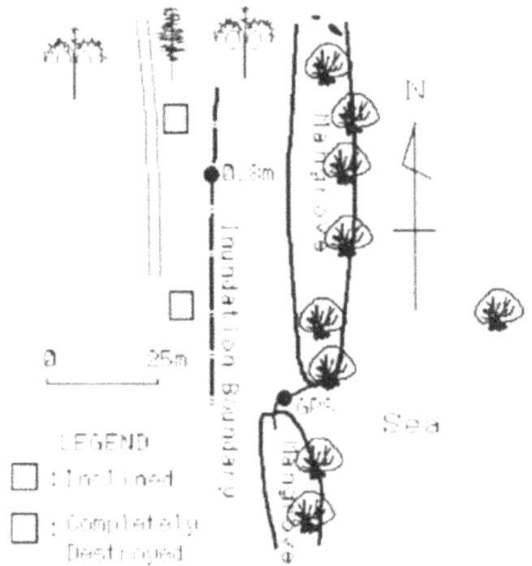

Figure 39
Inundation area at Konga on the southern coast of Flores Island.

(b) Konga, GPS 8° 26′ 36.5″S, 122° 47′ 04.8″E

Sea water inundated a width of 20–30 m from the shoreline, and many timbers drifted ashore. Evidence of a submergence limit could be detected by salt deposits on the sandy shore, which was corroborated by native people. The height at that point was 80 cm above the sea (Fig. 39).

(c) Ende

Ende City is located on the south coast of the central part of Flores Island. A sea-surface rise of about 50 cm was observed by the citizens.

4.9 Tsunami Evidence at Distant Locations

Kalaotoa Island is located about 150 km northwest of Maumere City and belongs to Selayar Regency in South Sulawesi Province. In the entire Selayar Regency, 19 persons perished and 130 houses were swept away resulting from the tsunami. Seventy of these houses were settled on Kalaotoa Island.

In Permana village on Buton Island, in Southwest Sulawesi Province 600 high floor houses entirely gave in or washed away and 3 persons were killed by the tsunami.

It is not clear whether the damage at Latuna on Pantar Island (see Fig. 9) was caused by the tsunami. Some tsunami damage was expected on the north coast of the Adonara and Lomblen Islands, but we could not observe this unambiguously.

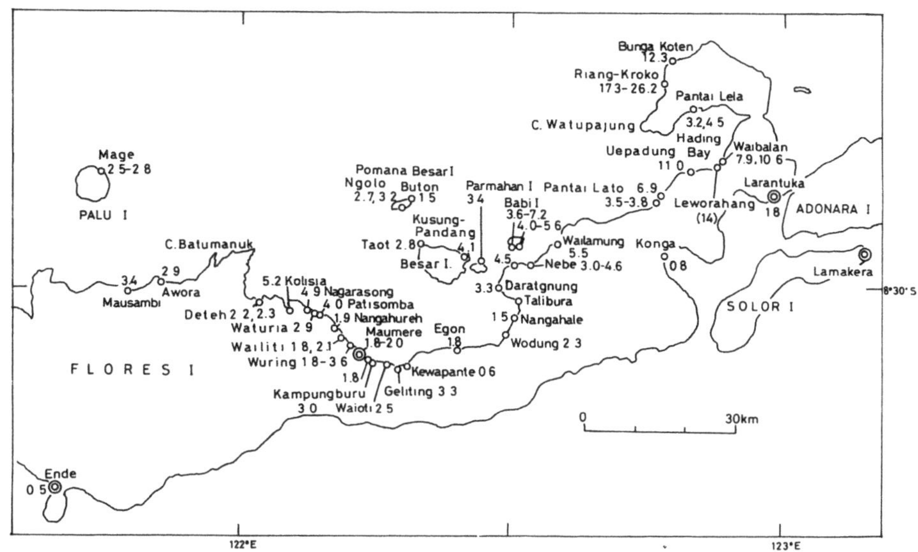

Figure 40

Distribution of tsunami heights on the coast of Flores Island. Numbers show runup height above mean sea level (m).

4.10 Distributions of Tsunami Heights

By using the result of our field survey we obtain a map of the tsunami height as shown in Figure 40. A clear contrast in the tsunami heights between the coasts west of Wailamung and east of Pantai Lato can be noticed. Tsunami heights on the western coast did not exceed 7 m, while those of the eastern coast generally exceeded 10 m. It is difficult to explain this variation in the tsunami height distribution simply by one fault model. It is probable that landslide(s) on the coastal area, and/or on the sea bed were the cause(s) of the huge tsunami heights observed on the eastern coast.

5. Human and House Damage

5.1 Statistics of Human and House Damage

In addition to our survey, we obtained statistics of human and building damage from the Military Commander of Maumere, and from the Sikka Regency Office at Maumere (see Appendix 2). These statistics are tabulated for each of the Regencies. In addition, we gathered articles describing the damage from the newspaper "Kompas", dated from December 13 to 31, 1992, in which reports of individual villages occasionally appeared. In addition to those accounts, we could use the

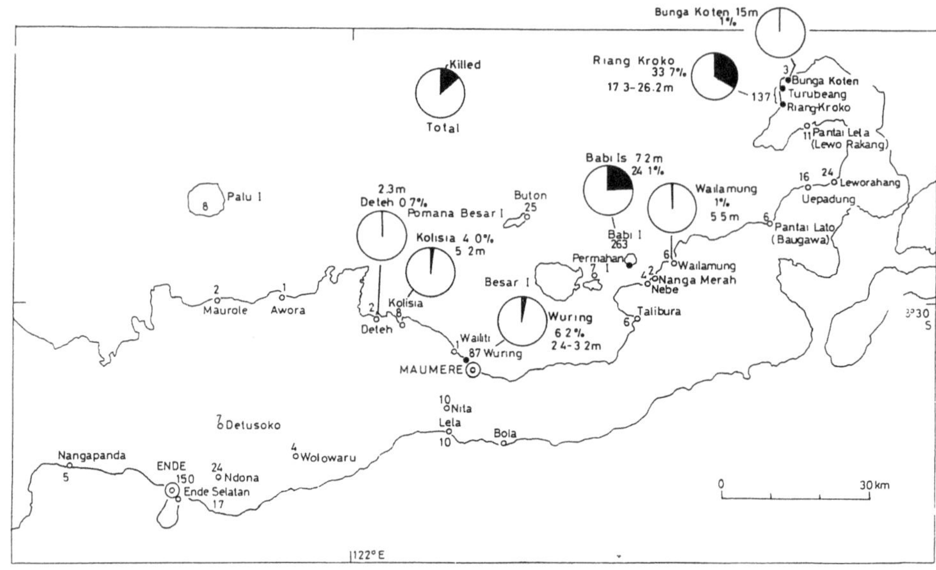

Figure 41
Numbers of deaths from the Flores Island earthquake and tsunami by villages. Mortality in seven villages struck by the tsunami is shown by circle graphs.

eyewitness accounts from our survey to help document the amount of damage and total number of houses and population in each village. Thus, we obtained the map for the distribution of the number of lives taken in the eastern part of Flores Island as shown in Figure 41. In Figure 41, numbers located inland and on the southern coast of Flores Island refer to victims of the earthquake, while those on the north coast are probably mostly of victims of the tsunami.

5.2. *House Damage*

The information regarding damage covers four regencies (Ngada, Ende, Sikka and Flores Timur) on Flores Island and two archipelagoes belonging to South Sulawesi and Southeast Sulawesi provinces. Figure 42 shows the number of swept away and completely collapsed houses, including both the damage due to the earthquake and that due to the tsunami. The total for the entire Flores Island is 16,967 houses. Most house damage in Ende and Ngada Regencies seem to have been caused by the earthquake, while that in Flores Timur was caused by the tsunami. Aggregating the amount of house damage, caused by the tsunami on both Sulawesi Provinces, we determined the total number of collapsed houses throughout Indonesia to be 17,697. But we should note the following fact: As mentioned above, the number of washed-away houses at Riang-Kroko and Bunga-Koten villages was about 200 and 100, respectively. Additionally house damage also was sustained at

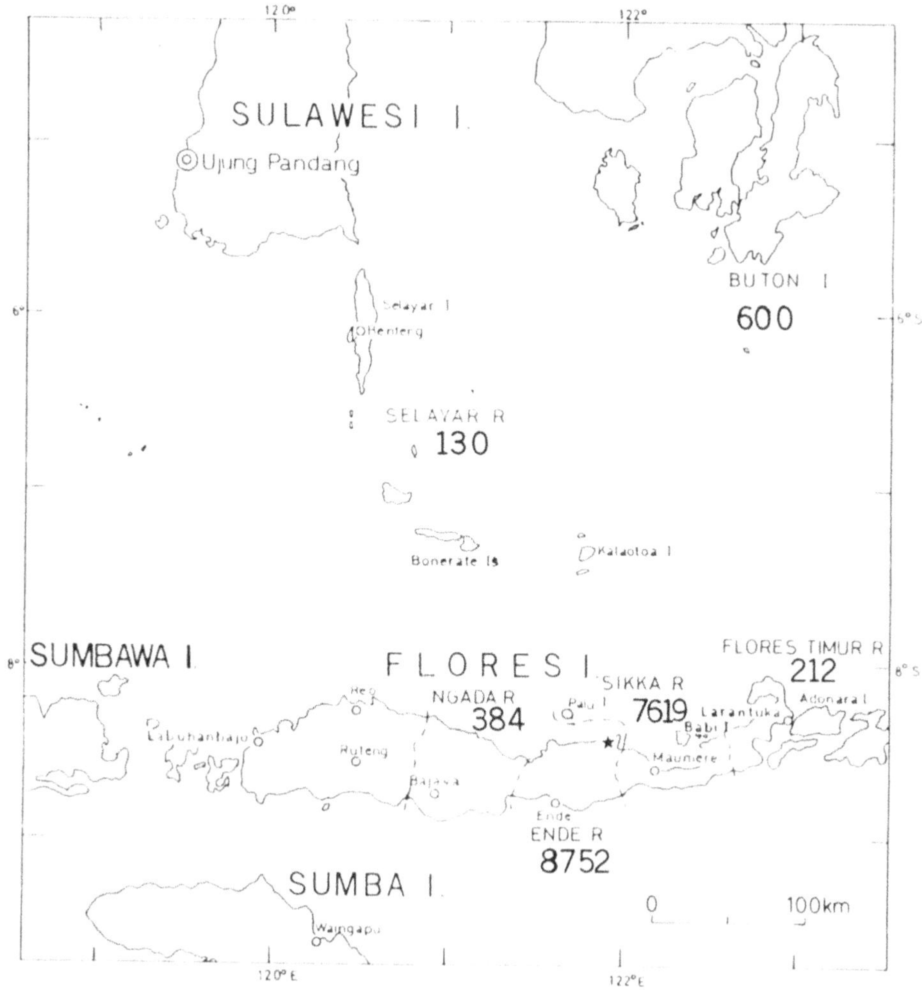

Figure 42

Number of houses totally destroyed (including washed away) by the earthquake and tsunami by regencies.

Pantai Lato, Pantai Lela, and other villages, therefore we must conclude that the damage statistics for the Flores Timur Regency (212) is substantially underestimated.

5.3 Number of Killed Persons

The number of deaths is given in Figure 43, which includes victims of both the earthquake and the tsunami. The total number of deaths on Flores Island is 1,690. Adding 22 deaths in South Sulawesi and Southeast Sulawesi Provinces, 1,702 people were killed in the entire Indonesian territory.

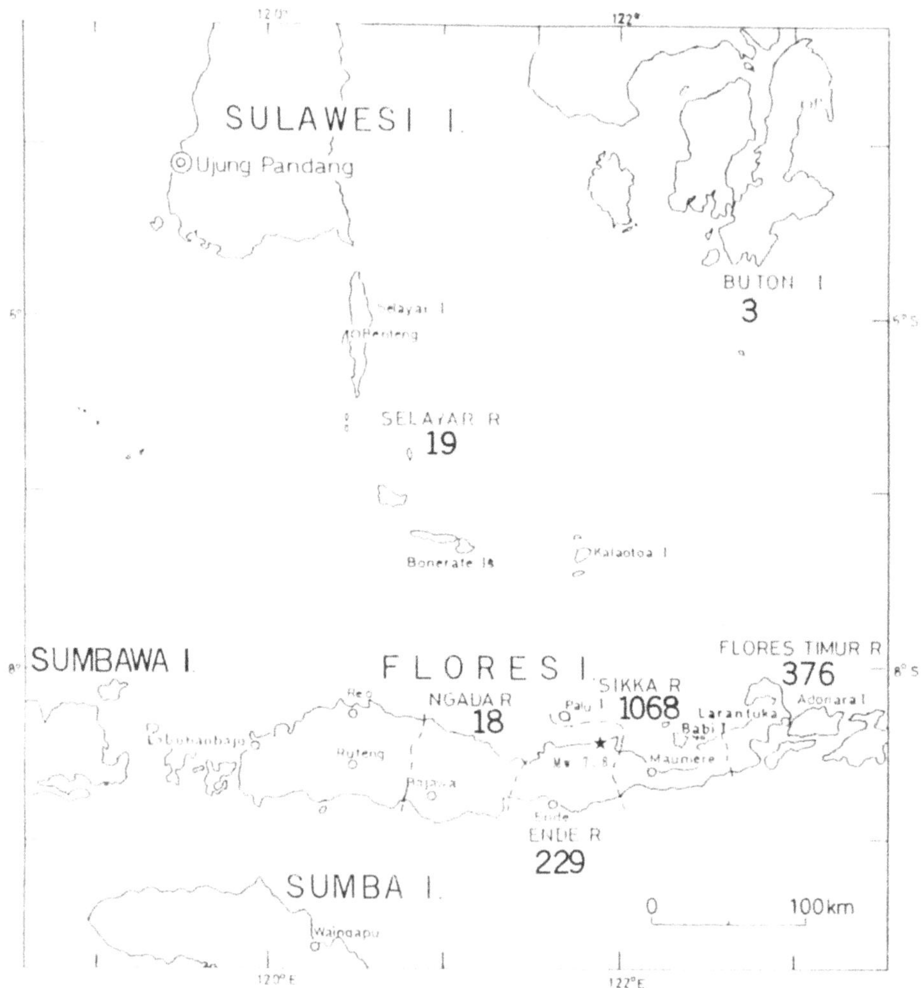

Figure 43
Number of deaths by regencies.

Comparing Figure 42, we can calculate the ratio of lives lost to the number of completely damaged houses for each regency. The ratios are 0.047 and 0.022 for Ngade and Ende regencies, respectively, where the damage was predominately caused by the earthquake. On the other hand, the ratios for Sikka and Flores Timur regencies are 0.145 and 1.774, respectively, where the damage was caused mainly by the tsunami. From this we can conclude that people were more swiftly killed by the tsunamis than by the earthquake.

5.4 Relationship between Tsunami Height and the Ratio of Mortality for Villages

We obtained information pertaining to the casualties suffered in coastal villages from articles in the newspaper, KOMPAS, and from other sources. We calculated mortalities, the ratio of victims to population, for seven villages on the damaged coast (circle graphs in Fig. 41). Approximately one fourth of the population was killed on Babi Island, and one third perished at Riang-Kroko village. In cases in which tsunami height exceeded 5 m some villages suffered mortality of several tens of percent. This fact also was recognized in the case of the MEIJI (1896) and SHOWA (1933) Sanriku tsunamis in Japan. If these people had had advance knowledge of the tsunamis, and had taken appropriate action upon feeling the earthquake, the mortality rate could have been substantially reduced.

6. Discussions and Conclusion

IMAMURA *et al.* (1995) demonstrated that it is difficult for fault models to fit the distribution of tsunami height even if two faults are assumed to occur separately in the eastern and western parts of the epicentral area, and if the amount of dislocation of the eastern plane is several times larger than that of the western. Historical data show that landslides sometimes generate tsunamis and/or increase the magnitude of the tsunami.

In Japan, examples of tsunamis induced by landslides are minimal. We know of only one example: the 1792 Ariake Bay Tsunami was induced by a large-scale landslide on the east slope of Mayuyama Hill, Shimabara Peninsula, Kyushu Island. For the Flores Island event, it appeared that landslides took place in the coastal region of Uepadung in Hading Bay. Extraordinarily high tsunami runup heights observed at Riang-Kroko village and its vicinity suggest that for the present event, landslides in the coastal and sea areas are partially the cause.

We examined the historical data for tsunami damage in Japan, for example, relative to the 1854 Ansei Tokai and Nankai Tsunamis, the 1896 and 1933 Great Sanriku Tsunamis, and others. We found that it is rare in the tsunami history of Japan that most houses were swept away by tsunami waves with heights of less than 5 m. Perhaps Indonesian carpenters can learn from the Japanese regarding the reinforcement of wooden houses against the horizontal forces, such that coastal houses of Indonesia would be safer from tsunamis. Specifically, wooden houses of Indonesia should use more diagonal bars.

At some coastal areas rocks fell along the slopes and sea water reached close to the foot of the slope of the hill behind the villages. In those cases, falling rocks may have prevented the inhabitants from evacuating safely. At Inaho village on Okushiri Island, Japan, falling rock also prevented the inhabitants from moving to higher ground at the time of the 1993 Hokkaido-Nansei-Oki Tsunami. In planning emergency shelters in the higher fields behind coastal villages, the possibility of rockfalls should be taken into account.

It is valid to protect coastal villages from the damage of tsunamis by planting trees in front of the residential areas. A zone of arranged trees, a greenbelt, cannot prevent sea water from flowing into the villages, but we can expect it to effectively dissipate the energy of the incident waves of the tsunamis, and to reduce the number of victims. For the Flores Island event, a row of palm trees had been planted in front of several villages, however it seemed that the row of trees was not very effective in dissipating tsunami energy. In most of these cases, the greenbelt was formed only by a line of palm trees. A palm tree generally has few low branches. If we want to improve the efficiency of dissipating the tsunami energy, then we should select varieties of trees that have many low branches with a high density of leaves.

We also noticed, in the case of Wuring and the villages of Babi Island, that a wide coral reef is an effective dissipation of tsunami energy. TSUJI and HINO (1993) noticed in the case of the 1792 Ariake tsunami, that tsunami height diminishes at a coast with a wide reef. This happens because the maximum height of a solitary wave is limited up to seventy percent of the depth, consequently if the wave height exceeds that limit, the wave energy will be lost by breaking at wave crests. It is worthwhile to consider the effect of a wide reef on the dissipation of tsunami energy when planning the locations of new coastal villages in tropical countries.

Acknowledgment

We wish to express our thanks to Dr. Karjoto Sontokusumo, the Director General of the Meteorological and Geophysical Agency (BMG) of Indonesia, Dr. R. Soetarjo, the secretary of BMG, and Mr. Hendrik Fernandez, the Governor of East Nusa Tenggara Province for their support of our survey. We also thank Drs. M. Hakuno, Jun'ichi Shibuya, Harry Yeh, Phillip Liu, Costas E. Synolakis, Prof. B. H. Choi, and Dr. S. Z. Shi for their support of our survey on Flores Island. We also thank Mr. Bambang, a student of the postgraduate course at the University of Tokyo, who translated the text of the questionnaires into the Indonesian language. Mr. E. Ruslan, the chief of the Maumere Meteorological Observatory, and Mr. E. Sariman, the chief of the Kupang Meteorological Observatory assisted us in the observation of aftershocks and in the field survey.

The Japan Disaster Relief Team of the Japan International Cooperation Team (leader: Ka. Abe) entered the damaged region five days earlier, and gave us ample useful information from which we could make a reasonable plan.

The present study was supported financially by a Scientific Grant-in-aid (B-4-4, No. 0436024) from the Ministry of Education, Science and Culture (Monbusho) of Japan.

Finally, we also thank Mr. Ibnu Purwana of BMG, and two reviewers of this paper for their services in polishing the text.

REFERENCES

HAKUNO, M. (1993), *Damage of Public Constructions and Life Lines due to the 1992 Flores Earthquake*, Kaiyo Monthly *25* (2), 782–785 (in Japanese).

IMAMURA, F., GICA, E., TAKAHASHI, T., and SHUTO, N. (1995), *Numerical Simulation of the 1992 Flores Tsunami: Interpretation of Tsunami Phenomena in Northeastern Flores Island and Damage at Babi Island* (in this book), Pure and Appl. Geophys. (this issue).

JAPAN DISASTER RELIEF TEAM (Leader, Ka. Abe) (1993), *Report of Japan Disaster Relief Team (Expert Team) on the Earthquake in Republic of Indonesia of December 12, 1992*, Japan International Cooperation Agency (JICA), 106 pp.

KATO, K., SUNARJO, and TSUJI, Y. (1993), *The Sumba Earthquake Tsunami of August 19, 1997*, Kaiyo Monthly *25* (12), 744–755 (in Japanese).

KOMPAS (Newspaper), issues dated December 13 to 30, 1992 (in Indonesian).

MIYOSHI, H. (1993), *A Blind Spot on Tsunami Study in the Indonesian Territory –"Tanah Longsor (Landslide)" Phenomena*, Kaiyo Monthly *25* (12), 797–803 (in Japanese).

MATSUTOMI, H. (1993), *Characteristics of the 1992 Flores Earthquake-Tsunami and its Damages*, Kaiyo Monthly *25* (12), 756–766 (in Japanese).

SHIBUYA, J. (1993), *Damages of Buildings Due to the 1992 Flores Earthquake*, Kaiyo Monthly *25* (12), 786–791 (in Japanese).

SOLOVIEV, S. L., and GO, C. N. (1974), *Catalog of Tsunamis in the Western Coast of the Pacific Ocean*, Nauka Press, Moscow, 308 pp. (in Russian).

SOLOVIEV, S. L., and GO, C. N. (1987), *Catalog of Tsunamis along the Coast of the Pacific Ocean, 1972 to 1986*, Nauka Press, Moscow (in Russian).

SUNARJO and TAJAN (1993), *The mitigation problem of tsunami hazard in Indonesia*, Proc. Indonesian Assoc. Geophys. (HAGI), 18th Annual Conv., Jakarta, 363–373 (in Indonesian).

TAKEO, M., and IDE, S. (1993), *Mechanism of the 1992 Flores Earthquake*, Kaiyo Monthly *25* (12), 767–770 (in Japanese).

TSUJI, Y., and MATSUTOMI, H. (1993), *Damages due to the Tsunami*, Report of the Field Survey on the Flores Island Earthquake-Tsunami of December 12, 1992, Rep. Grant in Aid, No. B-4-4, 70–87 (in Japanese).

TSUJI, Y., and HINO, T. (1993), *Damage and Inundation Height of the 1792 Shimabara Landslide Tsunami along the Coast of Kumamoto Prefecture*, Bull. Earthq. Res. Inst., Univ. Tokyo *68*, 91–176 (in Japanese).

UTSU, T. (1991), *Table of World's Earthquakes with Damages*, Earthq. Res. Inst., Univ. Tokyo, 243 pp. (in Japanese).

YEH, H., IMAMURA, F., SYNOLAKIS, C., TSUJI, Y., LIU, P., and SHI, S. Z. (1993), *The Flores Island Tsunamis*, EOS, Trans. Am. Geophys. Union *74* (33), 369, 371–373.

(Received October 30, 1994, revised April 17, 1995, accepted May 4, 1995)

Appendix 1

List of the measured tsunami heights of uncorrected and those of the tide (at measured time from mean sea level, MSL) corrected values. The standard port of the astronomical tide is Maumere. Six components of M2, S2, K1, O1, P1, and K2 are used for the tide calculation. The accuracy of heights of most of the points is estimated to be 0.1 m, except three points with marks "B" and "C" in the column of the reliability (R.B.), where B means slightly less reliable, and C means less reliable.

Location	GPS Measurement Lat. 8° S +	Long. 122° E +	date	Measured time	height no correction	Astrn. Tide	Tsunami height above MSL	R.B.
	′ ″	′ ″	y m d	h m	cm	cm	cm	
1. from Maumere to west								
Maumere port 1	37 03.0	13 09.0	92 12 29	15:16	156	+48	204	
Maumere port-2			12 29	16:00	119	+65	184	
Wuring-1 (tip)	36 56.0	12 14.0	12 30	08:45	–			
Wuring-2			12 30	08:45	322	−56	266	
Wuring-3	36 00.0	12 12.0	12 30	08:45	354	−56	298	
Wuring-4	36 04.0	12 10.0	12 30	08:45	373	−56	317	
Wuring-5	36 04.0	12 10.0	12 30	10:25	269	−89	180	
Wuring-6			12 30	08:45	364	−56	308	
Wuring-7			12 30	08:45	414	−56	358	
Wuring-8	36 14.0	12 03.0	12 20	08:45	384	−56	328	
Wuring-9			12 30	08:45	390	−56	334	
Wuring-10			12 29	14:40	210	+30	240	
Wuring-11					280	+30	310	
Wuring-Mosque			12 29	15:06	276	+44	320	
(3 traces)					259	+44	303	
					248	+44	292	
(Average of Wuring)							(296)	
Wailiti-1	35 16.2	11 02.8	12 31	10:35	279	−71	208	
Wailiti-2	35 21.8	10 56.6	12 31	10:35	249	−71	178	
Nangahureh	34 18.3	10 14.3	12 30	15:38	140	+50	190	
Patisomba	33 01.8	8 38.5	12 30	16:10	335	+64	399	
Waturia	32 48.8	7 57.2	12 31	11:00	362	−74	288	
Nagarasong	32 24.8	7 14.9	12 31	14:54	473	+15	488	
Kolisia	32 33.5	5 17.5	12 31	15:50	470	+47	517	
Magepandang-Deteh 1	31 49.1	2 10.6	12 31	17:20	152	+78	230	
Magepandang-Deteh 2	31 50.7	2 13.2	12 31	17:20	146	+78	224	
Awora	29 36.2	121 51 08.5	93 1 2	14:10	311	−23	288	
Mausambi	30 34.1	121 47 10.1	1 2	12:50	379	−42	337	
2. Islands, north offshore								
Palu I. Mage 1			1 5	9:45	278	+6	284	
Palu I. Mage 2			1 5	9:57	244	+8	252	

Appendix 1 (Contd)

Location	GPS Measurement Lat. 8° S +	Long. 122° E +	date	Measured time	height no correction	Astrn. Tide	Tsunami height above MSL	R.B.
Pomana B.I. Ngolo-1			92 12 30	11:20	350	−82	268	B
Pomana B.I. Ngolo-2			12 30	11:20	400	−82	318	
Pomana B.I. Buton			12 30	13:05	194	−42	152	
Besar I. Taot	25 30.7	20 09.3	12 30	17:00	203	+76	279	
Besar I., Kusung Pandang	26 48.4	24 32.1	12 31	11:50	486	−72	414	
Parmahan I.	27 20.0	26 41.0	93 1 2	15:30	320	+16	336	
Babi I., Moslem V-1 (Kampungbaru)	26 6.0	30 21.0	1 2	13:15	395	−38	357	
Babi I., Moslem V-2	(V. western end)		1 4	12:24	723	−9	714	
Babi I., Moslem V-3	(40 m west of V.2)		1 4	12:35	725	−10	715	
Babi I., Chrst. V-D (Pagaraman)	25 45.7	30 50.0	1 4	12:20	570	−7	563	
Babi I., Chrst. V-E	25 44.8	30 48.4	1 4	12:20	549	−7	542	
Babi I., Chrst. V-F	25 42.2	30 50.3	1 4	12:20	464	−7	457	
Babi I., Chrst. V-G			1 4	12:20	415	−7	408	
Babi I., Chrst. V-H			1 4	12:20	407	−7	400	
(Average of Babi I. Chrst. V.)							(474)	
3. East of Maumere								
Permatasari Beach Hotel	37 39.3	14.10.9	92 12 29	12:20	238	−56	182	
Kampungburu	37 49.0	14 34.0	93 1 1	10:00	335	−37	298	
Waioti	38 14.0	16 20.0	1 1	10:40	295	−49	246	
Geliting	38 32.0	17 24.0	1 1	10:55	380	−53	327	
Kewapante	38 08.0	18 38.0	1 1	11:20	120	−56	64	
Egon	36 41.0	24 19.0	1 1	12:15	235	−57	178	
Wodung	35 00.0	29 31.0	1 1	14.05	255	−23	232	
Nangahale	33 19.0	30 38.0	1 1	14:35	160	−7	153	
Talibura Market	31 38.0	31 12.0	1 1	15:25	215	+22	237	
Daratgnung	30 00.3	28 49.8	(not recorded)		281	+49	330	
Nebe-West	27 44.8	30 33.2	(not recorded)		466	−18	448	
Nebe-1	27 57.0	32 04.0	1 3	13:25	417	−26	391	
Nebe-2	28 10.0	31 58.0	1 3	13:51	385	−23	362	
Nebe-3			1 3	14:15	433	−17	416	
Nebe-4 (V. center)	27 50.0	32 24.0	1 3	15:17	400	+3	403	
Nebe-5			1 3	15:17	357	+3	360	
Nebe-6			1 5	12:02	292	+10	302	
Nebe-7	27 50.1	32 17.9	1 5	12:25	453	+8	461	
Nebe-8	27 49.7	32 28.5	1 3	11:30	350	−18	332	
(Average of Nebe-1 to 8)							(378)	
Wailamung	25 29.4	35 22.0	1 5	12:20	538	+12	550	
Pantai Lato-1	21 11.3	46 10.0	1 1	14:54	345	+4	349	
Pantai Lato-2			1 1	15:02	340	+8	348	
Pantai Lato-3			1 1	15:15	360	+16	376	
Pantai Lato-4			1 1	16:10	310	+47	357	
Pantai Lato-5	21 20.0	46 05.5	1 1	15:32	325	+26	351	
Pantai Lato-6			1 1	16:47	297	+64	361	

Appendix 1 (Contd)

Location	GPS Measurement Lat. 8° S +	Long. 122° E +	date		Measured time	height no correction	Astrn. Tide	Tsunami height above MSL	R.B.
Pantai Lato-7			93	1	1	17:10	277	+73	350	
(Average of P. Lato)									(356)	
Pantai Lato-North	21 30.9	45 57.1		1	1	16:35	630	+58	688	
Uepadung	17 42.6	49 56.8		1	3	11:50	1120	−22	1098	
Leworahang	17 23.1	52 39.2		1	2	12:50	14m*	–	14m*	C
Waibalan 1	16 51.9	53 13.3		1	2	13:30	622	−34	788	
Waibalan 2				1	2	13:40	1090	−32	1058	
Pantai Lela-1	11 21.9	50 08.6		1	3	10:47	347	−29	318	
Pantai Lela-2	11 08.2	50 14.7		1	2	11:06	483	−34	449	
Riang-Kroko 1				1	2	13:50	1865	−29	1836	
Riang-Kroko 2				1	2	13:50	1754	−29	1725	
Riang-Kroko 3	9 01.4	47 00.3		1	2	13:50	1778	−29	1749	
Riang-Kroko 4	9 01.9	47 00.8		1	3	12:45	2645	−27	2618	
(Average of R. Kroko)									(1982)	
Bunga-Koten	06 40.6	47 44.3		1	3	13:30	1250	−25	1225	
(= Bou Tanabeten)										
4. South Coast of Flores										
Larantuka, wharf	20 33.1	59 18.4		1	2	9:40	200	−25**	175	
Konga	26 36.5	47 04.8		1	1	13:03	90	−12**	78	C
Ende				1	5		50*	–	50*	

* Net height by eye measurement.
** The standard port of the astronomical tide for Konga and Larantuka is Lamakera on Solor Island.

Appendix 2

Statistics of human and building damage by regency. Reported by the Military Commander of Maumere on December 30, 1992. Number with parentheses are reported by Sikka and Ende Regency Offices on January 4, 1993, and December 22, 1992, respectively.

Regency	Population thousand	Deaths	Injured heavily	slightly	Houses totally collapsed
Sikka	247	1,068	252	1,256	7,619
		(1,103)	(392)	(1,597)	(8,057)
Ende	219	229	228	336	8,752
		(193)	(228)	(333)	(8,752)
Ngada	198	18	9	11	384
East Flores	265	376	21	31	212
Total	929	1,691*	510	1,634	16,967**
		(1,690)	(650)	(1,972)	(17,405)

* In addition 3 people were killed due to the tsunami on Buton Island in Southeast Sulawesi Province and 19 people were killed in Selayar Regency, South Sulawesi Province.
** In addition 130 and 600 houses were swept away on Kalaotoa Island, South Sulawesi Province and on Buton Island in Southeast Sulawesi Province, respectively.

PAGEOPH, Vol. 144, Nos. 3/4 (1995)

Coastal Sedimentation Associated with the December 12th, 1992 Tsunami in Flores, Indonesia

SHAOZHONG SHI,[1] ALASTAIR G. DAWSON,[2] and DAVID E. SMITH[2]

Abstract — This paper presents the result of a detailed granulometric investigation of sediments deposited by a modern tsunami, the 1992 tsunami in Flores, Indonesia. Eyewitness accounts indicate that sediments were deposited upon coastal lowlands over wide areas as a result of the tsunami inundation. Distinctive vertical and lateral variations in particle size composition are characteristic features of the tsunami deposits and these are intimately related to sedimentary processes associated with flood inundation. The geomorphological and sedimentary evidence is used here to establish a preliminary model of tsunami sedimentation. This information is believed to be of great value in understanding sedimentary processes associated with tsunami flooding and in the interpretation of palaeo-tsunami deposits.

Key words: Tsunami, coastal sedimentation, sorting processes, particle size, modal population, geomorphology, sediment.

1. Introduction

On 12th December, 1992, 1:30 pm (05:30 GMT), a major back-arc thrust earthquake of magnitude M_s 7.5 took place offshore approximately 50 km north of Maumere, the capital of Flores Island (Figure 1). Several minutes later a large tsunami struck the northern coastline of Flores. The earthquake and associated tsunami caused remarkable morphological changes to coastal areas of Flores and adjacent islands. The earthquake triggered extensive coastal landsliding at Lewora-hang (Figure 1) while conspicuous evidence of ground cracking and liquefaction was observed on coastal lowlands in other areas. However, the most widespread and dramatic changes in coastal morphology were caused by the erosion, transport and deposition of sediment during tsunami inundation. At Riangkrok, where the

---

[1] School of Applied Science and Computer Studies, Inverness College, IV1 1SA, U.K.
[2] Division of Geography, School of Natural and Environmental Sciences, Coventry University, CV1 5FB, U.K.

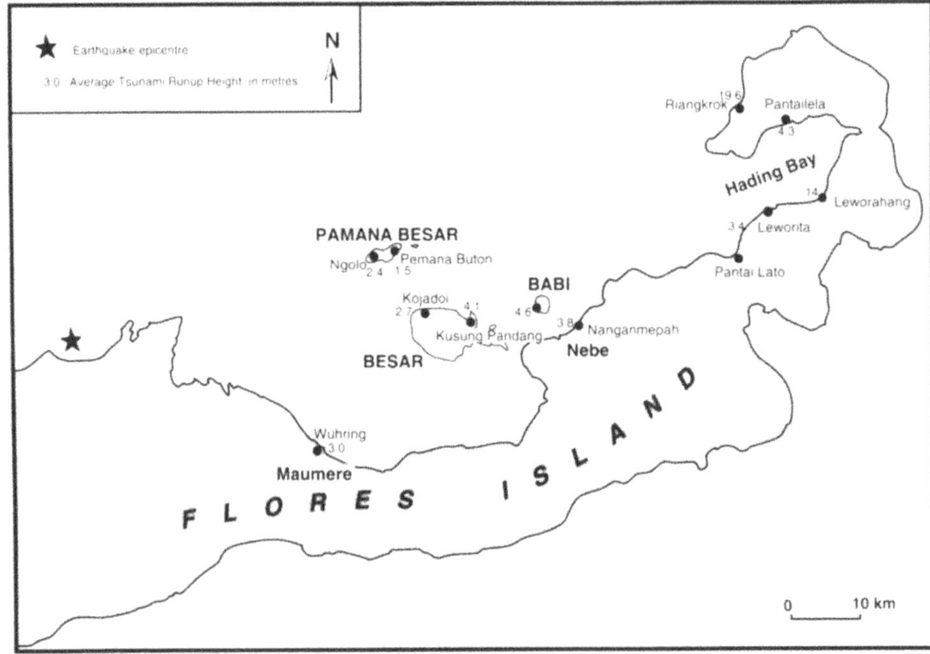

Figure 1

Map of eastern Flores, Indonesia showing locations mentioned in text. The numerical values denote the altitudes of tsunami runup (in metres) as measured by the International Survey Team (YEH *et al.*, 1993).

tsunami runup was as high as 26 m, widespread deposition of coral boulders indicates severe nearshore erosion of coral reefs off the coast (Figure 1). With the exception of Leworahang and Riangkrok, most affected areas of the coastline were flooded by the tsunami with runup values ranging between 1.5 and 4 m. In these areas, beaches and coastal lowlands were subject to widespread erosion, resulting in coastline retreat, lowering of coastal ground surfaces due to sediment removal, formation of ephemeral soil cliffs and gullies and the production of bare ground as a result of vegetation stripping (SHI *et al.*, 1993). Associated with the erosional features was widespread deposition of continuous and discontinuous sheets of sediment. In this study, the result of a detailed granulometric investigation of sediment sheets is described for two sites. Laboratory particle size analysis of these samples was undertaken using a laser granulometer (Malvern 2600c) (MALVERN INSTRUMENTS, 1990). Contiguous slicing of sediment core into samples was employed to detect vertical variations in particle size distribution with confidence.

2. Tsunami Sediment Characteristics

2.1 Nebe

2.1.1 Eyewitness Accounts

According to eyewitnesses at Nebe (Figure 1), the tsunami broke into a turbulent bore near the seaward edge of the coral reef approximately 500 m offshore and was associated with 3 to 5 separate episodes of tsunami inundation across the coastal lowlands. In this area, the sediment available offshore for tsunami transport was limited due to an absence of sediment on the coral reef and the occurrence of only a narrow beach flanked by mangrove trees.

2.1.2 Site Description

At Nebe the observed tsunami runup of *circa* 4 m was associated with the transport of sediment and debris including concrete blocks and artifacts. Tsunami flooding resulted here in coastline retreat of *circa* 10–20 metres with removal of sediment at least 10 m^3 per metre of coastline (SHI *et al.*, 1993). The newly formed coastline was typically crenulate in plan profile with distinctive soil cliffs up to 1 m

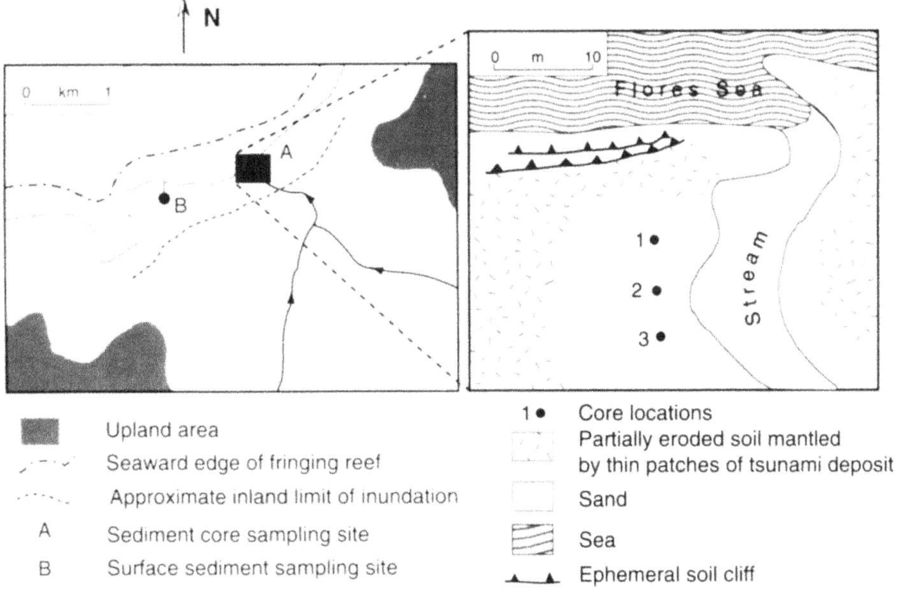

Figure 2

Map of the site at Nebe. Surface samples were obtained along a traverse line perpendicular to the local coastline at site B.

in height. Exposures along this cliffline disclose the presence of a coral platform overlain by dark sandy soil. These sediments are, in turn, mantled by an extensive and discontinuous sheet of sand sediment deposited by the tsunami. The tsunami sediment is composed of grey medium sand near the coast and of grey fine sand farther inland that forms sediment sheets ranging in thickness from 0.01 to 0.5 m. Cores of sediment were obtained in an area separated from the sea by a slightly eroded ground surface with soil cliffs seawards (Figure 2). Core 2 was examined visually and sliced into contiguous samples at 0.3 cm intervals. Nine surface samples were also obtained along a line from the coast inland (Figure 2).

2.1.3 Lithostratigraphy

Examination of core 2 revealed that a sharp uncomformity exists between the tsunami sediment and the underlying deposit, showing that erosion took place during tsunami inundation. This is indicated by an absence of part of the topsoil and vegetation in an area from the coastline to approximately 40 m inland at Location A (Figure 2), although there are no significant indications of severe surface erosion landward of the transect. The tsunami sediment is composed largely of medium and coarse sand. In general, it contains a significant amount of grit and pebbles at the base and is generally finer towards the top of the core although the grading is not distinctive. Coarse clasts including sandy intraclasts and grit occur intermittently at several levels. These intraclasts are composed of friable dark sandy soil that closely resembles the local soil.

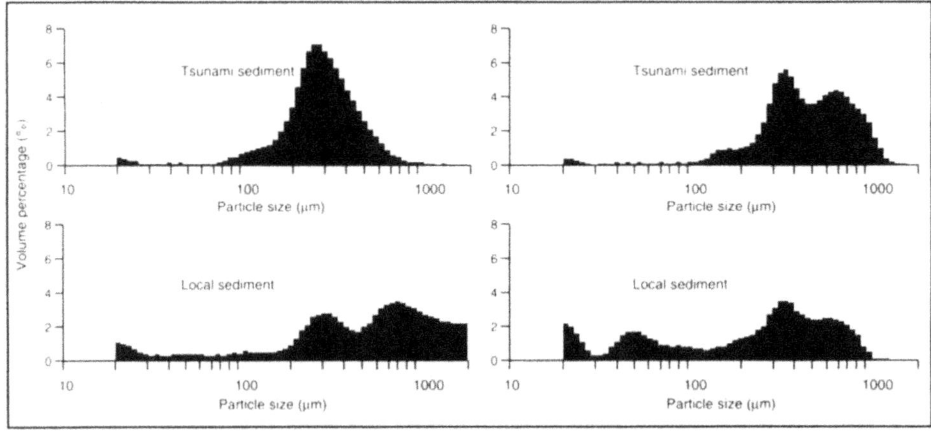

Figure 3

Particle size characteristics of tsunami sediment and local sediment at Nebe. Modal peaks of the tsunami sediment samples correspond to those of the local sediment, in particular at *circa* 300 and 600 µm on the size scale.

2.1.4 Particle Size Analysis: Downcore Variations

Representative particle size distributions obtained from core samples are shown in Figure 3. The tsunami sediment grains contained in the core largely occur in a size range between 100–1,100 μm with occasional grains coarser than 2 mm. Large variations in particle composition are illustrated by the percentage frequency curves that vary from relatively well-sorted to poorly-sorted distributions. The particle size distributions occur as three broad groups:

1) Size distributions that exhibit a dominant primary modal subpopulation with a near log-normal distribution (indicative of well-sorted sediments) in the sand range and a much less prominent tail in the finger range (silt and clay). The fine fraction generally comprises about 5% of the total composition and the primary modal position varies between *circa* 200–500 μm. About half of the samples are of this type.

2) Multimodal distributions with an additional coarse sand subpopulation, sometimes with small amounts of grit. The coarse sand subpopulation occurs in a size range of 600 μm–2 mm but sometimes is composed of a broad range of particle sizes with no distinct modal size. Although much less significant, a third subpopulation can be identified with a modal peak in the range of 100–150 μm.

3) Rarely, poorly-sorted broad distributions without distinct modal peaks occur. Size distributions of types 2 and 3 sometimes coincide with the presence of intraclasts or gravel.

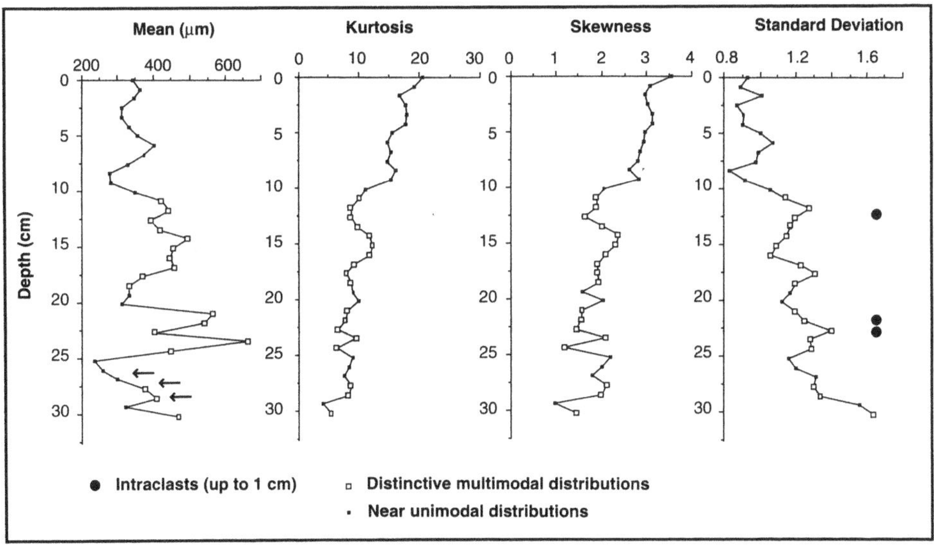

Figure 4

Vertical variations of parameters of tsunami sediment particle size distribution, Core 2, Nebe. Arrows denote particle size distribution histograms shown in Figure 5. The parameters are calculated with moment statistical formulae (McBride, 1971).

In summary, the tsunami sediments at Nebe are composed of three principal subpopulations, a) a major subpopulation at *circa* 300–350 µm, b) a finer component population in the range 100–150 µm, and c) a coarser subpopulation at *circa* 600 µm.

Vertical variations in textural property are illustrated by fluctuations in particle size composition. This is clearly shown by the saw-toothed curves of downcore variations in the values of mean particle size, skewness, kurtosis and standard deviation with depth (Figure 4). There appears to be a general-fining upward trend that corresponds with a general upcore increase in sorting. The detailed variations in the particle size characteristics consist of multiple fining-upward sequences, though these appear limited and there are occasional indications of inverse grading. Some of the fining-upward sequences are represented by better sorted, near unimodal particle size distributions, within which secondary subpopulations are much less

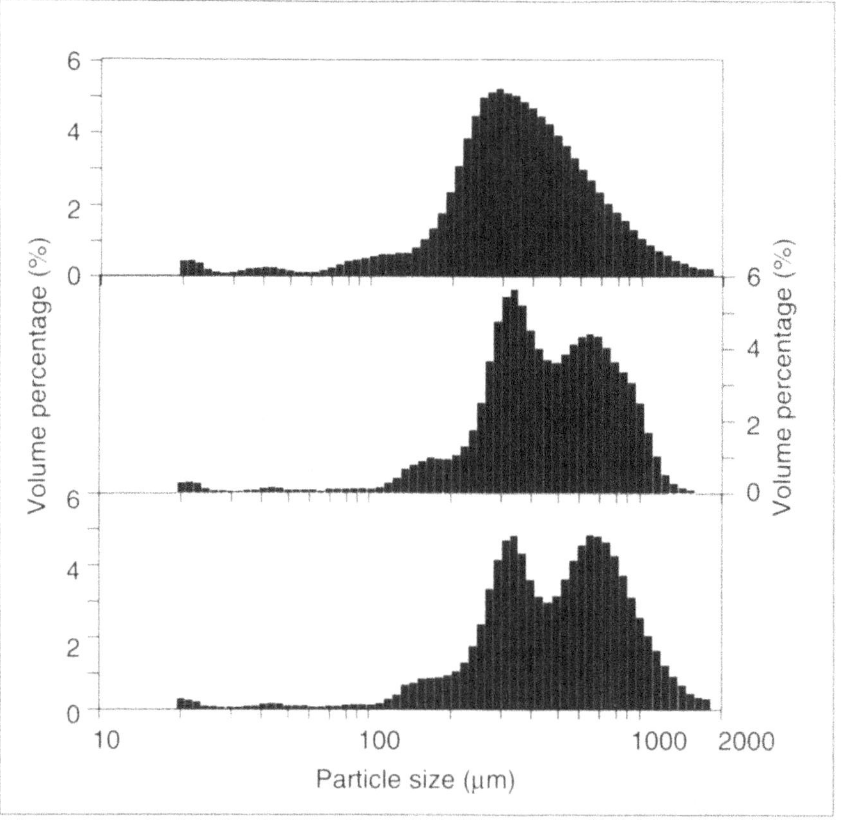

Figure 5

An example of upward progression of variations in multimodal distribution within some of multiple fining-upward sequences preserved in tsunami sediment at Nebe. Note that significant modification occurs at the coarsest end of the composite overall population. The positions of the samples are shown in Figure 4.

distinctive but often recognisable. Together with the fining-upward sequences of distinct multimodal populations, they form multiple sets of fining-upwards sequences through the whole of the sediment. Within each fining-upward sequence, conspicuous variations occur at the coarse end of the particle size distributions. This is well illustrated by progressive upcore compositional changes of the multimodal distributions (Figure 5). The modal positions of the subpopulations occur within consistent size ranges except the coarsest subpopulation fines and decreases in proportion upcore.

2.1.5 Particle Size Analysis: Spatial Variations

Surface samples were obtained along a line perpendicular to the coast in order to investigate spatial variations in particle size composition within the tsunami deposits (Figure 6). There is a progressive fining of mean particle size landwards that is locally interrupted and partly obscured by fluctuations in mean particle size. These fluctuations are restricted to an area near the coast adjacent to which an undulating topographic depression occurs.

2.1.6 Characteristics of the Local Soil

The local soil is composed of a mixture of sand, silt and clay and is characterised by multimodal and broad particle size distributions. It contains a large amount

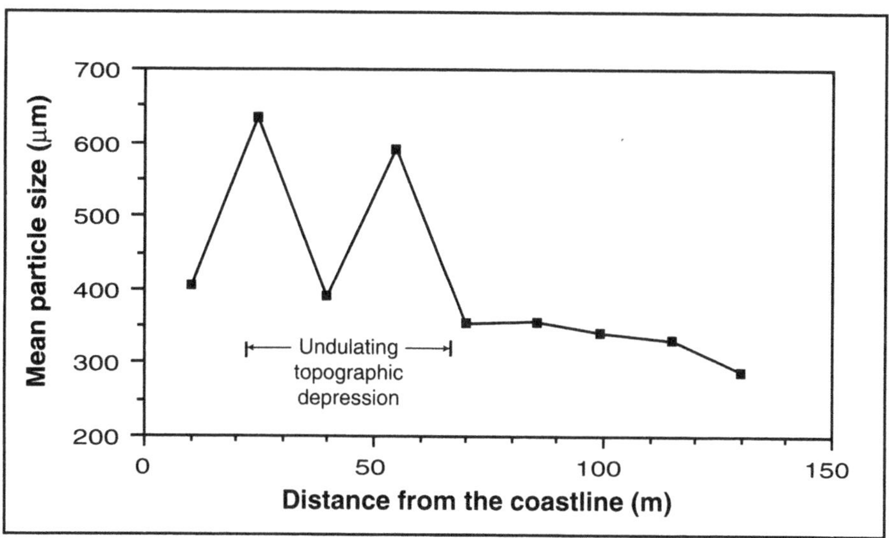

Figure 6

Variation in mean particle size along a 130 m traverse perpendicular to the coastline at site B, Nebe (see Figure 2).

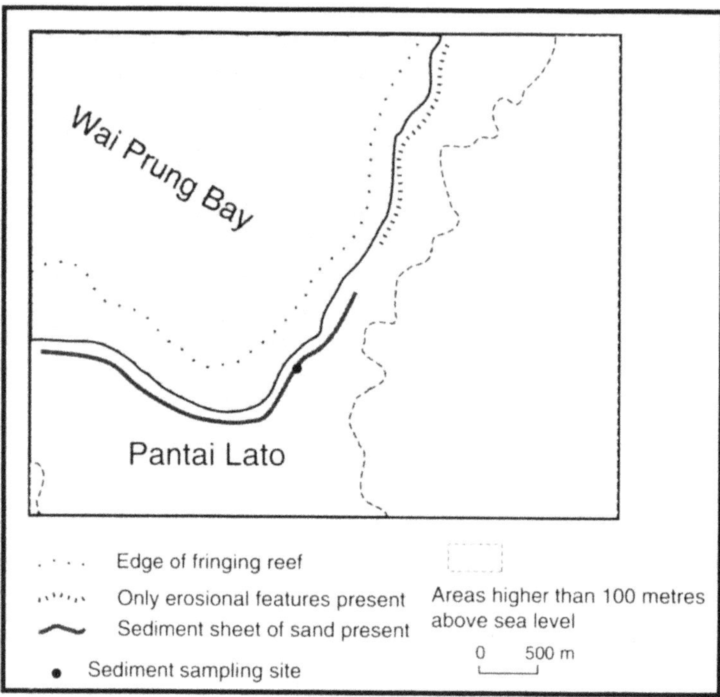

Figure 7
Location map of the site at Lato.

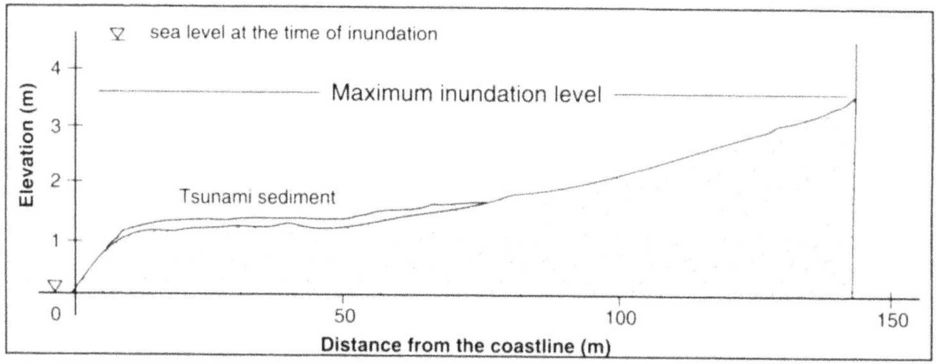

Figure 8
Schematic diagram showing the measured maximum inundation distance (*circa* 140 m), the altitudinal limit of runup (3.5 m) and the extent of a continuous sand sheet at Lato. Note that the thickness of the sand sheet is only a few centimetres and is exaggerated here. Farther inland from the sheet, only isolated patches of fine sand occur and their thicknesses are in the order of a few millimetres.

of clay and fine silt in contrast to the tsunami sediment which contains little (Figure 3). Whenever the deposit is multimodal, subpopulations of particle sizes are consistent in their respective ranges. The modal peaks of the sand subpopulations occur at *circa* 300 and 600 μm respectively, and clearly correspond to the respective subpopulations that are characteristic of the tsunami-deposited sand.

2.2 Lato

Interesting lateral variations in particle size distribution are illustrated by the results of analyses of samples obtained at Lato (Figures 1 and 7). Here, the

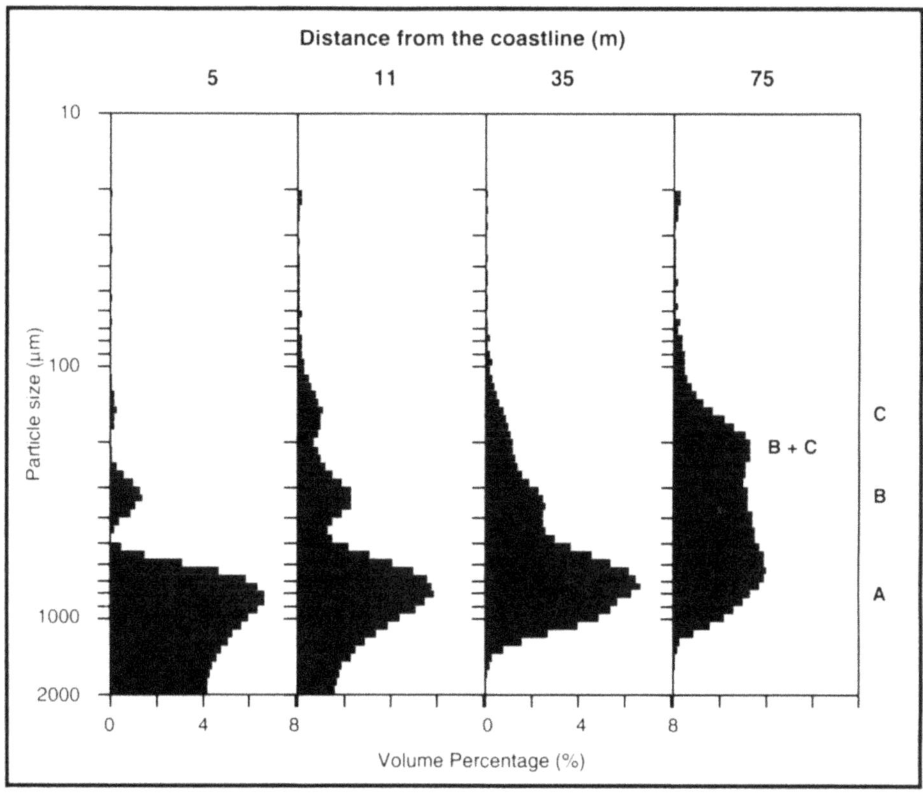

Figure 9
Lateral variations in the characteristics of particle size distribution at Lato. For convenience, the three subpopulations of particles are named population A, B and C, respectively from the coarsest to the finest. From the coast inland, the proportion of the coarsest population A decreases and its modal position shifts towards finer size range, while populations B and C increase and both their modal positions remain relatively constant at *circa* 300 and 150 μm on the size scale respectively for the first three samples. The most landward sample appears to exhibit a bimodal distribution with its finer modal position (*circa* 200 μm) resting between the modal positions of populations B and C. Clearly, it is very likely that an enhancement in the proportion of population C has led to a seemingly new population as a result of the merging of populations B and C. Unequivocally, the sediment fines landwards.

maximum runup was measured as *circa* 3 m with inundation reaching as far as *circa* 140 m inland (Figure 8). Tsunami sediments were deposited in this area as a thin continuous sheet restricted to an area 80 m inland from the coast. Four samples from a thin sediment sheet were obtained along a traverse line perpendicular to the coast. The particle size distributions are all multimodal and show clear lateral variations in sediment compostion that reflect changes in relative abundancy of the respective particle size subpopulations (Figure 9). A landwards fining trend is represented by a general decrease in the modal size and a decrease in the abundance of the coarsest subpopulation.

3. *Discussion*

3.1 *The Role of Sediment Sources*

Although there is difficulty in quantifying and interpreting the relative contributions of specific sediment sources, there are ample indications that the source of sediment for tsunami deposition was largely derived from local coastal soil at Nebe. Such indications include the especially light-grey colour of the tsunami sediment, local morphological changes and scarcity of sediment on the coral reef. The particle size distribution characteristics of the sediment are dependent on the nature of the source material and the processes of tsunami transportation and deposition. The source material (local soil) is composed of particles that cluster into several grain size subpopulations (Figure 3). The subpopulations in the sand range are readily identifiable in the tsunami sediment and this characteristic imposes a significant control upon the characteristics of the tsunami deposit. The significant difference in composition between the local soil and tsunami sediment indicates that clay and silt was probably removed seaward during periods of tsunami backwash flow. Modification of the transported sediment appears also to have occured as a result of differential transportation and sedimentation of grains.

3.2 *Tsunami Deposition and Sorting Processes*

The multimodal characteristic of the tsunami sediment clearly indicates that sedimentation took place at fast rates and that particles of a wide size range settled out simultaneously at different rates. This is also well illustrated by the sets of individual fining-upward sequences which possess the same sand subpopulations. This indicates that the modal grain size values of these subpopulations occur within consistent size ranges and that the coarsest fraction becomes finer and decreases in proportion upcore. Such variations indicate a unique settling process characterised by rapid rates of tsunami sedimentation resulting in the simultaneous deposition of fine and coarse particles.

The saw-toothed curves of particle size parameters highlight a positive correlation between a progressive upward-fining trend and an increase in the degree of sorting. This is interpreted as a result of the general decrease in energy of consecutive waves and the effects of prolonged sorting of some of the sediment, while it was in motion.

3.4 Tsunami Runup and Backwash

Tsunami inundation is characterised by both runup and backwash processes and both contribute to the reworking, transportion and deposition of sediment. The occurrence of multiple sets of fining-upwards sediment sequences clearly indicates that different episodes of runup and backwash have resulted in net sediment accumulations. Separate suites of sediment with different granulometric characteristics are identifiable, although it is currently difficult to tell which are due to backwash and which are attributable to runup.

3.5 Lateral Sorting

The general trend of fining landwards may partly be a result of differential transportation, which has transported more fine particles farther inland than coarser ones, and also may arise from the effects of differential settling velocities. At Nebe, such a trend is obscured due to the effects of complex topography and possibly also as a result of different episodes of inundation.

Landward transport of sediment associated with tsunami inundation is best recorded at Lato (Figure 9). This sediment sheet is interpreted as having been produced by the last tsunami inundation of the coastal lowland, owing to its very thin nature and its restricted presence near the coast, whereas surface soil shows signs of only slight erosion farther inland. The sediment was probably winnowed landwards and was subject to the influence of backwash. Nevertheless, lateral variations in the composition of particle sizes of the sediment provide important information on sediment dispersion processes associated with tsunami flooding. It seems likely that all sizes of particles were transported and dispersed at the same time but that finer particles travelled farther than coarser grains in the transporting water body. Moreover, it seems to be suggested by the results that lateral sorting was predominantly imposed upon the coarsest fraction of transported particles.

In summary, the sorting process of the source material by the Flores tsunami can be understood conceptually as having occurred in two different ways. Primary sorting is represented by the removal of a large amount of clay and silt by tsunami backwash. Secondary sorting resulted in further modification of the transported material during tsunami flooding. Variability is most pronounced at the coarsest end of particle size distribution within the tsunami sediment. At present, the distinction between the sediments deposited by tsunami runup and backwash is not fully understood. Nevertheless, the particle size distributions of the sediment and their variations probably reflect a unique process in coastal geomorphology.

4. Summary

The Flores study indicates that tsunami inundation is an ephemeral process and is associated with turbulent processes of sediment transport. As a result, sediment erosion is localised while sediment transport and deposition are major processes associated with the deposition of partially sorted and multimodal sediments. During the inundation of a tsunami, the energy regime changes from turbulence to relative calmness, and thus the transported grains settle out at a rapid rate as a result of a sudden decrease in transporting capacity. Sedimentation rates are so high that the tsunami sediment is frequently composed of several populations of particles in different size ranges. The interplay of turbulence, rapid sedimentation and the characteristics of transported material determines the resultant characteristics of the tsunami deposit. The analysis of the Flores tsunami sediments at Nebe and Lato therefore represents an attempt to understand the very complex processes of sediment transport and deposition associated with tsunami flooding.

Acknowledgements

Participation in the Flores International Survey Team is gratefully acknowledged to F. Imamura, P. J. Prih Hadjardi, R. Soetardjo and Mr. Sunarjo. Cartographic assistance was kindly provided by Ruth Gaskell and Shirley Addleton.

REFERENCES

MALVERN INSTRUMENTS (1989), *Series 2600 — User Manual*. Spring Lane South, Malvern, Worcs. WR *14 1*AQ, U.K.

MCBRIDE, E. F., *Mathematical treatment of size distribution data*. In *Procedures in Sedimentary Petrology* (ed. R. E. Carver) (Wiley, New York 1971).

SHI, S., DAWSON, A. G., and SMITH, D. E. (1993), *Geomorphological impact of the Flores tsunami of 12th December, 1992*. In Tsunami '93, Proc. IUGG/IOC Int. Tsunami Symp., Wakayama, Japan, August 23–27, 1993, pp 689–696.

YEH, H., IMAMURA, F., SYNOLKIS, C., TSUJI, Y., LIU, P., and SHI, S. (1993), *The Flores Island Tsunamis*, EOS, Trans. Am. Geophy. Union *74* (33), August 17, pp. 369, 371–373.

(Received July 23, 1994, revised January 13, 1995, accepted January 29, 1995)

PAGEOPH, Vol. 144, Nos. 3/4 (1995)

Modeling the Seismic Source and Tsunami Generation of the December 12, 1992 Flores Island, Indonesia, Earthquake

DANNIE HIDAYAT,[1] JEFFREY S. BARKER,[1] and KENJI SATAKE[2]

Abstract — On December 12, 1992 a large earthquake (M_s 7.5) occurred just north of Flores Island, Indonesia which, along with the tsunami it generated, killed more than 2,000 people. In this study, teleseismic P and SH waves, as well as PP waves from distances up to 123°, are inverted for the orientations and time histories of multiple point sources. By repeating the inversion for reasonable values of depth, time separation and spatial separation, a 2-fault model is developed. Next, the vertical deformation of the seafloor is estimated from this fault model. Using a detailed bathymetric model, linear and nonlinear tsunami propagation models are tested. The data consist of a single tide gauge record at Palopo (650 km to the north), as well as tsunami runup height measurements from Flores Island and nearby islands. Assuming a tsunami runup amplification factor of two, the two-fault model explains the tide gauge record and the tsunami runup heights on most of Flores Island. It cannot, however, explain the large tsunami runup heights observed near Leworahang (on Hading Bay) and Riangkroko (on the northeast peninsula). Massive coastal slumping was observed at both of these locations. A final model, which in addition to the two faults, includes point sources of large vertical displacement at these two locations explains the observations quite well.

Key words: Earthquake source, body waves, moment tensor, tsunami modeling, submarine slumps, Indonesia.

Introduction

The December 12, 1992 Flores Island, Indonesia, earthquake (M_s 7.5) and the ensuing tsunami caused over 2,000 deaths and another 2,000 injuries (Figure 1). Tsunami runup heights along the northern shore of Flores Island vary from 2–5 m in the central portion of the island to as much as 26 m in the eastern portion (YEH *et al.*, 1993; TSUJI and MATSUTOMI, 1993; TSUJI *et al.*, 1995). The tsunami washed away entire villages on Flores Island and on the small Babi Island just offshore.

Indonesia is the site of many large earthquakes, with the Indo-Australian plate subducting northward beneath the Eurasian plate. The shallow depth of the Flores Island earthquake (36 km, PDE; 15 km, Harvard CMT; 9 km, this study), however,

---

[1] Department of Geological Sciences, State University of New York, Binghamton, NY 13902-6000, U.S.A.
[2] Department of Geological Sciences, University of Michigan, Ann Arbor, MI 48109-1063, U.S.A.

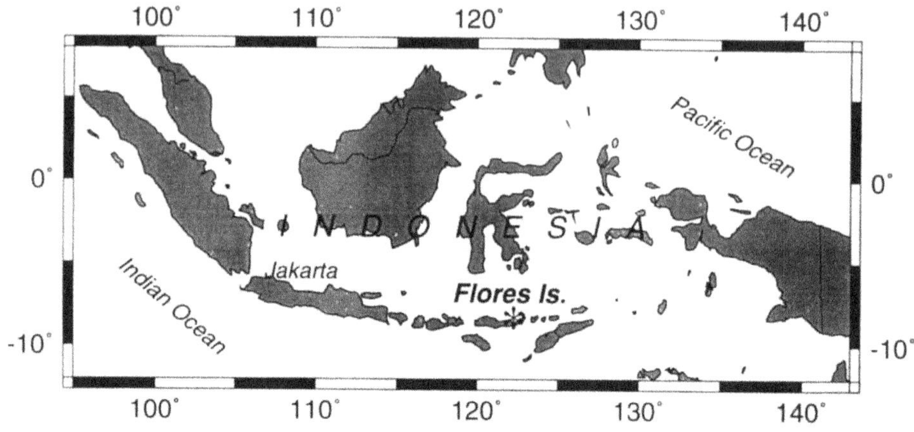

Figure 1
Map of Indonesia showing the location of the December 12, 1992 Flores Island earthquake.

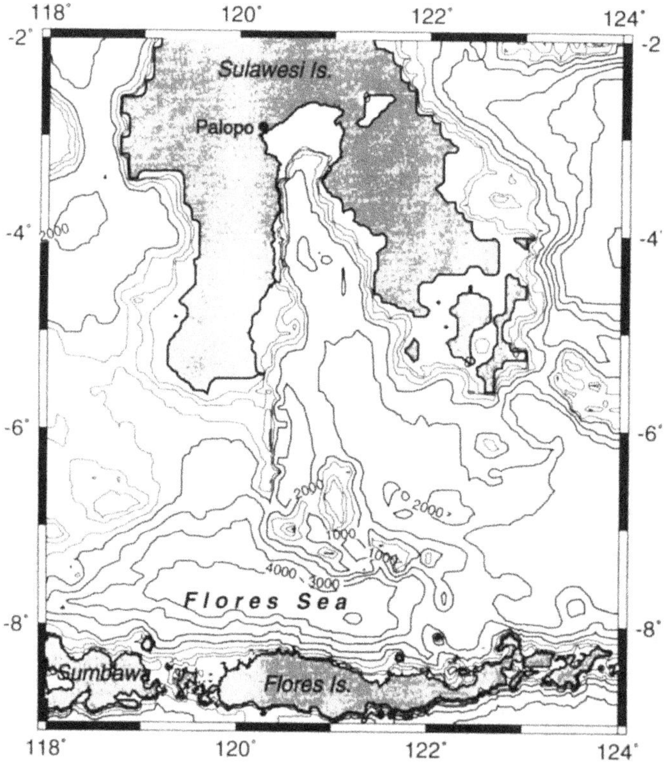

Figure 2
Regional bathymetry model based on edited ETOPO-5 data augmented with detailed information from nautical charts near the north coast of Flores Island and near the tide gauge at Palopo on Sulawesi Island. Contours show bathymetric depth in meters.

indicates that it occurred in the backarc region of the Eurasian lithosphere above the subducting slab. Other studies (HARVARD CMT; IMAMURA and KIKUCHI, 1994; BECKERS and LAY, 1995) as well as this study, suggest a thrust mechanism on a shallow-dipping fault plane dipping to the south, consistent with backarc tectonics (for example, PLAFKER and WARD, 1992).

In this study, we invert teleseismic, broadband P and SH waves, as well as PP waves (from ranges up to 123°) for the seismic moment rate tensor. We resolve two independent subsources, with the second located to the east of the first. From the seismological model, we estimate the vertical displacement of the seafloor. Using linear and nonlinear methods, we then compute the tsunami generation and propagation. Only one tide gauge recorded the tsunami, at Palopo on southern Sulawesi, 650 km north of Flores Island (Figure 2). Our tsunami models are constrained to fit this record, while attempting to explain the tsunami runup heights measured on Flores Island. We find that substantial secondary (nonseismic) sources are required to explain the extremely large tsunami runups on the eastern portion of Flores Island, and model these in terms of landslides or submarine slumps.

Modeling the Seismic Source

Teleseismic broadband P- and S-wave seismograms (ranges 30–90°) were obtained from the IRIS Data Management Center. The S waves were rotated into the tangential component, then the P and SH waves were corrected for instrument gain and, where necessary, integrated to ground displacement. To improve the azimuthal distribution of data, PP waves from ranges up to 123° were also obtained and corrected to ground displacement. Nevertheless, azimuths from south to west are poorly sampled. The data were quite noisy at some stations, so phaseless (two-pass) Butterworth filters (high-pass for P waves, low-pass for SH waves) were applied. The data modeled in this study and details of filters applied are listed in Table 1.

Synthetic seismograms are computed using a propagator matrix method (KIKUCHI and KANAMORI, 1991). The seismic velocity structure at the source is assumed to consist of water and a four-layer oceanic crust overlying the mantle (Table 2). Similarly, the velocity structure near each receiver is assumed to be continental crust with two layers overlying the mantle. For the PP waves, the structure near the bounce point is assumed to be the same as the oceanic structure at the source. The synthetic seismograms are computed for a triangular far-field source time function with vertical strike-slip, vertical dip-slip and compensated linear vector dipole (CLVD) sources, since the response due to an arbitrarily-oriented shear dislocation may be obtained from the linear combination of these three fundamental dislocations (LANGSTON and HELMBERGER, 1975). An attenuation operator with t^* of 1.0 for P waves, 4.0 for SH waves, and 2.0 for PP waves is applied. The displacement instrument response is convolved and, if necessary,

Table 1

Seismic data and filter characteristics

Station	Distance (°)	Azimuth (°)	P	PP	SH
BJI	48.3	354.1	+		+
CHTO	35.1	320.0	+ +		+ +
ENH	40.1	343.2	+		
HIA	57.2	358.3	+		+
KIP	83.8	67.4	+ +[1]		
KMI	38.0	331.1	+		
KONO	108.2	330.4		+ +[1]	
LSA	47.9	323.2	+		
MAJO	47.1	18.0	+ +		
MDJ	53.0	6.9	+		+
PAB	122.8	312.1		+ +[1]	
PAS	119.3	53.8		+ +[1,3]	
RAR	76.2	109.8	+ +[1]		+ +[2]
SNZO	57.0	134.3	+ +		+ +
SPA	81.8	180.0	+ +		+ +[2]
SSE	39.1	359.0	+		+
TATO	33.0	359.3	+ +		+ +[2]
TAU	41.1	151.4	+ +		
YSS	58.0	16.7	+ +		+ +[2]

+ Used only in preliminary, 1-source inversions.

+ + Used in 2-source inversions.

[1] Low-pass Butterworth filtered at 0.05 Hz.

[2] High-pass Butterworth filtered at 0.02 Hz.

[3] In preliminary inversions, a stack of Terrascope *PP* waves was used; in final inversions, the observed *PP* wave at PAS was used.

low-pass or high-pass filters are applied to obtain the Green's functions used in the inversion.

Prior to inversion, *P*, *PP* and *SH* wave windows are defined in the data, and synthetic seismograms for a starting model are aligned in time with the observed waveforms. After defining the time window of interest, the waveforms are inverted for the deviatoric moment rate tensors and far-field source time functions of one or more sources, varying the depths, relative locations and origin times of the sources. The moment tensor inversion method is given in LANGSTON (1981), with applications to teleseismic body waves in BARKER and LANGSTON (1981, 1982, 1983). The source centroid depth is determined by repeating the inversion using Green's functions computed for different depths, and selecting the depth that minimizes the difference between observed and synthetic seismograms (as measured by the RMS fit).

One-source Model. An initial inversion was performed using as a starting model the Harvard CMT orientation, and a source time function consisting of 11 overlapping isosceles triangles of equal amplitude. Each triangle has a rise and fall

Table 2

Seismic velocity structures

Oceanic Structure (at source and PP bounce points)			
V_P (km/s)	V_S (km/s)	Density (g/cm^3)	Thickness (km)
1.50	0.00	1.03	1.5
2.00	1.15	1.80	1.0
4.75	2.75	2.30	3.0
6.52	3.75	2.50	3.0
7.00	4.00	2.80	5.0
8.00	4.60	3.10	h.s.

Continental Structure (at receivers)			
V_P (km/s)	V_S (km/s)	Density (g/cm^3)	Thickness (km)
6.00	3.46	2.60	15.0
7.00	4.04	2.80	18.0
8.00	4.62	3.10	h.s.

time of 5 sec. After five iterations a solution was obtained with orientation consisting of a shallow-dipping reverse mechanism at a depth of 9 km. This is substantially shallower than the depths listed in the PDE (36 km) and the Harvard CMT solution (15 km). The southward-dipping fault plane is preferred since this is consistent with the sense of motion expected in backarc compression (e.g., PLAFKER and WARD, 1992). The source time history has a duration of about 55 sec, and consists of two pulses, each of 20–25 sec duration. This suggests that the earthquake has a double source, which we could interpret as two distinct asperities. The moment magnitude is calculated as 7.9, with moment approximately equally distributed between the two pulses. In this case, a single-point source, which is implicitly assumed in the moment tensor inversion, may not be a valid model. We may, instead, consider two point sources, with distinct depths, orientations, and source time histories.

Two-source Inversion. In an attempt to distinguish the parameters of the two subsources, simultaneous inversions were performed, holding the depth of the first source at 9 km and varying the depth, horizontal separation and time separation of the second source. The starting orientation of each source was the one-source moment tensor, and the source time histories were parameterized by separate series of eight overlapping isosceles triangles, now with a rise and fall time of 3 sec. Since we are interested in the details of the source process, and since the data of the Chinese Digital Seismograph Network (CDSN) are sensitive only to lower frequencies, these stations were excluded from the 2-source inversions. There was no improvement in RMS fit when the second source was located at a different depth

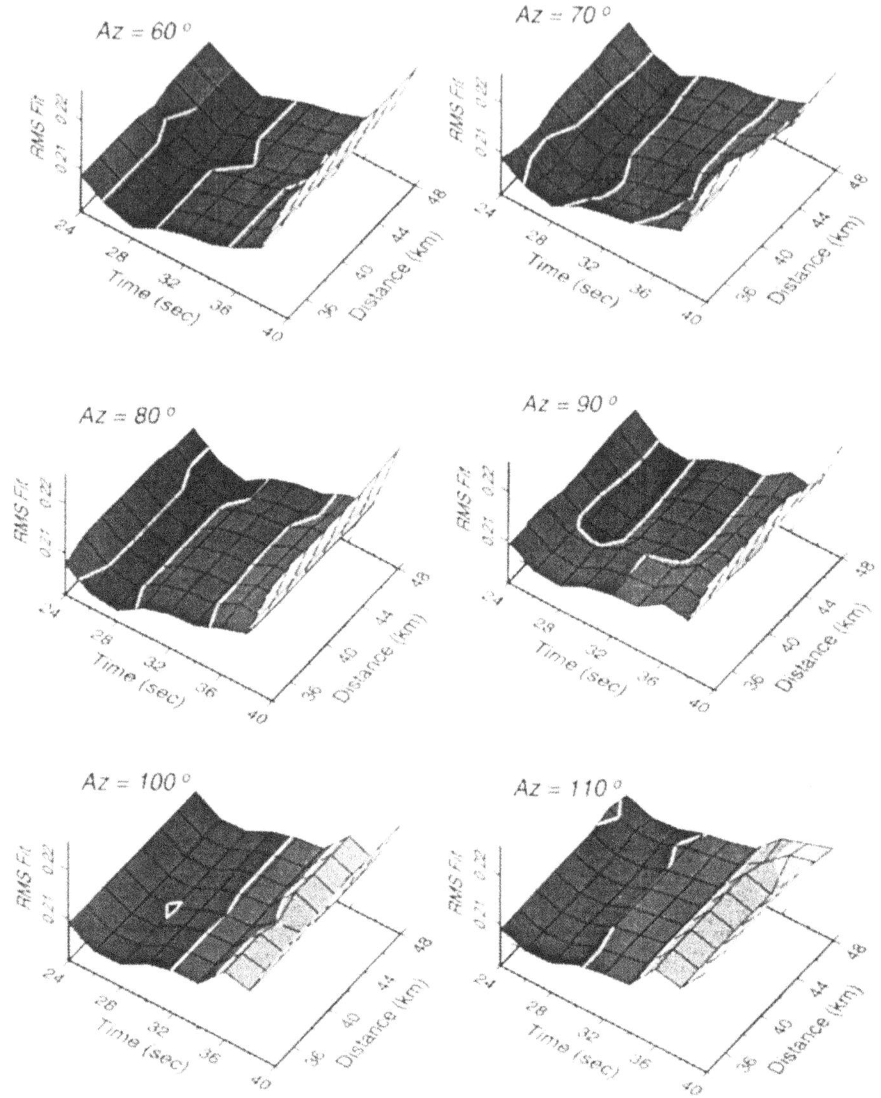

Figure 3

RMS fit (a measure of the residual between observed and synthetic seismograms) for various values of the distance, azimuth and time separation of the second source from the first. Time separation is well resolved at 28 sec. Azimuths from 60° to 90° yield good fits. Distance separation is poorly resolved, so we choose 40 km.

than the first, so we will assume both to be at 9 km depth. Inversions were performed for horizontal spatial separations of 34–48 km between sources, with the second located at azimuths of 60–110° from the first. Based on the one-source inversion result, time separations of 24–40 sec were considered. Plots of RMS fit for each of these inversions are shown in Figure 3. Time separation is well resolved

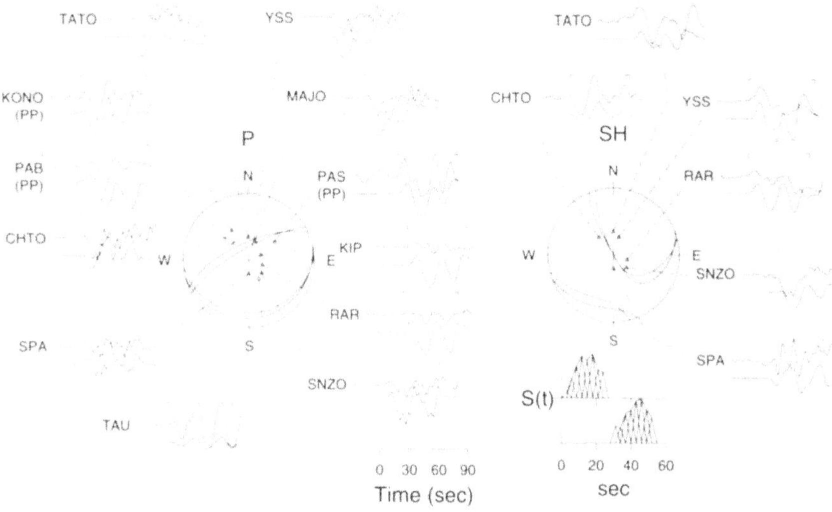

Figure 4

Inversion results for the preferred 2-source model. Observed (above) and synthetic (below) seismograms are shown plotted on common time and amplitude scales. The inversion time windows are indicated above the observed waveforms. Also shown are lower-hemisphere equal-area projections of the *P* and *SH* nodal surfaces for the major double couple of each source, with each station's position in the radiation pattern indicated. The two-source time histories are shown on the bottom right of the figure, plotted on common time and amplitude scales.

as 28 sec, but the spatial separation is not at all well resolved. For the distances chosen, a slight minimum occurs at 38 km for an azimuth of 70°, however the resolution is so poor that we can conclude only that the second source occurred at some distance along an azimuth of 60–90° from the first.

Figure 4 shows the waveforms and results for the two-source inversion, with the second source separated in time by 28 sec and in space by 40 km at an azimuth of 80° from the first. *P* and *PP* waves are shown on the left, and *SH* waves on the right, along with the nodal surfaces corresponding to the major double couple for each source, on lower-hemisphere equal-area projections. For each station, the observed waveform is shown above the final synthetic waveform, plotted on common time and amplitude scales. The inversion time window is indicated above each observed trace. Although some of the data are still rather noisy, significant details of the *P*, *PP* and *SH* waveforms are well modeled. In particular, nodal and small amplitude *P* waves to the north and west constrain the dip of the fault plane, while the *SH* polarity change between CHTO and TATO constrains the rake. *PP* waves at KONO and PAB are dilatational, although the Hilbert transform results in an upward first motion. The two-source model allows details of later arrivals in *P* waves (such as TATO, YSS, MAJO and SPA), *PP* waves (such as PAS and PAB), and *SH* waves (such as CHTO, YSS and SPA) to be well modeled. The two source time functions are plotted at the bottom right of the figure. Each source

Table 3

Seismic inversion results for the Flores Island earthquake

		First Source	Second Source
Depth		9 km	9 km
Time Separation			28 sec
Spatial Separation			40 km
			80° azimuth
Moment Tensor	NN	3.15	1.95
($\times 10^{27}$ dyne-cm)	EE	1.44	1.29
	NE	-0.016	0.0066
	ND	3.51	3.87
	ED	-2.16	-2.46
Seismic Moment ($\times 10^{27}$ dyne-cm)		5.8	5.4
Moment Magnitude		7.8	7.8
Major Double Couple Moment		5.61	5.22
($\times 10^{27}$ dyne-cm)			
Fault Planes:			
1	Strike(°)	85	77
	Dip(°)	23	15
	Rake(°)	108	106
2	Strike(°)	246	241
	Dip(°)	68	76
	Rake(°)	83	86
Time function duration		28 sec	30 sec

function has approximately equal duration and amplitude, indicating that the moment of the two sources is essentially the same. The final model parameters are listed in Table 3. Each source is a shallow-dipping reverse fault with moment magnitude 7.8.

Modeling Tsunami Generation

Using the seismic inversion results, we may model the tsunami generated by the Flores Island earthquake. As mentioned in the introduction, only a single tide gauge recorded a significant tsunami signal, and this was located 650 km away at Palopo, on the island of Sulawesi. Other tide gauges in the region either do not record continuously, or were shadowed by land (GONZÁLEZ *et al.*, 1993). An international survey team measured a number of tsunami runup heights along the northeast coast of Flores Island within three weeks of the earthquake (YEH *et al.*, 1993; TSUJI and MATSUTOMI, 1993; TSUJI *et al.*, 1995). They measured maximum tsunami runup heights and inundation areas at several villages. These measurements were supported by aerial photography and geodetic measurements (GONZÁLEZ *et al.*, 1993). At each location we consider the largest consistent measurement

Table 4

Observed and calculated tsunami runup heights (in m)*

#	Station Name	Latitude	Longitude	Observed	Linear model 2 faults	Linear model + 2 slumps	Nonlinear model 2 faults	Nonlinear model + 2 slumps
1	Mage, Palu Is.	−8.30	121.75	2.8	1.79	1.79	1.80	1.80
2	Mausanbi	−8.50	121.78	3.4	3.15	3.04	2.88	2.88
3	Awora	−8.48	121.85	2.9	2.90	2.74	2.68	2.68
4	Deteh	−8.53	122.03	2.3	5.88	5.88	5.46	5.46
5	Kolisia	−8.53	122.10	5.2	5.52	5.52	5.40	5.40
6	Nagasarong	−8.53	122.12	4.9	5.11	5.00	4.86	4.88
7	Waturia	−8.53	122.13	2.9	4.18	4.15	3.64	4.08
8	Patisomba	−8.55	122.15	4.0	4.11	4.61	4.06	4.06
9	Nangahureh	−8.55	122.17	1.9	3.44	3.30	3.30	3.30
10	Wailiti	−8.57	122.18	2.1	3.43	3.21	2.16	2.62
11	Wuring	−8.60	122.20	3.2	3.67	3.58	2.68	2.90
12	Maumere	−8.62	122.23	3.0	2.95	3.14	2.24	2.46
13	Waioti	−8.63	122.27	2.5	3.46	3.60	3.02	3.02
14	Geliting	−8.63	122.28	3.3	3.63	3.94	3.28	3.16
15	Kewapante	−8.63	122.30	0.6	3.02	3.13	2.70	2.70
16	Egon	−8.60	122.42	1.8	4.05	4.22	3.98	3.98
17	Wodung	−8.58	122.48	2.3	4.69	5.34	3.58	3.86
18	Nangahale	−8.55	122.50	1.5	3.58	5.26	3.70	3.38
19	Talobura	−8.52	122.52	2.4	5.52	6.11	4.92	4.74
20	Ngolo, Pomana Is.	−8.35	122.32	3.2	1.97	2.58	1.98	1.98
21	Buton, Pomana Is.	−8.33	122.33	1.5	2.23	2.23	2.24	2.24
22	Taot, Desar Is.	−8.87	122.35	2.8	1.33	2.02	1.20	1.32
23	Kusung, Besar Is.	−8.87	122.42	4.1	2.62	4.64	2.66	3.92
24	Permahan Is.	−8.45	122.45	3.4	3.21	4.46	3.16	2.74
25	Babi Is. N	−8.40	122.52	4.0	2.12	3.01	2.14	2.36
26	Babi Is. W	−8.42	122.50	7.1	2.40	3.62	2.40	2.40
27	Babi Is. S	−8.43	122.52	4.0	3.76	2.94	3.72	3.72
28	Babi Is. E	−8.42	122.53	5.6	2.69	2.94	2.66	2.66
29	Nebe	−8.45	122.53	4.6	5.11	5.40	3.04	3.04
30	Wailamung	−8.42	122.58	5.5	6.92	7.81	5.42	4.80
31	Larentuka	−8.37	122.98	1.8	1.97	1.96	1.70	1.70
32	Pantai Lato	−8.37	122.77	3.8	3.59	5.78	3.12	3.52
33	Uepadung	−8.30	122.83	11.0	3.44	10.27	3.12	5.66
34	Waibalen	−8.28	122.88	10.6	4.42	10.70	3.30	8.16
35	Pantai Lela	−8.20	122.83	4.5	3.83	6.52	3.52	4.38
36	Riangkroko	−8.15	122.78	18.4	3.31	11.14	3.24	14.20
37	Bunga	−8.10	122.80	12.3	2.70	8.00	2.70	2.94

* Calculated as 2.0 × computed tsunami height.

of tsunami runup height (see Table 4). These tsunami runup height measurements, combined with the tide gauge record at Palopo, serve as our data set. Tsunami runup heights are generally larger than tsunami wave amplitudes (as would be measured on tide gauges) by an amount that depends on the coastal topography and wave dynamics. Furthermore, computed tsunami heights depend on the grid size of the bathymetric model (SATAKE and TANIOKA, 1995). In this study, we assume a uniform amplification factor of two in modeling maximum tsunami runup heights on Flores Island.

Our model for the bathymetry is initially based on edited ETOPO-5 data, augmented with detailed bathymetry compiled from nautical charts near Flores Island and along the bay leading to Palopo. Particular attention is paid to coastlines and shallow water. The coastal water around Flores Island is generally deep (at least 20 m), whereas it is much shallower near the Palopo tide gauge station (6 m). The result is sampled onto a bathymetric grid with one minute increments from 118°E to 124°E longitude and 2°S to 9°S latitude (Figure 2).

The next step is to model the vertical deformation of the seafloor as a result of the earthquake, using the method by OKADA (1985). We use seismicity and geodetic observations to construct a fault model based on our seismic inversion results. Aftershock seismicity (located by the U.S.G.S.) ranges in depth from the surface to about 15 km depth. For the dip determined in the inversion, this yields a down-dip width of the fault of 40 km. If we assume an aspect ratio of 2, typical of large earthquakes (KANAMORI and ANDERSON, 1975; GELLER, 1976), the length of each source is assumed to be 80 km. If the second source is separated from the first by 40 km, as suggested by the seismic inversion, the two faults would overlap over half their area. On the other hand, noting that the spatial separation is poorly resolved in the inversion, the horizontal distribution of seismicity is better matched if the second source is moved so that it is 80 km from the first. Thus, fixing the bottom center of the first source at the location of the hypocenter published in the PDE, the surface projection of the fault plane occurs at the location of the heavy line on the left side of Figure 5. The center of the top of the second source is located 80 km

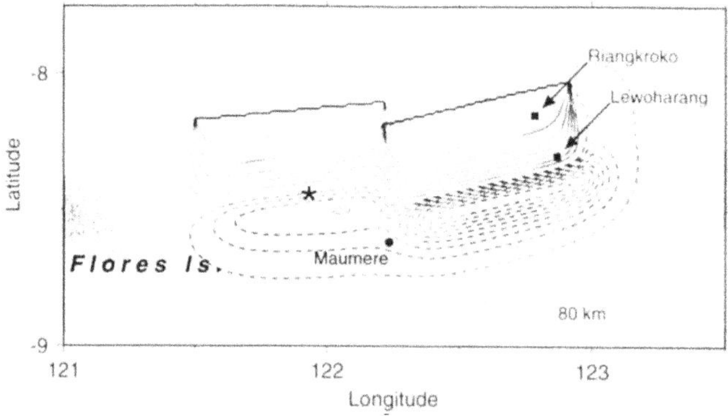

Figure 5

The vertical component of surface deformation estimated from the seismic inversion model. Contours of uplift (in 10 cm increments) are shown as solid curves, while contours of depression are shown as dashed curves. The surface projections of the two-fault planes may be seen where the uplift contours coalesce to form heavy lines. The epicenter (from the PDE) is shown by the asterisk. The city of Maumere is indicated, as well as the villages of Riangkroko and Leworahang.

from this point at an azimuth of 80° (the heavy line on the right). Since the strikes and dips determined from the inversion are slightly different, these two planes do not exactly meet at their edges.

Given these fault areas and the seismic moments determined in the inversion, and assuming the rigidity to be 3×10^{11} dyne/cm^2, the average slip on each fault plane should be 5.6 m. However, preliminary tsunami models computed with this amount of slip for the first source, generated tsunami heights that were too large on the central and western parts of the affected area of Flores Island. Therefore, in modeling tsunami generation, we presume the slip for the first source to be 3.2 m. This is the value assumed by YEH *et al.* (1993) in their initial model. An alternative would be to increase the area of the first fault. However, if the center of the fault is located at the hypocenter, this would result in significant tsunamis farther to the west than observed on Flores Island. Another alternative is to move both fault planes in a northerly or southerly direction, assuming that the hypocenter is either incorrect, or does not correspond to the bottom of the first fault plane. However, moving both planes either north or south by 5 minute increments (about 9 km) also generates tsunami amplitudes that are too large. The reason for this may be seen in the contours of vertical surface deformation in Figure 5. The shallow reverse mechanism generates uplift shown by the solid contours, and depression shown by dashed contours. The area of zero vertical deformation lies along much of the coast, so if the planes are shifted the amplitude of deformation (either positive or negative) increases. Finally, we note that geodetic measurements at Maumere indicated subsidence of 15 cm (GONZÁLEZ *et al.*, 1993), while the model indicates subsidence of slightly more than 20 cm, which is in reasonable agreement with that measured.

From the surface deformation and bathymetry models, tsunami propagation is computed using a finite-difference method. We use both a linear model (SATAKE and KANAMORI, 1990) and a nonlinear model (SATAKE, 1995). In the linear model, tsunami amplitude is proportional to the amount of slip on the fault. In the nonlinear model, the tsunami amplitude is controlled by nonlinear terms and bottom friction, in addition to the fault slip. It turns out that with a reasonable value of bottom friction, the results are very similar for this event; the maximum tsunami heights predicted by the nonlinear model are only slightly smaller than those predicted by the linear model. This is because the ocean depth is quite large around Flores Island.

Computations were made for a total duration of 6 hours, with a time increment of 5 sec. Snapshots of the linear tsunami model for the first six one-minute intervals are shown in Figure 6. Here we see that the seafloor uplift, due to the two faults, causes a large tsunami wave to be generated within the first minute. As the wave propagates into deeper water to the north, the amplitude decreases. However, in the shallow water along the coast of Fores Island, and particularly within Hading Bay, large amplitude tsunami waves continue for several minutes.

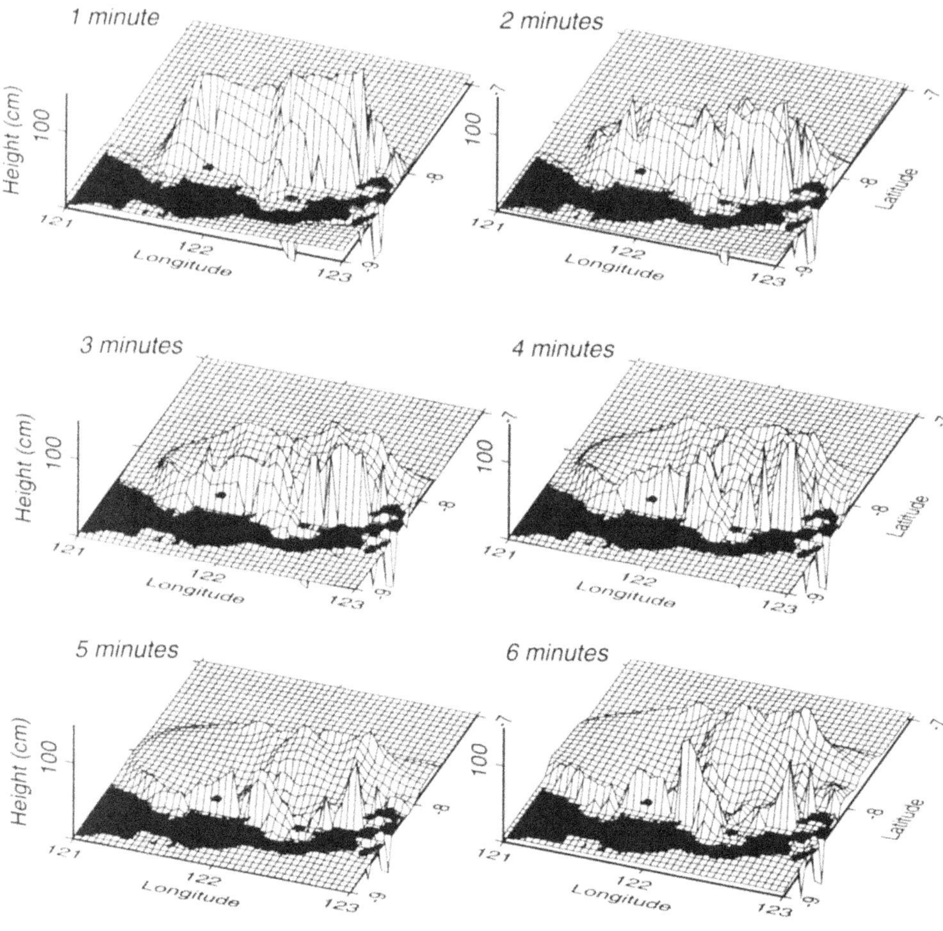

Figure 6

Snapshots of the first six minutes of tsunami propagation for the two-fault model. The large initial amplitude decreases rapidly as the tsunami moves into deeper water to the north, but remains large as shallow coastal areas are approached to the south. The wave meets Flores Island after three minutes, with large amplitudes (particularly near Maumere) continuing for minutes afterwards.

Tsunami waveforms were computed for the Palopo tide gauge station and for 37 sites on Flores Island and nearby small islands, where maximum tsunami runup measurements are available. At coastlines, total reflection is assumed in the computation (i.e., tsunami runup is not modeled directly). For these locations, we take the largest positive amplitude and multiply by two (the assumed tsunami runup amplification factor) in order to compare with the observed tsunami runup heights. A map comparing observed and computed tsunami runup heights is shown in

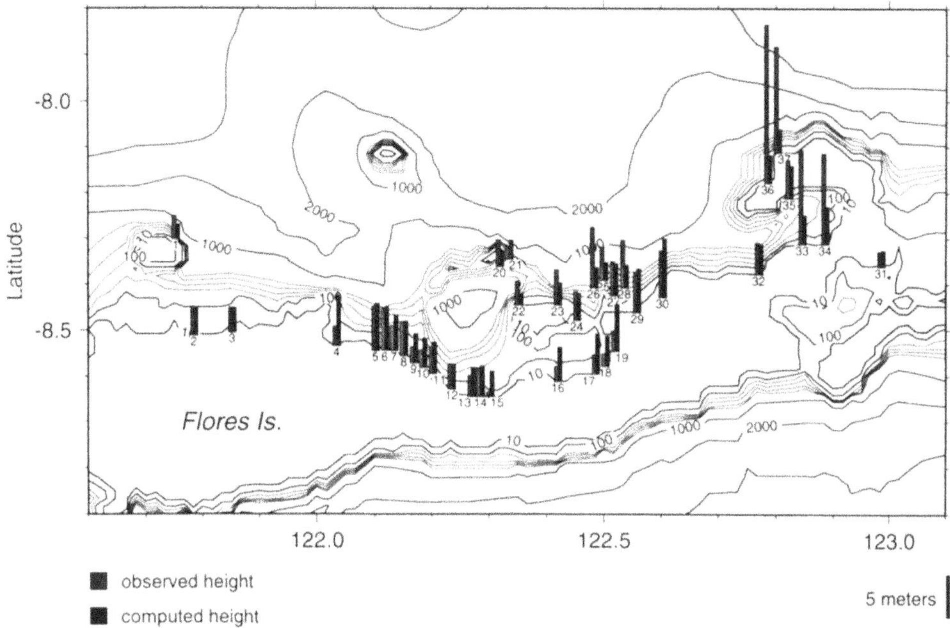

Figure 7

Bar graph showing observed (hatchured) and computed (solid) tsunami runup heights at various positions on and near Flores Island for the two-fault model. The location numbers and heights are listed in Table 4. A tsunami runup amplification factor of two is assumed for all locations. The model explains the observed tsunami runup heights well in the central and western portions of the affected area. However, it cannot explain the large observations near Leworahang (locations 33 and 34) and Riangkroko (locations 36 and 37).

Figure 7 (with locations and observed and predicted tsunami runup heights listed in Table 4). The results for the linear model, based on the two seismic sources, matches the tsunami runup heights on the central and western portion of the affected area of Flores Island quite well. However, this model substantially under-predicts the large tsunami runup heights measured at Riangkroko and Bunga on the northeastern peninsula (locations 36 and 37 in Figure 7), and at Waibalen and Uepadung, which are near Leworahang on the south edge of Hading Bay (locations 33 and 34).

A massive coastal slump was observed near Leworahang (YEH *et al.*, 1993). For purposes of modeling tsunami generation, we may model this as a very large vertical displacement over a very small area. The height of the scarp produced is about 7.5 m, and based on submerged coconut trees, the depth of the water immediately in front of the scarp is about 10 m (YEH *et al.*, 1993), so we assume a vertical displacement of 17.5 m located entirely within one grid point (about 1.2 km) at Leworahang. This unusual source will cause an unavoidable numerical dispersion in the finite-difference code. To investigate the effect of this error, we have also computed the tsunami generated by a vertical displacement of 10 m distributed over four grid points (approximately 5 km horizontal extent). The results are nearly identical, although the four-grid-point model may suffer from numerical dispersion as well. Physically, gravity waves from such a small source may not be modeled as long (shallow water) waves, but may be modeled as more dispersive deep water waves.

The largest observed tsunami runup, however, was at Riangkroko (with one measurement at 26 m!). A landslide with a horizontal extent of about 600 m was observed near this location (George Plafker, personal communication). Since the direction of fallen trees tend away from the landslide, it is reasonable to assume that large tsunami waves were generated by this landslide source. Another possible explanation is an anomalously large tsunami runup factor due to the topography of the area, which is influenced by a stream flowing into the sea. However, four different measurements gave an average tsunami runup height of 19.6 m at Riangkroko, while 6 km away at Bunga the tsunami runup height was 12.3 m. Lacking topographic data and appropriate programs for computing runup, we will presume a submarine slump by including a second single-grid-point slump source (also with a vertical dislocation of 17.5 m) located at Riangkroko. The results for this combined faulting and slumping model are plotted on the map in Figure 8 and

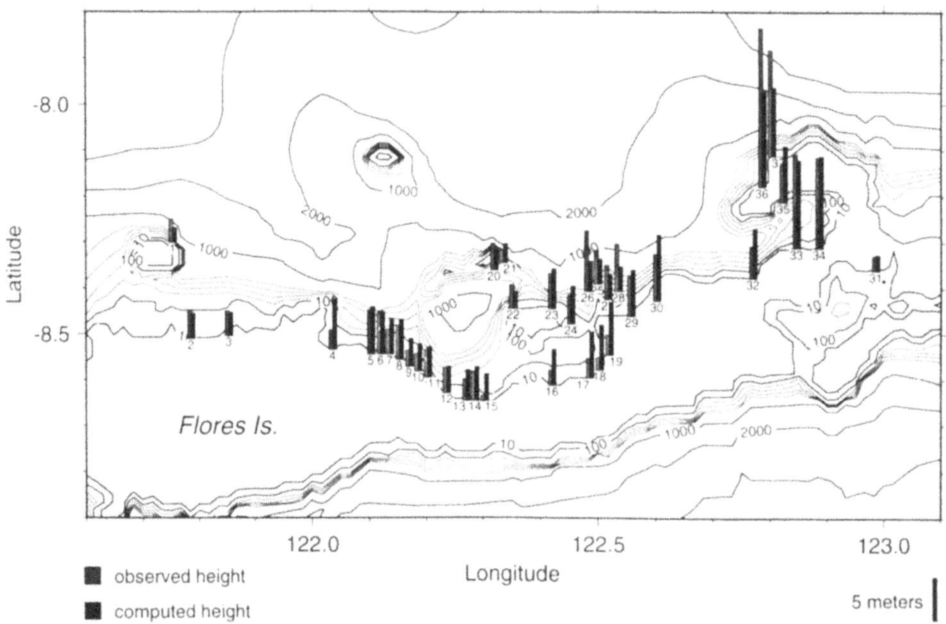

Figure 8
Bar graph showing observed (hatchured) and computed (solid) tsunami runup heights for the final model, which includes two faults and two slumps (one near Leworahang and one near Riangkroko). The heights are, once again, listed in Table 4. This model explains most of the observed tsunami runup heights, although the computed height at Riangkroko is still less than observed.

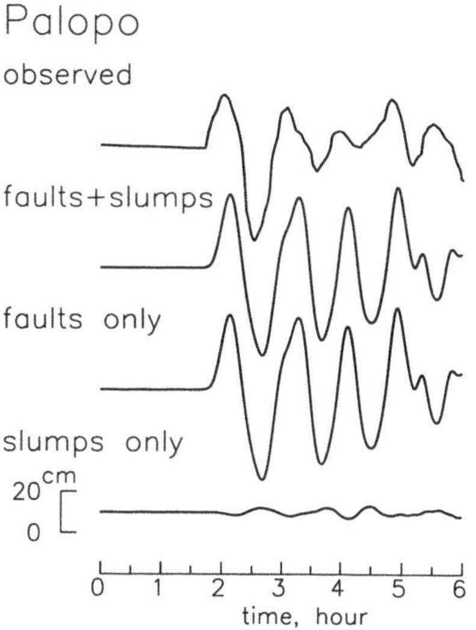

Figure 9

Observed and computed tide gauge records at Palopo. The observed waveform is shown on the top. The second trace is that computed for the final model. The third trace is that computed for the two-source model (without slumps). The bottom trace is that computed for a model consisting only of the two slumps. The slump sources have little influence on the computed waveform at Palopo. The two-fault model explains the observed waveform well, particularly in the first three half-cycles. The large later arrivals in the computed waveform limits the amplitude of slip on the second source.

listed in Table 4. The tsunami runup heights on the western and central portions of the affected area of Flores Island are, once again, well modeled. In fact, the slumping sources generate large tsunamis only in the immediate vicinity of the slump. The model explains the observed tsunami runup heights in Hading Bay (locations 33 and 34) very well, indicating that the coastal landslide near Leworahang was responsible for the large tsunamis experienced there. Our slumping source at Riangkroko also explains over half of the extremely large tsunami runup height observed there and at Bunga (locations 36 and 37). If a slumping source is responsible for the large amplitudes there, either a larger area of slumping or greater vertical displacement is required.

Finally, the most rigorous test of the tsunami models is the fit to the tide gauge record at Palopo. This is shown in Figure 9, with maximum and minimum tsunami wave amplitudes listed in Table 5. Due to the shallow coastal water around Palopo, we have chosen to display the tsunami waveforms computed with the nonlinear model, although the results are nearly identical for the linear model. Displayed at the top of the figure is the observed tide gauge record for the first six hours after the earthquake (digitized from GONZÁLEZ *et al.*, 1993). Below this, and plotted on the same time and amplitude scales, are the computed tsunami waveforms for the final model, for the initial 2-fault model, and for the two slumping sources alone.

Table 5

Observed and predicted tsunami amplitude at Palopo (latitude $-2.98°$, longitude $120.22°$)

	Observed	Linear model		Nonlinear model	
		2 faults	+2 slumps	2 faults	+2 slumps
Maximum (cm)	25.0	43.9	44.9	40.5	39.9
Minimum (cm)	−46.6	−46.8	−47.6	−44.2	−43.1

We see initially that the slumping sources have very little effect on the tsunami wave observed at such a large distance. The waveform for the final model is nearly identical to that for the two faulting sources. For the first three half-cycles, these fit the observed tide gauge record extremely well. If the slip were increased on either of the faults, whether in an attempt to explain the large tsunami runups on the eastern end of Flores Island, to reduce the assumed fault areas, or to be in better agreement with the seismic moment determined in the seismic inversion, the amplitude of the Palopo tsunami record could not be as well modeled. This is particularly true of the second seismic source which, perhaps due to slightly less complicated bathymetry along the path to Palopo (see Figure 2), is primarily responsible for the large-amplitude oscillations that continued for nearly four hours in the computed waveforms. Large amplitudes are observed at Palopo extending to six hours after the earthquake, but the amplitude of oscillation is not as large or consistent as in the computed waveforms. If the slip were increased on the second fault, these later oscillations would be even larger. With a more detailed bathymetric model around the Palopo tide gauge station, it may be possible to improve the later details of the computed tsunami waveforms. However, our final model, with two seismic sources and two slumping sources, provides a reasonable fit to the observed tide gauge record, given the bathymetric data available.

Conclusions

Inversion of teleseismic P and SH waves, as well as PP waves at ranges up to 123°, indicates that the Flores Island earthquake occurred with reverse motion on a shallow-dipping fault with two distinct asperities. The centroid depth of both sources is quite shallow (9 km), certainly a contributing factor to the tsunami generation potential of this earthquake. The second source occurred 28 sec after the first and is located to the east, although the spatial separation is poorly resolved. The inversion indicates a similar seismic moment for each source, with a moment magnitude for each of 7.8. The duration of each source is also similar (28–30 sec), therefore using the relationship with seismic moment and source duration (COHN *et al.*, 1982), the average stress drops for the first and second source are 49 and 37 bars, respectively.

Using a detailed bathymetry model and estimating surface deformation from the seismic inversion results, we obtained linear and nonlinear models of tsunami generation and propagation. Using as observed data a single tide gauge record at Palopo (650 km away) and maximum tsunami runup heights measured on and near Flores Island, we find that the slip of the first source must be reduced (to 3.2 m) in order to fit the tsunami runup heights on the western portion of the affected area of Flores Island. The slip of the second source cannot be substantially larger than 5.6 m, or the predicted tsunami amplitude at Palopo will be too large. The two-source seismic model cannot explain the large tsunami runup heights measured near Leworahang on Hading Bay, or near Riangkroko on the northeast peninsula of Flores Island. A large coastal slump was observed at Leworahang, and a landslide was observed at Riangkroko. When we model these as point sources of vertical displacement, the tsunami runup heights near these locations are well modeled. It is clear that the two-source seismic model can explain the tsunami height observed at Palopo and the tsunami runup heights observed on most of Flores Island. However, the large tsunami runups observed near Leworahang and Riangkroko require localized, large secondary sources such as coastal or submarine slumps.

Acknowledgements

We would like to thank Frank González, Yoshinobu Tsuji and Fumihiko Imamura for providing information and data prior to publication. In addition, we appreciate the comments of two anonymous reviewers. Seismic data were obtained from the IRIS Data Management Center, and the figures were generated using GMT. This research was supported in part by the U.S. Geological Survey (grant 1434–93–G–2277) at SUNY Binghamton and by the National Science Foundation (grant EAR–9117800) at U. Michigan.

REFERENCES

BARKER, J. S., and LANGSTON, C. A. (1981), *Inversion of Teleseismic Body Waves for the Moment Tensor of the 1978 Thessaloniki, Greece, Earthquake*, Bull. Seismol. Soc. Am. 71, 1423–1444.

BARKER, J. S., and LANGSTON, C. A. (1982), *Moment Tensor Inversion of Complex Earthquakes*, Geophys. J. Roy. Astr. Soc. 68, 777–803.

BARKER, J. S., and LANGSTON, C. A. (1983), *A Teleseismic Body-wave Analysis of the May 1980 Mammoth Lakes, California, Earthquakes*, Bull. Seismol. Soc. Am. 73, 419–434.

BECKERS, J., and LAY, T. (1995), *Very Broadband Seismic Analysis of the 1992 Flores, Indonesia Earthquake ($M_u = 7.9$): Observations on Back-arc Tectonic Setting and Tsunami Generation*, J. Geophys. Res., in press.

COHN, S. N., HONG, T.-L., and HELMBERGER, D. V. (1982), *The Oroville Earthquakes: A Study of Source Characteristics and Site Effects*, J. Geophys. Res. 87, 4585–4594.

GELLER, R. (1976), *Scaling Relations for Earthquake Source Parameters and Magnitudes*, Bull. Seismol. Soc. Am. 66, 1501–1523.

GONZÁLEZ, F., SUTISNA, S., HADI, P., BERNARD, E., and WINARSO, P. (1993), *Some Observations Related to the Flores Island Earthquake and Tsunami.* In *Proc., Intl. Tsunami Symposium, TSUNAMI '93*, 23–27 August 1993, Wakayama, Japan.

IMAMURA, F., and KIKUCHI, M. (1994), *Moment Release of the 1992 Flores Island Earthquake Inferred from Tsunami and Seismic Data*, Sci. Tsunami Hazards *12*, 67–76.

KANAMORI, H., and ANDERSON, D. L. (1975), *Theoretical Basis of Some Empirical Relations in Seismology*, Bull. Seismol. Soc. Am. *65*, 1073–1095.

KIKUCHI, M., and KANAMORI, H. (1991), *Inversion of Complex Body Waves—III*, Bull. Seismol. Soc. Am. *81*, 2335–2350.

LANGSTON, C. A. (1981), *Source Inversion of Seismic Waveforms: The Koyna, India, Earthquake of September 13, 1967*, Bull. Seismol. Soc. Am. *71*, 1–24.

LANGSTON, C. A., and HELMBERGER, D. V. (1975), *A Procedure for Modeling Shallow Dislocation Sources*, Geophys. J. Roy. Astr. Soc. *42*, 117–130.

OKADA, Y. (1985), *Surface Deformation due to Shear and Tensile Faults in a Halfspace*, Bull. Seismol. Soc. Am. *75*, 1135–1154.

PLAFKER, G., and WARD, S. N. (1992), *Backarc Thrust Faulting and Tectonic Uplift along the Caribbean Sea Coast during the April 22, 1991, Costa Rica Earthquake*, Tectonics *11*, 709–718.

SATAKE, K. (1995), *Linear and Nonlinear Computations of the 1992 Nicaragua Earthquake Tsunamis*, Pure and Appl. Geophys., this issue.

SATAKE, K., and KANAMORI, H. (1991), *Abnormal Tsunamis Caused by the June 13, 1984, Torishima, Japan, Earthquake*, J. Geophys. Res. *96*, 19,933–19,939.

SATAKE, K., and TANIOKA, Y. (1995), *Tsunami Generation of the 1993 Hokkaido Nansei-oki Earthquake*, Pure and Appl. Geophys., this issue.

TSUJI, Y., and MATSUTOMI, H. (1993), *Damages due to the tsunami.* In *The Report of the Field Survey of the Flores Island Earthquake-Tsunami of December 12, 1992* (ed. Tsuji, Y.) (in Japanese with English abstracts) pp. 70–87.

TSUJI, Y., MATSUTOMI, H., IMAMURA, F., TAKEO, M., KAWATA, Y., MATSUYAMA, M., TAKAHASHI, T., SUNARJO and HARJADI, P. (1995), *Damage of Coastal Villages due to the 1992 Flores Island Earthquake Tsunami*, Pure and Appl. Geophys., this issue.

YEH, H., IMAMURA, F., SYNOLAKIS, C., TSUJI, Y., LIU, P., and SHI, S. (1993), *The Flores Island Tsunamis*, EOS, Trans. Am. Geophys. Union *74*(33), 369, 371–373.

(Received October 3, 1994, revised March 7, 1995, accepted March 14, 1995)

PAGEOPH, Vol. 144, No. 3/4 (1995)

0033–4553/95/040555–14$1.50 + 0.20/0
© 1995 Birkhäuser Verlag, Basel

Numerical Simulation of the 1992 Flores Tsunami: Interpretation of Tsunami Phenomena in Northeastern Flores Island and Damage at Babi Island

FUMIHIKO IMAMURA,[1] EDISON GICA,[1] TOMOYUKI TAKAHASHI[2] and NOBUO SHUTO[2]

Abstract —Numerical analysis of the 1992 Flores Island, Indonesia earthquake tsunami is carried out with the composite fault model consisting of two different slip values. Computed results show good agreement with the measured runup heights in the northeastern part of Flores Island, except for those in the southern shore of Hading Bay and at Riangkroko. The landslides in the southern part of Hading Bay could generate local tsunamis of more than 10 m. The circular-arc slip model proposed in this study for wave generation due to landslides shows better results than the subsidence model, It is, however, difficult to reproduce the tsunami runup height of 26.2 m at Riangkroko, which was extraordinarily high compared to other places. The wave propagation process on a sea bottom with a steep slope, as well as landslides, may be the cause of the amplification of tsunami at Riangkroko. The simulation model demonstrates that the reflected wave along the northeastern shore of Flores Island, accompanying a high hydraulic pressure, could be the main cause of severe damage in the southern coast of Babi Island.

Key words: The 1992 Flores earthquake tsunami, numerical simulation, landslide, hydraulic pressure, tsunami damage.

1. Introduction

On 12 December 1992, an earthquake of 7.5 M_s and an accompanying destructive tsunami struck the northeastern coast of Flores Island (Figure 1). There were 1,713 casualties and 2,126 people were injured. Half of these were due to tsunami (TSUJI *et al.*, 1993). An extremely large tsunami runup height of 26.2 m was measured at Riangkroko, in the northeastern peninsula of Flores Island. There are only two other examples in this century—the 1993 Sanriku earthquake tsunami and the 1993 Hokkaido Nansei-Oki earthquake tsunami—in which more than 20 m tsunami runup heights were observed.

---

[1] School of Civil Engineering, Asian Institute of Technology, G.P.O. Box 2754, Bangkok 10501, Thailand.
[2] Disaster Control Research Center, Tohoku University, Aoba, Sendai 980-77, Japan.

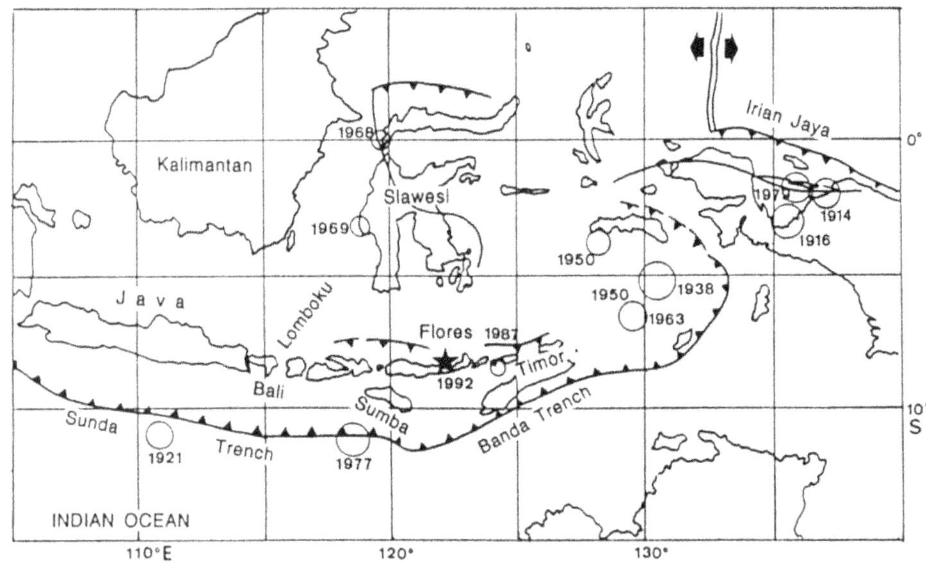

Figure 1

Map of the tectonics and epicenter of the earthquakes in eastern Indonesia. Flores Island is located in the back arc of eastern Sunda and western Banda thrusts.

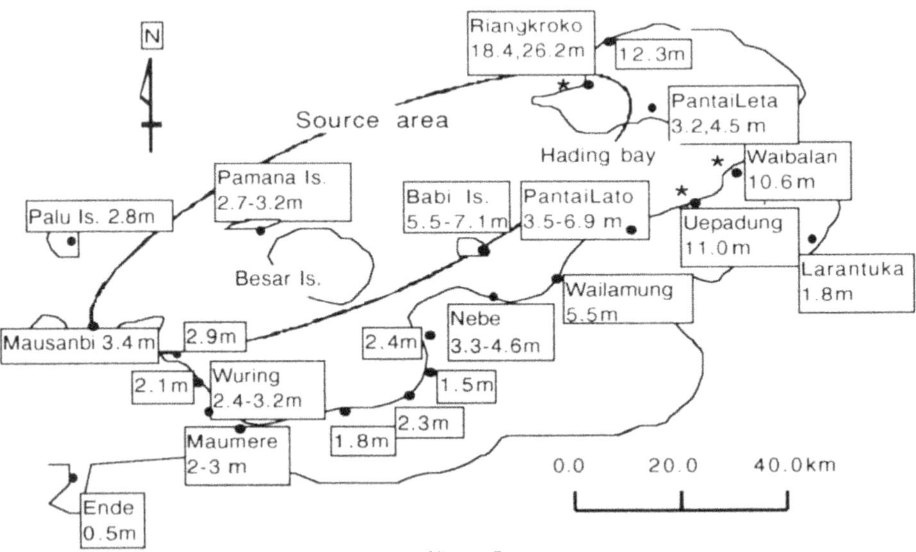

Figure 2

Measured tsunami runup heights in meters on eastern Flores Island and the tsunami source (TSUJI *et al.*, 1993). A general trend of increasing runup height from west to east in this region is significant. The asterisks indicate the location of the landslide in Hading Bay.

An International Tsunami Survey Team (ITST) consisting of engineers and scientists from Indonesia, Japan, and the U.K., Korea and the U.S. conducted a field survey along the northeastern coast of Flores Island and several smaller offshore islands. Tsunami runup heights were measured as shown in Figure 2 and problems relative to the damage and the generation mechanism were pointed out. In this paper two problems associated with the 1992 Flores tsunami are investigated, because of their important role in preventing tsunami disasters in the future. The first problem concerns the mechanism which generated a giant tsunami with runup heights at Riangkroko, Uepadung, and Waibalan in Hading Bay of 10.6–26.2 m, which are much larger than those on the western part of Flores Island (Figure 2). The second point of concern is to investigate an underlying cause of the damage at Babi Island, where two villages located along the southern coast of Babi Island were completely destroyed due to a large wave force, even though the source lay to the north.

2. Tsunami Source Model

Our initial tsunami model uses one fault model based upon the quick Harvard CMT solution [event file M121292Y]. Taking into account the tectonics in this region (thrust in a back arc, HAMILTON, 1988), a shallow dip thrust fault as the source with (strike, dip, slip) = (61°, 32°, 64°) shown in Table 1, is selected from two mechanism solutions. According to the distribution of the aftershocks determined by the United State Geological Survey (USGS), the fault plane is approximately 100 km long and 50 km wide. An average dislocation of 3.2 m can be determined by using a rigidity of 4.0×10^{11} dyne/cm^2.

A numerical simulation of tsunami generation and propagation, the TUNAMI-N1 code (SHUTO et al., 1990; IMAMURA et al., 1993a), was carried out using one fault model. The lack of detailed topographical data in shallow regions and on the land made it very difficult to carry out runup simulations. The computed maximum water level along the coastline modeled as a vertical wall is associated with the computed runup height. Both are correlated as long as the slope of the shoreline is steeper than around 0.01 in this case. Therefore, in this model, the

Table 1

Fault parameter of the 1992 *Flores earthquake*

	M_0 ($\times 10^{27}$ dyn-cm)	Depth (km)	Strike (deg)	Dip (deg)	Slip (deg)	Length (km)	Width (km)	Dislocation (m)
East fault	4.8	3	61	32	64	50	25	9.6
West fault	1.6	3	61	32	64	50	25	3.2

Table 2

Computational condition

Governing equation	Shallow water theory (nonlinear theory)
Spatial grid size	300 m
Time step	1 sec
B.C. of coastal line	Perfect reflection condition (vertical wall)
Reproduction time	1 hour

computed maximum levels along the coastline are directly compared with the measured runup heights. The computational condition is summarized in Table 2. In Figure 2, the computed results are smaller than the measured runup heights in the eastern part of Flores Island, suggesting that the slip value on the eastern side of the fault might have been larger than that on the western side. Aftei several trials, a composite fault model with different siip values; 3.2 m in the west and 9.6 m in the east, was proposed (IMAMURA and KIKUCHI, 1994). In this model, the seismic moment of 6.4×10^{27} dyne-cm is kept and both segments have the same fault area of 50 km long and 25 km wide (Figure 3). The crustal movements estimated by the composite fault model with a depth of 3 km coincide with the crustal deformation measured by TSUJI *et al.* (1993). The computed maximum water levels using this model are compared to the measured runup heights in Figure 4. The numerical model with the composite fault model reproduces the distribution of the measured tsunami runup heights along the northeastern coastline of Flores Island, except for the southern shore of Hading Bay and Riangkroko.

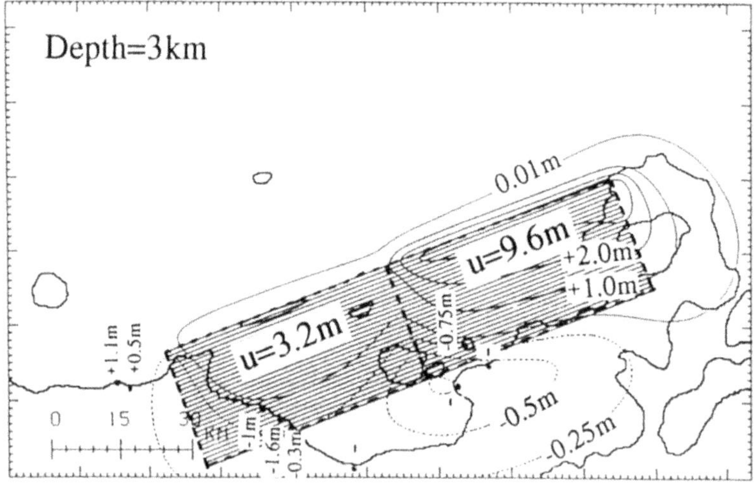

Figure 3

Composite fault model (IMAMURA and KIKUCHI, 1994) and its vertical displacement of ground. Moment release in the eastern part is larger than that in the western part.

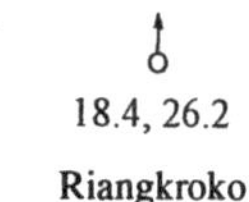
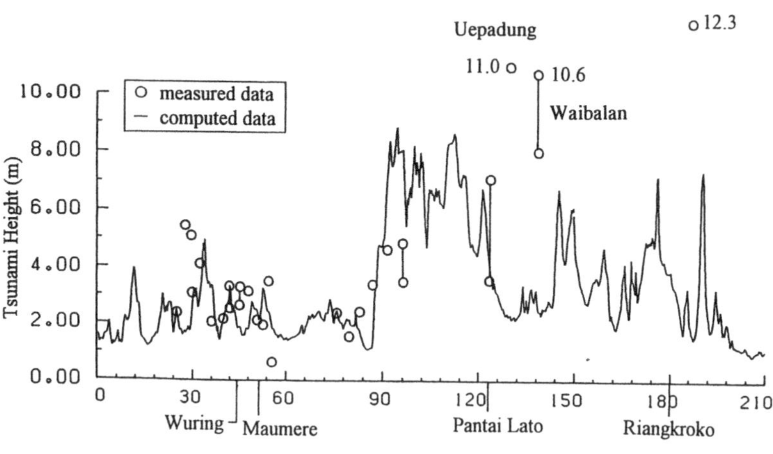

Figure 4

Comparison between the measured runup heights and the computed maximum water level along the northeastern coast of Flores Island.

3. Tsunami in Hading Bay and Riangkroko

Landslide in Hading Bay

As shown in Figure 4, the measured runup heights of 10.6 m at Waibalan and 7.6–11.0 m at Uepadung on the southern shore of Hading Bay, were considerably higher than elsewhere in this bay such as Pantai Lato and Pantai Leta. It is possible that abnormal runup heights were generated by other geological agents. ITST noticed large landslides along the southern shore of Hading Bay (YEH *et al.*, 1993), and reported that the landslide shear planes are almost vertical. Their planes form a steep and vertical cliff along the present shoreline approximately 150 m wide and 2 km long (Photo 1). The observed large runup heights are limited to the area of subaqueous slumps, suggesting the possibility that they were caused by landslide-generated waves rather than tectonic tsunami waves. A model which includes wave generation due to landslides may be required to reproduce the measured tsunami heights in the Flores tsunami event.

Photo 1

Landslide in Waibalan taken by the second field survey team in 1994. The cliff is 150 m wide and 2 km long. The average height of the cliff is around 10 m.

Wave Generation Model due to Landslides

There are several models for wave generation due to landslides. They range from the inflow volume method and unit-width discharge method (MING and WANG, 1993), to two-dimensional boxes model dropped vertically at the end of a semi-infinite channel (NODA, 1970). None of these methods can be applied directly to the present case, because higher tsunami heights were observed, not on the opposite side of the landslide area, but along the landslide and its surrounding area. In addition, the observed runup was higher than the level of the top of the landslide. In the present study, the subsidence model and the circular-arc slip model (see Figure 5) are proposed, since the cavity formed by subsidence could generate a wave propagating toward the coast (GICA, 1994). For wave generation due to landslides, modeling with runup condition is used. The averaged depth of 20 m and land height of 8 m are estimated from the field survey, while the radius of circular-arc slip of 30 m, and properties of the soil are assumed because no data pertinent to them are available. Assuming the same maximum vertical displacement in both models, the computed results demonstrate that the wave runup height (3.9 m) of the subsidence model is half that of the circular-arc slip model (8.2 m).

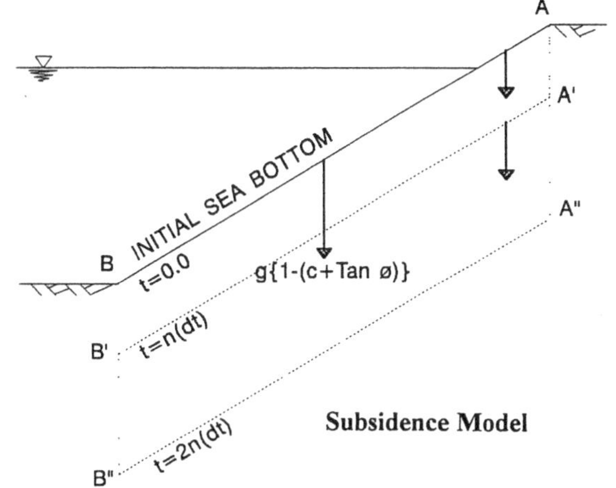

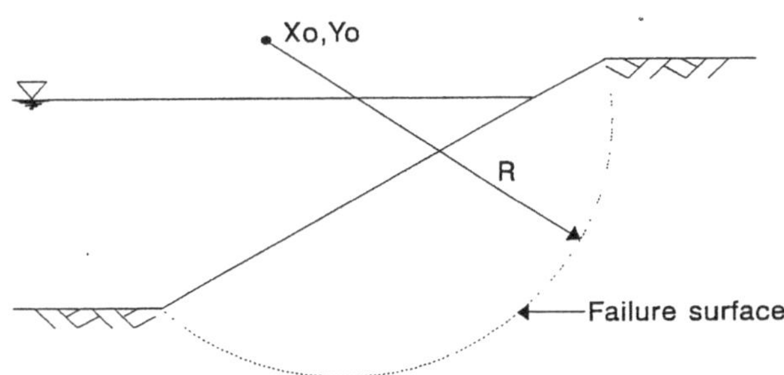

Figure 5

Subsidence (upper) and circular-arc slip (lower) models for the wave generation due to landslides. The deformation of sea bottom is calculated at every step based upon two models. c: soil cohesion, ϕ; soil friction angle, g; gravitational acceleration.

This is the reason that a deformation of the sea bottom near the toe of the landslide mass in the circular-arc slip, increasing the disturbances of the water surface, causes a wave propagation with a larger amplitude toward the coastal line. The maximum runup height of the circular-arc slip model is observed as identical to the ground level. The above results suggest that the circular-arc slip model more suitably reproduces the tsunami of more than 10 m observed in Flores. Unfortunately, the lack of field data such as landslide scale under the sea and soil properties does not allow us to compare the computed values with those measured in Hading Bay.

Riangkroko

An extremely large tsunami runup was measured in the small village of Riangkroko (Figure 6), where 137 people from a population of 406 prior to the

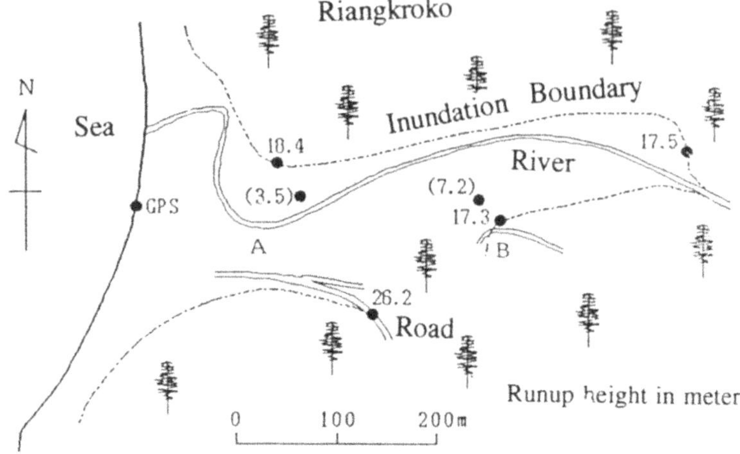

Figure 6

Inundation area and tsunami runup at Riangkroko. A maximum tsunami height of 26.2 m was measured. All houses were completely washed away by the tsunami. Numerals in parenthesis are the inundated tsunami heights.

event lost their lives. The village is located at the mouth of a small river, the Nipar River, with its northwestern side facing the Flores Sea. A remarkably steep sea bottom existed offshore of the village (Figure 7), similar to the case of Okushiri Island in the Hokkaido Nansei-Oki earthquake tsunami of 1993. The maximum

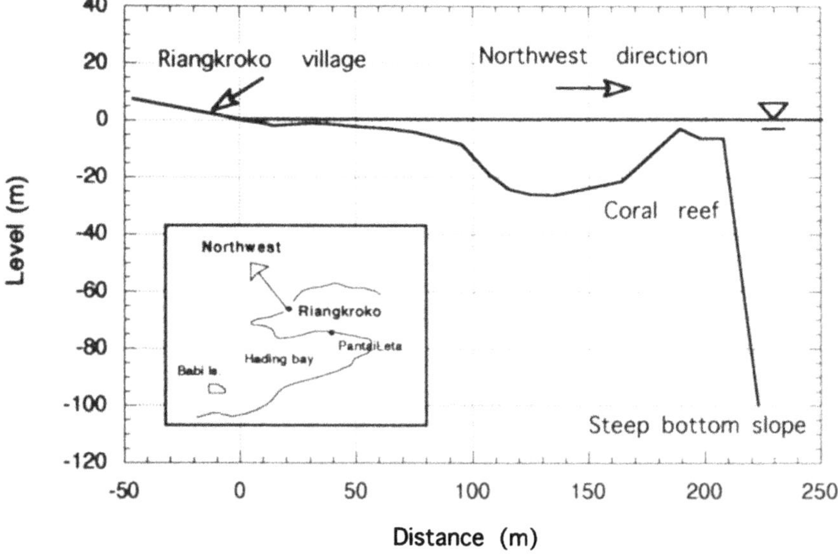

Figure 7

Topography of sea bottom offshore Riangkroko. A steep bottom slope after passing the coral reef is significant, suggesting large amplification on the slope and energy trapping on the coral reef in the shallow region.

Photo 2
Huge landslide nearby Riangkroko taken by the second field survey team in 1994. Its height was estimated to be more than 30 m.

runup height measured at Riangkroko was 26 m and the average of heights based on four different tsunami marks was 19.6 m. This indicates that the magnitude of the tsunami runup is probably not an isolated local phenomenon like wave splash-up.

Figure 4 shows the comparison of computed and measured tsunami runup heights, indicating that the computed results are much smaller than those measured at Riangkroko. The present model with the fault motion only, could not simulate the observed data. This discrepancy is not caused by the runup condition assumed in this model, since the ratio of the measured to the computed values exceeds 5, which is quite large. A propagation process on the steeply sloping sea bottom might cause the significant amplification of a wave, or another geophysical phenomena such as a landslide might have generated a big wave. A second survey conducted in December 1994 found a huge landslide of more than 30 m in height along the coast, about 1 km from Riangkroko (Photo 2), whose location is shown by an asterisk in Figure 2. The effect of a landslide here may also be signficant. For further analysis, a field investigation in this area and a numerical analysis with more detailed offshore topography and bathymetry are required.

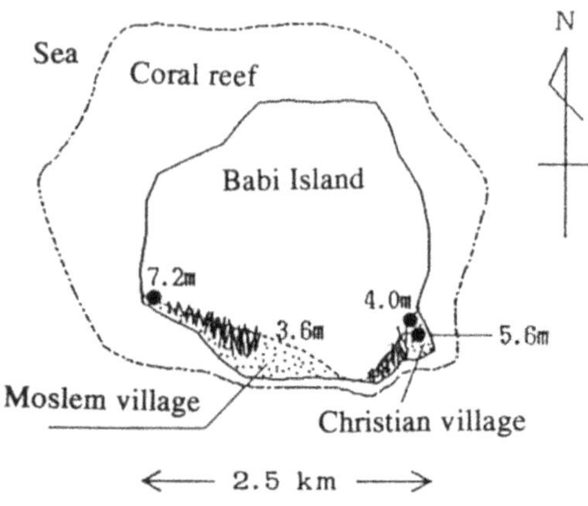

Figure 8
Map of Babi Island. There are two villages on the southern coast of this island. The observed tsunami
heights are not extremely high and yet damage to the two villages was considerable.

4. Damage in Babi Island

Babi Island

Babi Island (Figure 8) sustained significant damage with 263 casualties out of a
population of 1,100. All houses were completely destroyed. Maximum tsunami
runup heights were 5.6 m in the Christian village and 3.6 m in the Moslem village.
A maximum height of 7.2 m on Babi Island was measured on a cliff in the western
side. Although the runup heights were not as high as those at other damaged places,
the damage due to tsunamis was quite severe.

Babi Island is about 5 km offshore of Flores Island. It has a conical shape with
a 351 m elevation at the summit and is approximately 2 km in diameter. The
northern shore faces the Flores Sea and possesses a wide coral reef, while the
southern shore, where the two villages were located, has a much narrower reef.
Even when strong wind waves and swells of the Flores Sea attack from the north,
wave conditions on the southern side are usually calm, because most incoming wave
energy is dissipated on the wide coral reef on the northern side (YEH *et al.*, 1993).
However, in this event the southern part of the island sustained severe damage due
to tsunami currents rather than waves.

Reflected Waves and Hydraulic Pressure

In order to explain why the southern coast of Babi Island suffered severe
damage due to the tsunami, a numerical simulation with the same condition as
shown in Table 2 was carried out, focusing specifically on Babi Island. Some of the

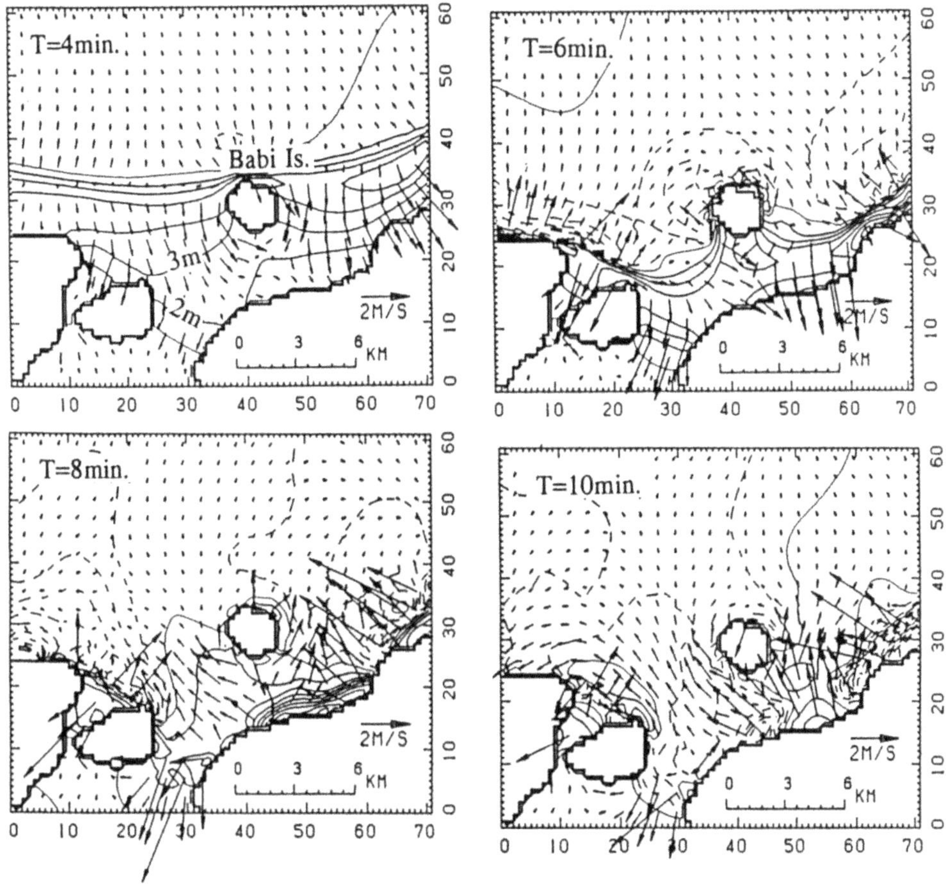

Figure 9

Water elevation contours with intervals of 1 m and current vectors around Babi Island. The first wave was reflected along the northern shore of Flores Island 8 minutes after the earthquake and attacked the southern coast of Babi Island.

numerical results are shown in Figures 9 and 10. Figure 9 shows the water elevation contours and the current vector distributions around this island from 4 minutes to 10 minutes after tsunami generation. Figure 10 shows the time histories of water level, velocity, and hydraulic pressure in front of the Moslem village. The hydraulic pressure is defined as follows (AIDA, 1977);

$$[\text{Hydraulic pressure } (m^3/s^2)] = [\text{Tsunami inundated height } (m)]$$

$$\times [\text{Current velocity } (m/s)]^2. \tag{1}$$

HATORI's (1984) empirical result of the relationship between damage to houses and several hydraulic properties, demonstrated that hydraulic pressure is the most suitable parameter to estimate the degree of house damage. Figures 9 and 10

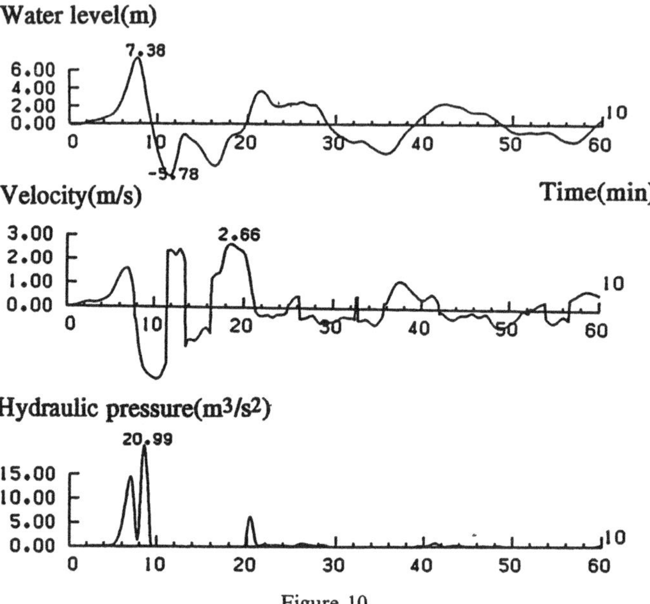

Figure 10

Time histories of water level, velocity and hydraulic pressure. The positive values of the velocity component indicate that the tsunami propagated from the northern side to the southern.

indicate that the first wave attacked the island from the north without high hydraulic pressure. The same wave was reflected off the coast of Flores Island and again attacked Babi Island on the southern part, accompanied by high hydraulic pressure, which is consistent with the eyewitnesses' account. In addition, the island is located at the nodal point of the standing wave, suggesting that the wave height is small whereas the current is large.

Measured and Computed Hydraulic Pressure

We could estimate hydraulic pressure and velocity at Wuring near Maumere by measuring the tsunami traces on the wall of the Mosque (MATSUTOMI, 1993; IMAMURA *et al.*, 1993b). Here most of the wooden houses were destroyed due to the tsunami. Applying the Bernoulli equation, by assuming energy conservation between the front and back of the Mosque, we computed the velocity to be 2.7–3.6 m/s and the hydraulic pressure to be 6.2–15.2 m^3/s^2. This estimation agrees with HATORI's (1984) criterion that hydraulic pressure over 5–9 m^3/s^2 corresponds to damage of over 50%. The present simulation yields a velocity of 2.38 m/s and a hydraulic pressure of 7.70 m^3/s^2, which agrees well with the measured data, and supports the accuracy and reliability of this simulation model. From the numerical model, it is determined that the locations with hydraulic pressure beyond 10 m^3/s^2 are Babi Island, Waibalan, Pantai Lato and Riangkroko, where severe damage was reported. Surprisingly, the hydraulic pressure at Riangkroko is over 30 m^3/s^2.

5. Conclusions

The landslide found on the southern shore of Hading Bay could generate local waves which are substantially higher than those obtained by the tsunami simulation from the fault model only. A new model for wave generation due to landslide, the circular-arc slip model, is proposed. This model reproduced higher runup heights along the coastline than the subsidence model. The tsunami runup height at Riangkroko was extraordinarily high compared to other locations. The wave propagation process on a steep slope of sea bottom as well as other geological agents, such as submarine landslide, could be related to this data. The numerical model reveals that the reflected wave along the northeastern shore of Flores Island is the main cause of severe damage in southern Babi Island. The computed hydraulic pressure and velocity, which correspond well to the measured data at Wuring, are useful in estimating the area of damage due to the tsunami.

Acknowledgements

We express our thanks to the Meteorological and Geophysical Agency of Indonesia (BMG) for their support of our survey. The study was supported by a Grant-in-Aid No. 04306024 for Cooperative Research (A) from the Ministry of Education, Science and Culture, Japan.

REFERENCES

AIDA, I. (1977), *Numerical Experiment for the Tsunami Inundation — in the Case of Susaki and Usa in Kochi Prefecture*, Bull. Earthq. Res. Inst. Univ. Tokyo *52*, 441–460 (in Japanese).

GICA, E. (1994), *A Study on the 1992 Flores Indonesia Earthquake Tsunami; Numerical Model on the Wave Generation due to Landslide*, Master Thesis, Asian Institute of Technology, 67 pp.

GONZÁLEZ, F., SUTISNA, S., HADI, P., BERNARD, E., and WINNARSO, P. (1993), *Some Observations Related to the Flores Island Earthquake and Tsunami*, Proc. Int. Tsunami Symp. in Wakayama, 789–801.

HAMILTON, W. B. (1988), *Plate Tectonics and Island Arcs*, Geological Soc. Am. Bull. *100*, 1503–1527.

HATORI, T. (1984), *On the Damage to House due to Tsunamis*, Bull. Earthq. Res. Inst. Univ. Tokyo *59*, 422–439.

IMAMURA, F., SHUTO N., IDE, S., YOSHIDA, Y., and ABE, Ka. (1993a), *Estimate of the Tsunami Source of the 1992 Nicaraguan Earthquake from Tsunami Data*, Geophys. Res. Lett. *20*, 1515–1518.

IMAMURA, F., MATSUTOMI, H., TSUJI, Y., MATSUYAMA, M., KAWATA, Y., and TAKAHASHI, T. (1993b), *Field Survey of the 1992 Indonesia Flores Tsunami and its Analysis*, Proc. of Coastal Eng. in Japan *40*, 181–185 (in Japanese).

IMAMURA, F., and KIKUCHI, M. (1994), *Moment Release of the 1992 Flores Island Earthquake Inferred from Tsunami and Teleseismic Data*, Sci. Tsunami Hazards *12*, 67–76.

MATSUTOMI, H. (1993), *Tsunami and Damage in the Northeast Part of Flores Island*, Kaiyo Monthly *25*, 756–761. (in Japanese).

MING, D., and WANG, D. (1993), *Studies on Waves Generated by Landslide*, Proc. XXV Congress of IAHR, Tokyo, Tech. Session C, 1–8.

NODA, E. K. (1970), *Water Waves Generated by Landslide*, J. Waterways, Harbors and Coastal Eng. Div., ASCE *96*, 835–855.

SHUTO, N., GOTO, C., and IMAMURA, F. (1990), *Numerical Simulation as a Means of Warning for Near-field Tsunami*, Coastal Eng. in Japan, *33*, 2, 173–193.

TSUJI, Y., IMAMURA, F., KAWATA, Y., MATSUTOMI, H., TAKEO, M., HAKUNO, M., SHIBUYA, J., MATSUYAMA, M., and TAKAHASI, T. (1993), *The 1992 Indonesia Flores Earthquake Tsunami*, Kaiyo Monthly *25*, 735–744 (in Japanese).

YEH, H., IMAMURA, F., SYNOLAKIS, C., TSUJI, Y., LIU, P., and SHI, S. (1993), *The Flores Island Tsunamis*, EOS, Trans. Am. Geophys. Union *74* (33), 371–373.

(Received August 18, 1994, revised March 23, 1995, accepted April 8, 1995)

PAGEOPH, Vol. 144, Nos. 3/4 (1995)

Laboratory Experiments of Tsunami Runup on a Circular Island

MICHAEL J. BRIGGS,[1] COSTAS E. SYNOLAKIS,[2] GORDON S. HARKINS,[1]
and DEBRA R. GREEN[1]

Abstract—Laboratory experiments of a 7.2-m-diameter conical island were conducted to study three-dimensional tsunami runup. The 62.5-cm tall island had 1 on 4 side slopes and was positioned in the center of a 30-m-wide by 25-m-long flat-bottom basin. Solitary waves with height-to-depth ratios ranging from 0.05 to 0.20 and "source" lengths ranging from 0.30 to 7.14 island diameters were tested in water depths of 32 and 42 cm. Twenty-seven capacitance wave gages were used to measure surface wave elevations at incident and four radial transects on the island slope. Maximum vertical runup measurements were made at 20 locations around the perimeter of the island using rod and transit. A new runup gage was located on the back or lee side of the island to record runup time series.

Key words: Tsunamis, tsunami runup, laboratory experiments, physical models, three-dimensional models, tsunami simulation, solitary waves, wavemakers, tsunami evolution, instrumentation.

1. Introduction

Recently, tsunamis in Indonesia and Japan caused millions of dollars in damages and killed thousands of people. On December 12, 1992, a 7.5-magnitude earthquake off Flores Island, Indonesia, killed nearly 2,500 people and washed away entire villages (YEH *et al.*, 1993; 1994). Field surveys found an average runup height near Riangkrok of 19.6 m, with a maximum height of 26 m. Reflection off Flores Island may have been partially responsible for the catastrophe at Babi Island, where 750 people were killed due to tsunami waves running up to 7.3 m above SWL. On July 12, 1993, a magnitude 7.8 earthquake off Okushiri Island, Japan, triggered a devastating tsunami with recorded runup measurements as high as 30 m. This tsunami resulted in larger property damage than any 1992 tsunamis, and it completely inundated an entire village with overland flow. Property damage was $600 million.

When a tsunami approaches an island from deep water, it undergoes refraction, diffraction, breaking, and wave trapping. The tsunami increases in height and

---

[1] USAE Waterways Experiment Station, Coastal Engineering Research Center, Vicksburg, MS 39180-6199, U.S.A.

[2] Department of Civil and Aerospace Engineering, University of Southern California, Los Angeles, CA, U.S.A.

steepness with complicated currents and multiple wave trains. Edge waves may even develop depending on beach slope and bathymetry, coastline irregularity, and incident wave direction. Reflections from adjacent shorelines may affect the number of tsunami waves and their amplitudes around the perimeter of the island.

Several numerical models have been developed to solve the linear, mild-slope equations for regular periodic waves approaching a circular island (SMITH and SPRINKS, 1975; JONSSON and SKOVGAARD, 1979). However, none of these models calculate wave runup, which is the most devastating hazard associated with tsunamis. The only available experimental data on wave runup on a circular island were obtained by PROVIS (1975). His conical island had a diameter of 3 m and a slope of 1 on 10. It was positioned in a 5.55-m-wide and 5.80-m-long wave basin. The water depth was only 15 cm, which caused the data to be dominated by laboratory scale effects (SPRINKS and SMITH, 1983). They found that viscous damping and standing waves between the wavemaker and the island contaminated the experimental results.

Field surveys of tsunami damage on both Babi and Okushiri Islands showed unexpectedly large runup heights, especially on the back or lee side of the island. Interestingly, numerical simulations by different international teams of the wave runup for both tsunamis produced results which differed substantially from the field measurements, often by factors of ten. Recognizing the need for a better understanding of the important physical parameters involved in three-dimensional tsunami runup, the National Science Foundation funded a three-year study beginning in 1992. This joint research study includes participants from Cornell University, Harvard University, University of Washington, University of Southern California (USC), and the U.S. Army Engineer Waterways Experiment Station (WES).

One of the goals of this project is to develop large-scale experimental databases for verification and modification of numerical models. Previous laboratory studies focused on tsunami wave runup on a plane 1 on 30 beach (BRIGGS *et al.*, 1993). A conical island was selected for study because of its mathematical simplicity and realistic geometry to actual islands (i.e. Babi Island, Okushiri Island, Hawaiian Islands). These large-scale experiments were conducted at WES during 1993 and 1994.

LIU *et al.* (1994) obtained very good agreement between this experimental data and their nonlinear, shallow-water model with bottom friction for the free surface displacements and maximum runup heights. The numerical model uses a staggered explicit finite difference leapfrog scheme to solve the governing equations. The numerical model was then used to examine several other important processes including velocity field, wave trapping around the island, and beach slope effects on runup heights.

In this paper, results from the wave height evolution and runup measurements on the island are presented. First, the experimental design including the physical model, wavemaker, and instrumentation are described. Next, a description of the tsunami wave simulation using solitary waves is given. Thereafter, amplitude evolution is briefly described. Finally, results from the measurements of maximum runup around the perimeter of the island and runup histories are presented and discussed.

2. Laboratory Experiments

2.1 Physical Model

A physical model of a conical island was constructed in the center of a 30-m-wide by 25-m-long flat-bottom basin at WES (Figure 1). The island had the shape of a truncated, right circular cone with diameters of 7.2 m at the toe and 2.2 m at the crest. The vertical height of the island was approximately 62.5 cm, with 1 V on 4 H beach face (i.e. $\beta = 14°$). The surface of the island and basin were constructed with smooth concrete. Tests were conducted at two water depths, 32 and 42 cm, to vary the relative waterline diameter of the island.

The X axis (X) of the right-hand, global coordinate system was perpendicular to the wavemaker and the Y axis (Y) was parallel to the wavemaker. The origin was

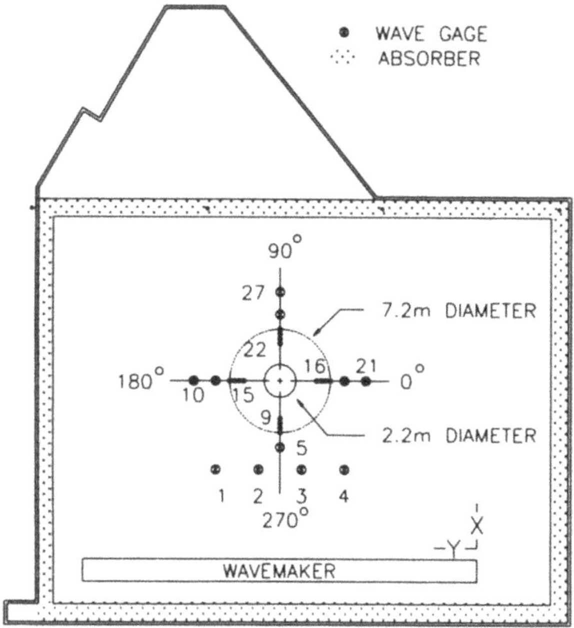

Figure 1
Schematic of island and wave gages.

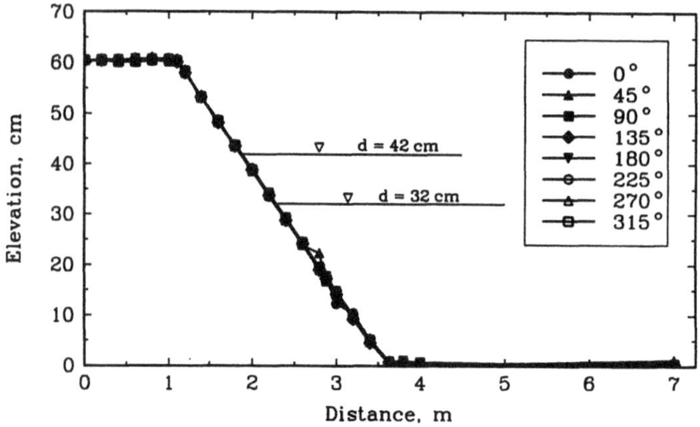

Figure 2
Circular island bathymetry for eight radial transects.

located at the end of the wavemaker, in line with the front surface of all paddles at their rest position. The center of the island was located at $X = 12.96$ m and $Y = 13.80$ m. A local coordinate system (x, y) was located at the center of the island. Angles increase counterclockwise (polar convention) from the x axis (x) pointing in the 0° direction (see Figure 1).

Bathymetric surveys of the island at eight radial transects showed it to be very uniform (Figure 2). The largest variation was a slight "bump" on the 45° transect at an elevation of approximately 22 cm. A bathymetric survey of the basin revealed a maximum elevation of 2.8 cm, a minimum of -1.4 cm, and a standard deviation of 7.3 mm. The basin sides and rear were lined with wave absorber to minimize wave reflections and cross-basin seiche. The irregular shape of the rear wall minimized reflections into the study area.

2.2 Wavemaker

A directional spectral wave generator (DSWG), designed and built by MTS Systems Corporation, was used to generate tsunami waves. Figure 3 shows a wave shoaling and refracting around the island with the DSWG in the background. The electronically controlled DSWG is 27.4-m-long and consists of 60 paddles, 46 cm wide and 76 cm high. The paddles are grouped in four modules of 15 paddles. Each of the 61 paddle joints is independently driven in piston mode by a 3/4-HP closed-loop dc servometer. The paddles are connected in a continuous chain with flexible polyethylene seals to produce smooth wave forms using the "snake principle" without spurious waves from end effects. Maximum stroke of the DSWG is 30.5 cm.

Figure 3
Overhead photograph showing runup around island.

Digital and analog circuits comprising the DSWG control console were located in a nearby climate-controlled room. This MTS console supplies digital wave-board control signals for input to 61 Preston digital-to-analog (D/A) signal converters. Minicomputers (a) perform D/A conversion for the 61 paddles at run time, (b) monitor paddle displacement and feedback, (c) calibrate wave gages, (d) digitize data, (e) update the control signals, and (f) analyze collected data (BRIGGS and HAMPTON, 1987).

2.3 Instrumentation

Twenty-seven capacitance wave gages were used to measure surface wave elevations (see Figure 1). The first four gages were located parallel to the wavemaker to measure incident wave conditions. Prior to each run, these gages were moved seaward from the toe of the island a distance equivalent to half-a-wavelength (i.e. L/2) of the wave to be generated. This procedure insured that the tsunami wave was always measured at the same relative stage of evolution.

A circular measurement grid of six concentric circles covers the island to a distance 2.5 m beyond the toe. Measurement points were located at the intersection of these concentric circles and the 90° radial lines. The spacing between the grid points was a function of the water depth. The shallowest gage was located in an 8-cm depth and the deepest gage was located over the toe. Two gages were evenly spaced between these points along each 90° transect. Two additional gages were spaced in the deepwater portion at distances of 1.0 m and 2.5 m from the toe (except for the 270° transect). Table 1 lists the X, Y, and Z coordinates for each of the 27 gages for both water depths.

A unique aspect of these tests was the measurement of runup time histories using a new digital runup gage (Figure 4). YEH *et al.* (1989) used a digital runup gage embedded in a model beach to study runup velocities of a bore propagating up the beach. His 35-cm-long gage consisted of 8 rods spaced 5 cm apart with tips which projected no more than 1 mm above the beach surface. The runup gage used in this study possesses some special features which were newly developed. Rather

Table 1

Wave gage locations

Gage ID	X, m	Y, m	Z (cm)	Comments
		$d = 32$ cm		
1	$f(L/2)$	16.05	32.0	Incident gage
2	$f(L/2)$	14.55	32.0	Incident gage
3	$f(L/2)$	13.05	32.0	Incident gage
4	$f(L/2)$	11.55	32.0	Incident gage
5	8.36	13.80	32.0	270° transect
6	9.36	13.80	31.7	270° transect
7	9.76	13.80	22.5	270° transect
8	10.08	13.80	14.7	270° transect
9	10.36	13.80	8.2	270° transect
10	12.96	19.93	32.0	180° transect
11	12.96	18.43	32.0	180° transect
12	12.96	17.43	31.5	180° transect
13	12.96	17.00	22.5	180° transect
14	12.96	16.68	14.6	180° transect
15	12.96	16.40	7.9	180° transect
16	12.96	11.22	7.9	0° transect
17	12.96	10.92	15.2	0° transect
18	12.96	10.60	21.9	0° transect
19	12.96	10.25	30.1	0° transect
20	12.96	9.17	32.0	0° transect
21	12.96	7.67	32.0	0° transect
22	15.56	13.80	8.3	90° transect
23	15.84	13.80	15.7	90° transect
24	16.16	13.80	22.8	90° transect
25	16.59	13.80	31.7	90° transect
26	17.59	13.80	32.0	90° transect
27	19.09	13.80	32.0	90° transect

Table 1 (*Cont.*)

Gage ID	X, m	Y, m	Z (cm)	Comments
		$d = 42$ cm		
1	$f(L/2)$	16.05	42.0	Incident gage
2	$f(L/2)$	14.55	42.0	Incident gage
3	$f(L/2)$	13.05	4.20	Incident gage
4	$f(L/2)$	11.55	4.20	Incident gage
5	8.36	13.80	42.0	270° transect
6	9.36	13.80	42.0	270° transect
7	9.81	13.80	30.7	270° transect
8	10.27	13.80	19.3	270° transect
9	10.72	13.80	8.0	270° transect
10	12.96	19.93	42.0	180° transect
11	12.96	18.43	42.0	180° transect
12	12.96	17.43	42.0	180° transect
13	12.96	16.98	30.7	180° transect
14	12.96	16.52	19.3	180° transect
15	12.96	16.07	8.0	180° transect
16	12.96	11.53	8.0	0° transect
17	12.96	11.08	19.3	0° transect
18	12.96	10.62	30.7	0° transect
19	12.96	10.25	42.0	0° transect
20	12.96	9.17	42.0	0° transect
21	12.96	7.67	42.0	0° transect
22	15.20	13.80	8.0	90° transect
23	15.65	13.80	19.3	90° transect
24	16.11	13.80	30.7	90° transect
25	16.59	13.80	42.0	90° transect
26	17.59	13.80	42.0	90° transect
27	19.09	13.80	42.0	90° transect

than a continuous wire or rod placed along the bottom, this new prototype gage consisted of a series of 32 surface-piercing, vertical rods which are turned on or off by water contact. Gage resolution is limited by the 1-cm minimum spacing between rods. Software is used to convert the wetted rod number to the appropriate vertical runup or rundown. The gage is positioned along the transect so that the still-water level is approximately midway among the rods, enabling measurement of runup and rundown. The advantage of this design is that runup can be measured in the laboratory for uneven bottom conditions, such as rubble-mound breakwaters. For these tests, the runup gage was located only at the 90° transect on the back or lee side of the island. Based on the success of these tests, four new runup gages with 64 rods were designed. They will be used in future tests to cover each of the 90° transects (i.e., 0°, 90°, 180°, and 270°) around the perimeter of the island.

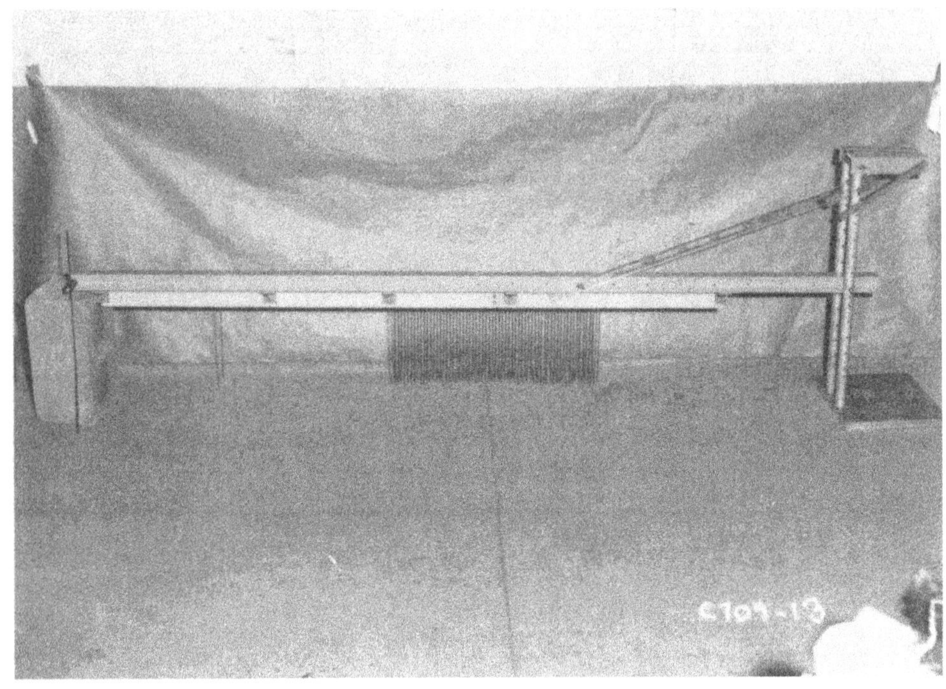

Figure 4
Prototype runup gage.

3. Tsunami Wave Simulation

3.1 Solitary Waves

Tsunami waves were simulated using solitary waves because they model some of the coastal effects of tsunamis well. Although the solitary wave is a single wave, it consists of a complex spectrum of frequencies that allows for elegant analysis and reliable generation in the laboratory. Also, it propagates over constant depth without appreciable changes, allowing for consistent referencing of its offshore or incident wave height. SYNOLAKIS (1987) and others have used the height-to-depth ratio $\mathbf{H} = H/d$ to describe solitary waves. The surface profile $\eta(x, t)$ for a wave centered at $x = X_1$ and time $t = 0$ is defined as

$$\eta(x, 0) = \mathbf{H} \operatorname{sech}^2 \gamma(x - X_1) \tag{1}$$

where $\gamma = (0.75\,\mathbf{H})^{1/2}$. A measure of the wavelength L can be defined in terms of $\mathbf{H}$ and water depth d as

$$L = \frac{2d}{\gamma} \operatorname{arccosh} \sqrt{20} \tag{2}$$

so that it is equal to the distance between the two end points in the symmetric profile where the height is 5 percent of the height at the crest H.

3.2 Target Parameters

Table 2 lists the target solitary wave parameters for the different H values used for each water depth. Due to stroke limitations of the DSWG, a maximum $H = 0.20$ was used for these tests. Because of the flat offshore region and relatively steep island slopes, these waves were nonbreaking until final stages of transformation near the shoreline where gentle spilling occurred.

As mentioned previously, two water depths ($d = 32$ and 42 cm) were used to change the effective island diameter D and beach exposed to the tsunami wave runup. Different DSWG lengths S (i.e., number of paddles) were used to vary the source length of the incoming tsunami wave. Both symmetric and eccentric source lengths were investigated. Symmetric cases were centered about the center of the DSWG and eccentric cases were offset from the center of the island a distance D_x along the x-axis and D_y along the y axis (i.e., waves were not generated directly at the island). Corresponding dimensionless parameters $\mathbf{S_D}(= S/D)$, $\mathbf{D}_x(= D_x/d)$, and $\mathbf{D}_y(= D_y/d)$ are listed in Table 3 for the different symmetric and eccentric cases as a function of the number of modules m and their associated paddle locations. Not all cases were run for each H.

The solitary wave control signal was imbedded in a longer control signal, which included a long ramp time and wait time before and after the main solitary wave to allow the water to still. Since solitary waves are generated with a single positive stroke, the wavemaker was ramped back to its minimum excursion (i.e., largest negative stroke) to enable use of its full stroke capability. The entire control signal was converted to an analog signal with a D/A rate of 20 Hz.

Table 2

Target solitary wave parameters

No.	H	Height (cm)	Length (m)	Period (sec)
		$d = 32$ cm		
1	0.05	1.60	7.20	7.01
2	0.10	3.20	5.09	4.90
3	0.20	6.40	3.60	3.41
		$d = 42$ cm		
4	0.05	2.10	9.46	8.03
5	0.08	3.36	7.48	6.31
6	0.10	4.20	6.69	5.62

Table 3

Dimensionless wavemaker lengths

m	Description	No. Paddles	Paddle ID		S (cm)	S_D		D_x		D_y	
			From	To		$d = 32$	$d = 42$	$d = 32$	$d = 42$	$d = 32$	$d = 42$
					Symmetric Cases						
0.25	1/4 Module Center	4	29	32	137.2	0.30	0.36	40.50	30.86	0.0	0.0
0.5	1/2 Module Center	8	27	34	320.0	0.69	0.83	40.50	30.86	0.0	0.0
1	1 Module Center	15	24	38	640.1	1.38	1.67	40.50	30.86	0.0	0.0
1.5	1.5 Module Center	23	19	42	1005.8	2.17	2.62	40.50	30.86	0.0	0.0
2	2 Module Center	30	16	45	1325.9	2.86	3.45	40.50	30.86	0.0	0.0
3	3 Module Center	45	8	52	2011.7	4.34	5.24	40.50	30.86	0.0	0.0
4	4 Module Center	61	1	61	2743.2	5.91	7.14	40.50	30.86	0.0	0.0
					Eccentric Cases						
0.25e	1/4 Module End	4	1	4	137.2	0.30	0.36	40.50	30.86	40.7	31.0
0.5e	1/2 Module End	8	1	8	320.0	0.69	0.83	40.50	30.86	37.9	28.8
1e	1 Module End	16	1	16	685.8	1.48	1.79	40.50	30.86	32.1	24.5
2e	2 Module End	31	1	31	1371.6	2.96	3.57	40.50	30.86	21.4	16.3

Notes:
1. Island diameter at toe, cm = 720 cm
2. Island diameter at $d = 32$ cm waterline = 464 cm
3. Island diameter at $d = 42$ cm waterline = 384 cm

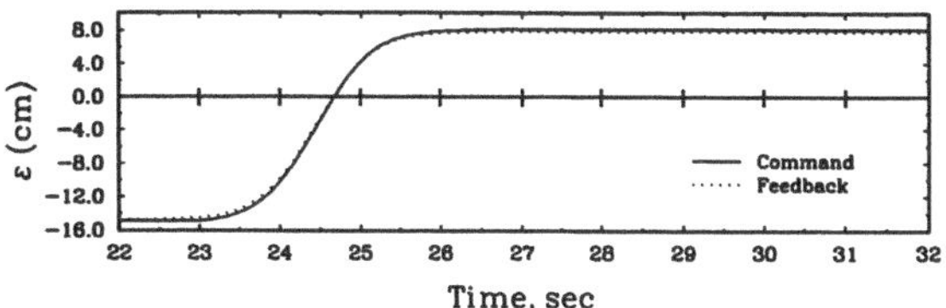

(a) Command and feedback control signal.

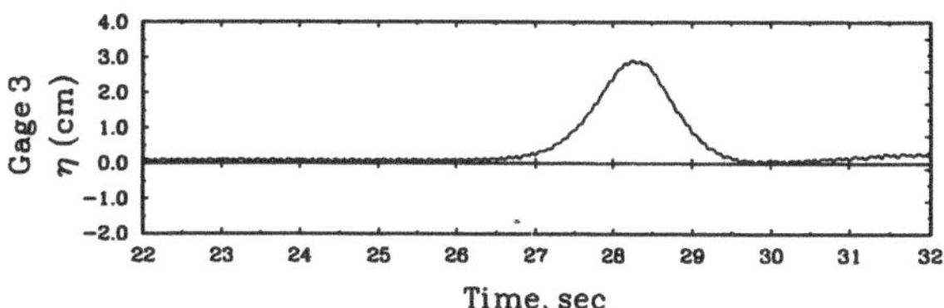

(b) Measured surface elevation for gage 3.

Figure 5
Surface elevation time series for $\mathbf{H} = 0.10$ at $d = 32$ for $m = 4$.

As an example, the top panel of Figure 5 shows the command and feedback control signal at the DSWG and the bottom panel shows the surface elevation for incident gage 3 for the target $\mathbf{H}_{tgt} = 0.10$, $m = 4$ modules, $d = 32$ cm condition in the basin. All data were collected at a sampling frequency $f_s = 25$ Hz. The command and feedback signals are nearly identical. The incident solitary wave profile is relatively clean, without spurious harmonics. Other cases manifested similar patterns. This information is important for future numerical model simulations/verifications of this laboratory data.

3.3 Measured Wave Heights

The shelf width is the distance (i.e., 9.36 m) between the DSWG and the toe of the island. For source lengths smaller than this value (i.e., less than $m = 1.5$ modules), the wave front is not uniform in the longshore direction due to radiation from the ends before it reaches the island. The incident wave front was very uniform when larger source lengths were used, however.

Measured wave heights are listed in Table 4 for each of the different cases at the two water depths. Both symmetric and eccentric source length cases are given for each target $\mathbf{H}$. The measured wave heights are an average of incident gages 2 and 3 (see Figure 1) for all runs for each case. Also listed are $\mathbf{H}_{meas}$ and the ratio of measured to target wave height $\mathbf{H}_{meas}/\mathbf{H}_{tgt}$.

In general, measured wave heights were smaller than target wave heights. For the symmetric cases when two or more modules were used, the measured wave height was approximatley 90 percent of the target value. The decrease in measured wave height from the target was due to losses in the mechanical generation of the solitary waves resulting from gaps between the floor and the wavemaker. For the smaller source lengths less than $m = 1.5$, the radiation condition from the end of the wave was an important factor in decreasing measured wave height. For $m = 1$, the average measured wave was only 60 percent of the target value. For $m = 0.25$ and 0.50, the measured wave heights were proportionally smaller.

For the eccentric cases ($m = 0.25e$ to $2e$), listed values reflect the fact that the incident gages were offset from the source. These small values indicate the size wave experienced by the island as opposed to the size wave actually generated. Actual measured values would have been the same as the corresponding symmetric case if

Table 4

Measured wave heights

H	m	Wave Height, cm		H_{meas}	H_{meas}/H_{tgt}
		Target	Meas		
		$d = 32$ cm			
0.05	0.25	1.6	—	—	—
	0.5	1.6	—	—	—
	1	1.6	0.98	0.03	0.61
	1.5	1.6	—	—	—
	2	1.6	1.42	0.04	0.89
	3	1.6	1.47	0.05	0.92
	4	1.6	1.44	0.05	0.90
	1	1.6	—	—	—
	2	1.6	—	—	—
0.10	0.25	3.2	0.55	0.02	0.17
	0.5	3.2	1.07	0.03	0.34
	1	3.2	1.92	0.06	0.60
	1.5	3.2	—	—	—
	2	3.2	2.74	0.09	0.86
	3	3.2	2.82	0.09	0.88
	4	3.2	2.90	0.09	0.90
	1	3.2	0.63	0.02	0.20
	2	3.2	—	—	—
0.20	0.25	6.4	1.02	0.03	0.16
	0.5	6.4	2.13	0.07	0.33
	1	6.4	3.92	0.12	0.61
	1.5	6.4	—	—	—
	2	6.4	5.68	0.18	0.91
	3	6.4	5.81	0.18	0.90
	4	6.4	5.78	0.18	0.90
	1	6.4	1.10	0.03	0.17
	2	6.4	3.03	0.09	0.47

Table 4 (*Cont.*)

H	*m*	Wave Height, cm		H_{meas}	H_{meas}/H_{tgt}
		Target	Meas		
			$d = 42$ cm		
0.05	0.25	2.1	0.37	0.01	0.17
	0.5	2.1	0.75	0.02	0.36
	1	2.1	1.28	0.03	0.61
	1.5	2.1	—	—	—
	2	2.1	1.83	0.04	0.87
	3	2.1	1.91	0.05	0.91
	4	2.1	1.91	0.05	0.91
	1	2.1	0.46	0.01	0.22
	2	2.1	0.47	0.01	0.22
0.08	0.25	3.4	0.53	0.05	0.59
	0.5	3.4	1.10	0.03	0.33
	1	3.4	2.00	0.05	0.59
	1.5	3.4	—	—	—
	2	3.4	2.92	0.07	0.87
	3	3.4	3.06	0.07	0.91
	4	3.4	3.07	0.07	0.91
	0.25	3.4	0.17	0.00	0.05
	0.5	3.4	0.37	0.01	0.11
	1	3.4	0.70	0.02	0.21
	2	3.4	1.62	0.04	0.48
0.10	0.25	4.2	0.70	0.02	0.17
	0.5	4.2	1.37	0.03	0.33
	1	4.2	2.41	0.06	0.57
	1.5	4.2	—	—	—
	2	4.2	3.66	0.09	0.87
	3	4.2	3.82	0.09	0.91
	4	4.2	3.83	0.09	0.91
	0.25	4.2	0.27	0.01	0.07
	0.5	4.2	0.55	0.01	0.13
	1	4.2	0.94	0.02	0.22
	2	4.2	—	—	—

the incident gages had been located symmetrically with the source. Shorter and longer eccentric source lengths were proportionately smaller or larger than their symmetric counterparts.

4. Amplitude Evolution

When tsunami waves approach the island they undergo complicated nonlinear transformations. Amplitude evolution for the four radial transects is shown in

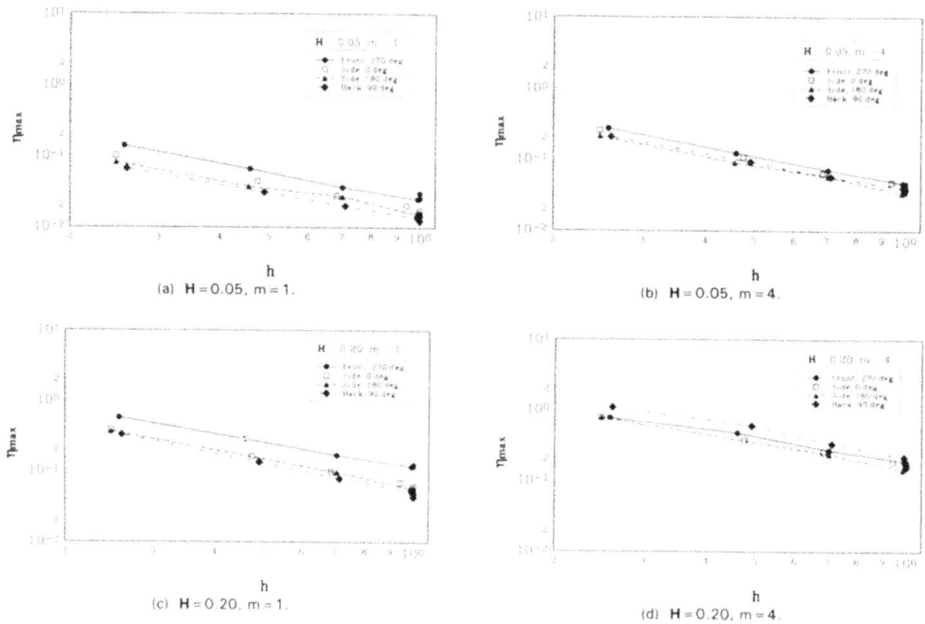

Figure 6
Normalized wave height evolution versus local water depth for $\mathbf{H} = 0.05$ and 0.20, $m = 1$ and 4, $d = 32$ cm.

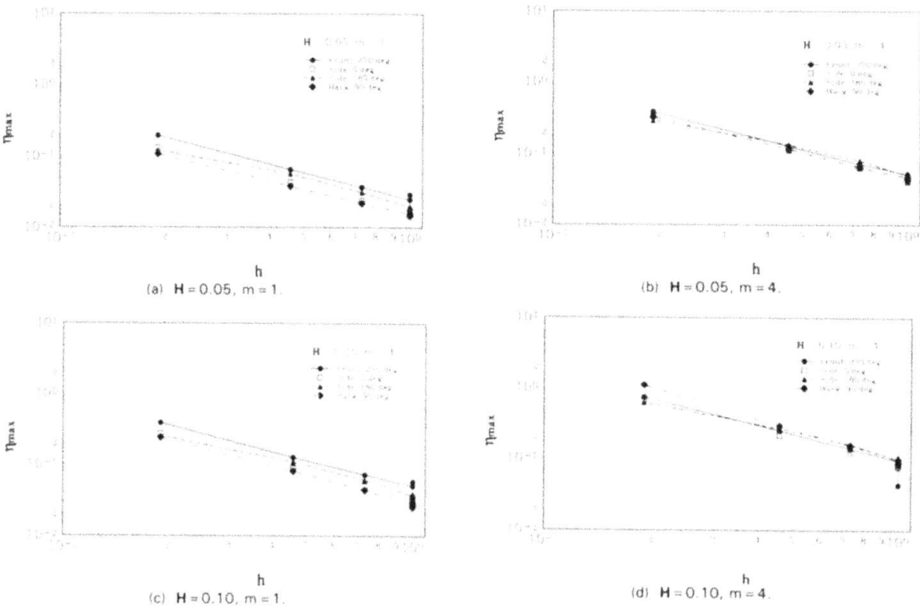

Figure 7
Normalized wave height evolution versus local water depth for $\mathbf{H} = 0.05$ and 0.10, $m = 1$ and 4, $d = 42$ cm.

Figure 6 for $d = 32$ cm for $\mathbf{H} = 0.05$ and 0.20 and $m = 1$ and 4 modules. The corresponding amplitude evolution with cross-shore distance for the deeper $d = 42$ cm cases is shown in Figure 7 for $\mathbf{H} = 0.05$ and 0.10. Dimensionless η_{max} (wave height at each gage η_{max} in the cross-shore radial transect normalized by the undisturbed water depth h at that gage) was plotted versus dimensionless gage depth $\mathbf{h}$ ($= h/d$).

In Figure 6a, the largest wave heights are on the front side of the island and the smallest are on the back side. For the larger source length of $m = 4$ in Figure 6b, there is a slight difference among the four transects in wave height, although the front side remains largest. For the larger $\mathbf{H} = 0.20$ in Figure 6c, the front side waves are still noticeably larger than the back side waves. However, for $m = 4$ in Figure 6d, the back side waves are larger because of the constructive effects of the edge waves. The same phenomenon occurs for the deeper depth cases in Figure 7.

5. Runup Measurements

In this section, results from maximum vertical runup measurements with a rod and transit and runup time series with the prototype runup gage are presented and discussed.

5.1 Maximum Vertical Runup Heights

Maximum vertical runup R_v was measured at twenty locations around the perimeter of the island. Sixteen were evenly spaced every 22.5° around the perimeter. Four radial transects with uneven spacing were located on the back side of the island (i.e., 90°) to improve the resolution in this critical area. At the conclusion of each run, maximum runup along each transect was manually located. A surveyor's rod and transit were then used to measure vertical runup at each transect.

Changes in runup shape and magnitude R were investigated by varying the water depth, wave height, source length (number of modules), and eccentricity of the source. Figures 8a–d are polar plots of maximum vertical runup for $d = 32$ cm, $\mathbf{H} = 0.05$ and 0.20 for $m = 1$ and 4 modules, respectively. Figures 9a–d are analogous plots for the deeper water $d = 42$ cm cases for $\mathbf{H} = 0.05$ and 0.10 for $m = 1$ and 4 modules, respectively. The tsunami wave propagated from the bottom in each panel of the figure. The island crest, waterline, and toe are shown for reference. Two or three runs of each case are overlain, demonstrating excellent repeatability.

In Figure 8a for $\mathbf{H} = 0.05$ and $m = 1$, runup is fairly uniform around the perimeter of the island. In Figure 8b for $m = 4$ modules and the same wave height, runup is higher on the front side. For the larger $\mathbf{H} = 0.20$ and $m = 1$ in Figure 8c, runup is still larger on the front side of the island; however, a distinctive pattern of

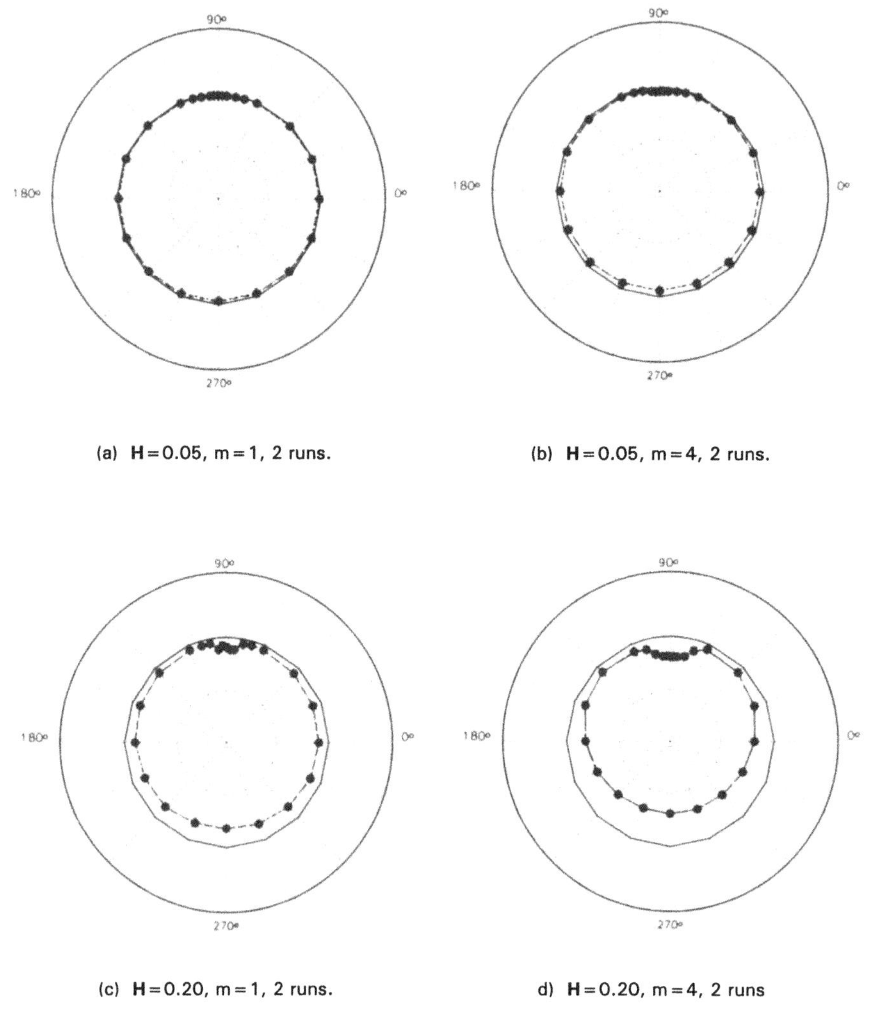

(a) **H** = 0.05, m = 1, 2 runs.   (b) **H** = 0.05, m = 4, 2 runs.

(c) **H** = 0.20, m = 1, 2 runs.   d) **H** = 0.20, m = 4, 2 runs

Figure 8
Maximum runup measurements for $\mathbf{H} = 0.05$ and 0.20, $m = 1$ and 4, $d = 32$ cm.

runup due to the edge waves propagating around the island from the symmetric source begins to become apparent. By Figure 8d for $m = 4$ modules, the runup on the back side is almost as large as that on the front side of the island. Similar patterns are suggested by Figure 9, except that runup on the back side is more pronounced. For $\mathbf{H} = 0.10$ and $m = 4$ in Figure 9d, runup on the back side is slightly larger than that on the front side. Refraction and diffraction cause the wave to bend around the island as edge waves. Because the island and source were symmetric, the wave wraps evenly around the island and produces relatively large runup on the back side. This is a very interesting phenomenon since most people

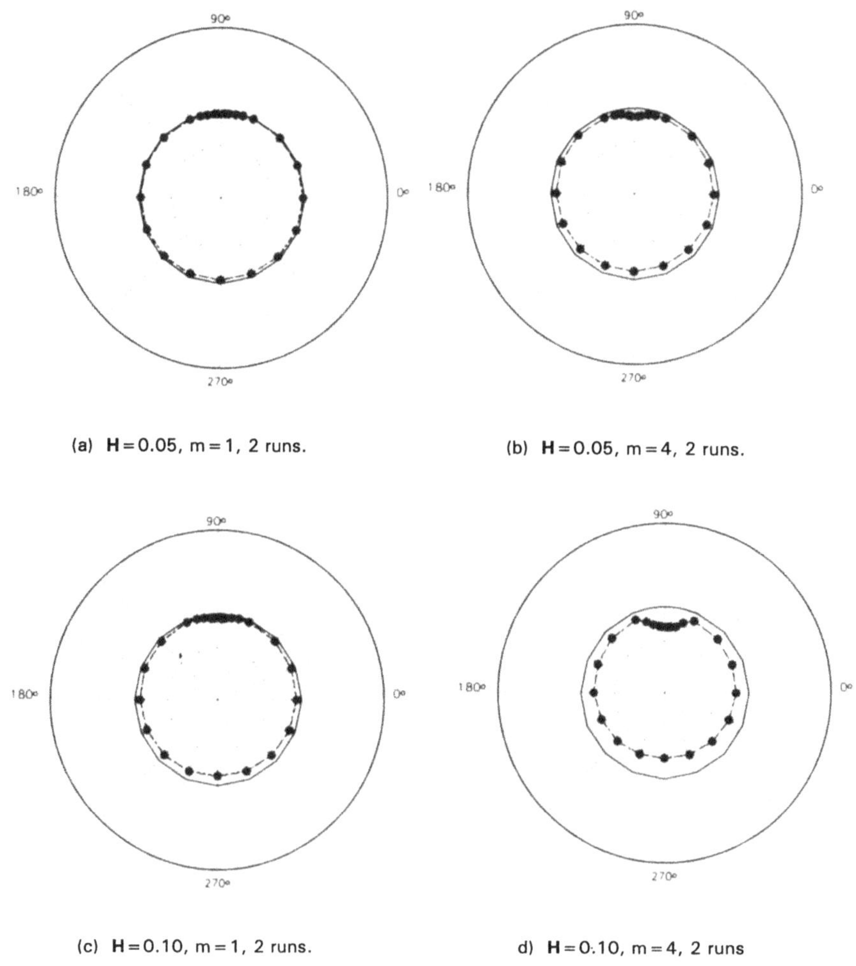

(a) **H** = 0.05, *m* = 1, 2 runs.

(b) **H** = 0.05, *m* = 4, 2 runs.

(c) **H** = 0.10, *m* = 1, 2 runs.

d) **H** = 0.10, *m* = 4, 2 runs

Figure 9

Maximum runup measurements for **H** = 0.05 and 0.10, *m* = 1 and 4, *d* = 42 cm.

would feel "safe" on the back side of an island. When the source is offset from the island center (i.e., eccentricity effects), runup is largest on the island quadrant closest to the source between 0° and 270°, decreasing linearly around the perimeter to the opposite side. Figures 10a–d illustrate this effect for both water depths for the largest wave height of **H** = 0.20 and 0.10 for *m* = 1*e* and 2*e*, respectively.

5.2 Maximum Runup versus Source Length

Maximum runup **R** versus length S_D values on the front, side and back of the island for the two water depths are shown in Figure 11. Values for the side of the

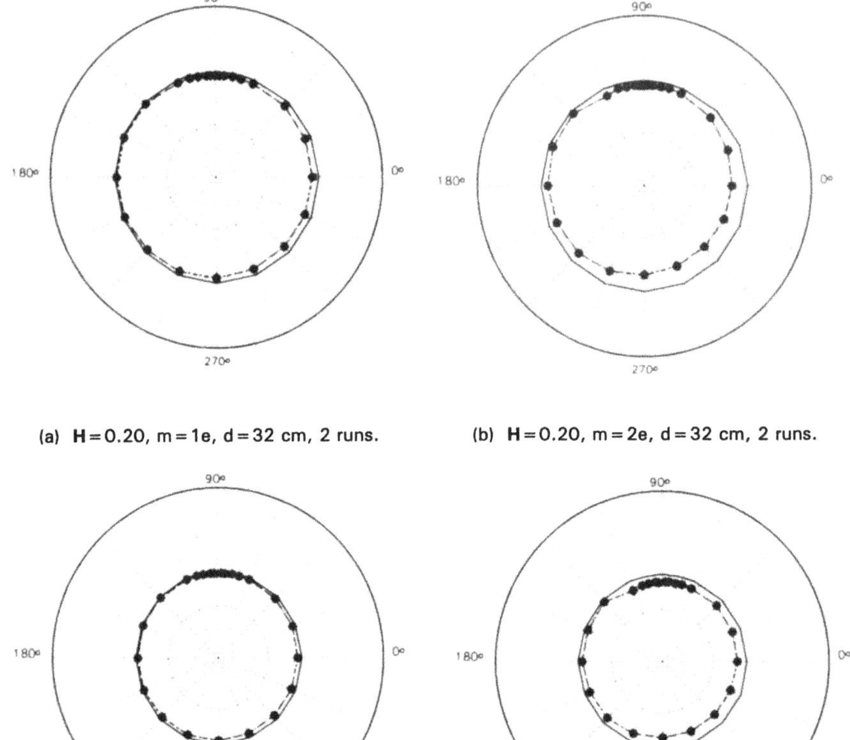

(a) **H** = 0.20, m = 1e, d = 32 cm, 2 runs.   (b) **H** = 0.20, m = 2e, d = 32 cm, 2 runs.

(c) **H** = 0.10, m = 1e, d = 42 cm, 2 runs.   d) **H** = 0.10, m = 2e, d = 42 cm, 2 runs

Figure 10
Maximum runup measurements for eccentric source, $\mathbf{H} = 0.20$ and 0.10, $m = 4$, $d = 32$, 42 cm.

island are an average from both $0°$ and $180°$ sides. Because the prototype runup
gage was located on the back side of the island on the $90°$ transect for $d = 42$ cm,
no values were collected. Least-square fit lines for each of the three target
normalized wave height **H** values are also shown. A second order polynomial fit
produced the highest correlation coefficient r^2 (i.e., best fit) for all cases. Quadratic
equation coefficients a, b, and c corresponding to an equation of the form

$$\mathbf{R} = a + b\mathbf{S_D} + c\mathbf{S_D^2} \tag{3}$$

and the associated r^2 are listed in Table 5 for both the 32-cm and 42-cm water
depths. These empirical equations can be used to estimate the runup on an island

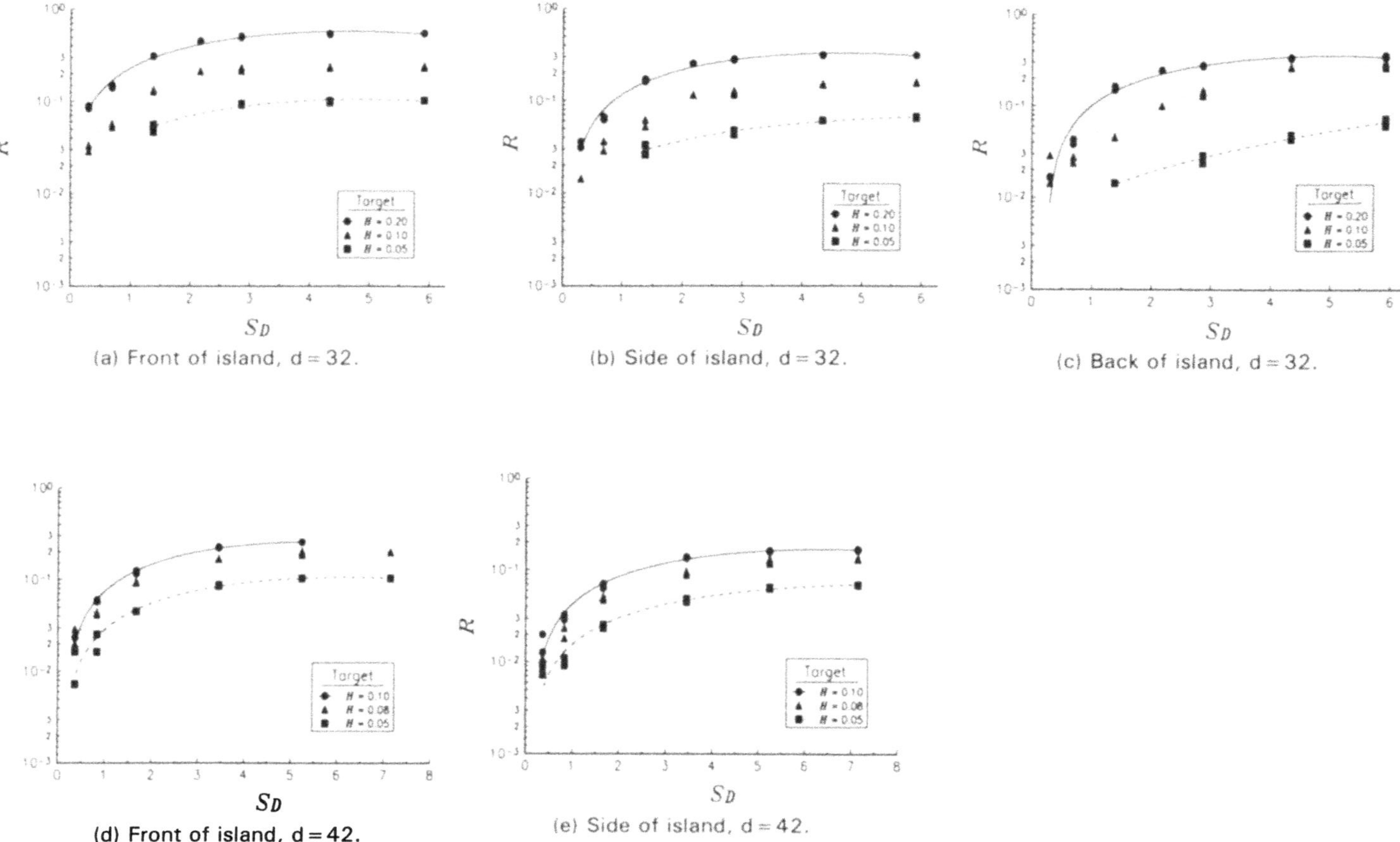

Figure 11

Maximum runup around island as a function of source length for two water depths.

Table 5

Least square parameters for R vs S_D

H	a	b	c	r^2
		Front, $d = 32$ cm		
0.20	0.0125	0.2444	−0.0264	0.99
0.10	−0.0079	0.1192	−0.0134	0.98
0.05	0.0031	0.0421	−0.0044	0.95
		Side, $d = 32$ cm		
0.20	−0.0137	0.1463	−0.0152	0.99
0.10	−0.0072	0.0620	−0.0056	0.99
0.05	−0.0079	0.0170	−0.0011	0.97
		Back, $d = 32$ cm		
0.20	−0.0358	0.1526	−0.0152	0.99
0.10	−0.0147	0.0608	−0.0016	0.96
0.05	0.0045	0.0057	0.0008	0.97
		Front, $d = 42$ cm		
0.10	−0.0137	0.0972	−0.0088.	1.00
0.08	−0.0020	0.0654	−0.0053	0.99
0.05	−0.0036	0.0352	−0.0028	0.99
		Side, $d = 42$ cm		
0.10	−0.0078	0.0548	−0.0043	0.99
0.08	−0.0070	0.0383	−0.0026	0.99
0.05	−0.0013	0.0180	−0.0011	0.99

for the range of conditions studied. However, given the small coefficient of the second-order term, it is conjectured that a linear relationship between runup and source length, analogous to that found by BRIGGS *et al.* (1993) for the plane beach, could be used without loss of accuracy for empirical predictions.

5.3 Runup Time History

Measurement of maximum runup is a labor-intensive effort which only gives one value of runup and no information about rundown. A runup time series showing both runup (positive values) and rundown (negative values) is a more useful measurement for verification of numerical models and prediction of subsequent runup waves due to bathymetry variations and reflections from adjacent shorelines. Measurements were only made at the 90° transect on the back side of the island for selected deeper water cases. As previously stated, future tests are planned at each 90° transect around the perimeter of the island with four new runup gages.

Figure 12 is an example runup time series for $\mathbf{H} = 0.10$ and $m = 1$ module. The first runup and rundown wave is usually the largest. The maximum runup of 2.9 cm agrees very well with the manual measurement of 2.7 cm at the adjacent 87.5° and

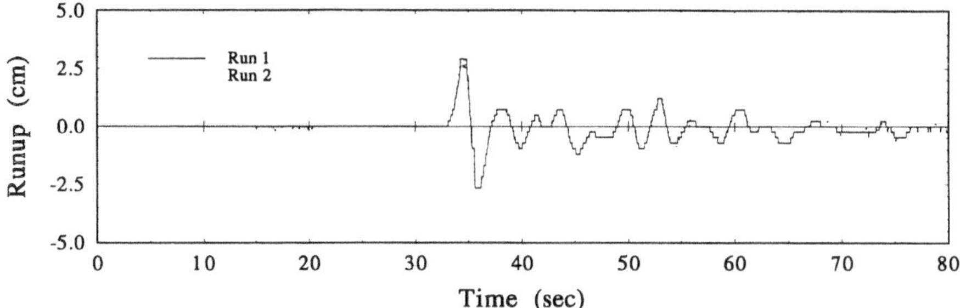

Figure 12
Repeatability for runup time histories for **H** = 0.10.

Table 6

Comparison of maximum vertical runup

H	m	Test ID	Meas. Ave	Runup Gage
0.05	0.25	5	0.55	0.24
	0.50	6	0.70	0.49
	1	1	0.99	0.97
	2	2	2.00	2.18
	3	3	3.35	3.40
	4	4	4.19	4.37
	0.25e	A	—	—
	0.50e	B	—	—
	1e	7	0.86	0.49
	2e	8	1.37	1.21
0.08	0.25	5	0.53	0.49
	0.50	6	0.93	0.73
	1	1	1.91	1.94
	2	2	5.30	5.58
	3	3	9.16	—
	4	4	10.42	10.91
	0.25e	A	—	—
	0.50e	B	0.23	0.24
	1e	7	0.72	0.73
	2e	8	2.67	2.91
0.10	0.25	5	0.61	0.49
	0.50	6	1.14	1.21
	1	1	2.67	2.91
	2	2	8.99	—
	3	3	10.53	10.91
	4	4	10.73	10.91
	0.25e	A	0.46	0.49
	0.50e	B	0.67	0.73
	1e	7	1.03	1.21
	2e	8	4.04	4.37

92.5° transects (the gage made it impossible to manually measure runup at the 90° transect). The stair-step pattern is due to the resolution of 1 cm (0.24 cm in the vertical) between gage rods. Two runs are overlain, demonstrating excellent repeatability.

Table 6 compares maximum vertical runup between manually measured and runup gage values in centimeters. Measured values are averages of the 87.5° and 92.5° values. Both measured and runup gage values were averaged over two or three runs. In most cases, the agreement is good. The runup gage should read slightly

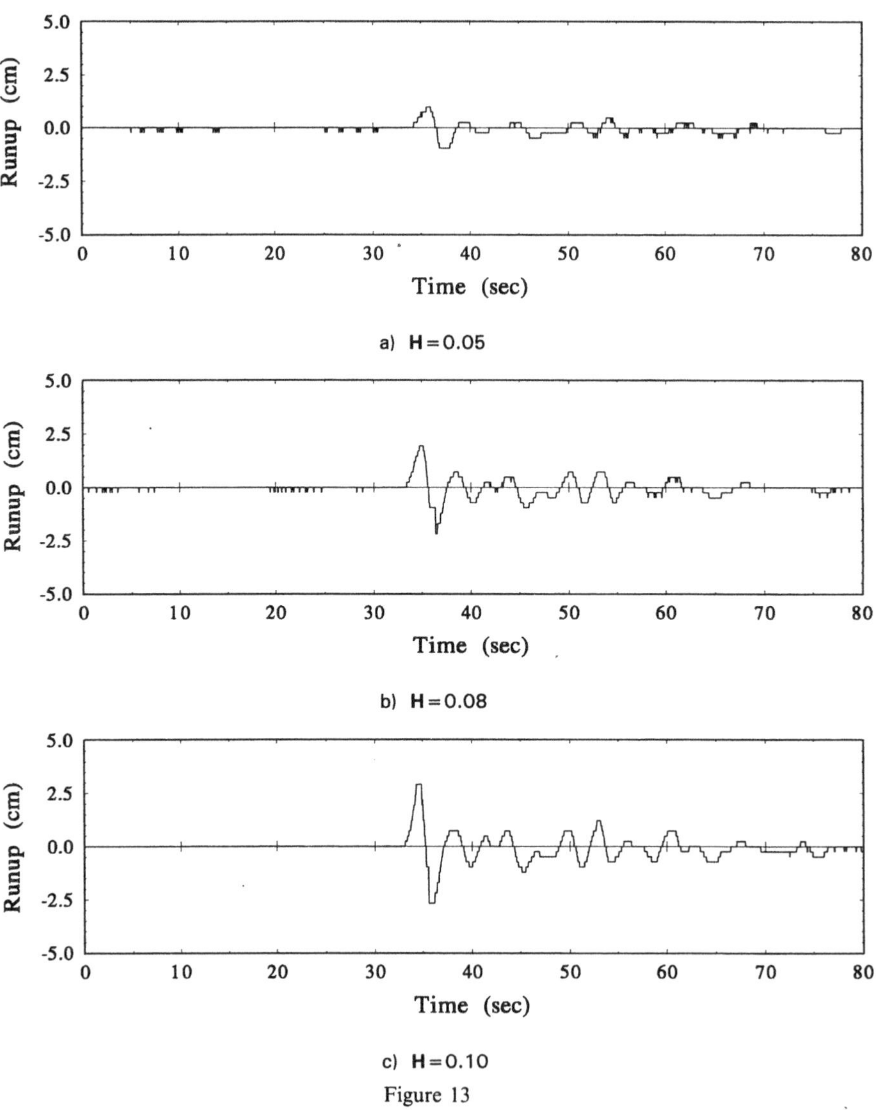

a) H = 0.05

b) H = 0.08

c) H = 0.10

Figure 13

Runup time histories for $m = 1$.

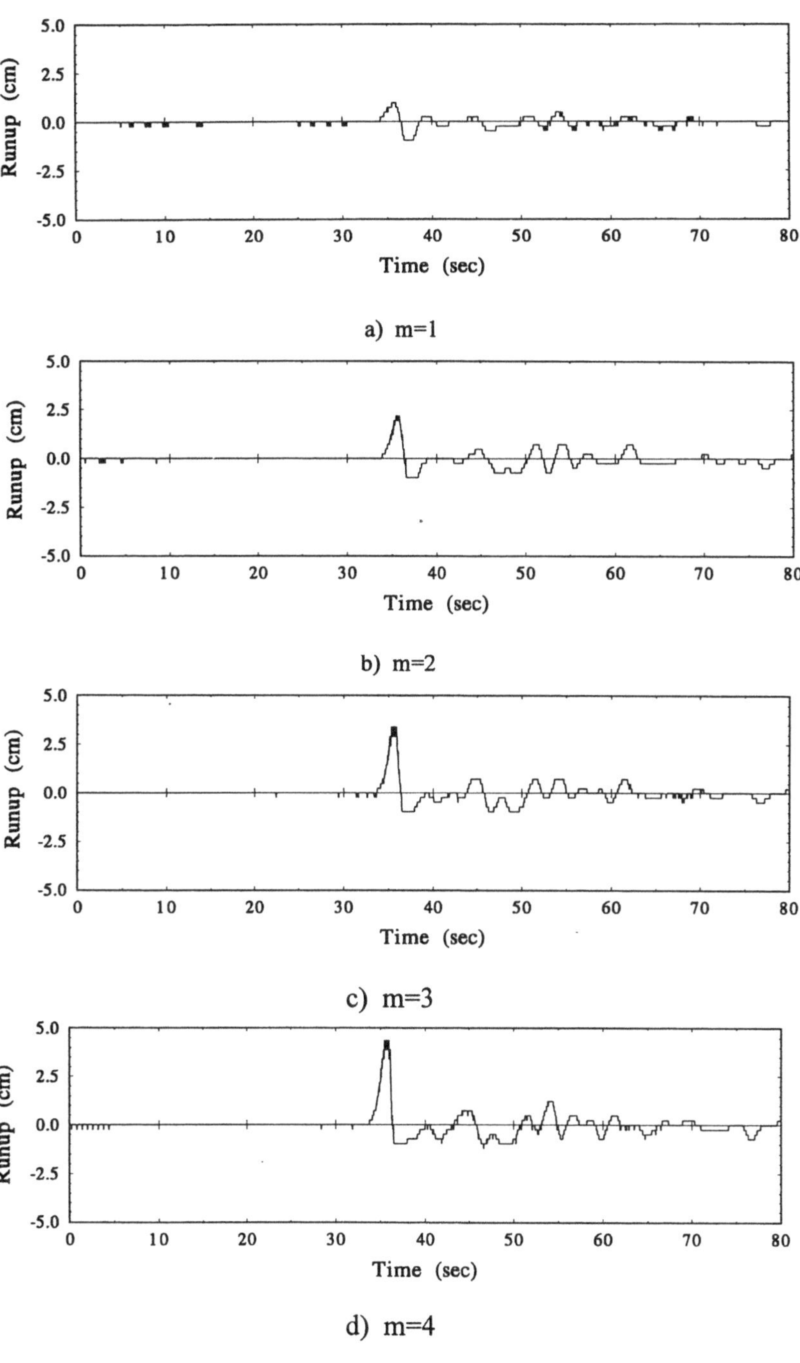

a) m=1

b) m=2

c) m=3

d) m=4

Figure 14

Runup time histories for a range of source lengths at $H = 0.05$.

higher than the measured values because the gage is over the $90°$ transect where runup is expected to be the largest. The minimum and maximum runup gage readings were 0.24 and 10.91 cm, respectively. Because of rod spacings of 1 or 2 cm, the resolution varied between ± 0.24 and ± 0.48 cm. In light of this fact, the low readings are within tolerance levels of measured values. Thus, the runup gage is accurate enough to be used in future tests in lieu of manual measurements of maximum runup, a considerable time and cost savings.

Figure 13 shows how the runup time history varies for one symmetric module (i.e., $m = 1$) for a range of wave heights, $H = 0.05$, 0.08, and 0.10. The amplitude of both the runup and rundown increases as H increases. The runup portion appears to increase more than the rundown portion as the number of modules increases (Figure 14) from $m = 1$ to 4 for a fixed $H = 0.05$.

6. Summary and Conclusions

This paper presents results from three-dimensional, laboratory tests of tsunami wave runup on a conical island. The 7.2-m diameter, 62.5-cm tall island had 1 on 4 side slopes and was located in the center of a 30-m-wide by 25-m-long flat-bottom basin. Solitary waves with height-to-depth ratios ranging from 0.05 to 0.20 and "source" lengths ranging from 0.30 to 7.14 island diameters were tested in water depths of 32 cm and 42 cm. Maximum vertical runup measurements were made at 20 locations around the perimeter of the island. Runup on the back side of the island can be higher than the front side, depending on the tsunami wave. Runup time series measurements from a new runup gage show great promise.

Acknowledgements

The authors wish to acknowledge the Office, Chief of Engineers, U.S. Army Corps of Engineers, and the University of Southern California for authorizing publication of this paper. It was prepared as part of the "Three-dimensional Tsunami Runup" study funded by the National Science Foundation through the U.S. Army Engineer Waterways Experiment Station and the University of Southern California (NSF Grants BCS-9205134 and BCS-9201326, respectively). We would especially like to thank the following individuals for their assistance and participation in this project: Mr. David Daily, Mr. Utku Kanoglu, and Mr. Allen Collidge.

REFERENCES

BRIGGS, M. J., and HAMPTON, M. L. (1987), *Directional Spectral Wave Generator Basin Response to Monochromatic Waves*, WES Technical Report, CERC-87-6, USAE Waterways Experiment Station, Vicksburg, MS, 1–90.

BRIGGS, M. J., SYNOLAKIS, C. E., and HUGHES, S. A. (1993), *Laboratory Measurements of Tsunami Runup*, Tsunami '93 Proceedings, Wakayama, Japan, 585–598.

JONSSON, I. G., and SKOVAARD, O. (1979), *A Mild-slope Equation and its Application to Tsunami Calculations*, Marine Geodesy 2,41–58.

LIU, P.L.-F., CHO, Y.-S., BRIGGS, M. J., and SYNOLAKIS, C. E. (1994), *Runup of Solitary Waves on a Circular Island*, JFM, in revision 1995.

PROVIS, D. G. (1975), *Propagation of Water Waves near an Island*, Ph.D. Thesis, U. Essex.

SMITH, R., and SPRINKS, T. (1975), *Scattering of Surface Waves by a Conical Island*, JFM, 72, 373–384.

SPRINKS, T., and SMITH, R. (1983), *Scale Effects in a Wave-refraction Experiment*, JFM 129, 455–471.

SYNOLAKIS, C. E. (1987), *The Runup of Solitary Waves*. JFM 185, 523–545.

YEH, H., GHAZALI, A., and MARTON, I. (1989), *Experimental Study of Bore Runup*, JFM 206, 563–578.

YEH, H., IMAMURA, F., SYNOLAKIS, C., TSUJI, Y., LIU, P., and SHI, S. (1993), *The Flores Island Tsunamis*, EOS Transactions, AGU 74, 33, Aug. 17, 369–373.

YEH, H., LIU, P., BRIGGS, M., and SYNOLAKIS, C. (1994), *Tsunami Catastrophe on Babi Island*, Nature Magazine, November.

(Received September 19, 1994, revised December 29, 1994, accepted January 10, 1995)

PAGEOPH, Vol. 144, Nos. 3/4 (1995)

0033-4553/95/040595-25$1.50 + 0.20/0
© 1995 Birkhäuser Verlag, Basel

Tsunami Trapping near Circular Islands

STEFANO TINTI[1] and CESARE VANNINI[1]

Abstract—Trapping of long water waves that are induced by submarine earthquakes and that attack circular islands is studied by applying a theoretical model (TINTI and VANNINI, 1994) that is based on the linear shallow water approximation. The solution is computed as the superposition of the eigenmodes of the water basin. The tsunami trapping is seen in terms of the capability of the source to excite the "trapped" eigenmodes of the basin. The bottom depth dependence around the island is shown to be quite important in determining the trapping capability of the island: a depth profile that is downwardly concave as the distance from the island coasts increases is substantially more efficient in amplifying the incoming waves and in trapping their energy than a profile exhibiting an upward concavity.

Key words: Analytical model, circular island, shallow water approximation, tsunami, wave amplification.

Introduction

The impact of tsunamis against islands can be devastating, as recent examples such as the 1992 Flores tsunami investing Babi Island (see YEH *et al.*, 1993) have shown. This engenders a motivation to investigate the interaction of tsunamis with ocean islands with special attention. In this paper we are interested in studying a peculiar aspect of this interaction, namely the amplification and trapping of tsunami waves that can be observed in the sea region encircling the island, especially when the water wavelength and the island have a comparable scale. Tsunamis generated by submarine earthquakes often have horizontal source dimensions that are in the range of tens of kilometers and that, from the viewpoint of the present paper, can be considered as strictly larger than, but nevertheless comparable with the typical size of most ocean islands and seamounts. This study is both an application and an extension of the theoretical model of long-wave propagation developed by TINTI and VANNINI in 1994, making use of the free oscillation modes of a closed water basin and of their linear combination. In this paper the main

---

[1] Dipartimento di Fisica, Settore di Geofisica, Università di Bologna, Italy.

concern regards the trapping of tsunamis near circular islands and the dependence of the island trapping capability upon the geometrical configuration of the ocean bottom near the island region. The study highlights the special importance of the depth profile shape in proximity of the island: indeed, downwardly concave profiles prove to be potentially more capable of locally capturing tsunamis than profiles with upward concavity. In the following sections illustrative key examples of wave propagating in the ocean with different depth profiles are analyzed and discussed, with particular emphasis being posed on the wave system generated near the island by the passing wavefront.

The Basic Geometry and Model

The model that is extensively explained elsewhere (TINTI and VANNINI, 1994) takes into account an axisymmetric ocean basin, with a circular island in the center, the bathymetry $h = h(r)$ being only a function of the radial distance r from the basin centre. The basin is closed by an outer boundary $r = r_2$ where a null wave elevation condition is imposed and by island coasts $r = r_1$ that are purely reflective. An important feature of the basin is that the outer boundary must be distant enough from the island coasts to be considered at infinity: this condition is practically fullfilled if the waves back-reflected by the outer boundary reach the internal region of interest, i.e., the region around the island, at a considerably longer time than the typical time of interaction of the wavefront with the island itself. The inviscid linearized shallow water equations are taken to model the tsunami evolution, which permits (see SATAKE and SHIMAZAKI, 1987) expression of the water elevation above the mean sea surface $\zeta(\bar{x}, t)$ as a superposition of the basin normal eigenmodes $e_{ikq}(\bar{x})$. In the case of a circular axisymmetric basin, the eigenfunctions:

$$e_{ikq}(r, \varphi) = \rho_{ik}(r)\alpha_{iq}(\varphi) \qquad (1)$$

are factorisable in the product of a linear combination of Bessel functions $\rho_{ik}(r)$ of complicated argument times a trigonometric function $\alpha_{iq}(\varphi)$ and are identified by an angular wavenumber i, by a radial wavenumber k, by a parity number $q = 1, 2$ and by an eigenfrequency ω_{ik}, that is independent from q. In the expression (1) r and φ denote the polar coordinates. The quantum numbers i and k are associated with the number of zeros, i.e., of nodal lines, of the eigenfunction e_{ikq} in the coordinates φ and r respectively, while the parity number q identifies the eigenfunctions with even and odd dependence upon the coordinate φ. As a rule, eigenfunctions with higher quantum numbers exhibit more oscillations, smaller wavelength and higher frequency. The solution is unequivocally determined by the initial condition $\zeta(r, \varphi, t_0)$: the decomposition of $\zeta(r, \varphi, t_0)$ over the set of the eigenmodes results in the computation of the coefficients γ_{ikq} satisfying identically the relation

$$\zeta(r, \varphi, t_0) = \sum_{ikq} \gamma_{ikq} e_{ikq}(r, \varphi). \qquad (2)$$

These coefficients are in turn used in computing the solution as a weighted synthesis of the eigenmodes (TINTI and VANNINI, 1994):

$$\zeta(r, \varphi, t) = \sum_{ikq} \gamma_{ikq} e_{ikq}(r, \varphi) \cos[\omega_{ik}(t - t_0)]. \tag{3}$$

It is trivial to verify that the above expression reduces identically to $\zeta(r, \varphi, t_0)$ at $t = t_0$ and that it matches the boundary conditions since all the eigenmodes e_{ikq} share this common property.

An Island in a Constant Depth Basin

In order to fully understand the dynamics of the interaction between a wavefront and an island, the easiest example to treat is that of a wave propagating in a uniform-depth floor basin, since it permits ruling out the effects related to the variable bathymetry and isolates and emphazises only those connected to the circular island system: in this case the island is simply a cylinder with vertical walls

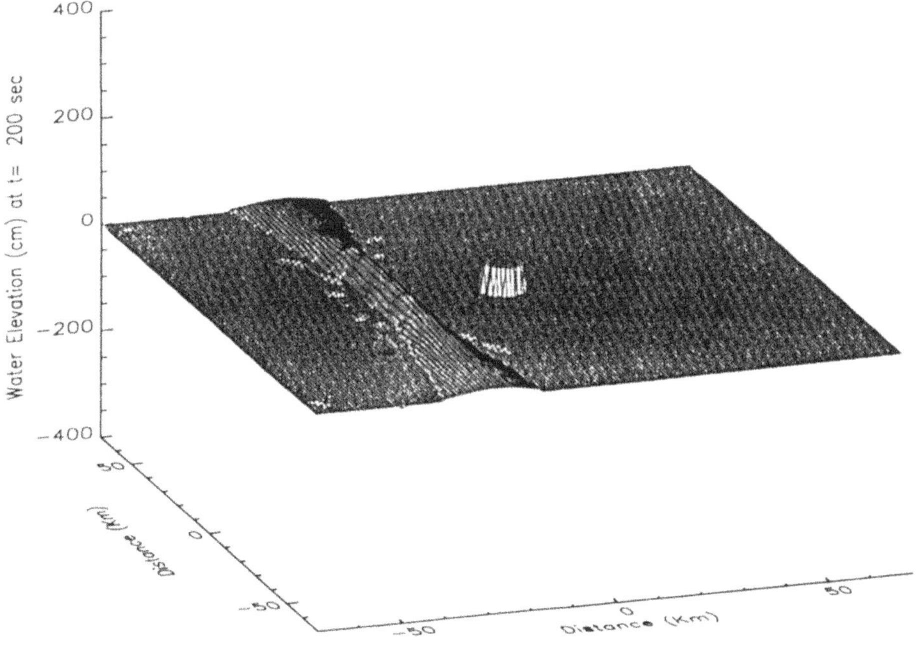

Figure 1

Positive wave propagating (left to right) towards a cylindrical island of radius $r_1 = 5$ km emerging from a flat-bottom ocean. Computations are performed in a circular basin of radius $r_2 = 200$ km, while the sketch shows the ocean surface in the central 140 km × 140 km rectangular space window. 200 seconds after the initial time the wavefront is at about 25 km from the island coasts.

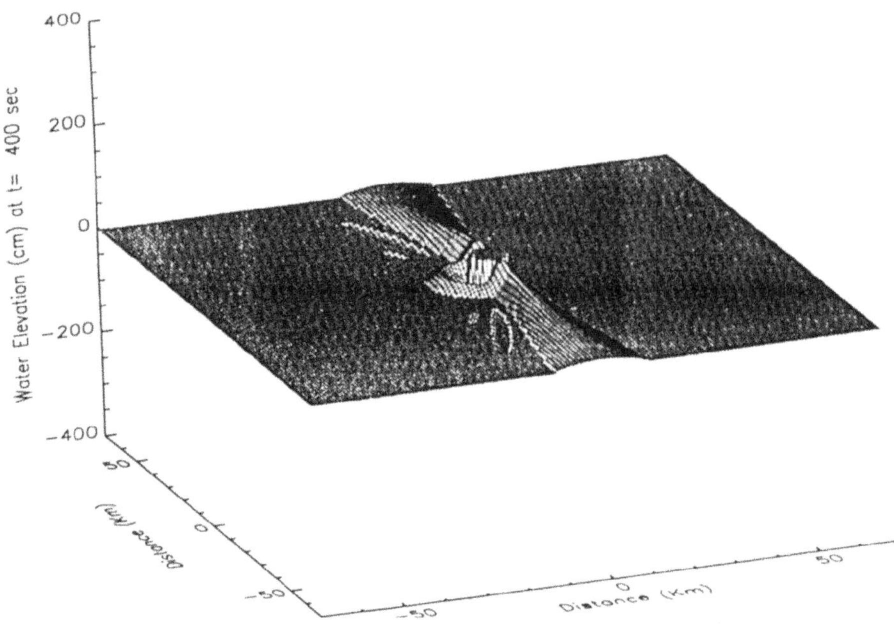

Figure 2

The forcing wave interacts with the island vertical walls. The main front is passing beyond the island, whereas a diffracted front is forming around the island coasts.

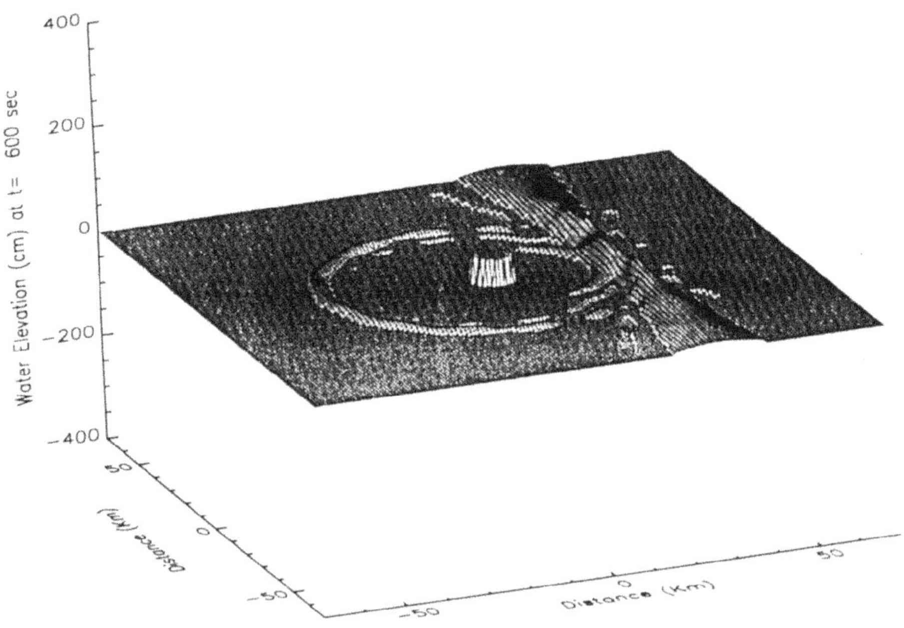

Figure 3(a)

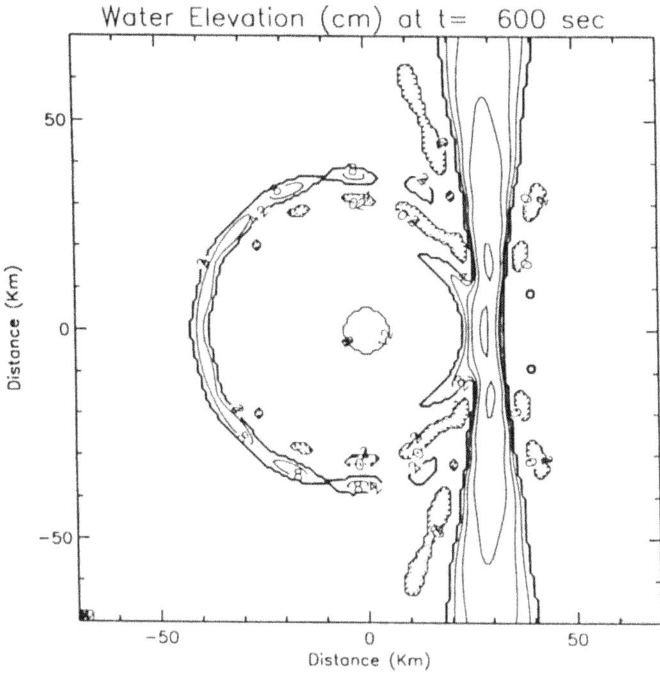

Figure 3(b)

Figure 3

The main linear front has surpassed the island, originating a well-defined circular diffracted wave that is leaving the island at the same propagation velocity as the main wave. Prospective view (a) as well as a planar contour-line plot (b) are given.

placed in the middle of a flat circular basin. The first two simulations regard a basin characterized by the following geometrical parameters: $r_1 = 5$ km, $r_2 = 200$ km and $h = 2000$ m. The total number of eigenmodes used is 2000.

In the first example the initial perturbation consists of a wavefront with the shape of a bulge that is travelling toward the island, as is depicted in Figure 1. In Figure 2 the main forcing wave has already reached the island coasts and has given rise to a diffracted wavefront: one can observe that the wave amplitude near the island coasts is greater, because of the constructive interference between the diffracted wave, that is radiating outward, and the main wave. At a later stage, visible in Figures 3a and 3b, the forcing wave propagates beyond the island, having approximately taken its initial shape again, while the diffracted wave is radiating outward in the form of a circular front that is closely centered on the first impact point on the island coasts and that shows somewhat of a detachment tail in proximity of the forcing bulge. It is worthwhile remarking that a similar diffraction pattern was also computed by PINETTES (1980) for a similar basin, through an analytical model, making use of nonlinear shallow water equations.

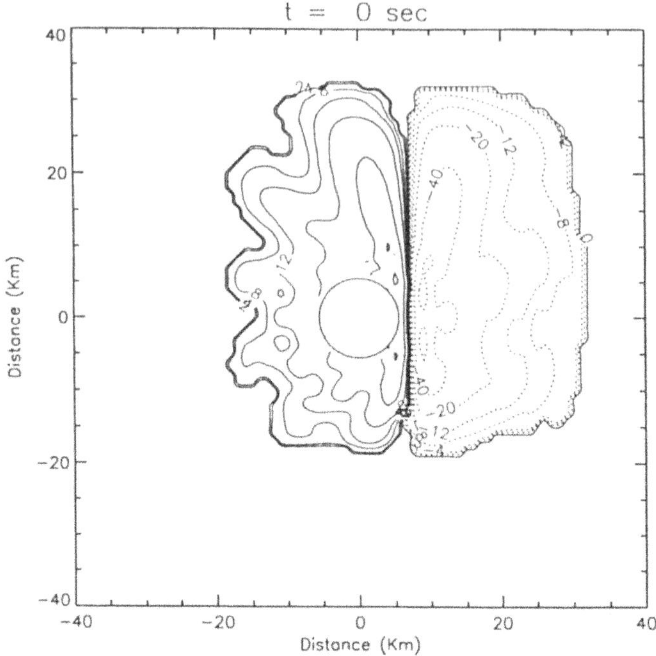

Figure 4
Initial sea-surface disturbance (in cm) produced by a dip-slip vertical fault that is placed in the proximity of a circular island in a flat-bottom basin. The maximum vertical displacement is about 50 cm. The downlifted region is completely offshore, while the uplift region involves the island itself. The initial wave is a dipole (crest-trough) system. For this and all the following graphs, contour curves are solid and dotted lines respectively for postitive and negative label values.

A further exemplification is given with a more complex initial perturbation induced by a seismic fault placed near the island in an asymmetric position. A pure dip-slip mechanism on a vertical fault is used with the uplifting block involving the island, and the subsiding region being offshore. Figure 4 shows the initial surface displacement of the basin, characterized by a dipolar wave system. The initial perturbation splits into two dipoles with opposite directions, one travelling to the left towards the island coasts and one propagating towards the right. At $t = 150$ sec (Figure 5) one can distinctly observe a circular diffraction wave, as in the example treated above, in addition to the two dipolar forcing waves that continue to travel in opposite directions and exhibit a shape that has not been significantly deformed by the interaction with the island. At $t = 250$ sec practically no residual wave remains near the island coasts, all the energy having been radiated outward (Figure 6). The detachment tails are visible in both dipolar waves, but the one that is associated with the wave propagating toward the left is more conspicuous, for this experienced a more relevant interaction with the island.

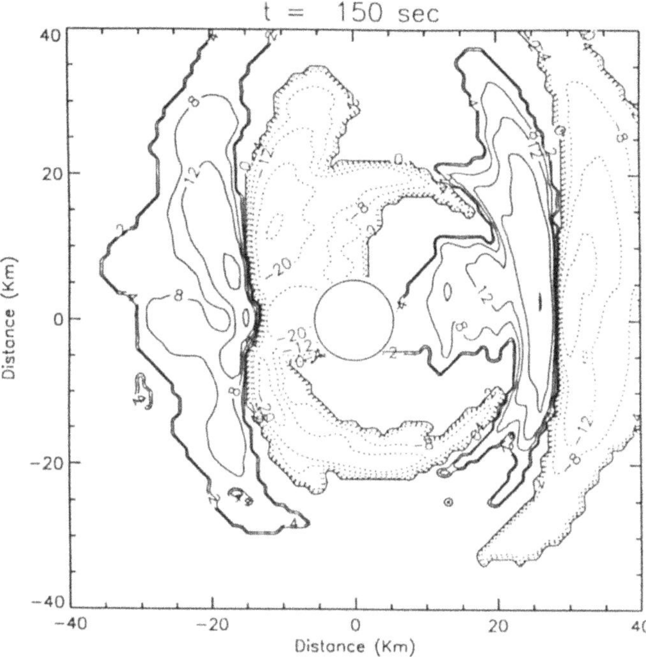

Figure 5

The initial dipole splits into two equal dipole systems, travelling one right and the other left. Both dipoles interact with the island, though the larger interaction pertains to the dipole advancing leftward. At the interaction stage, the dipole surpasses the island, generating a circular diffracted wavefront. Water elevations are in centimeters.

Variable Bathymetry: Upwardly-concave Radial Depth Profile

After learning that in a flat bottom basin the chief characteristic of the wave propagation is that the exciting wave passes nearly undeformed beyond the island and that concurrently it generates a circular diffracted wavefront, we turn our attention to a case with a more complicated bathymetry. Our goal is to investigate the further effects of wave refraction, induced by a variable-depth ocean bottom. In fact, since the long-wave phase velocity is equal to $\sqrt{gh}$, where g is the gravity constant and h is the water depth, then the wavefronts are expected to be slowed and bent toward shallower waters as observed for a uniformly sloping beach by URSELL (1952) and for a uniformly sloping shelf of finite width by MYSAK (1968). A simulation has been carried out for a circular basin, differing from one taken into account in the previous cases, in that ocean depth increases monotonically from the island shore value (10 m) to the offshore value (2000 m) that is reached at a 20 km distance from the basin center: the radial depth profile is assumed to exhibit an upward concavity in the region with variable depth representing the submerged

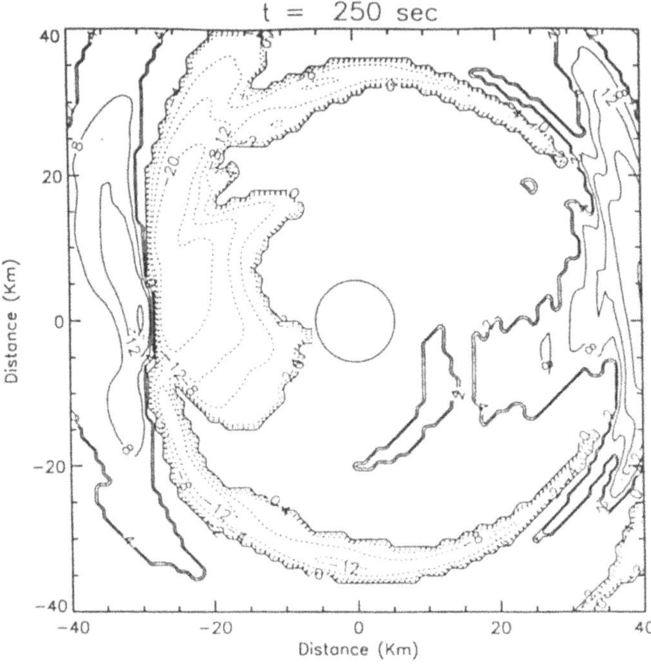

Figure 6
The dipole-island interaction is completed: the dipole has surpassed the island and the diffracted wave
system is radiating towards the open sea. No wave energy remains around the island after the passage
of the main forcing system. Water elevations are in centimeters.

base of the island (see Figure 7). In the following, we will find it convenient to refer
to this bathymetry as a bathymetry of type UC (Upward Concavity). A seismic
source equal to that used in the flat-bottom example is used to produce the initial
perturbation illustrated in Figure 4. A similar case was presented in TINTI and
VANNINI (1994) where a smaller basin however was used (with an outer boundary
placed only at $r_2 = 60$ km vs. the actual $r_2 = 180$ km), which permitted computation
of the wave-island interaction only during the first stages of the process. The choice
in the present paper allows us to prolong the simulation through the end of the
interaction. One question could arise as to whether the solution is affected by the
basin size: since the solution is calculated as a synthesis of the eigenmodes of the
basin, if we change the basin size, we change the basic set of eigenmodes and
consequently the contribution of each term in the summation of the expression (3)
is changed as well. From the physical point of view, however, the outer boundary
of the basin has no influence on the interaction of the wave with the central island
(as long as the back-reflected wave can be neglected) and the final solution is
expected to be invariant. We have tested our model and have verified that it satisfies

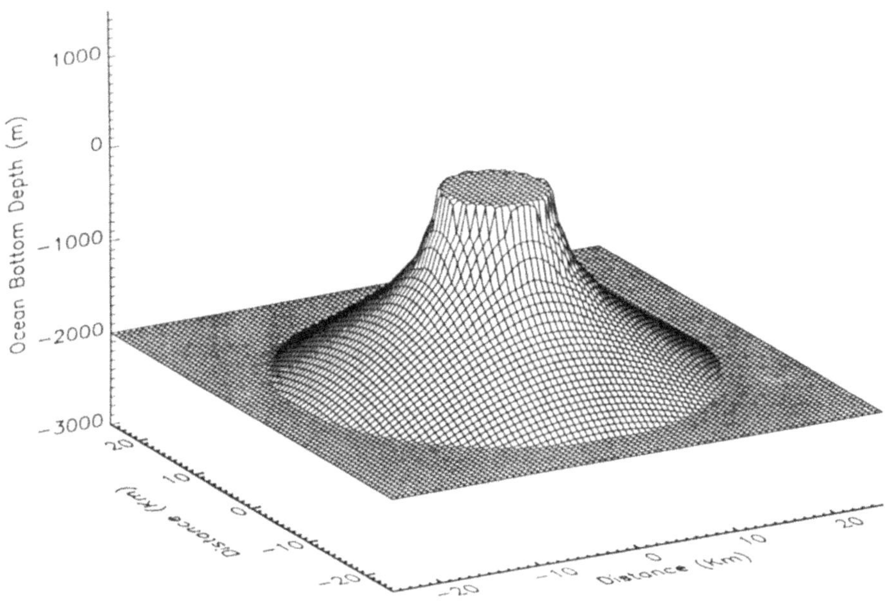

Figure 7

Sketch of the variable-depth sea floor used in the simulation with the radial depth profile of type UC. The island base has a 20 km radius, rising from a flat-bottom 2000 m-deep ocean.

this physical requirement, by comparing the solutions computed for the smaller and the larger basin.

If we examine the case of the initial dipole system in the vicinity of the island (see Figures 8 and 9), we can observe that initially the wave evolution seems to be basically similar to the constant-depth case: a separation of the disturbance into two equal dipoles is occurring, but with the important difference that presently the wave system approaching the island is deflected toward the coasts by the decreasing depth floor. A diffracted wave system of circular shape is further generated, radiating from the island towards the open ocean. Accounting for the real difference in this case and worthy of emphasis is the generation of a local system of edge waves that persists near the island coasts long after the tsunami passage (TINTI and VANNINI, 1994). The progression of these waves occurs as a rotation around the island with a gradual leakage of energy towards the open sea. At $t = 600$ s the waves have already lost much of their initial energy. We stress once more that this system of local waves is a feature that is absent in the constant-depth case, since the physical mechanism of their generation is founded on the refraction and hence on the variable sea-depth in the coastal island region.

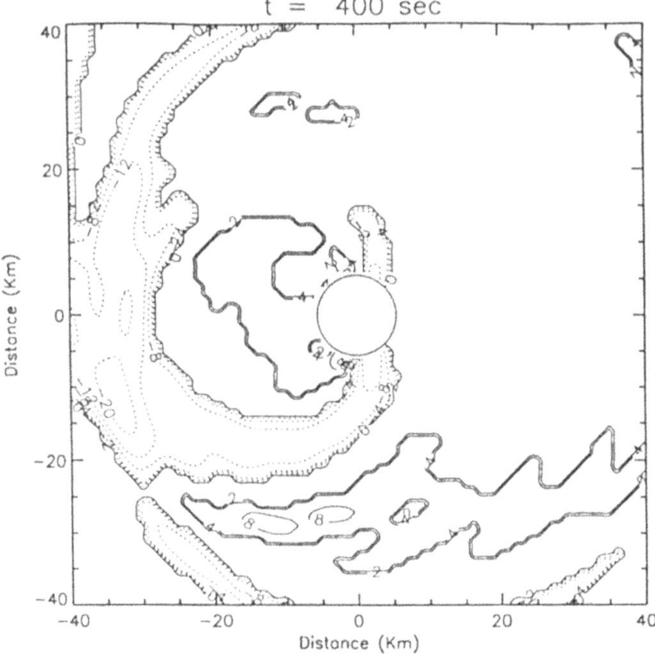

Figure 8
Wavefield computed 400 seconds after the initial dipolar disturbance (shown in Figure 4). At this time, the main wave system has already left the island. Two edge-wave systems have been produced that respectively rotate clockwise and counterclockwise around the island coasts, the latter being much stronger. Contour lines are given in centimeters.

Variable Bathymetry: Downwardly-concave Radial Depth Profile

The simulation described in this section concerns a radial depth profile showing a downward concavity near the island, as is illustrated in the sketch of Figure 10. This bathymetry will be hereafter called a type DC (Downward Concavity) bathymetry. The basin horizontal sizes ($r_1 = 5$ km and $r_2 = 180$ km) as well as the initial perturbation (see Figure 4) are the same as those used in the previous experiment, in order to allow a more expeditious comparison between the two cases. As many as 1200 eigenmodes are used in synthesizing the solution.

The evolution of the wave system can be briefly highlighted as follows with the aid of the sequence of Figures 11a–e covering a time span of 2000 seconds. The initial condition evolves normally, giving rise to two dipole waves, one approaching and the other leaving the island (see Figure 11a). Wavefronts travelling towards the island are intensely deflected towards the coasts since they feel a quite strong depth variation.

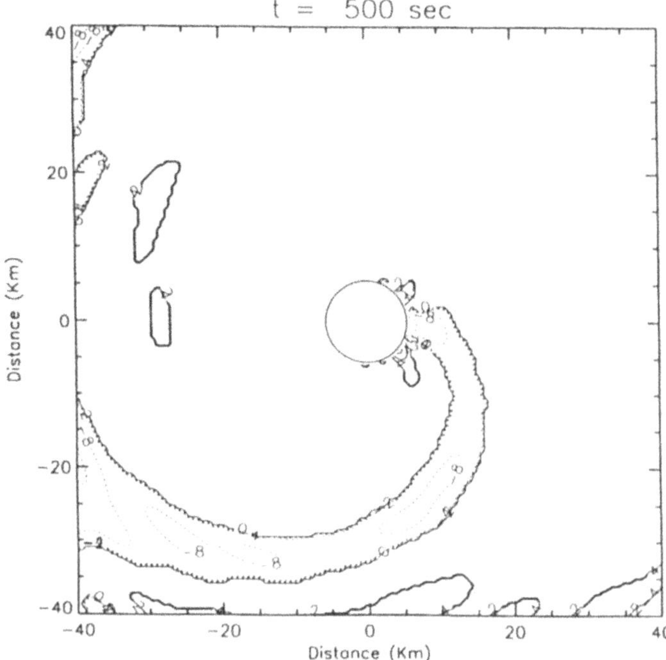

Figure 9

At a time of 500 seconds after the tsunami excitation, the edge-wave system propagating counterclockwise is practically the only wavefront involving the shallow water region around the island. Note that the front has an arc-like shape and is so remarkably long as to affect the deep ocean waters. Contour lines are in centimeters.

Here is the key difference with the upwardly-concave profile case: refraction is now a dominant feature and is able to considerably deform the wavefronts that assume a well pronounced arc-like shape, being intensely decelerated near the island (Figure 11b). At $t = 400$ s (Figure 11c) the diffraction system of waves which is radiating outward from the island coasts is already discernable. A strong refraction is expected to produce a robust system of local edge waves rotating around the island with a weak leakage of energy. This is confirmed by Figures 11d and 11e which delineate the wave fields at $t = 1000$ s and at $t = 2000$ s. The residual fronts emanate from the edge waves which continue travelling around the island with a significant amplitude, while the forcing dipole and the diffracted waves are already distantly far in the open sea. These edge waves probably persist considerably longer, however we cannot compute their evolution further in time with the present-size basin, because of the interaction with the waves reflected by the outer boundary. We notice that the edge waves can be distinguished into two different systems, one rotating clockwise and the other counterclockwise, the latter being remarkably stronger than the former. This is simply the effect of the initial

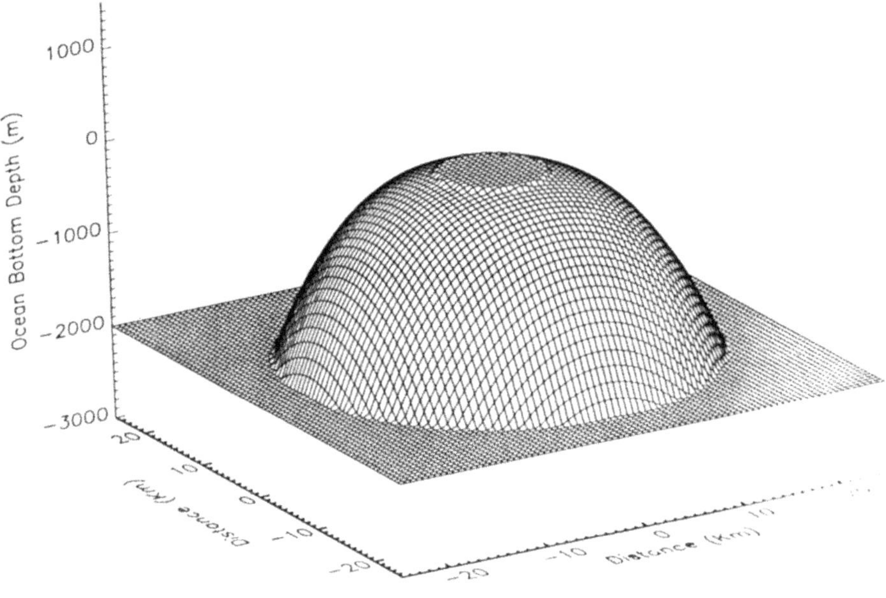

Figure 10
Sea-floor sketch concerning the tsunami simulation in the case of a radial depth profile of type DC. The island emerges from a 2000-m deep ocean with a 20-km radius axisymmetrical root.

disturbance asymmetry. Let us refer to the planar view of Figure 4. We can see that the initial perturbation is offset upward with its center placed in the first Cartesian quadrant. This implies that the most energetic waves initially attack the island against the coasts located in the upper half-plane and are progressively forced to a counterclockwise rotation. On the other hand, an initial perturbation predominantly placed in the fourth Cartesian quadrant would have produced a prevalent system of edge waves rotating clockwise.

In order to appreciate the level of interaction between the attacking wave and the island, we can compute the maximum amplitude reached by the wave during the entire interaction stage at any coastal points. If this amplitude is normalized over the amplitude of the initial disturbance, the wave amplification for different coastal positions ensues. Curves of such wave amplifications vs. position, which is identified by an azimuthal angle, are a common way to show the effect of an incoming wave against a circular island and have been used in prior investigations concerning plane monochromatic waves (see for example, LAUTENBACHER, 1970 and JONNSON *et al.*, 1976). Figure 12 shows the curve computed in the case of the DC bathymetry. The position on the coast is measured in degrees and increases clockwise, likewise the angular coordinate φ. The curve is asymmetric, reflecting the asymmetry of the selected initial perturbation which is placed predominantly in the first Cartesian quadrant. The curve of Figure 12 is smooth, but nevertheless it exhibits small-scale

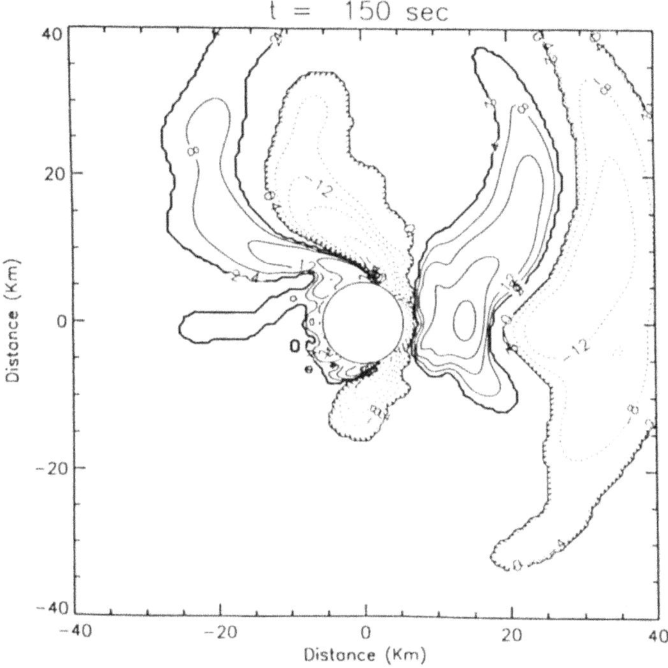

(a)

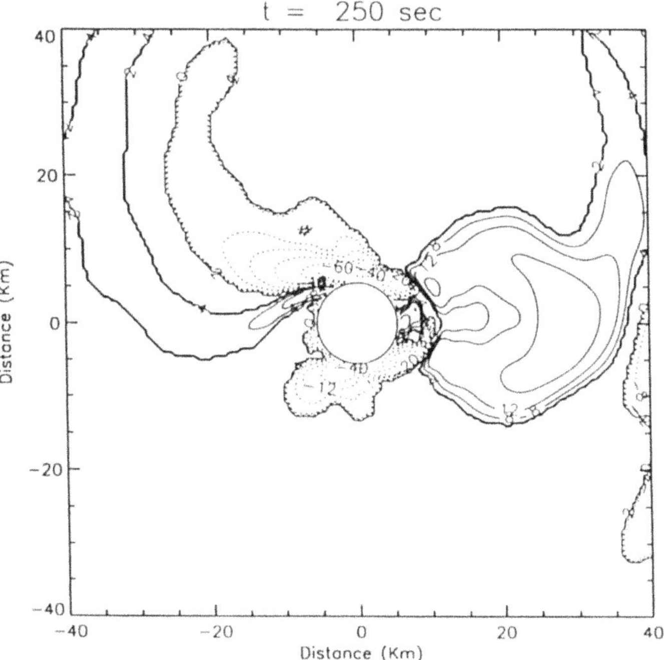

(b)

Figure 11(a,b)

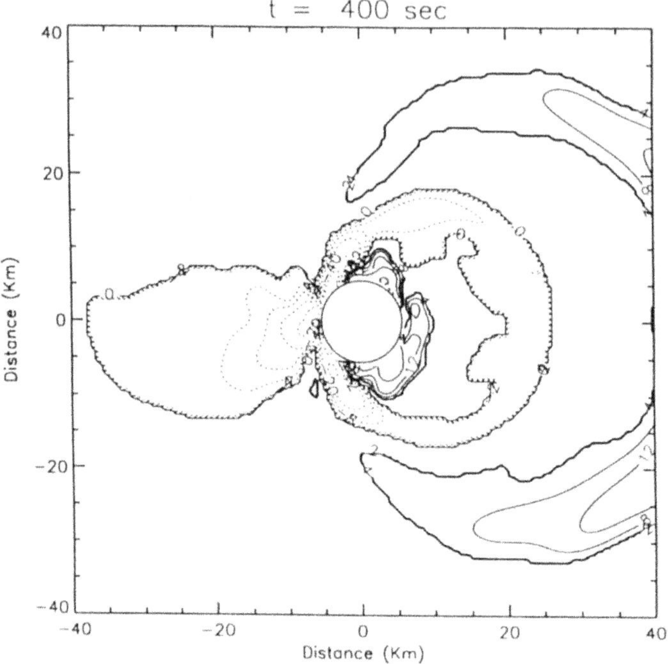

(c)

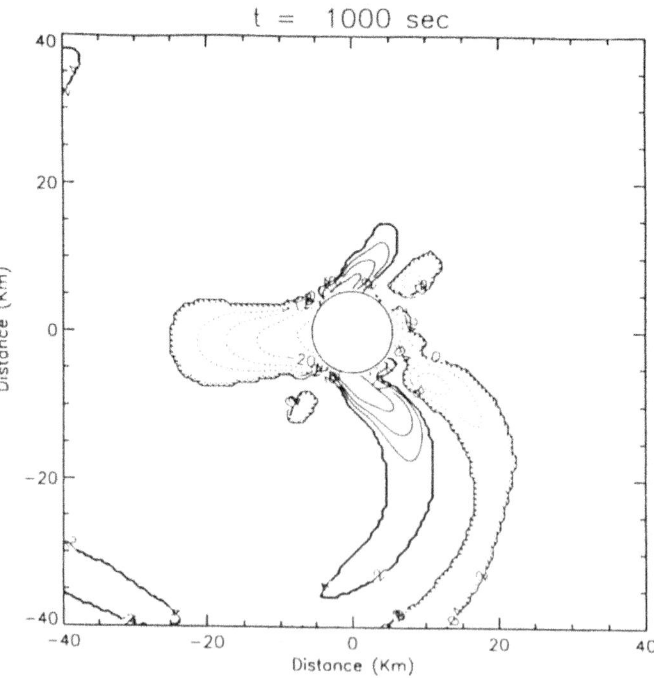

(d)

Figure 11(c,d)

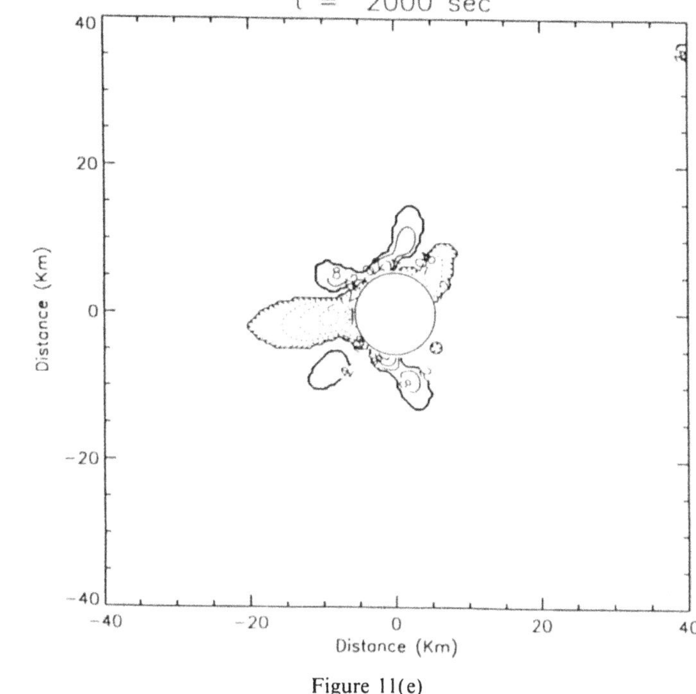

(e)

Figure 11(e)

Figure 11

Wavefield time evolution for a tsunami of seismic origin with the initial perturbation given in Figure 4. Contour plots with isolines given in centimeters are shown for different times: 150 sec (a), 250 sec (b), 400 sec (c), 1000 sec (d) and 2000 sec (e). At the first stages (a) and (b) the left dipole tends to embrace the island, the upper arm being stronger than the lower. At the intermediate stage (c), the main dipole systems have left the island, while the diffracted wavefront is forming. At the later stages (d) and (e), only the edge-wave systems persisting around the island coasts are observable. The system rotating counterclockwise is stronger. Notice the wave amplification in the lee of.the island where the two systems interfere. Notice further that the wave amplitude decays substantially along the front as the distance from the island increases.

perturbations which are the mathematical consequence of the solution (2) being computed through a truncated summation involving only a finite number of eigenfunctions e_{ikq}, which however in no way affects the significance of the results. It is interesting to note that a point facing the seismic fault ($\varphi = 0°$) does not experience the maximum amplification, which instead is higher in the lee of the island ($\varphi = 180°$) and attains its highest value around the position $\varphi = 120^c$. This effect is not surprising since it was previously observed in the case of an incoming plane wave (that is for a disturbance that exists at any time and in any spatial point and is characterized by infinite-length wavefronts) of wavelength comparable with the island scale. Nonetheless, it is important to remark that the effect is present even for a finite-size initial disturbance such as the ones we are able to take into account with our analysis.

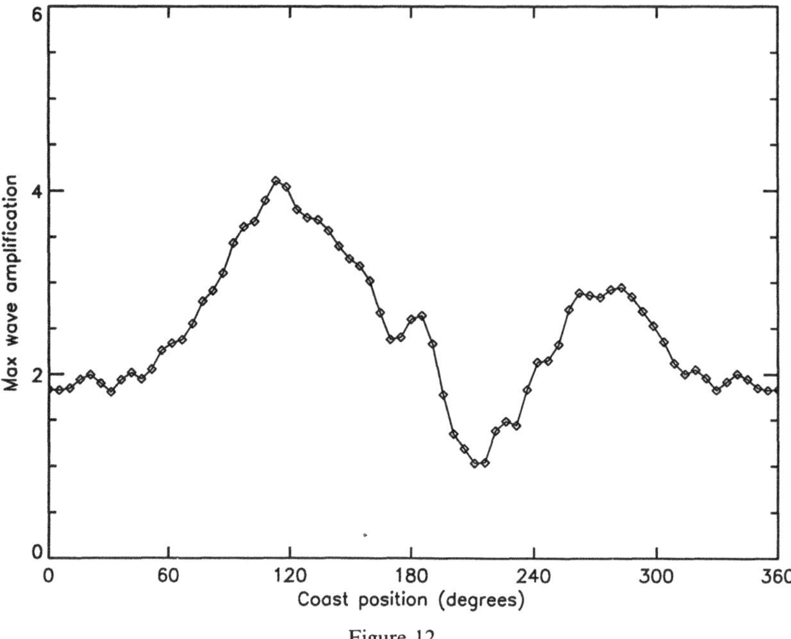

Figure 12

Maximum wave amplification vs. angular position on the coast measured counterclockwise from the point facing the seismic fault ($\varphi = 0°$). The amplification is defined as the ratio of the maximum amplitude the elevation takes at a given point during the wave evolution over the maximum amplitude of the initial disturbance, which in our experiments is about 50 cm.

The influence of the depth variation on the wave behavior was already emphasized by LONGUET-HIGGINS (1967) who was the first to observe that for axisymmetrical bathymetries where the depth $h(r)$ increases with the distance r from the center faster than r^2, i.e., where $d_r(h/r^2) > 0$, waves can be trapped near the island coasts. Our experiments can be compared with Longuet-Higgins's theory. The axisymmetric bathymetry of type DC we have used in the present section increases as $r^{2.86}$ in the belt comprised between $r_1 = 5$ km and $r = 20$ km, remaining constant from $r = 20$ km up to $r_2 = 180$ km, and is associated with a well-formed, long-duration system of trapped edge waves which agrees with Longuet-Higgins statement. Conversely, the type UC profile corresponds to a depth function $h(r)$ that is piecewise proportional to powers of r with an exponent smaller than 2. In this case while Longuet-Higgins would have predicted no trapping, we have observed as well the generation of a local system of edge waves, though they loose their energy rather quickly (see Figure 9), and certainly faster than in the case with downward concavity. It seems difficult, however, to define the meaning of trapping in a quantitative manner or in other words to define one discriminating parameter or a set of parameters on the basis of which it would be possible to discern what is and what is not a trapped wave. In this section we will not address this classification problem, but we will limit ourselves noting that both type UC and type DC axisymmetric bathymetries are able to induce a local system of waves, though with life duration and amplitude that are strongly influenced by the bathymetry features. In the next section we will investigate further the characteristics of these local waves, paying special attention to understanding their typical spectrum and amplitude.

Wave Amplification Near the Coasts

In the foregoing we have seen examples of edge waves induced by a tsunami attacking an ocean island, with the purpose of pointing out that the same source (namely a vertical dipolar displacement of the sea bottom causing a similar disturbance at the sea surface) produces quite a different system of waves in the coastal region, depending on the local bathymetry. No edge waves originate in a flat-floor ocean. In a variable-depth environment edge waves can be expected, and they are larger for type-DC radial profiles than for type UC. In this section, we will

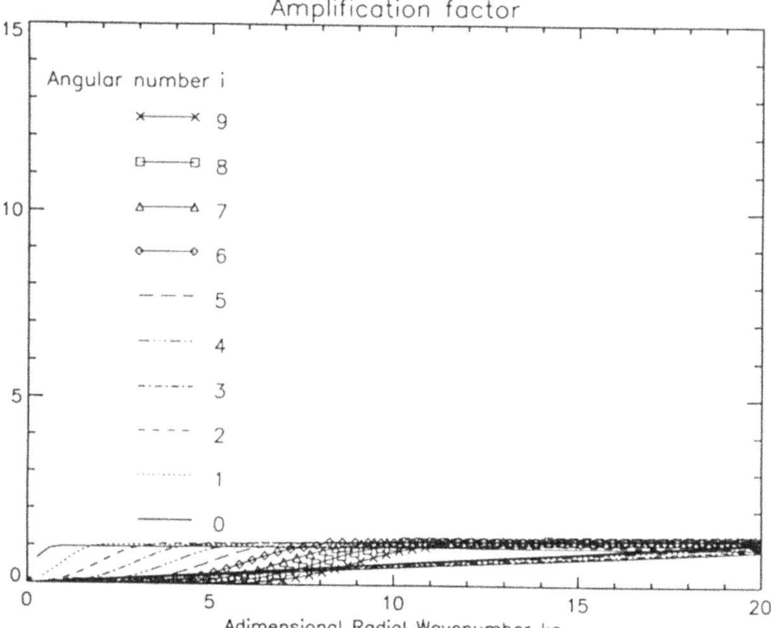

Figure 13

Amplification factor curves for the flat-bottom basin. No peaks are present, which means that no edge waves can be excited in this basin.

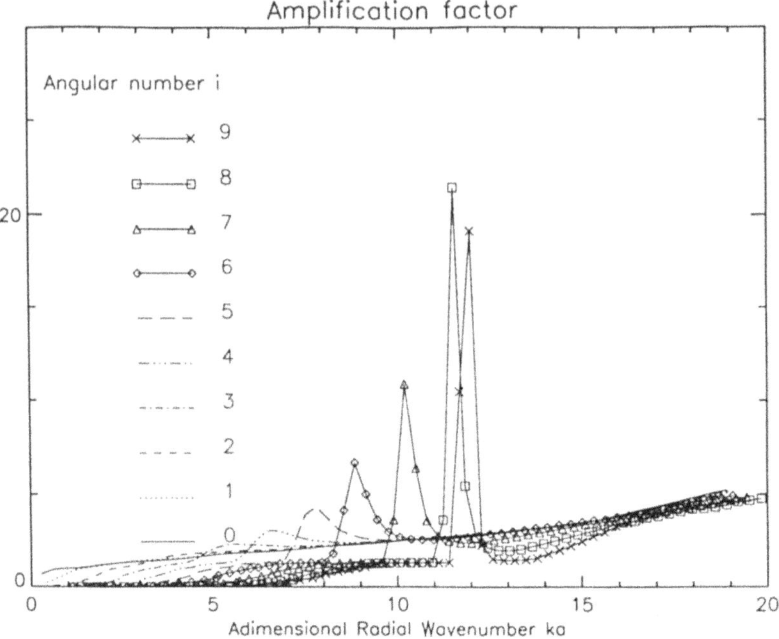

Figure 14
Amplification factor curves for the basin with type-UC bathymetry. The most relevant peaks have values
about 20 and are found for radial wavenumbers k_a larger than 10.

analyze the solution of our shallow-water linear problem, regarding it as the result
of the response of a physical linear system (i.e., the ocean basin surrounding the
island) to an external excitation (namely the initial perturbation). In this respect,
the solution depends on the source as well as on the physical system itself, whose
properties are fully contained in the corresponding set of eigenmodes. This consideration implies that, if a system is distinctly able to generate significant edge waves,
there must exist at least some eigenmodes of this system that are prone to edge
wave generation. Since in our basins with axisymmetrical bathymetries edge waves
have larger amplitude in the coastal region and are weak offshore, we introduce an
amplification factor A_{ikq} for each eigenmode e_{ikq}. This is defined as the ratio
between the maximum amplitude of the eigenmode in the inner region and its
maximum amplitude in the outer region. In our variable-depth examples the inner
region is obviously taken as the region of the coastal waters corresponding to the
underwater basis of the island ($5 \text{ km} \leq r \leq 20 \text{ km}$), whereas the outer region is the
deep flat-bottom ocean ($r > 20 \text{ km}$). Notice that we assume the same basin partition
even for the case with uniform depth, in order to make our comparative analysis
more straightforward. By this definition it descends that A_{ikq} depends only on the
properties of the radial factor ρ_{ik} of the eigenmode e_{ikq} (see the expression (1)) and

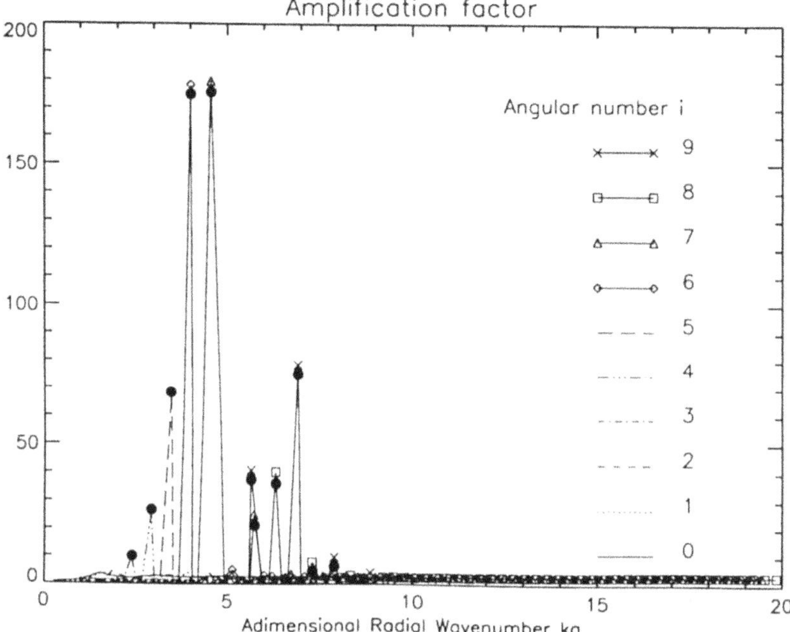

Figure 15

Amplification factor curves for the basin with type-DC bathymetry. The highest peaks exceed the value of 150 and are in the low wavenumbers range. Solid circles mark the modes that satisfy Longuet-Higgins's trapping condition (see discussion in the text).

therefore it does not depend upon the parity number q. This enables us to drop the subscript q and to denote the amplification factor simply as A_{ik}. Let us now designate with R the characteristic distance delimiting the inner and the offshore region, that is $R = 20$ km. It is then convenient to introduce the adimensional radial wavenumber $k_{a,ik}$ defined as

$$k_{a,ik} = \frac{\omega_{ik} R}{\sqrt{gH}} \tag{4}$$

where ω_{ik} is the eigenmode frequency and $H = 2000$ m is the offshore depth. We have computed the values A_{ik} for all three basins we took into account in the previous section in the range $0 \le i \le 9$ and $1 \le k \le 100$. For graphical convenience, these are plotted in the form of curves, each corresponding to a specific value of the angular wavenumber i, vs. the adimensional radial wavenumber k_a.

Let us first examine Figure 13 where the amplification factor curves corresponding to a constant-depth ocean are shown. All curves have small values (less than 1) for small wavenumbers k_a and tend to a flat platform of about unit value for higher wavenumbers. Evidently all eigenmodes have near-coast amplitudes smaller than or at most comparable with their offshore amplitudes. Let us now consider the case of a radial depth profile of type UC, whose amplification factors are depicted in Figure 14. The significant peaks observable in some curves indicate that there are special eigenmodes with coastal amplitude much larger than the corresponding open-ocean value. If we look at Figure 15 concerning our third case with type DC radial depth profile, we see that there exist very sharp peaks in the curves, two of which having values exceeding 150. Eigenmodes exhibiting such a conspicuous amplification must contain oscillations that are practically significant only in the inner region, being nearly negligible outside. The oscillation pattern of the eigenmodes corresponding to the heighest peaks of Figure 15 are displayed in the contour plots of Figures 16a and 16b. They have a multilobal structure consisting respectively of 4 and 5 wavelengths in the angular direction and, most importantly, their motion is virtually confined or "trapped" in the inner coastal region at a distance from the center shorter than $R = 20$ km.

The above observations support the idea that the eigenmode amplification factor A_{ik} can be used as an indicator for evaluating the degree of confinement or trapping of a given eigenmode. Eigenmodes with lower A_{ik} are less trapped or, we can say, less edge-wave-like than those associated with higher A_{ik}. This statement cannot be turned into a binary classification criterium because we are not able to specify a critical value of A_{ik}, delimiting two distinct classes of trapped and

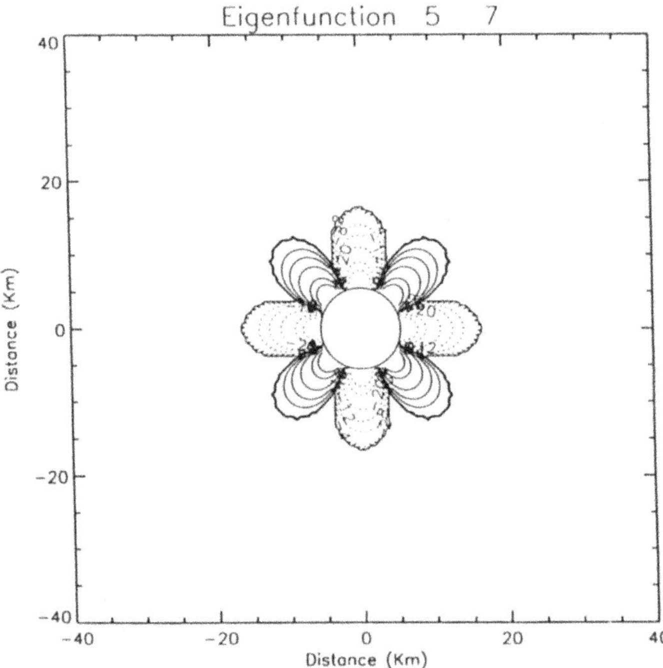

Figure 16(a)

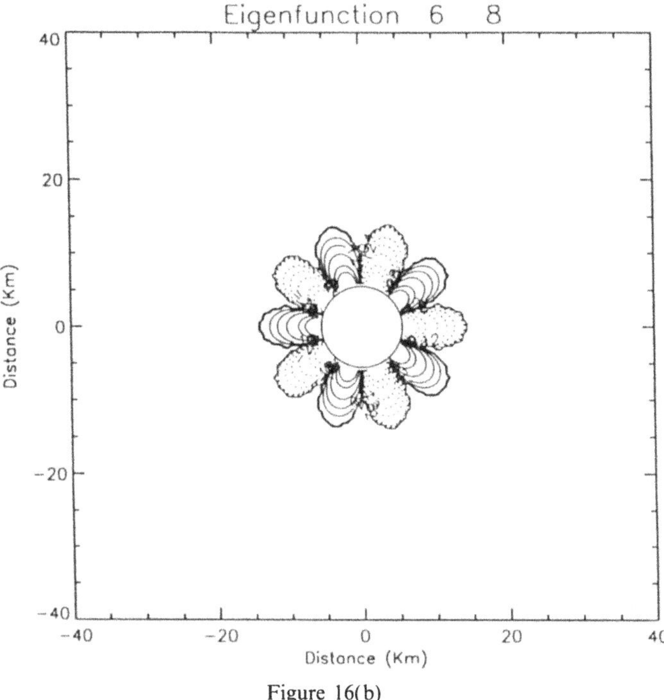

Figure 16(b)

Figure 16

Contour plots of the eigenmodes of the type-DC bathtmetry basin with the two highest amplification factors: (a) mode with $i = 5$ and $k = 7$; (b) mode with $i = 6$ and $k = 8$. Both modes are strongly trapped in the region, with variable depth forming the underwater base of the ocean island.

untrapped eigenmodes. We simply affirm that the trapping character of an eigenmode is the more marked, the larger is A_{ik}. At the lower extreme, where A_{ik} is about 1 or smaller, there is no trapping at all. Furthermore, if an amplification factor curve shows high and narrow peaks, the uppermost eigenmodes of the peaks are considerably more trapped than their neighbors; and therefore, assuming this point of view, it can be stated that peaks identify the trapped eigenmodes of the basin. Following this idea, from inspecting Figure 13 we can conclude that a flat basin has no trapped eigenmodes, whereas from Figures 14 and 15, we deduce that both basins either with UC- or with DC-type bathymetry do possess trapped eigenmodes, though with quite different degree of trapping.

The LONGUET-HIGGINS (1967) approach to distinguish trapped modes appears inappropriate for our examples. In a basin with a simple bathymetry consisting of a flat sill rising from a deeper flat ocean with depth H, he proposed that a necessary condition for a mode to be trapped is that the critical mode radius r_{ik} be greater than the typical radius r_t. In this inequality r_t was taken as the sill radius, while r_{ik} was defined as $r_{ik} = i\sqrt{gH}/\omega_{ik}$, where i is the angular wavenumber. We have

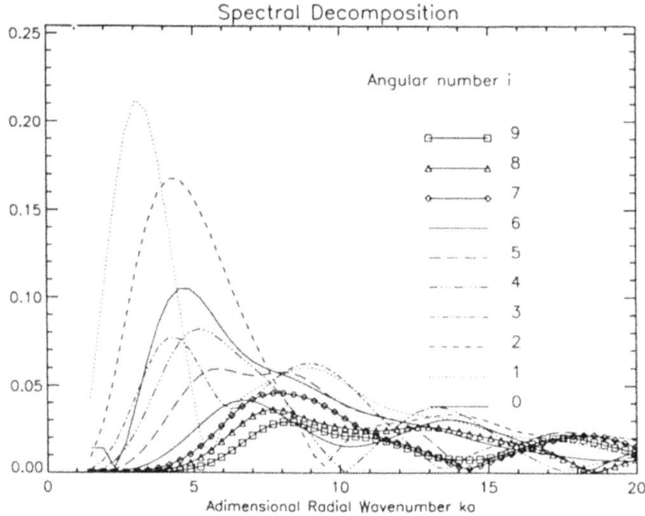

Figure 17

Spectral decomposition of the initial tsunami disturbance given in Figure 4 for the flat-bottom basin. Square root values Γ_{ik} computed from the spectral coefficients γ_{ikq} (see expression (5) in the text) are shown. Each curve corresponds to a given constant value of the angular wavenumber i and is plotted vs. the radial wavenumbers suitably adimenionalized.

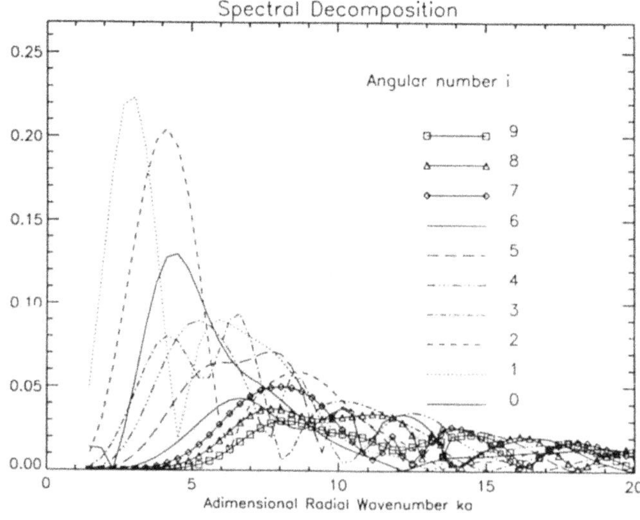

Figure 18

Spectral decomposition as in Figure 17, but for the basin with bathymetry of the type UC.

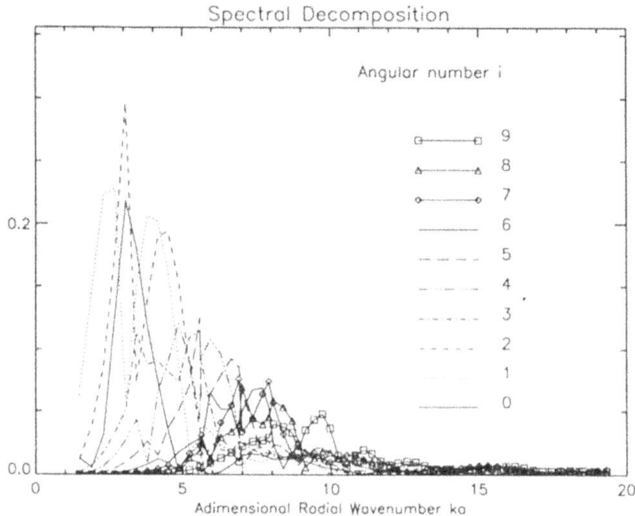

Figure 19
Spectral decomposition as in Figure 17, but for the basin with bathymetry of the type DC.

applied Longuet-Higgins's condition to the eigenmodes of our variable-depth basin, taking as the typical radius r_t the radius $R = 20$ km delimiting the inner from the outer regions. No eigenmode of the UC-type bathymetry is found to match this condition, which means that Longuet-Higgins's method would conclude that no energy trapping is possible for such a basin. Conversely, several eigenmodes fulfilling the condition are found in the DC-type bathymetry case. They are marked with a solid circle in Figure 15. As may be seen, Longuet-Higgins's modes coincide with the modes at the top of the peaks in the shown amplification factor curves, which means that the two criteria, Longuet-Higgins's and ours, are in agreement in this case.

In the foregoing discussion we have substantiated the idea that each water basin may contain a number of free oscillation modes that are the most outstandingly trapped. To distinguish the well-formed edge waves in the final solution of our linear problem, the trapped eigenmodes must be excited by the initial sea-surface disturbance. In Figures 17–19 we have plotted the spectral decomposition of the seismically-induced initial perturbation that is displayed in Figure 4 and that was used in the numerical experiments described in this paper. As spectral decomposition we mean the square-root values Γ_{ik} of the sum of the squared coefficients γ_{ikq}, $q = 1, 2$ defined in the expression (2), properly normalized, that is

$$\Gamma_{ik} = \frac{\sqrt{\gamma_{ik1}^2 + \gamma_{ik2}^2}}{\Sigma_{ikq}\gamma_{ikq}^2}.$$ (5)

Like the amplification factors A_{ik}, the spectral coefficients Γ_{ik} are likewise plotted as curves of constant wavenumber i vs. the adimensional radial wavenumber

k_a. By inspecting the Figures, it may be observed that all spectral curves decay substantially as k_a increases, which i) is consistent with the fact that the summations (2) and (3) must be convergent and moreover ii) enables us to truncate the summations to a limited, though sufficiently high, number of eigenmodes. Essentially for our discussion is however the spectral energy confinement in the lower radial wavenumber region. Viewing Figures 18 and 19, we see that most of the spectral power is in the range of $k_a < 10$, which means that the initial condition we have selected is capable of exciting those trapped eigenmodes falling in that wavenumber interval. Since in the type-UC bathymetry basin all significantly trapped modes correspond to higher values of k_a (see Figure 14), this explains in a simple and direct way why only weak and short-duration edge waves are generated in this case. A source of smaller size, with a spectral power peaked over higher wavenumbers would have more suitably given rise to stronger edge waves. The case of the DC-type bathymetry is the most favorable for edge wave generation, because the source spectrum is especially energetic just in that wavenumber range where the most conspicuous trapped modes can be found.

Conclusions

The main conclusion that can be drawn from our work is that tsunami wavefronts attacking ocean islands produce local systems of waves rotating around the island coasts as the effect of the strong wave refraction caused by the considerable depth changes. The characteristics of these rotating edge waves, such as their amplitude, their life time and their dominant periods, derive from the joint combination of the basin bathymetric features and the source features. From a mathematical point of view, the former control the set of the basin eigenmodes determining what eigenmodes are trapped and the degree of trapping. Oppositely, the source spectrum acts as a selective operator, exciting the trapped modes differentially so as to determine those which will be predominant in the final solution.

Acknowledgements

This work has been partly financed by funds from MURST (the Italian Ministero dell' Università e Ricerca Scientifica e Tecnologica) and partly by means of funds from the European Community: EVCV-CT92-0175.

REFERENCES

JONNSON, I. G., SKOVGAARD, O., and BRINK-KJAER, O. (1976), *Diffraction and Refraction Calculations for Waves Incident on an Island*, J. Mar. Res. 34, 469–496.

LAUTENBACHER, C. C. (1970), *Gravity Wave Refraction by Islands*, J. Fluid Mech. *41*, 655–672.

LONGUET-HIGGINS, M. S. (1969), *On the Trapping of Wave Energy Round Islands*, J. Fluid Mechanics *29*, 781–821.

MYSAK, L.A. (1968), *Edge Waves on a Gently Sloping Continental Shelf of Finite Width*, J. Mar. Res. *26*, 24–33.

PINETTES, M. J. (1980), *Diffraction d'une onde solitaire par une île circulaire*, J. Eng. Math. *14*, 207–218.

SATAKE, K., and SHIMAZAKI, K. (1987), *Computation of Tsunami Waveforms by a Superposition of Normal Modes*, J. Phys. Earth *35*, 409–414.

TINTI, S., and VANNINI, C. (1994), *Theoretical Investigation on Tsunamis Induced by Seismic Faults near Ocean Islands*, Marine Geodesy. *17*, 193–212.

URSELL, F. (1952), *Edge Waves on a Sloping Beach*, Proc. Roy. Soc. *214*, 79–97.

YEH, H., IMAMURA, F., SYNOLAKIS, C., TSUJI, Y., LIU, P., and SHI, S. (1993), *The Flores Island Tsunamis*, EOS, Trans. Am. Geophy. Union *74*, 369–373.

(Received August 10, 1994, revised November 24, 1994, accepted December 28, 1994)

PAGEOPH, Vol. 144, No. 3/4 (1995)

0033–4553/95/040621–11$1.50 + 0.20/0
© 1995 Birkhäuser Verlag, Basel

Source Model of Noto-Hanto-Oki Earthquake Tsunami of 7 February 1993

KUNIAKI ABE[1] and MASAMI OKADA[2]

Abstract — A source model was discussed for a small tsunami accompanied by the Noto-Hanto-Oki earthquake (M_s 6.6), striking Japan on 7 February, 1994. Assuming a fault model under the sea bottom, we estimated the focal parameters jointly, using synthesized tsunami source spectra as well as the tsunami numerical simulation. The fault proposed by this study consists of a plane sized 15×15 km, dipping N47°W with the dip angle of 42°, which is almost pure reverse fault (slip angle 87°) with a dislocation of 1 meter. The numerical simulation shows that the shallow sea in the source region caused a comparatively long recurring tsunami (the periods are 12–18 minutes) in spite of its small size. The model fault is corresponding to an aftershock area of this earthquake.

Key words: Noto-Hanto-Oki earthquake tsunami, source model, spectral synthesis, numerical simulation.

Introduction

AIDA (1978) showed that it is effective to use a deformation of sea bottom estimated by the fault model for modeling a tsunami source. MANSINHA and SMYLIE's (1971) theory assumes to take a rectangular plane as the fault plane and the displacement of sea bottom due to the fault motion at an arbitrary point are given by position, size and direction of the fault. However, there is no method to find the best parameters uniquely. However, some constraints derived from seismic wave and tsunami, yield better fault model solutions. Systematically comparing the waveform in the numerical simulation with the observed one we can construct a tsunami source model. Positively, ABE (1993) proposed a spectral synthesis to reproduce the observed spectra. The value is found in the use of a predominant frequency which is sensitive to the fault size. This advantage counterbalances a disadvantage of spectral-transform process of direct records. On the other hand, numerical simulations simulate waveforms as integrated results of initial and boundary conditions. This method is inadequate to examine an effect of

---

[1] Niigata Junior College, Nippon Dental Univ., Hamauracho 1–8, Niigata, 951, Japan.
[2] Meteorological Research Institute, Nagamine 1-1, Tsukuba, 305, Japan.

each fault parameter. Therefore, a combined method of numerical simulation with the spectral synthesis is proposed to construct a source model and applied to the small tsunami observed at the Japan Sea on 7 February, 1993, accompanied by the Noto-Hanto-Oki earthquake (M_s 6.6) in the central part of Japan.

Observed Tsunami

The earthquake (M_s: 6.6), which had an epicenter of 37°39'N and 137°18'E at the continental shelf in the Japan Sea off Noto Peninsula, Central part of Japan, and the origin time of 13 h 27 min 43.7 s (UT), 7 February 1993, was accompanied by the small tsunami recorded at several tide stations. The map near the Noto Peninsula including the epicenter is shown in Figure 1. The maximum double amplitude observed was 0.5 m at the nearest tide station, Wazima. The first tsunami waves were observed as a sharp rise at Wazima and as a gentle rise at Naoetsu.

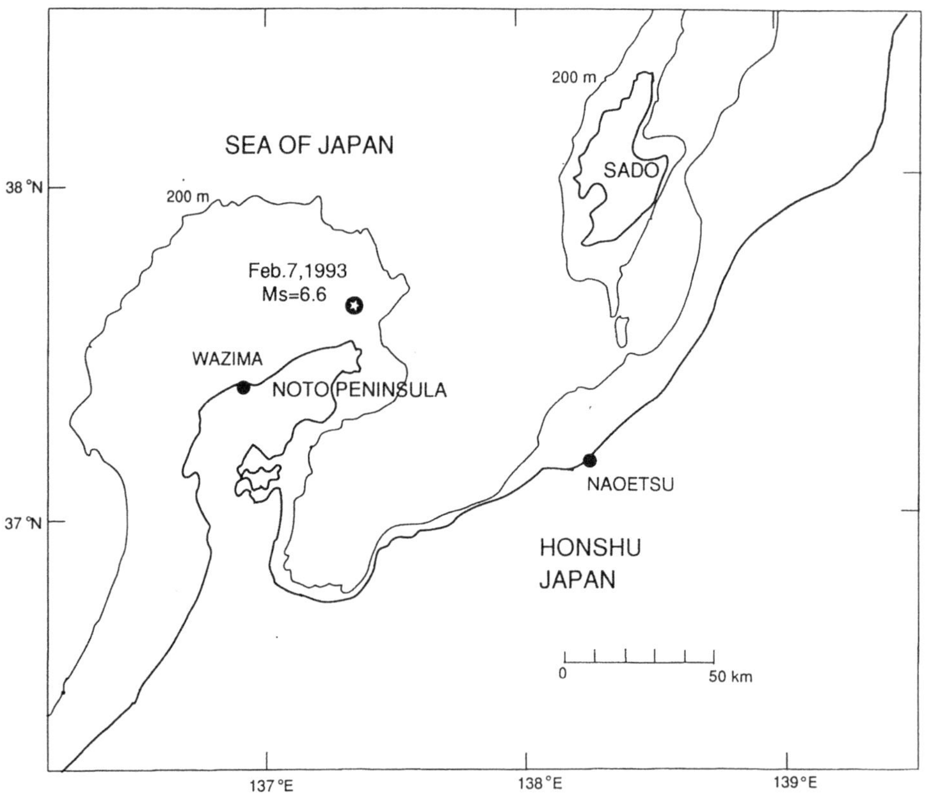

Figure 1

Computation area of numerical simulation. The epicenter of main shock and tide stations are shown with star and solid circles, respectively.

However, at the other distant stations the tsunamis were masked in the background noise such that the first waves were barely identified. Therefore, it was one of the smallest tsunamis which was detectable.

The period can be estimated by a ratio of wavelength to phase velocity. The observed periods of 18 and 12 minutes in the first waves at Wazima and Naoetsu, respectively, were explained from wavelengths of 37 and 25 km, by assuming a water depth of 120 m at the source.

Tsunami waveforms were obtained by removing the astronomical tide from the original records, and the spectra for 6 hours from the origin time of the earthquake were calculated for comparisons.

Numerical Simulation

The linearized shallow water theory is used as the governing equation and is descretized by the leap-frog scheme. The initial condition of the tsunami is given by vertical displacement of ground surface calculated from the fault model (MANSINHA and SMYLIE, 1971), and assumed to be completed through rise time of 10 s. Total reflection and transmission are assumed at land-sea boundary and open-sea boundary, respectively. The computational area, shown in Figure 1, is divided into a grid space of 2 km and 1 km intervals and water level is followed with a time step of 5 s and 3 s, respectively. Using the coarse mesh we studied the strike and depth dependence of the model, and for the best model we computed waveforms using a fine mesh of 1 km. Time histories of water elevations were computed at Wazima and Naoetsu for comparisons with tidal records.

Spectral Synthesis

The observed tsunami spectrum is explained by a synthesized spectrum of the source with shelf response (ABE, 1993). In the synthesis, tide gage response and propagation effects from source to shelf are assumed to be flat to frequency. Generally, tide gage decays shorter period components than a critical period. The tide gage responses at Wazima and Naoetsu were measured by SATAKE *et al.* (1988) and the recovery times for 1 m difference were 255 and 450 s, respectively. Taking into consideration the 0.1 m rise of this tsunami, we can estimate recovery times of 85 and 150 s, respectively, applying their formula. This condition shows that the critical period is less than about 3 minutes. Thus, we can expect a flat response for the period range (frequency range between 0 and 0.002 Hz).

In the synthesis, source spectrum is analytically expressed by YAMASHITA and SATO (1980), and shelf response by ABE and ISHII (1980). Modeling of the shelf response assumes a flat shelf of depth h_1 with a width of l accompanying a ramp of

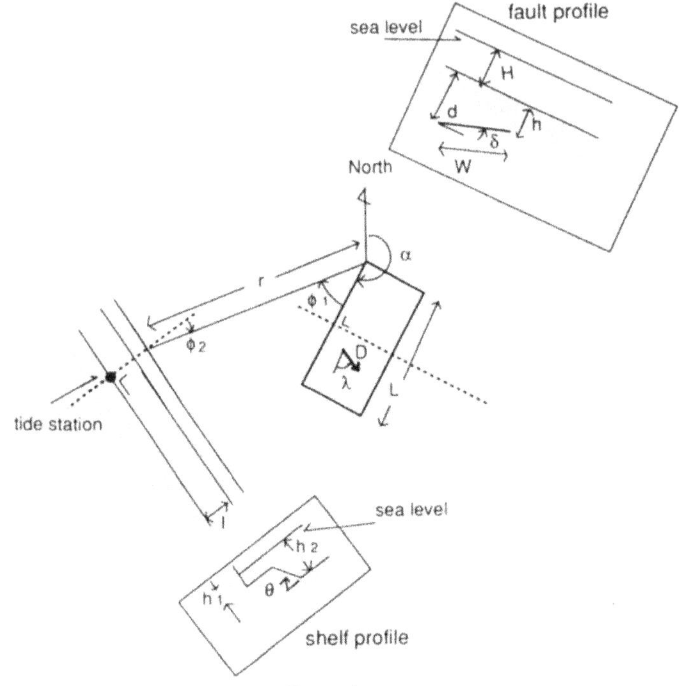

Figure 2
Model fault, model shelf and the relative position with definitions of parameters.

a constant slope θ from the open sea with a constant depth of h_2, and an incident angle of ϕ_2. The definitions of source and shelf models are shown in Figure 2. In the application of source spectrum unilateral fault motion is assumed identical to the original paper by YAMASHITA and SATO (1980).

Comparison of Various Models with Observations

1. Size Effect

The length and width of the fault affect the tsunami spectra. Source spectra, which are expected at the margin of the shelf off Wazima (propagation distance, r, 57 km and azimuth angle, ϕ_1, 10°), were computed for various fault sizes and shown in Figure 3. For the computation, fault parameters of directions are assumed identical to those obtained from the CMT solutions of seismic wave, as described later. These are 42° in dip angle, 87° in slip angle and 223° in azimuth angle. Fault sizes are reflected to periods of tsunami. As a range of fault size, 10 km to 20 km, is assumed from sizes of wavelength predicted from observed periods of the first

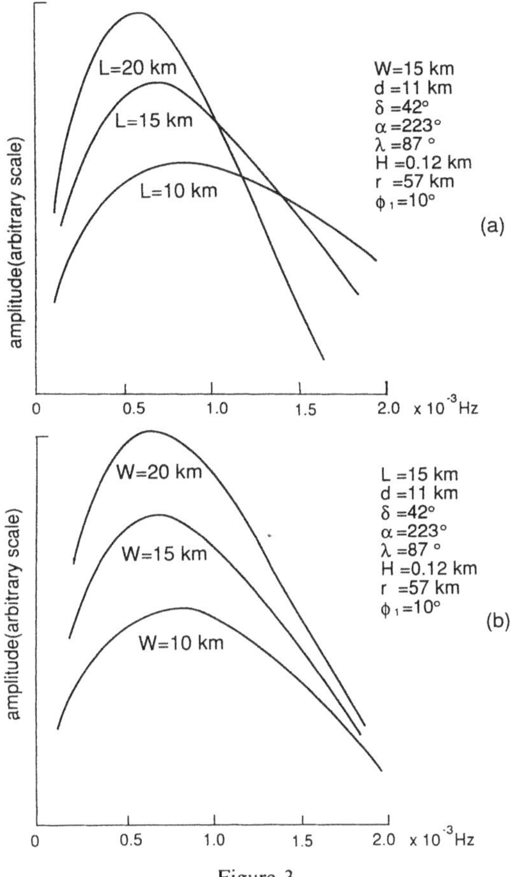

Figure 3

Fault size dependence of source spectra expected at sea off Wazima. Effects of fault length (a) and width (b).

waves. The source spectra are estimated as various combinations of the fault length and width within the range. The result indicates that the increase of fault length contributes to a large predominant period and the width accelerates a sharpness of the spectral peak. The relation is understood from the azimuth direction of Wazima and located in the direction of the long axis of the source. Thus it is concluded that the size of 15 km × 15 km is a reasonable one because both the observed spectra are explained in the general trend from the synthesized spectra as shown in Figure 7. As for a difference of peak frequency between observation and synthesis at Wazima, it is explained that the peak frequency component is a wave radiated in the direction normal to the fault strike and propagated along a curved path, due to a gentle slope of sea bottom to the open sea. In the spectral synthesis this component is not explained because the source is assumed to be located on a flat sea bottom.

2. *Azimuthal Effect*

Azimuthal effect was studied in the numerical simulation varying the strike angles. Three cases around the strike angle of 223°, derived from the CMT solutions, were considered as shown in Figure 4a, and waveforms are shown in Figure 4b. Fault sizes and other parameters are kept constant through the process. It is pointed out that the azimuth dependence of the computed waveform is lacking clarity but discrepancies of arrival times are broad between model and observation for Model A to Wazima and Model C to Naoetsu. The result suggests that model B best explains the observed waveforms, and the strike of 223° is estimated.

3. *Depth Effect*

Numerical simulation reveals that a shallow fault generates a sharp rise of the tsunami and is a good approximation of the observed waveforms as displayed in Figures 5a and 5b. Three cases are treated for the depth. Those are 1, 5 and 10 km as upper depth of the fault. The shallowest case produces the best agreement with

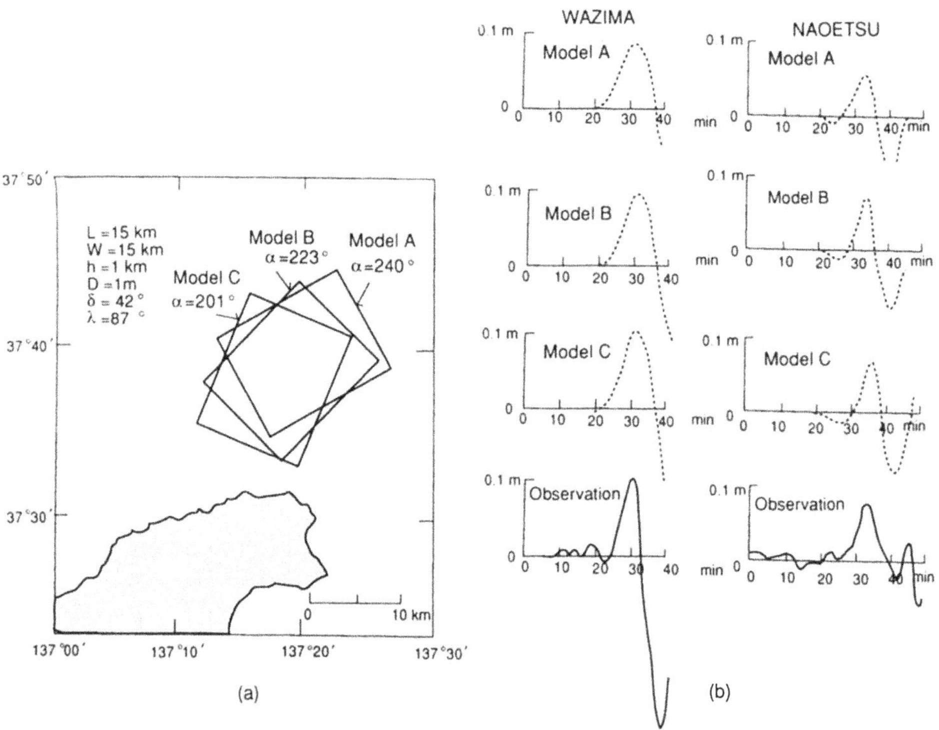

Figure 4
Fault strike dependence of waveforms. Positions of Models A, B and C (a) and the computed waveforms
from the models with observed waveforms (b).

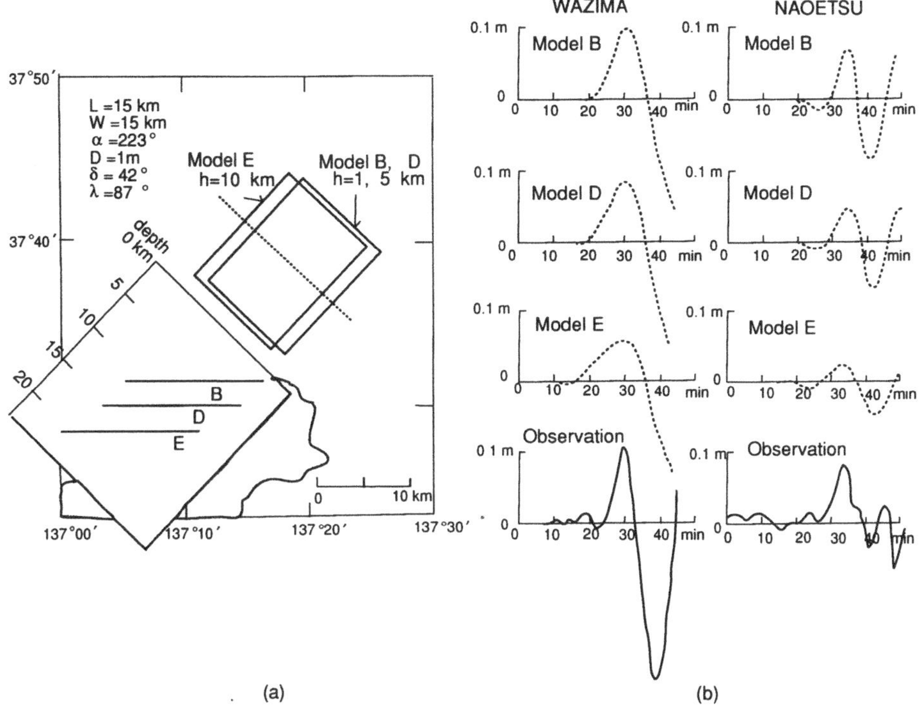

Figure 5

Fault depth dependence of waveforms. Positions of Models B, D, E and the cross section (a) and the computed waveforms from the models with observed waveforms (b).

the observed waveforms. This result suggests that the fault nearly reached the sea floor.

4. Other Parameters

Fault sizes of 15 km × 15 km and fault upper depth of 1 km were determined in the above consideration. Unit dislocation (1 m) was presumed through this simulation. Since dislocation is reflected in tsunami amplitude, it is estimated from a comparison of amplitude. The linear basic equation makes it possible to invert the dislocation. In the current comparison there is no reason to change the value, because the assumed dislocation explains the observed amplitude. Accordingly, a dislocation of 1 m is reasonable to the observed amplitude.

Dip angle of 42° and slip angle of 87° are taken from seismic source mechanisms as Centroid Moment Tensor solutions determined by Harvard University and Earthquake Research Institute, Tokyo University. The assumed parameters show no conflict with observed waveforms. The slip angle of 87° gives us a good approximation of pure reverse fault. The dip angle of 42° is almost equal to a critical angle, which is a changing point of subsidence area at the ground surface from a lower side to an upper side on the center line of a pure dip-slip fault (e.g., MATSUURA and SATO, 1975). Since the Naoetsu tide station is located in the direction normal to the strike direction, the polarity of initial wave is sensitive to the variation of dip angle. If the dip angle is larger than this critical angle, we can expect a definite upward motion of the first arrival and inversely we expect down motion. The observed one is not definite regarding up or down and the estimated dip angle is the approximately critical value.

Source Model

The best model is represented by a fault with the length of 15 km, width of 15 km, upper depth of 1 km, dip angle of 42°, slip angle of 87°, strike angle of 223° and dislocation of 1 m. The comparison was conducted on the results with a 2 km mesh, and the best Model B was determined. For the best model, a computation

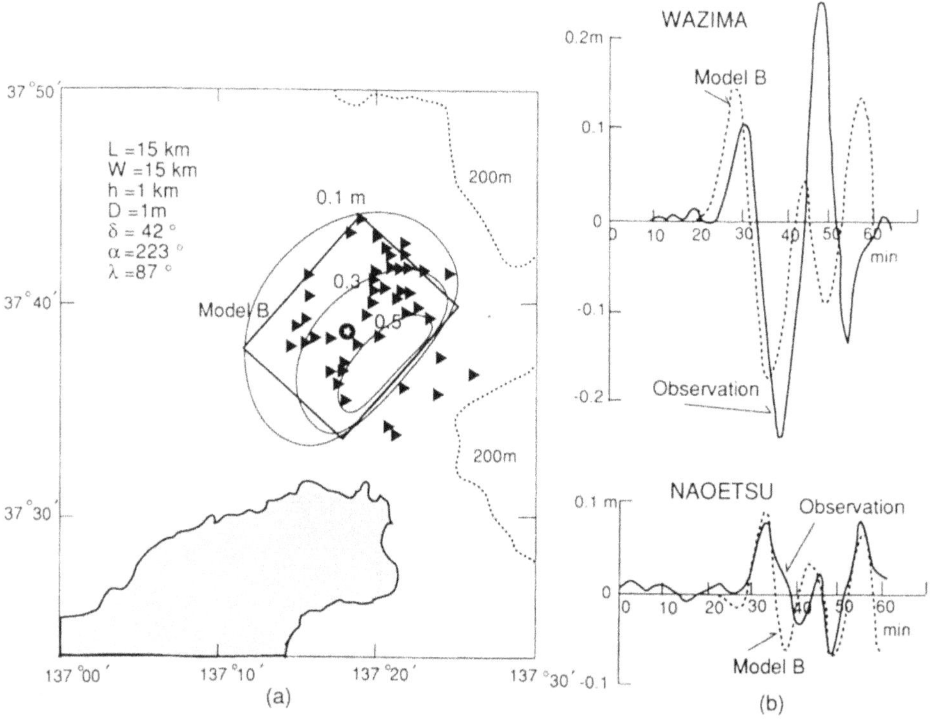

Figure 6
The location of the best Model B with aftershock distribution (triangles) observed by Japan Meteorological Agency from 22:27 to 24:00, Feb. 7, 1993 (a) and comparison of computed waveforms with observed waveforms at Wazima and Naoetsu (b).

was carried out using a finer mesh of 1 km and time interval of 3 s for the same area as the former coarse one. The model is shown in Figure 6a with the aftershocks obtained by the Japan Meterological Agency (HASHIMOTO *et al.*, 1993) and the waveforms computed from the model are shown in Figure 6b with the ones observed. The figure shows that the resolution was improved. The result reaffirmed our conclusion that the Model B explains the observed amplitudes and periods in the initial stages of the waveforms. A small phase shift between observation and computation is within the propagation time for one mesh at the coast. The coincidence of aftershock area with the fault plane supports the idea that the tsunami and aftershocks are related to the same fault. The improved result suggests that a descretization error was reduced because of a mesh interval 1 km smaller than one twentieth of wavelength, which is nearly equal to twice the fault length (e.g., IMAMURA *et al.*, 1990). However, the former result of the coarse mesh is valid for the comparison due to observations of the relative effect.

Spectrum synthesis was conducted for the Model B and the results were compared at Wazima and Naoetsu as shown in Figures 7a,b, respectively. The

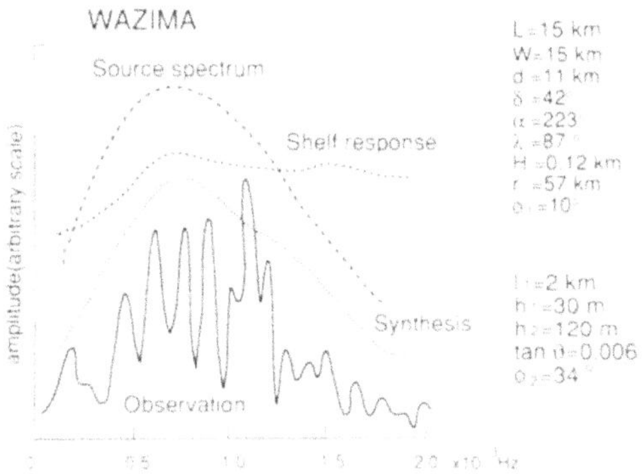

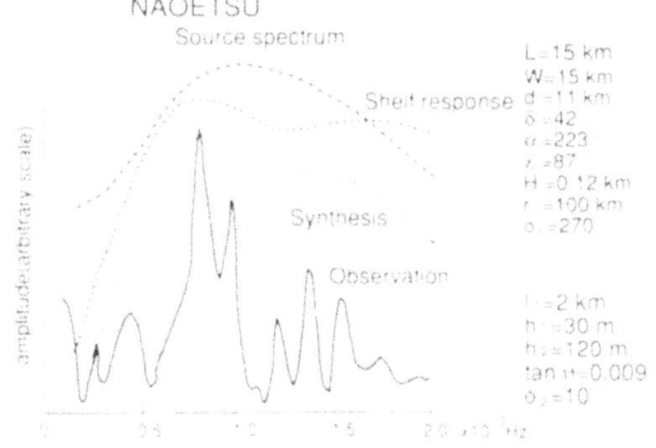

Figure 7

Synthesized spectra at Wazima (upper) and Naoetsu (lower) and the comparison with observed spectra.

synthesis consists of the source spectra and shelf responses corresponding to Wazima and Naoetsu. It is demonstrated that the shelf responses contribute to the modification of the source spectra and general trends of the observed spectra follow the modification. A specifically observed spectrum at Wazima is explained by a general trend of the synthesized spectrum, and Model B is also supported by this agreement.

Discussion

It is affirmed that an earthquake of magnitude, 6.6, and the associated tsunami were modelized from a fault formation. The tsunami propagation was explained within a linear shallow water theory. The size of 15 km is long enough in comparison with a water depth of 0.12 km. Accordingly, long wave approximation in the shallow water theory was valid and the nonlinear effect was neglected from the small amplitude.

It is known that there is a directivity in the radiation of the tsunami and a large amplitude is expected in the direction normal to fault strike (e.g., YAMASHITA and SATO, 1980). In this case the normal direction is towards the Naoetsu tide station. The fact that the tsunami amplitude of the first wave of Naoetsu was comparable to that of Wazima, in spite of longer propagation distance is harmonious with the directivity of tsunami propagation.

Regarding the tide gage records, the observed initial heights did not reach the initial level (0.6 m) generated by the coseismic deformation, which means that the tsunami did not develop itself. The decrease of amplitude is possibly explained by the small size of the source and the geometrical spreading. It is noted that the period is comparatively long despite the small fault size. The period of 12–18 minutes is approximated by a period of wave motion derived from the wavelength (nearly equal to twice the fault length or width/cos δ) and the long wave's phase velocity at the source.

In this model the seismic moment is 6.75×10^{18} Nm, assuming a rigidity of 3×10^{10} N/m. Recently MASUI *et al.* (1994) explained a strong ground motion due to the 1993 Noto-Hanto-Oki earthquake from the fault model of 15 km × 15 km with the seismic moment of 6.75×10^{18} Nm.

Conclusion

A tsunami source model was derived for the 1993 Noto Peninsula earthquake (M_s: 6.6) in the joint use of numerical simulation and spectral synthesis. The observed tsunami was explained with the model in waveforms and spectra. The source model was represented by a fault with a length of 15 km, width of 15 km, upper depth of 1 km, dip angle of 42°, slip angle of 87°, strike of 223° and a dislocation of 1 m. The source is located at the shallow shelf off the Noto Peninsula and the comparatively long-perioded tsunami was generated in spite of the small size of the fault plane. The estimated fault plane coincides with the aftershock area. The seismic moment is 6.75×10^{18} Nm with an assumption of rigidity of 3×10^{10} N/m^2.

Acknowledgment

We thank personnel at the Wazima and Naoetsu tide stations for providing the records. We also are grateful to Dr. Y. Tsuji for his cooperation in collecting tide gage records and Dr. Y. Ishikawa for his fruitful discussion.

REFERENCES

ABE, Ku. (1993), *Tsunami spectrum as a synthesis of source spectrum and shelf response.* In Proceeding of the IUGG/IOC International Tsunami Symposium, pp. 151–163.

ABE, Ku., and ISHII, H. (1980), *Propagation of Tsunami on a Linear Slope between Two Flat Regions. Part II Reflection and Transmission,* J. Phys. Earth 28, 543–552.

AIDA, I. (1978), *Reliability of a Tsunami Source Model Derived from Fault Parameters,* J. Phys. Earth. 26, 57–73.

HASHIMOTO, T., ISHIKAWA, Y., and UHIRA, J. (1993), *The 1993 Noto-Honto-Oki Earthquake and its Tectonic Implication (Part 1): A Rupture Process,* 1993 Joint Conference of Seismology in East Asia, Proceedings, 142.

IMAMURA, F., SHUTO, N., and GOTO, C. (1990), *Study on Numerical Simulation of the Transoceanic Propagation of Tsunami. Part 2: Characteristics of Tsunami Propagating over the Pacific Ocean,* Zisin 2 (43), 389–402 (in Japanese).

MASUI, T., TAKEMURA, M., and KAMATA, M. (1994), *Simulation of Strong Ground Motion during the 1993 Noto-Hanto-Oki earthquake by a Semi-Empirical Method,* Zisin 2 (47), 375–382 (in Japanese).

MANSINHA, L., and SMYLIE, D. E. (1971), *The Displacement Fields of Inclined Faults,* Bull. Seismol. Soc. Am. 61, 1433–1440.

MATSUURA, M., and SATO, R. (1975), *Displacement Fields due to the Faults,* Zisin 2 (28), 429–434 (in Japanese).

SATAKE, K., OKADA, M., and ABE, Ku. (1988), *Tide Gauge Response to Tsunamis: Measurements at 40 Tide Gauge Stations in Japan,* J. Marine Res. 46, 557–571.

YAMASHITA, T., and SATO, R. (1980), *Geneartion of Tsunami by a Fault Model,* J. Phys. Earth 22, 415–440.

(Received August 10, 1994, revised January 19, 1995, accepted January 27, 1995)

PAGEOPH, Vol. 144, Nos. 3/4 (1995)

0033-4553/95/040633-15$1.50 + 0.20/0
© 1995 Birkhäuser Verlag, Basel

Two 1993 Kamchatka Earthquakes

JEAN M. JOHNSON,[1] YUICHIRO TANIOKA,[1] KENJI SATAKE,[1] and LARRY J. RUFF[1]

Abstract—Two earthquakes occurred in 1993 off southern Kamchatka. They have similar surface wave magnitudes, focal mechanisms, and depths, but have distinctly different characteristics. The November earthquake is a standard or "impulsive" $M7$ underthrusting event. The June earthquake is a tsunamigenic or "low-stress-drop" event with several unusual characteristics, including a large, diffuse aftershock zone, directivity, and a long source time function. The 1993 earthquakes ruptured a segment of the Kamchatka Arc which has not ruptured since 1904. The 1993 earthquakes seem to signal the midpoint in the southern Kamchatka seismic cycle. .

Key words: Earthquake parameters, tsunamis, earthquake cycle.

Introduction

In 1993, two $M \sim 7$ earthquakes occurred immediately adjacent to each other off the east coast of southern Kamchatka. The first occurred on 8 June (51.218°N, 157.829°E, 13:03 GMT, $M_s = 7.2$) the second on 13 November (51.934°N, 158.647°E, 1:18 GMT, $M_s = 7.1$). The focal mechanisms of the two earthquakes (Figure 1) are similar and suggest that these are typical underthrusting events occurring at the down-dip edge of the coupled plate interface, which in this region is 38–40 km deep (TICHELAAR and RUFF, 1993). While the November event is in every respect a standard event of its kind, the June earthquake has certain unusual characteristics. First, this event produced anomalous aftershock activity, far up dip of the main shock epicenter. Second, the seismic moment of the earthquake as determined by CMT inversion is larger than expected from the surface wave magnitude. Third, and most importantly, the June earthquake generated a tsunami that was observed on the east coast of Kamchatka and on tide gauges on several Pacific islands. The waveforms recorded at the IRIS Pasadena station (Figure 2) show the different characters of the two events, in particular, the long period nature of the waves from the June event.

---

[1] Department of Geological Sciences, University of Michigan, Ann Arbor, MI 48109-1063, U.S.A.

1993 Kamchatka Earthquakes

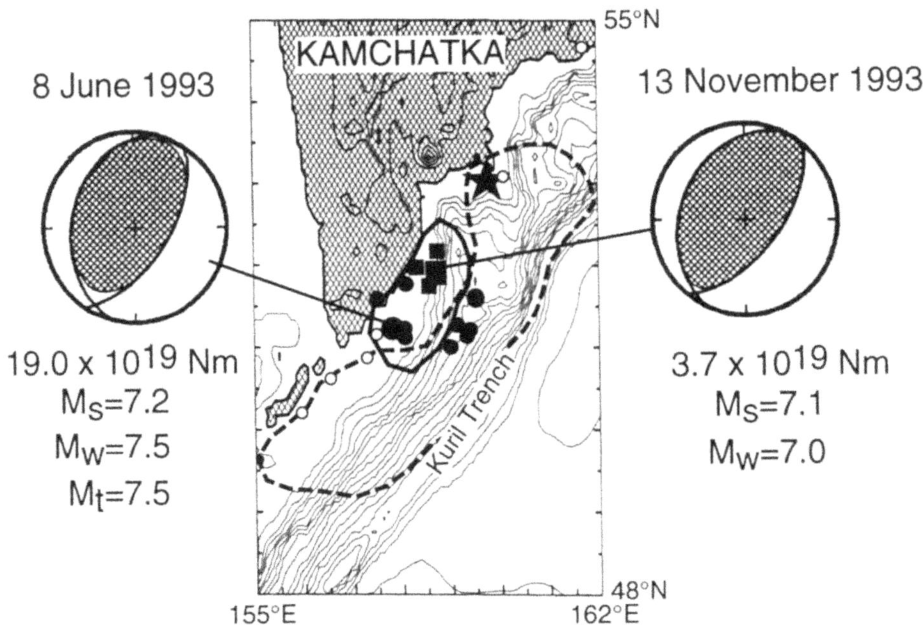

Figure 1

Aftershock zones of Kamchatka earthquakes. One-month aftershocks of the June 1993 earthquake are solid circles, solid squares are aftershocks of the November 1993 earthquake. Data are from NEIC. The dashed line outlines the aftershock zone of the 1952 earthquake, the large star is the epicenter (KELLEHER and SAVINO, 1975), the solid line outlines the inferred aftershock zones of the 1904 earthquakes (FEDOTOV *et al.*, 1982). Open circles are epicenters of earthquakes defining the down-dip edge of the coupled plate interface (TICHELAAR and RUFF, 1993).

We analyzed both the June and November earthquakes using several seismological methods. We determined the best focal mechanism, depth, and source time function for each. We also analyzed the tsunami of the June event in an attempt to determine the generating mechanism. The results suggest that the June Kamchatka earthquake may be an example of a low-stress-drop earthquake.

This portion of the Kuril-Kamchatka Arc last ruptured in the great 1952 Kamchatka earthquake (4 November 1952, 52.75°N, 159.5°E, $M_w = 9.0$). The 1952 zone, however, did not rupture at depth in the area of the 1993 earthquakes. This part of the plate interface last ruptured in a series of three earthquakes in June 1904. These three earthquakes were all of similar magnitude, approximately $M_s = 7.2$ (PACHECO and SYKES, 1992). This makes them directly comparable to the 1993 events. The relationship between all these events is sketched in Figure 1. We discuss the relationship of the 1993 earthquakes to both the 1904 earthquakes and

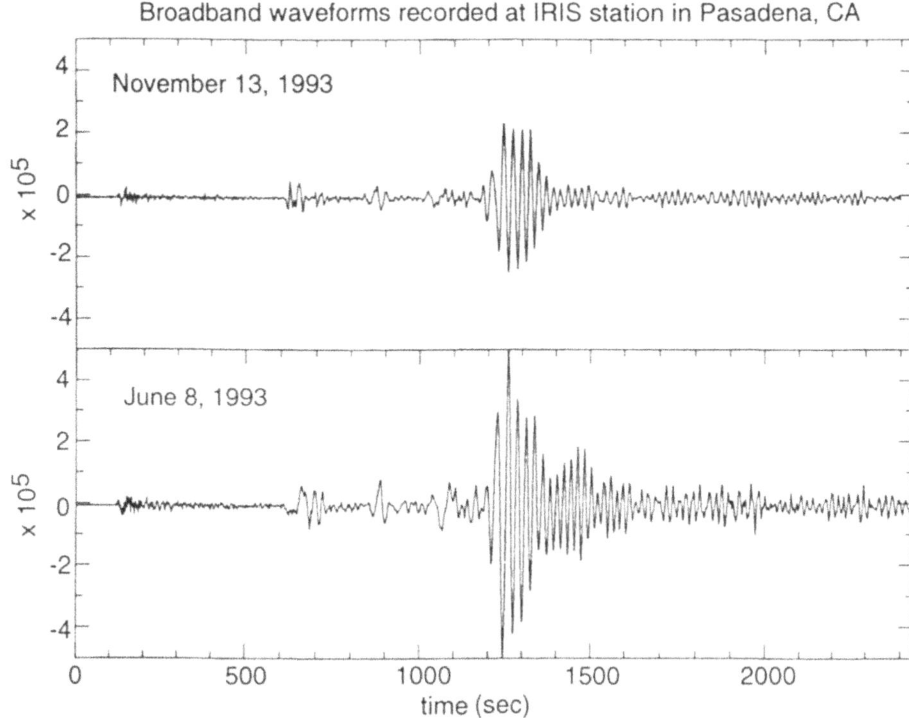

Figure 2

Broadband waveforms recorded at the IRIS network station in Pasadena, CA for the June and November 1993 Kamchatka earthquakes. The time axis is seconds after the beginning of the record.

the great 1952 Kamchatka earthquake. We also attempt to discern the basic features of the seismic cycle in this portion of the Kamchatka Arc.

Seismic Data Analysis

We performed CMT inversion (DZIEWONSKI *et al.*, 1981) for both events, using ten stations of the IRIS network for the June event and eight stations for the November event. The focal mechanism and seismic moment can be seen in Figure 3. The Harvard CMT (DZIEWONSKI *et al.*, 1981) and U.S.G.S. (SIPKIN, 1986) focal mechanisms are also shown for comparison and are listed in Table 1. The strike, dip, and rake of the presumed fault plane for the June earthquake are 191°, 27°, and 66°. For the November event, they are 204°, 31°, and 77°. These fault planes are consistent with the direction of subduction in the Kuril-Kamchatka trench. The seismic moment of the November event, 3×10^{19} Nm, or $M_w = 7.0$, is compatible with the surface wave magnitude of 7.1. The moment of the June event,

8 June 1993 Kamchatka Earthquake

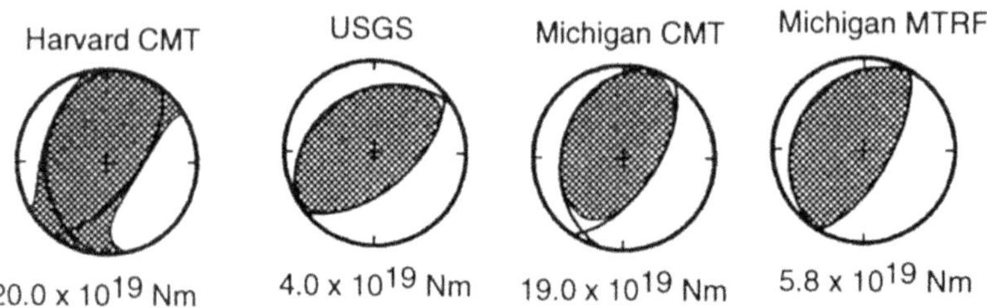

13 November 1993 Kamchatka Earthquake

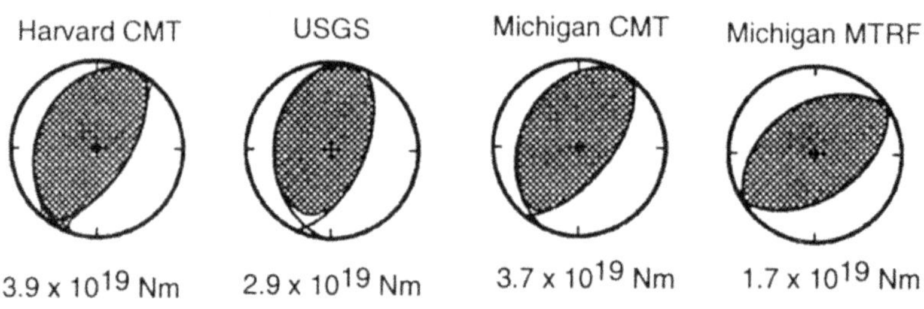

Figure 3
Focal mechanisms and seismic moments of the 1993 earthquakes.

19.0×10^{19} Nm, however, is over five times greater than the November earthquake. The moment magnitude of the June earthquake is 7.5, higher than the surface wave magnitude of 7.2.

We also performed body wave inversion for the moment tensor rate functions (MTRF) (RUFF and MILLER, 1994) to determine the best depth and focal mechanism for both earthquakes (see Figures 3 and 4). The strike, dip, and rake of the June earthquake are 212°, 24°, and 92°; for the November earthquake, 236°, 42°, and 90°. To determine the best depth, we searched over a range of possible depths to find the highest correlation coefficient, which measures the fit of the synthetics to the observed waveforms. The best depth of the June event is 40 km, which is significantly shallower than the NEIC published depth of 70 km. The best depth of the November event is more equivocal, but the highest correlation occurs at 45 km. The correlation declines significantly below 50 km for both events. This places both events at the down-dip edge of the coupled plate interface and certainly not deeper. As seen in Figure 1, both events lie along a line of earthquakes which have occurred in the past thirty years which define the edge of the coupled plate interface (TICHELAAR and RUFF, 1993). It is interesting to note that there is a definite gap in moderate size earthquakes at the southern edge of the Kamchatka Arc which the 1993 events filled.

Table 1

Location and focal mechanisms of 1993 earthquakes and principal aftershocks

Date	Location	Strike	Dip	Rake	Moment $\times 10^{19}$ Nm
8 June 13:03:36.48 $M_s = 7.2$	51.24°N 157.80°E	207	29	79	20
10 June 12:04:56.41 $m_b = 5.5$	51.170°N 159.097°E	205	32	69	0.014
10 June 12:58:59.49 $m_b = 5.8$	51.115°N 159.272°E	206	35	71	0.014
12 June 20:33:25.70 $m_b = 5.9$	51.259°N 157.692°E	214	32	92	0.27
13 November 01:18:04.18 $M_s = 7.1$	51.952°N 158.796°E	206	31	83	3.9
17 November 11:18:51.62 $m_b = 6.1$	51.81°N 158.659°E	208	36	73	0.12

In addition to finding the best depth, we also determined the source time function of each earthquake. The source time function for the November earthquake (Figure 5) consists of a single pulse of moment release of about 10 sec duration. This is typical of many $M7$ earthquakes which occur in subduction zones (SCHWARTZ and RUFF, 1987). On the other hand, the source time function of the June event has a duration of approximately 30 sec. This long duration of moment release is of special interest when the aftershock activity and the tsunami are considered. Further discussion relating to this point can be found below in the section dealing with tsunami data analysis.

In attempting to determine the source time function for the June earthquake, we discovered that it was quite difficult to fit all the data using a single source time function. This led us to consider whether there was any directivity affecting the data. We divided the stations used in the body wave inversion into several groups based on their azimuthal distribution. We then inverted the waveforms from each group. The results are shown in Figure 6. In general, the stations to the east of the earthquake have a compressed source time function of approximately 20 sec, while those to the west exhibit a lengthened source time function of about 30 sec. This indicates that the rupture propagated from west to east, or that rupture initiated at

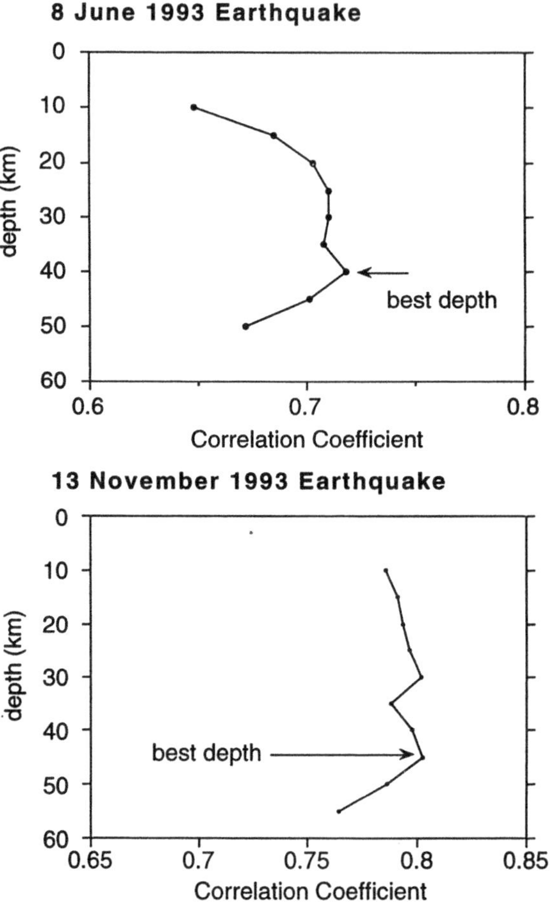

Figure 4

Correlation coefficient measuring the fit between the observed and synthetic as a function of depth.

the down-dip edge of the plate interface and proceeded updip. From the individual groups of source time functions, we estimate the length of faulting by using the variation in timing of the truncation of the source time functions relative to the onset. The observed delay time T_i at the ith station relative to the epicentral arrival time T_0 is

$$T_i - T_0 = X\left(\frac{1}{v_r} - p \cos \theta\right)$$

where X is the distance to the termination of faulting, v_r is the rupture velocity, p is the ray parameter, and θ is the angle between the rupture azimuth and the station azimuth. A full discussion of directivity analysis can be found in BECK and RUFF (1984). To simply estimate the length of faulting, we assume the rupture velocity is

Source time functions of Kamchatka Earthquakes

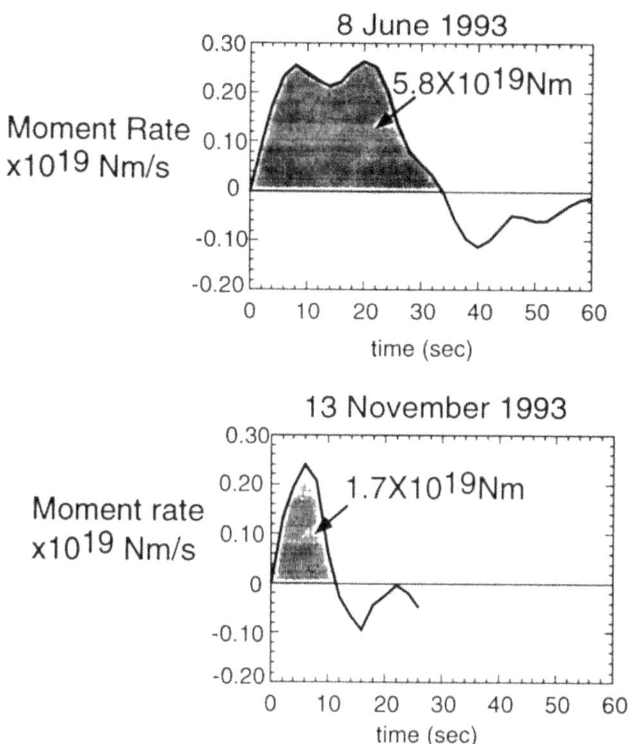

Figure 5

Source time functions of 1993 Kamchatka earthquakes from inversion of *P* waves.

3.0 km/sec and the rupture azimuth is perpendicular to the fault strike. A time delay of approximately 10 seconds produces a fault length of about 100 km, which is consistent with the aftershock zone.

To complete our study of the seismic data for both earthquakes, we analyzed the aftershocks. The focal mechanisms of the larger aftershocks are listed in Table 1. They are similar to the focal mechanisms of the main shocks. As seen in Figure 1, however, the aftershock distributions of the June and November earthquakes are quite different. The aftershocks of the November event are clustered close to the main shock. The aftershocks of the June event are spatially more diffuse. Also, there is a significant cluster of aftershocks to the east of the main shock. Considering the unusually long duration of the source time function, as described above, and the length of faulting as estimated by the directivity, this distribution is not surprising. The differences between the two aftershock zones will be discussed in more detail later.

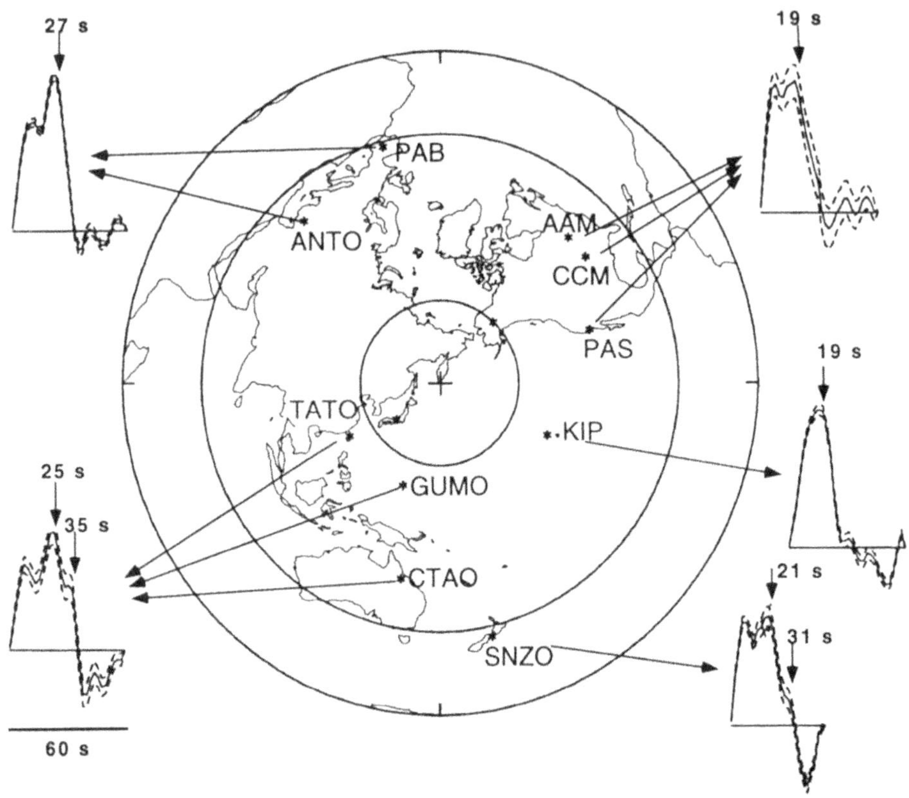

Figure 6
Directivity results. Arrows indicate the stations used in the inversion for each of the source time functions. The cross indicates the epicenter of the June 1993 earthquake.

In summary, the June and November earthquakes are distinctly different, despite the similarity in focal mechanism, depth, and surface wave magnitude. The November earthquake is a standard underthrusting earthquake such as commonly occurs at the down-dip edge of the coupled plate interface. The June event is more unusual, but the seismic data are all consistent with a longer duration of faulting on a longer fault.

Tsunami Data Analysis

One of the most interesting aspects of the June event was the tsunami it generated. It was observed on Kamchatka (10 cm at Petropavlosk-Kamchatshiy, V. K. Gusiakov, written communication) and was also recorded on tide gauges at

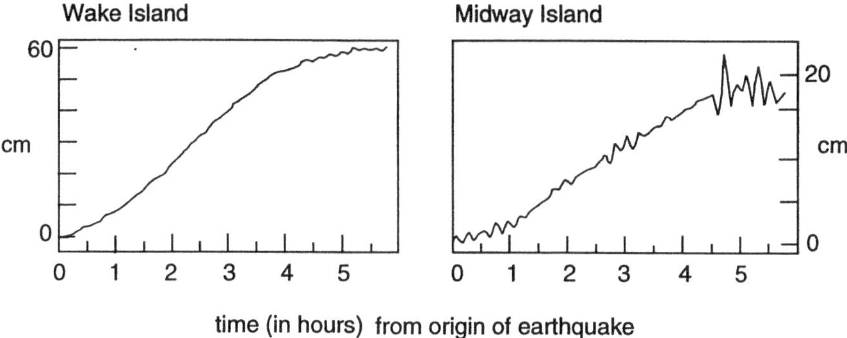

Figure 7

Digitized marigrams of the June 1993 tsunami recorded at Wake and Midway.

Midway (8 cm), Wake (5 cm), Hawaii (12 cm), and Shemya (10 cm) in the Aleutians. The tsunami magnitude (ABE, 1979) as determined by the maximum height of the tsunami on tide gauge records is 7.5. This is higher than the surface wave magnitude of 7.2. While it is not a tsunami earthquake as defined by ABE (1979), the discrepancy between the surface wave magnitude and the tsunami magnitude does point out the interesting character of this event. Luckily, several tsunami waveforms are available for this earthquake (Figure 7); unfortunately, the sampling rate is only once every six minutes, making the records of limited use. One of them, however, from Midway, can be used to constrain the tsunami generation mechanism.

In constructing models for the tsunami generation, we attempted to remain within the constraints imposed by the seismic data. Thus, for each model, the seismic moment is held constant. This determines the slip on each fault. Also, we attempted to choose faulting parameters consistent with the focal mechanism determined by the CMT inversion. The four models are shown in Figure 8.

In the first model, we assume the moment is released uniformly along the entire aftershock zone. The fault is 110 km downdip × 50 km along strike; the depth to the top edge of the fault is 18 km, the dip is 13°, and the slip amount is 90 cm of dip-slip motion. Although the focal mechanism shows that the dip of the fault plane is 25–30°, the subduction of the Pacific Plate occurs at a dip of 13° (TICHELAAR and RUFF, 1993); therefore, we chose the 13° value as the dip to be consistent with the actual plate geometry.

In the second model, we assume that faulting continued all the way to the trench. This is possible if we consider the maximum directivity time delay allowed by the seismic data. If it is 16 sec, rather than 10 sec, then rupture could have propagated to the trench. Although there are no aftershocks near the trench, this is not unusual, as the trench is filled with soft sediments unlikely to sustain large aftershocks. The fault for this model is 165 km downdip × 50 km along strike, the

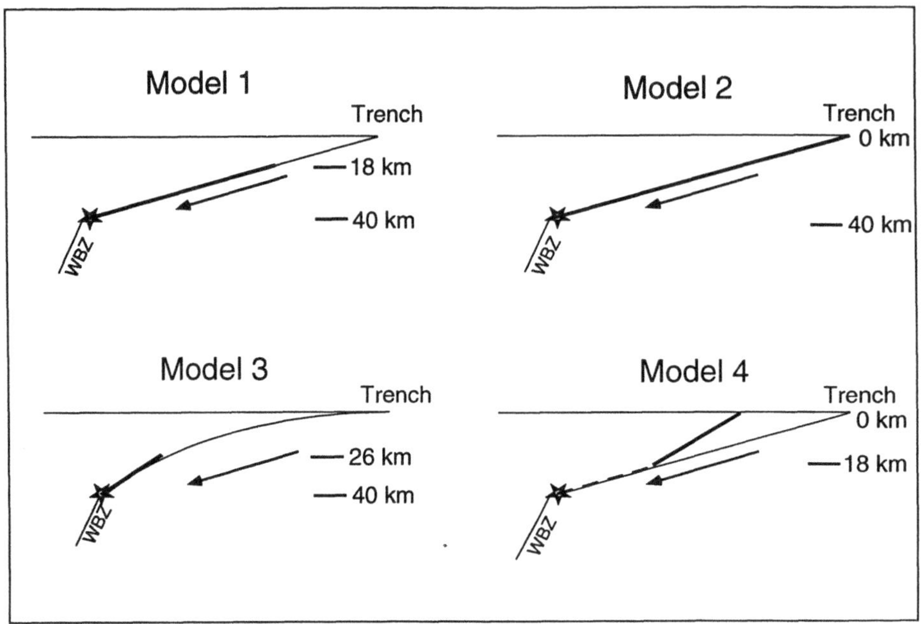

Figure 8
Models used to determine the initial condition of the June tsunami.

depth to the top of the fault is 0.5 km, the dip is 13°, and the slip amount is 60 cm dip-slip.

In the third model, we assumed that all the moment release occurred in the deep section of the aftershock zone and that no moment was released in the area of the second cluster of aftershocks. The fault parameters are: 50 km length along strike, 35 km width down-dip, depth to top of faulting 26 km, dip 30°, and slip amount 290 cm. The dip is consistent with the focal mechanism.

In the fourth model, we allow all the moment to be released on a subsidiary fault in the overlying plate. This corresponds to the cluster of aftershocks which are located to the east of the main shock. The faults parameter are: length 50 km, width 50 km, and slip 200 cm. We chose the dip to be 30° to be consistent with the focal mechanism. The faulting was allowed to reach the surface.

Using each of these fault models, we computed the ocean-bottom deformation applying the equations of OKADA (1985) for an elastic half-space. Using this as the initial condition for the tsunami, we calculated the tsunami propagation as a linear-long wave by a finite-difference method using the actual bathymetry of the Pacific Ocean. In the source area and in the northern Pacific, we used 5′ × 5′ grid size. In the area around Midway, however, we used a grid size of 1′ × 1′ to ensure the accuracy of the synthetic waveform.

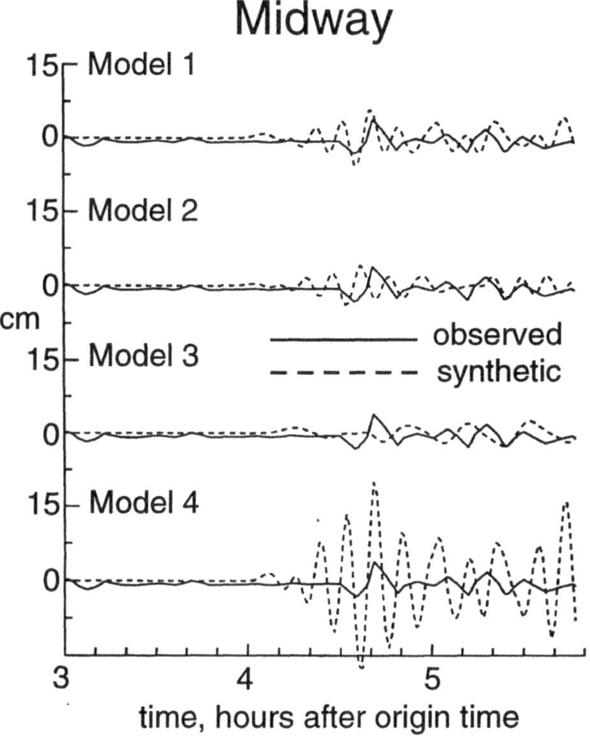

Figure 9
Observed and synthetic tsunami waveforms.

Figure 9 shows the results of forward modeling. The observed waveform is compared to the synthetic for each model. It is immediately obvious that none of the models matches the observed waveform perfectly. Both Model 1 and Model 2 have approximately the correct amplitude. The largest pulse of Model 2 arrives too early, as would be expected, as the slip is allowed to go to the trench. The arrival time of each of the synthetics does not match the observed first pulse, but this could be due to the aliasing caused by the undersampling. The amplitude of Model 3 is too small to match the observed, and the amplitude of Model 4 is too large.

It is not possible to uniquely determine the tsunami generation mechanism from a single tide gauge record. We can only make a qualitative estimate of the tsunami source. Of the four models we used, Models 3 and 4 are ruled out, but the problem of aliasing in the original tide gauge record makes it difficult to make an absolute determination between Models 1 and 2. Model 1, however, is slightly preferable as it agrees more closely on the whole with the seismic data. Unfortunately, nothing more can be determined from the tsunami data analysis.

Low-stress-drop earthquakes

Perhaps more interesting than the mechanism of faulting resulting in a tsunami, is the question of why the June earthquake is so different from the adjacent November earthquake. One possible answer can be suggested by looking at other well-studied earthquakes in the Kuril Trench. The work of SCHWARTZ and RUFF (1987), which examines large earthquakes ($M \geq 7.0$) in the southern Kuril Trench, offers some explanation for these types of events.

There were several earthquakes in the southern Kurils similar to the June 1993 Kamchatka earthquake. These events had large aftershock areas which extended to the trench, they generated tsunamis, and they had unusually long source time functions. SCHWARTZ and RUFF refer to these as "low-stress-drop" events. Other earthquakes in the Kurils were similar to the November 1993 Kamchatka earthquake. They occurred at the down-dip edge of the coupled plate interface, their aftershocks were tightly clustered near the main shock, and their source time functions were of short duration. These events are referred to as "impulsive" events. There is even one instance of a pair of adjacent earthquakes in the Kurils that exhibits the opposing characteristics seen in the Kamchatka earthquakes. The first earthquake (21 December 1946, $M_s = 7.2$, 44.1°N, 148.2°E) had an unusually large aftershock zone, much like the June 1993 event. Though this 1946 Kuril earthquake did not generate a tsunami (or at least none is reported) it appears to be one of the low-stress-drop earthquakes. Immediately adjacent to this event, an impulsive earthquake occurred four months later (14 April 1947, $M_s = 7.0$, 44.0°N, 148.5°E), similar to the November 1993 Kamchatka event.

To determine if the June event is a low-stress-drop earthquake, we estimate the stress drop from the seismic moment determined by the MTRF inversion m and the duration of the source time function 2τ (YOSHIDA *et al.*, 1992). We use the moment determined by the body wave inversion, 5.8×10^{19} Nm, and the half duration of the time function, 15 sec. The stress drop is estimated as $\Delta\sigma = 2.5\, m/(v_r\tau)^3$, where v_r is the rupture velocity. We assume the rupture velocity is 3.0 km/s. The stress drop we estimate is 16 bars. We also estimate the stress drop of the November event to be 125 bars, where $m = 1.8 \times 10^{19}$ Nm and $\tau = 5$ sec. This confirms the different character of the two events and lends credence to the supposition that the June event is a "low-stress-drop" earthquake and the November earthquake is an "impulsive" event.

As the parallels we have have already drawn between the southern Kuril Arc and the Kamchatka Arc have proven valid, we can perhaps proceed further and speculate on the controlling factors for these types of events. SCHWARTZ and RUFF showed that the locations of the low-stress-drop and impulsive events were controlled by the asperities which ruptured in large or great earthquakes. The smaller events are generally adjacent to the asperities. If the 1993 earthquakes are also controlled by the asperities in the Kamchatka region, we can hypothesize on the location of at least one asperity of the 1952 great earthquake. Assuming that an asperity is associated with the location of the epicenter, we can fix one asperity to the northeast of the 1993 earthquakes. Unfortunately, this can only be speculation at present because the slip distribution of the 1952 earthquake is unknown. Hopefully, in the future, tsunami waveform inversion can be performed, such as has been done for the 1957 Aleutian earthquake (JOHNSON and SATAKE, 1993), which will allow us to test our hypothesis.

Discussion

As noted in the introduction, the two 1993 Kamchatka earthquakes occurred within the rupture area of three earthquakes which occurred in 1904. Little is known about these events; only their magnitudes are known, being estimated as $M_s = 7.1-7.3$ (PACHECO and SYKES, 1992). The rupture area of the 1904 events, while not well determined, apparently did not rupture in the great 1952 Kamchatka earthquake which has a moment magnitude of $M_w = 9.0$ (KANAMORI, 1977). The aftershock zone of the 1952 earthquake (KELLEHER and SAVINO, 1975; FEDOTOV *et al.*, 1982) wraps around the 1904/1993 rupture areas with little overlap between the two.

The historical seismicity of the 1993 rupture zone does predate 1904. The 1952 zone, on the other hand, has a longer history. Tsunami evidence suggests that the 1952 area ruptured previously in $M9$ earthquakes in 1737 (IIDA *et al.*, 1967) and in 1841 (ABE, 1979). If the rupture zones for each of these earthquakes correspond, as the evidence suggests, to the 1952 zone, then this segment of the arc has a fairly regular recurrence history with an interval of about 108 years.

There are several different possible scenarios which could define the rupture cycle in this region. The first is that the southern Kamchatka Arc is one integrated, complex system with interacting parts: the 1952 zone and the 1993 zone. Each part ruptures approximately every 100 years, but out of phase by one-half cycle, as appears from the 1841–1904–1952–1993 sequence. Another possibility is that the two systems are essentially independent of each other. This is slightly difficult to explain as the 1904/1993 zone is down-dip of the larger event and is almost encompassed by it. It seems likely that the larger system must have some effect on the down-dip system. On the other hand, this may explain the difference in rupture styles; that is, the 1904/1993 zone appears to rupture in a series of $M7$ earthquakes while the 1952 zone ruptures in a single great event. A final possibility is that the 1904/1993 zone ruptures about every 50–60 years and was a part of the 1737 and 1841 zones. Why the 1904/1993 zone apparently did not rupture in 1952 remains unexplained.

It is impossible to determine if the 1904/1993 rupture area was involved in the 1737 or 1841 earthquakes, but the 1841–1904–1952–1993 sequence does suggest an interesting rupture cycle in this region. The 1904 earthquakes occurred 63 years after the 1841 earthquake and 48 years before the 1952 earthquake. The 1993 earthquakes occurred 41 years after the 1952 earthquake. The 1904/1993 zone seems to be rupturing at approximately the midpoint of the 1952 cycle. The 1993 earthquakes could be considered a "flag" that the entire system is continuing to operate as it has in the last seismic cycle.

Conclusions

The 1993 Kamchatka earthquakes, while having similar surface wave magnitudes, focal mechanisms, and depths, have distinctly different characteristics. The November earthquake is a standard or "impulsive" $M7$ underthrusting event. The June earthquake is a tsunamigenic or "low-stress-drop" event with several unusual characteristics. The 1993 earthquakes ruptured a segment of the Kamchatka Arc which has not ruptured since 1904. The 1993 earthquakes seem to signal the midpoint in the southern Kamchatka seismic cycle.

Acknowledgements

This work was supported by the National Science Foundation (EAR91–17800 and EAR90–19003).

REFERENCES

ABE, K. (1979), *Size of Great Earthquakes of 1873–1974 Inferred from Tsunami Data*, J. Geophys. Res. *84*, 1561–1568.

BECK, S., and RUFF, L. (1984), *The Rupture Process of the Great 1979 Colombia Earthquake: Evidence for the Asperity Model*, J. Geophys. Res. *89*, 9281–9291.

DZIEWONSKI, A. M., CHOU, T. A., and WOODHOUSE, J. H. (1981), *Determination of Earthquake Source Parameters from Waveform Data for Studies of Global and Regional Seismicity*, J. Geophys. Res. *86*, 2825–2853.

FEDOTOV, S. A., CHERNYSHEV, S. D., and CHERNYSHEVA, G. V. (1982), *The Improved Determination of the Source Boundaries for Earthquakes of $M \geq 7.75$, of the Properties of the Seismic Cycle, and of Long-term Seismic Prediciton for the Kurile-Kamchatka Arc*, Earthq. Predict. Res. *1*, 153–171.

IIDA, K., COX, D. C., and PARARAS-CARAYANNIS, G. (1967), *Preliminary Catalogue of Tsunamis Occurring in the Pacific Area*, Hawaii Institute of Geophysics, University of Hawaii.

JOHNSON, J. M., and SATAKE, K. (1993), *Source Parameter of the 1957 Aleutian Earthquake from Tsunami Waveforms*, Geophys. Res. Lett. *20*, 1487–1490.

KANAMORI, H. (1977), *The Energy Release in Great Earthquakes*, J. Geophys. Res. *82*, 2981–2987.

KELLEHER, J., and SAVINO, J. (1975), *Distribution of Seismicity before Large Strike Slip and Thrust-type Earthquakes*, J. Geophys. Res. *80*, 260–271.

OKADA, Y. (1985), *Surface Deformation due to Shear and Tensile Faults in a Half-space*, Bull. Seismol. Soc. Am. *75*, 1135–1154.

PACHECO, J. F., and SYKES, L. R. (1992), *Seismic Moment Catalog of Large, Shallow Earthquakes, 1900–1989*, Bull. Seismol. Soc. Am. *82*, 1306–1349.

RUFF, L. J., and MILLER, A. D. (1994), *Rupture Process of Large Earthquakes in the Northern Mexico Subduction Zone*, Pure Appl. Geophys. *142*, 101–172.

SCHWARTZ, S. Y., and RUFF L. J. (1987), *Asperity Distribution and Earthquake Occurrence in the Southern Kuril Islands Arc*, Phys. Earth Planet. Inter. *49*, 54–77.

SIPKIN S. A. (1986), Estimation of Earthquake Source Parameters by the Inversion of Waveform Data: Global Seismicity, Bull. Seismol. Soc. Am. *76*, 1515–1541.

TICHELAAR, B. W., and RUFF, L. J. (1993), *Depth of Seismic Coupling along Subduction Zones*, J. Geophys. Res. *98*, 2017–2037.

YOSHIDA, Y., SATAKE, K., and ABE, K. (1992), *The Large Normal-faulting Mariana Earthquake of April 15, 1990 in Uncoupled Subduction Zone*, Geophys. Res. Lett. *19*, 297–300.

(Received August 10, 1994; revised November 11, 1994; accepted January 3, 1995)

PAGEOPH, Vol. 144, Nos. 3/4 (1995)

Field Survey of the 1993 Hokkaido Nansei-Oki Earthquake Tsunami

N. SHUTO,[1] and H. MATSUTOMI[2]

Abstract—Runup data in Hokkaido and in three prefectures in the Tohoku District are described with a few witnessed arrival times and with comments of tide records. The highest runup of 31.7 m was found at the bottom of a narrow valley on the west coast of Okushiri Island. In order to explain high runups of 20 m at Hamatsumae in the sheltered area, roles of edge waves, refraction of the Okushiri Spur and tsunami generation by causes other than the major fault motion should be understood. An early arrival of the tsunami on the west coast of Hokkaido suggests another tsunami generation mechanism in addition to the major fault motion. ·

Key words: Tsunami, runup, arrival time, edge wave, Japan Sea.

1. Introduction

At 22:17, July 12th (local time), 1993, an earthquake of $M_w = 7.8$ generated a giant tsunami. The tsunami reached the maximum runup height of 31.7 m. This was the highest runup of the century in Japan while human casualties were the largest of the last 50 years.

Shortly after the event, the authors formed two teams, composed mostly of members from Tohoku and Akita Universities; one team for numerical simulation and the other for field survey. The simulation team of 4 members began the work by purchasing detailed maps and charts. They then digitized the maps, obtained the fault parameters from TBB (Tsunami Bulletin Board) on e-mail communication, carried out the simulation and made a computer-graphics video animation.

The survey team of 19 members arrived in Hokkaido on July 15th and proceeded to Okushiri Island on July 18th as soon as the ferry transportation resumed. The team continued the survey until July 23rd, maintaining close contact with the numerical simulation team. Every morning and night the two teams exchanged their results, and determined the locations and density of measurement points.

Thereafter runup heights were measured along the Japan Sea coast of three prefectures in the Tohoku District.

[1] Disaster Control Research Center, Faculty of Engineering, Tohoku University, Sendai 980-77, Japan.

[2] Department of Civil and Environmental Engineering, Faculty of Mine Engineering, Akita University, Tegata Gakuen 1-1, Akita 010, Japan.

2. Areas and Types of Field Survey

The survey covers the entire coast of Okushiri Island, 60 km long; the western coast of Hokkaido, 310 km long from Furubira to Matsumae; and the coasts of Aomori, Akita and Yamagata Prefectures, 910 km long in the Tohoku District. Figure 1 shows the area of the survey and the position of the epicenter.

Major objectives of the survey were:

(1) Measurement of height and position of runups. Maps and GPS (Global Positioning System) were used to determine the positions. Locations of runup measurements were marked with stakes at sites by the survey team. All the measurements are in the Japan Sea where the tidal range is of the order of 20 cm. The runup is expressed in meters above mean sea-water level.

(2) Characterization of the tsunami such as arrival time, initial movement of water surface (i.e., whether the tsunami began with ebb or flood), incident direction, tsunami profile, the time at which the highest and/or maximum wave appeared, tsunami wave period, and so on.

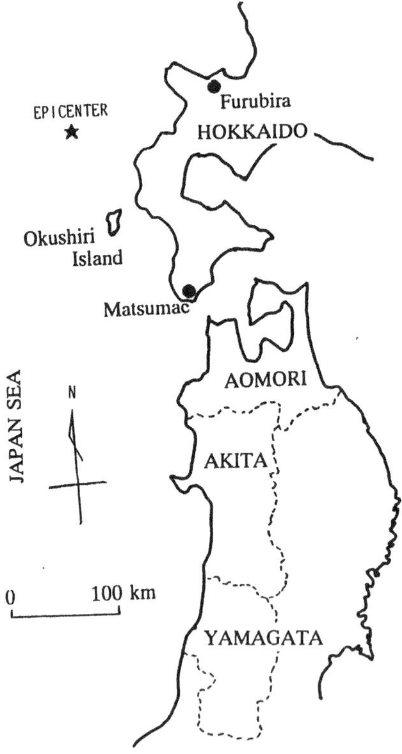

Figure 1
Area of survey and the epicenter.

(3) Inspection of damage to coastal structures.

The present paper reports the results of items (1) and (2).

About 130 runups were measured in Okushiri Island, about 70 in Hokkaido, and approximately 80 (60 measured by the authors' team and 20 measured by local authorities) in Aomori, Akita and Yamagata Prefectures in the Tohoku District.

Daily communication between the two teams was reflected in: (1) a dense measurement along the east coast of Okushiri Island, (2) a detailed survey at Aonae and Hamatsumae on the south coast of Okushiri Island, (3) efforts to access the coastal cliff near Horonai on the west coast of Okushiri Island, and (4) attempts to determine the arrival time along the southwestern coast of Hokkaido.

3. The Tsunami in Okushiri Island

It was night when the tsunami hit the island. Neither photos nor videos were taken and no witnesses could give a clear image of the tsunami. Figure 2 shows the runup distribution along the island.

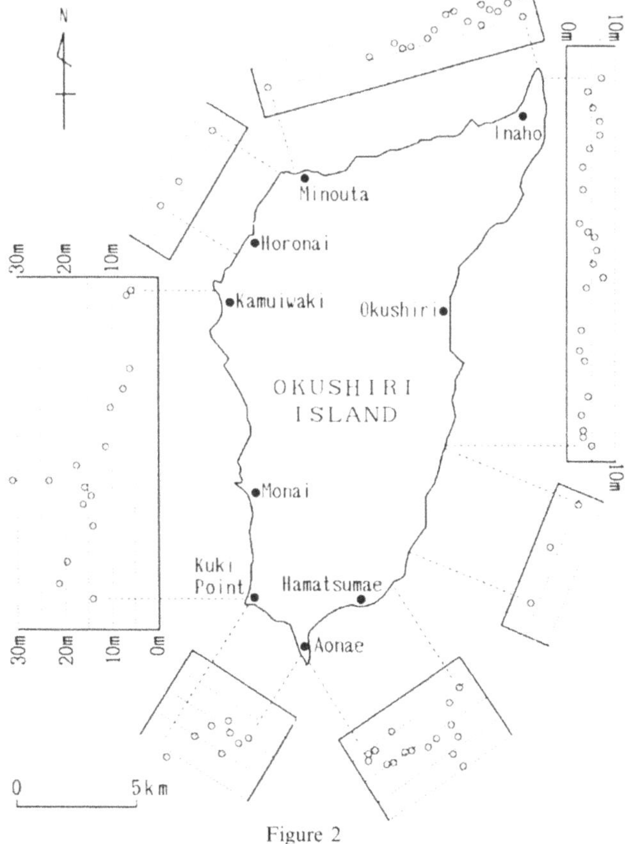

Figure 2
Runup distribution in Okushiri Island.

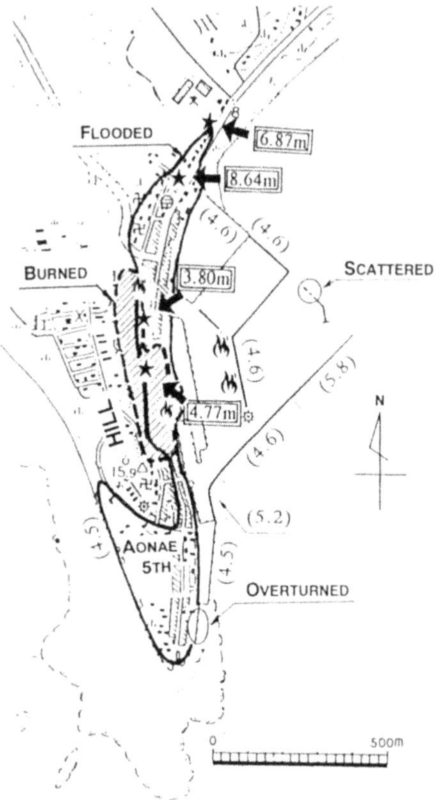

Figure 3

Tsunamis and damages in the town of Aonae. The fifth section of Aonae is on a sand spit in the south of a hill, and the first to fourth sections are along the eastern foot of the hill. Numerals in double rectangles are runup heights measured at the locations indicated by stars. Numerals in parentheses are the crown heights of structures. The area surrounded by the solid line was flooded by the tsunami. The hatched area was burned by fires.

At the southern tip of the island, the tsunami hit the town of Aonae. This town, composed of five sections, is located at the foot of a hill 20 m high as shown in Figure 3. The first to fourth sections are located on a beach along the eastern foot of the hill. Their waterfront is used as a fishing harbor, protected by breakwaters. The fifth section of Aonae is on a southern sand spit 3 m high. This fifth section was hit by the tsunami from the west, 4 to 5 minutes after the earthquake. The entire fifth section was swept away by the first tsunami, although the section was protected by sea walls 4.5 m high.

This first tsunami did not damage the first to fourth sections of the town, because the tsunami was not high enough to overflow the hill. Approximately ten minutes after the first tsunami, the second tsunami struck these sections from the east.

At the northern tip of the island, the tsunami hit the village of Inaho from the north.

Along most of the eastern coast of Okushiri Island, runups are not high because the coast is sheltered by the island itself against the tsunami source. A spatial fluctuation of runup distribution along it, however, may suggest the tsunamis behaved as edge waves.

Special attention should be directed to the fact that runups 20 m high were measured at Hamatsumae on the southeastern coast of the island even though it is located in the sheltered area.

The highest runup of 31.7 m was measured on the western coast directly facing the tsunami source. It was found at the bottom of a small valley north of Monai. The tsunami arrived at Monai 4 minutes after the earthquake.

The total length of the coastline on which runups are higher than 10 m is about 13 km on the west and south coasts of Okushiri Island.

4. The Tsunami on the Southwestern Coast of Hokkaido

Figure 4 displays the runup distribution. Runup heights are less than 10 m, except in a few locations.

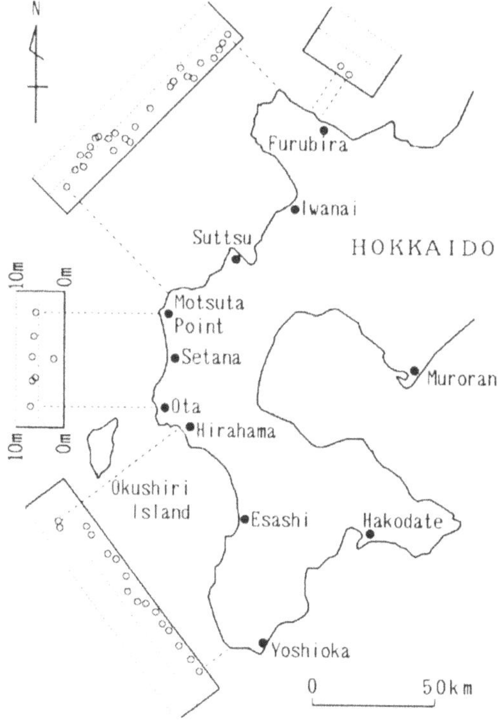

Figure 4
Runup distribution in Hokkaido Island.

Spanning the coast from Motsuta Point to Furubira, runups decrease northward from 5 m. The tsunami recorded at Iwanai began with a rise of the water surface.

Along the coast from Motsuta Point to Ota which protrudes westward and is parallel to the major axis of the estimated fault, runups are higher than 5 m. According to witnesses, the first tsunami arrived 4 to 5 minutes after the main shock. It began with a fall of the water surface, and the first or second wave was the highest.

Along the coast from Hirahama down to Esashi, runups are less than 5 m. The tsunami recorded at the Esashi tide gauge station began with a fall of the water surface. At Yoshioka on the southern end of the Oshima Peninsula, the tsunami on tide records began with a rise of the water surface. The maximum tsunami height measured above the estimated astronomical tide was 0.82 m, induced by the seventh wave which lasted a period of 8 minutes.

5. The Tsunami along the Tohoku District

5.1. Aomori Prefecture

Figure 5 shows the runup distribution along the coast of Aomori, measured by the local authorities.

The tsunami did not affect the Pacific coast.

In the middle of the Tsugaru Strait, a maximum runup of 1 m was obtained at Oma, at the northern end of the Shimokita Peninsula.

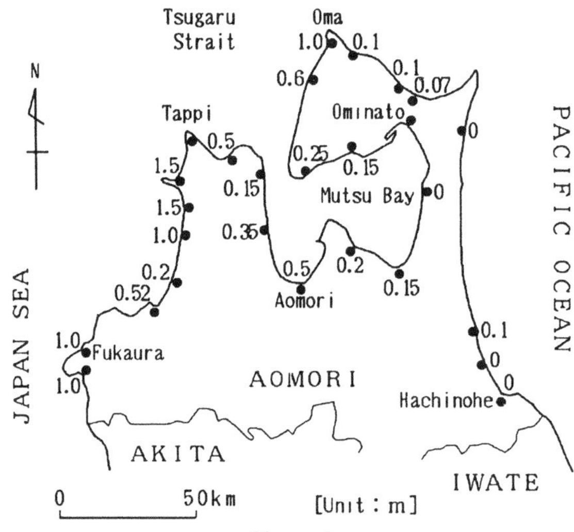

Figure 5
Runup distribution in Aomori Prefecture (measured by local authorities).

In Mutsu Bay, a maximum runup of 0.5 m was measured at Aomori, at the southern end of the bay. At Ominato on the northeastern side of the bay, the tsunami was recorded by the tide gauge but it proved difficult to determine the arrival time of the tsunami. The record here shows an oscillation with a period of 30 to 40 minutes, which is estimated to be close to the natural oscillation period in the bay. The maximum wave height was estimated as 0.57 m.

The tide record obtained at Tappi at the western entrance of the Tsugaru Straight reveals that the tsunami began with a rise of the water surface at 22:44, similar to the record at Yoshioka on the opposite side of the straight.

At Fukaura near the southern border of Aomori Prefecture, the tsunami recorded by the tide gauge began with a rise of the water surface at 22:43. The tsunami height of the first wave was the highest, 0.25 m. The first and fourth waves reached a wave height of 0.48 m. The wave period slightly exceeded 10 minutes.

5.2. Akita Prefecture

Figure 6 shows the runup distribution. Runup heights were on the order of 2 m. The highest runup of 3.47 m in Akita Prefecture was found at the left bank of the Mizusawa River in the village of Minehama, where a maximum runup of about

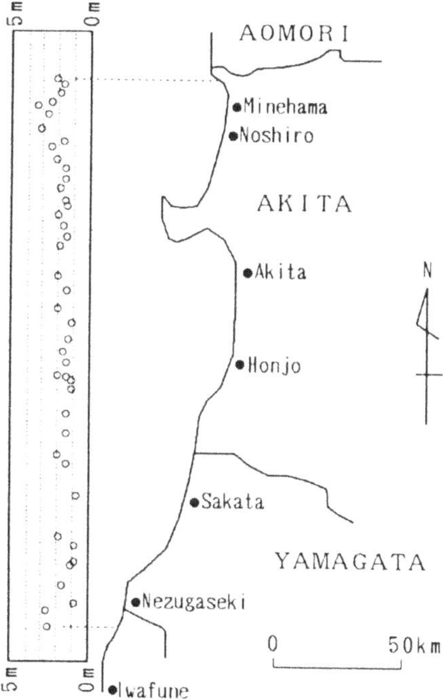

Figure 6
Runup distribution in Akita and Yamagata Prefectures.

15 m had been measured 10 years ago, at the time of the 1983 Nihonkai-Chubu earthquake tsunami.

At the tide station in Noshiro Harbor, the tsunami began with a rise of the water surface at 23:00. The tsunami reached a maximum height of 0.72 m at 1:12 and maintained the tsunami height of about 0.5 m until 7:00 on the 13th. The oscillation in the harbor continued until 7:00 on the 14th when the wave period was about 30 minutes. The wave period in the early stage could not be determined because of the very irregular time history.

The tide record at Akita Harbor demonstrates that the tsunami began with a rise of the water surface at 23:18, and reached the maximum tsunami height of 0.35 m at 0:35 on the 13th. The maximum wave height of 0.63 m was observed twice, at 0:35 and 11:00 on the 13th. The dominant wave period was about 20 minutes around 0:35 on the 13th.

5.3. *Yamagata Prefecture*

Figure 6 also shows the runup distribution in this prefecture. Runups are on the order of 2 m, and increase southward to 3 m.

The tide record at Sakata Harbor shows that the tsunami began at 23:23 with a rise of the water surface. The maximum tsunami height of 0.58 m was due to the 9th wave at 4:18 on the 13th, which registered a wave period of about 33 minutes.

At Iwafune in Niigata Prefecture, just south of the border, the tsunami began with a rise of the water surface. The maximum tsunami height of 0.70 m was due to the first wave. The maximum wave height of 1.36 m was also due to the first wave. The wave period exceeded 17 minutes.

6. *Special Characteristics of the Tsunami*

6.1. *Early Arrival*

A few minutes after the earthquake, the tsunami hit Okushiri Island and Hokkaido. Table 1 summarizes the results of the authors' team, the UJNR team (HOKKAIDO TSUNAMI SURVEY GROUP, 1993), and the arrival time obtained from tide records. Although there are doubts concerning the accuracy of witnessed arrival times owing to the nighttime arrival, it seems certain that the tsunami arrived very early at the west and south coasts of Okushiri Island and at the coasts of the towns of Taisei and Setana on Hokkaido.

The early arrival 4 to 5 minutes, on the west coast of Okushiri Island, is no wonder, because the coast is located close to the estimated tsunami source. The first tsunami was very large. For example, the fifth section of the town of Aonae was completely destroyed by the first tsunami as mentioned in Section 3.

Table 1

Arrival time and other remarks

Hokkaido	Arrival Time	Remarks
SHAKOTAN TOWN		
Shakotan Point	20 min.	The highest first wave began with an ebb
Kamui Point	5–6 min.	Ditto
Numamae Point	25–35 min.	Ditto
KAMOENAI VILLAGE		
Kawashiro Point	10 min.	
Sannai	10 min.	
Ryujin Point	5–10 min.	
IWANAI TOWN		
Iwanai	15 min.	
Iwanai Tide Gauge	22:37	Began with a flood
SHIMAMAKI VILLAGE		
Enoshima	5 min.	
SETANA TOWN		
Sukki	about 3 min.	The first wave was the highest
Shimauta	less than 5 min.	Ditto. Began with an ebb
		From the northwest
Setana	5 min.	
Futoro	less than 5 min.	The first wave from the northwest. Begin with an ebb. The second or third was the highest
TAISEI TOWN		
Ota	5 min.*	
Miyano	5 min.	The highest second wave at 22:27 or 22:28
Hirahama	5 min.	Began with an ebb from the west
		The second wave was the highest
Esashi Tide Gauge	22:28	Began with an ebb.

Okushiri Island	Arrival Time	Remarks
WEST COAST		
Hoyaishikawa P.S.	22:23*	
Monai	22:21	
SOUTH COAST		
Aonae 5th	4–5 min.	From the west
Aonae 1st–4th	22:37, 22:38*	From the east
Hamatsumae	22:22	

* Data with * from UJNR (1993).

On the other hand, the early arrival at Taisei and Setana towns on Hokkaido would be one of the riddles of the present tsunami. They are located far from the estimated faults, but the tsunami arrived about 5 minutes after the earthquake. This arrival time is of the same order as that to the west coast of Okushiri Island. In addition, there were many witnesses who confirmed that the first tsunami was small. These facts suggest that there might be another tsunami generation mechanism in addition to the major fault movement.

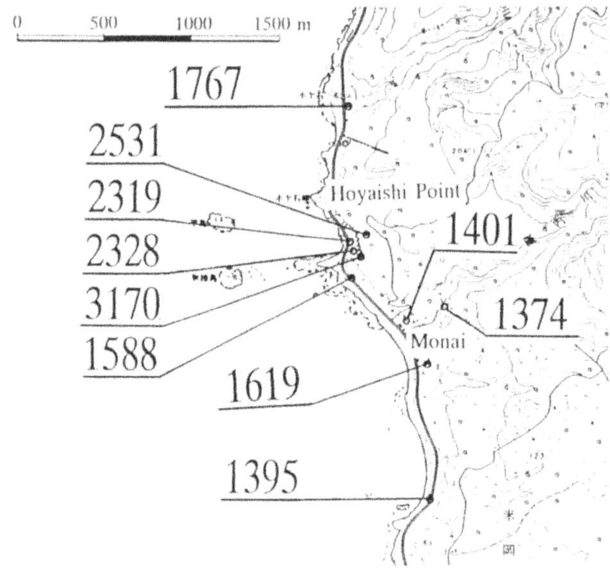

Figure 7
Runup in cm at and near the site of the highest runup of 31.7 m.

Photo 1
A small valley where the highest runup was measured.

6.2. The Highest Runup of 31.7 m

This very high runup was measured at a small pocket beach, north of Monai and south of Hoyaishi Point, on the west coast of Okushiri Island which directly faces the major fault. Figure 7 shows the map and the measured runups in cm.

The north and south ends of the pocket beach, which is 250 m long, are bounded by rocky shallows which continue to small islands and exposed rocks offshore. From the shoreline toward land, there are a narrow pebble beach, a coastal road of two lanes, and high coastal cliffs. A small sharp valley which has two branches is open at the pocket beach. Its entrance width is about 50 m. The northern branch is wider and shorter than the southern branch. The runup height was about 23 m on the south cliff at the entrance of the valley, 25.3 m at the bottom of the northern branch, and 31.7 m at the bottom of the southern branch (see Photo 1).

The authors consider that the maximum runup height of 31.7 m is a very local value and is not a representative value of the present tsunami. The representative runup of the present tsunami would be 23 m to 24 m, marked on the coastal cliff.

The authors also consider that if one tries to simulate this very high runup, he would need a map of very fine resolution, and must design a simulation with a spatial grid finer than 5 m.

6.3. Tsunami at Aonae

Aonae, which is located at the southern end of Okushiri Island, was most severely damaged by the tsunami and the tsunami-induced fires (see Fig. 3).

The island of Okushiri consists of hilly areas and narrow beaches. Extending to the southern end of the island, a hill 20 m high continues and it ends with a sudden decrease in height. On the west coast of the hill, there are very narrow beaches with no residents. From the southern end of the hill, a sand spit 3 m high, 500 m long and 250 m wide extends. The fifth section of the town of Aonae was located on this sand spit, protected by surrounding sea walls 4.5 m high. Along the eastern foot of the hill, there is a sandy beach about 150 m wide, on which the first to fourth sections of the town of Aonae were located. The land in front of the town area was recently reclaimed to construct Aonae Fishing Harbor. The crown height of breakwaters and sea walls at the harbor is 4.6 m.

The tsunami hit the fifth section first from the west 4 to 5 minutes after the earthquake. The tsunami height is estimated to be 7 m to 10 m. The sea walls 4.5 m high were inundated and left almost intact, but the entire 5th section was swept away by this first tsunami.

This first tsunami, however, did not influence and damage the first to fourth sections of the town because they were protected by the hill. About ten to fifteen minutes after the first tsunami, the second tsunami struck these sections from the east. The second tsunami overflowed a sandy hill on the northern part of the area, engulfed sea walls and breakwaters of the harbor, invaded the first to fourth sections of the town of Aonae, destroyed about half of these sections, and ignited a fire. Another fire started about two hours after the second tsunami. The two fires burnt the first to fourth sections except for a few houses in the fourth section. No fire engine could reach the fire, because the road was covered by debris transported by the tsunami.

It is certain that the second huge tsunami originated from the east. The time interval between the first and second waves is ten to fifteen minutes. There are three possibilities which explain the development of the second tsunami. The first possibility is that the tsunami generated by the north fault, rounded the northern tip of Okushiri Island and propagated along the eastern shore. There is evidence that the tsunami which hit the east coast of the island emanated from north. The

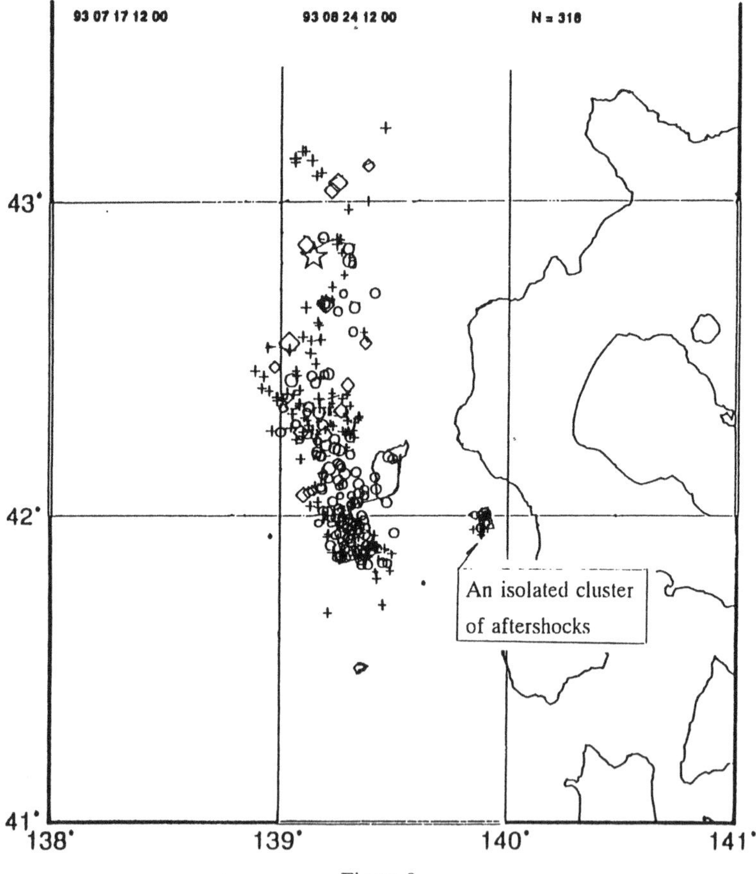

Figure 8
Aftershock distribution (RESEARCH GROUP, 1993) with the epicenter shown by star.

breakwaters under construction and parts of seawalls in Okushiri Harbor were overturned southward. The second possibility is that the tsunami was generated by the south fault, refracted by the Okushiri Spur, a very wide shallow in the sea as an extension of the Aonae Point, and hit the Aonae from the east after propagating on a long path due to refraction. The third possibility is that the tsunami might be generated not by the major fault motion, but by another crustal motion which is indicated in Figure 8 (RESEARCH GROUP, 1993), by an isolated cluster of after-shocks between Okushiri Island and Hokkaido.

6.4. High Runup at Hamatsumae

There are few locations where runup height exceeds 20 m. They are only found on Okushiri Island (see Fig. 2). Along the west coast from Monai down to the Kuki Point to the south, runup heights are from 20 m to 24 m. Moving eastward from the Kuki Point to the Aonae Point, runup heights gradually decrease to 10 m. This tsunami height was still high enough to sweep away the 5th section of Aonae, even though the 5th section was protected by seawalls 4.5 m high, which were built by taking into consideration the maximum heights of the 1983 Nihonkai-Chubu earthquake tsunami.

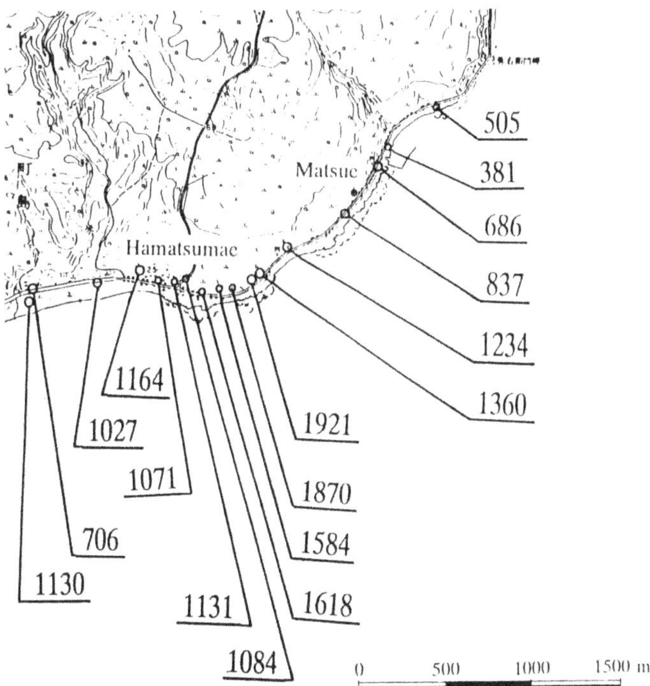

Figure 9
Runups in cm at and near Hamatsumae.

Further east of the Town of Aonae, high runups were found at Hamatsumae, in the area sheltered by the Aonae Point. When the field survey team found this high runup height, the authors decided to densely measure the runup height distribution in the neighborhood of Hamatsumae. The result is that, approaching Hamatsumae from both sides, runups become higher, as shown in Figure 9. Hamatsumae is located on a shore, the shape of which is convex to the sea. There is no special topography such as a V-shaped bay which suggests a local amplification effect.

This high runup may be (1) a result of the refraction effect of the Aonae Spur, (2) a result of the superposition of the tsunami trapped around the island, or (3) a result of the superposition of the tsunami generated by the crustal motion mentioned in Section 6.3 and the tsunamis which were generated by the major faults and trapped around the island.

6.5. Tsunami Trapped around the Island of Okushiri

A remarkable output of the authors' numerical simulation and CG animation was the entrapment of the tsunami around the island of Okushiri. In order to provide data for further study of a trapped tsunami, the field survey team was asked to increase the measurement points of runup heights on the east coast, even though damage was minor and the runups were less high there. The result was a spatial fluctuation of runup distribution as shown in Figure 2.

7. Conclusion

Since the 1993 Hokkaido Nansei-oki earthquake tsunami was generated at night, it was very difficult to collect visual data of the tsunami, such as photos and videos. The authors mostly collected tsunami data with a few witnessed arrival times and a few tide records. The weakness of runup data is that they do not impart their derivation.

On Okushiri Island, which was most devastated by the tsunami, there are no tide records. The time history of the tsunami is unknown. We must reconstruct the tsunami with the runup data, taking unreliable witnessed arrival times into consideration.

Special features of the tsunami are summarized in Section 6, which should be clarified by further study.

Acknowledgements

The authors would like to express their sincere thanks to Mr. S. Yamaki (INA Corporation) and Mr. M. Matsuyama (CRIEPI) for their vigorous cooperation in the field survey, to Mr. T. Nagai (Ohbayashi Co., Ltd.) and Mr. T. Takahashi (Tohoku University) for their efforts in the computer simulation, as well as to many students of Tohoku and Akita Universities for their earnest cooperation in the field and computer studies. A portion of this study was financially supported by a Scientific Grant-in-aid from the Ministry of Education, Science and Culture. The publication of the present study was financially supported by the Ogawa Commemorative Fund.

REFERENCES

HOKKAIDO TSUNAMI SURVEY GROUP (1993), *Tsunami Devastates Japanese Coastal Region*, EOS, Trans. AGU *74*, 417–432.
RESEARCH GROUP for aftershocks of the July 12 (1993), Hokkaido-nansei-oki Earthquake, 1993, *Geometry of the Aftershocks of the July 12, 1993, Hokkaido-Nansei-oki Earthquake*, Programme and Abstracts, Seismol. Soc. Japan *15* (in Japanese).

(Received August 1, 1994, revised February 15, 1995, accepted February 21, 1995)

PAGEOPH, Vol. 145, Nos. 3/4 (1995)

Field Survey Report on Tsunami Disasters Caused by the 1993 Southwest Hokkaido Earthquake

TOSHIHIKO SHIMAMOTO,[1] AKITO TSUTSUMI,[1] EIKO KAWAMOTO,[1] MASAHIRO MIYAWAKI[1] and HIROSHI SATO[1]

Abstract — Detailed field work at Okushiri Island and along the southwest coast of Hokkaido has revealed quantitatively (1) the advancing direction of tsunami on land, (2) the true tsunami height (i.e., height of tsunami, excluding its splashes, as measured from the ground) and (3) the flow velocity of tsunami on land, in heavily damaged areas. When a Japanese wooden house is swept away by tsunami, bolts that tie the house to its concrete foundation resist until the last moment and become bent towards the direction of the house being carried away. The orientations of more than 850 of those bent bolts and iron pipes (all that can be measured, mostly at Okushiri Island) and fell-down direction of about 400 trees clearly display how tsunami behaved on land and caused serious damage at various places. The true tsunami height was estimated by using several indicators, such as broken tree twigs and a window pane. The flow velocity of tsunami on land was determined by estimating the hydrodynamic force exerted on a bent handrail and a bent-down guardrail by the tsunami through *in situ* strength tests.

Contrary to the wide-spread recognition after the tsunami hazard, our results clearly indicate that only a few residential areas (i.e., Monai, eastern Hamatsumae, and a small portion at northern Aonae, all on Okushiri Island) were hit by a huge tsunami, with true heights reaching 10 m. Southern Aonae was completely swept away by tsunami that came directly from the focal region immediately to the west. The true tsunami height over the western sea wall of southern Aonae was estimated as 3 to 4 m. Northern Aonae also suffered severe damage due to tsunami that invaded from the corner zone of the sand dune (8 m high) and tide embankment at the northern end of the Aonae Harbor. This corner apparently acted as a tsunami amplifier, and tide embankment or breakwater can be quite dangerous when tsunami advances towards the corner it makes with the coast. The nearly complete devastation of Inaho at the northern end of Okushiri Island underscored the danger of tsunami whose propagation direction is parallel to the coast, since such tsunami waves tend to be amplified and tide embankment or breakwater is constructed low towards the coast at many harbors or fishing ports. Tsunami waves mostly of 2 to 4 m in true height swept away Hamatsumae on the southeast site of Okushiri Island where there were no coastal structures. Coastal structures were effective in reducing tsunami hazard at many sites. The maximum flow velocity at northern Aonae was estimated as 10 to 18 m/s (TSUTSUMI *et al.*, 1994), and such a high on-land velocity of tsunami near shore is probably due to the rapid shallowing of the deep sea near the epicentral region towards Okushiri Island. If the advancing direction, true height, and flow velocity of tsunami can be predicted by future analyses of tsunami generation and progagation, the analyses will be a powerful tool for future assessment of tsunami disasters, including the identification of blind spots in the tsunami hazard reduction.

Key words: 1993 Southwest Hokkaido earthquake, tsunami, tsunami hazard, Okushiri Island, tsunami hazard assessment.

---

[1] Earthquake Research Institute, University of Tokyo, 1-1-1 Yayoi, Bunkyo-ku, Tokyo 113, Japan.

Introduction

The 1993 Southwest Hokkaido Earthquake of Magnitude 7.9 (July 12, 22:17 JST) caused serious tsunami disasters in the southwestern part of Hokkaido, particularly on Okushiri Island (a tiny island off the southwest coast of Hokkaido with a population of about 4,500 at the time of earthquake). Of 230 casualties, including 28 missing, about 200 deaths are attributable to the tsunami. We have conducted detailed field surveys of tsunami disasters to learn lessons from this costly natural experiment for the future prevention of similar tsunami disasters. Our field work was conducted in four surveys totaling 39 days. During the first field survey (July 16 through July 21, 1994), we worked mostly on the estimation of the subsidence of Okushiri Island during the earthquake. Hence, our main work on tsunami disasters initiated from the second field survey (July 31 through Aug. 15, 1994).

Several groups have conducted detailed surveys of the distribution of tsunami runup height as measured from the level of sea water (TSUJI *et al.*, 1994a, b; MATSUTOMI and SHUTO, 1994; GOTO *et al.*, 1994). Such a precise runup height distribution is essential for characterizing tsunami, including its overall size. Indeed, the height distribution is the fundamental data for inferring earthquake source parameters through the simulation of tsunami generation (TAKAHASHI *et al.*, 1994; IMAMURA *et al.*, 1994; TSUJI *et al.*, 1994a; SATAKE and TANIOKA 1994; ABE, 1994; TANIOKA *et al.*, in review).

We have carried out our field work with a somewhat different scope from those of the above-mentioned workers and have been more concerned with tsunami disasters and their prevention. Thus we have worked only on heavily damaged areas due to tsunami on Okushiri Island and along the southwest coast of Hokkaido, trying to determine how the tsunami caused the heavy damage and to single out key factors for the prevention of tsunami disasters. More specifically, we have attempted to determine objectively (1) the advancing direction, (2) true height (height as measured from the ground), and (3) velocity of tsunami that run up on land in heavily damaged areas. We only depended slightly on the testimonies from the residents, since tsunami occurred late at night. The estimation of the true tsunami height near shore, rather than the tsunami runup height (i.e., the height of the highest point tsunami reached, as measured from the sea-water level), is needed to determine the height of coastal structures required to prevent tsunami. The velocity of tsunami will be useful to estimate the required strength of coastal structures to stop tsunami.

We have collected data to estimate the above three quantities just in time before the major reconstruction started, and raw data will be made open herein for detailed theoretical analyses. Such data are a completely new set of data to be compared with the simulation of tsunami generation and propagation in the future.

Advancing Direction of Tsunami in Heavily Damaged Areas

Indicators of the Advancing Direction of Tsunami

Now the question is what to use as reliable indicators of advancing direction of tsunami near shore and on land. Grass was overturned at many places that were swept away by tsunami. However, this overturned direction is not always a reliable indicator because the return flow of tsunami can easily turn the grass over to another direction. After searching for good indicators for several days during our first field survey, we finally found that iron bolts of about 10–15 mm in diameter, tying a Japanese wooden house to its concrete foundation, are an ideal indicator of the movement direction of tsunami. The houses are tied to their foundation only through these iron bolts, and the bolts resist the displacement of the houses by tsunami until the last moment. As a result, those bolts became bent down towards the direction of houses to be carried away (Figure 1a). The bolts are so stiff that

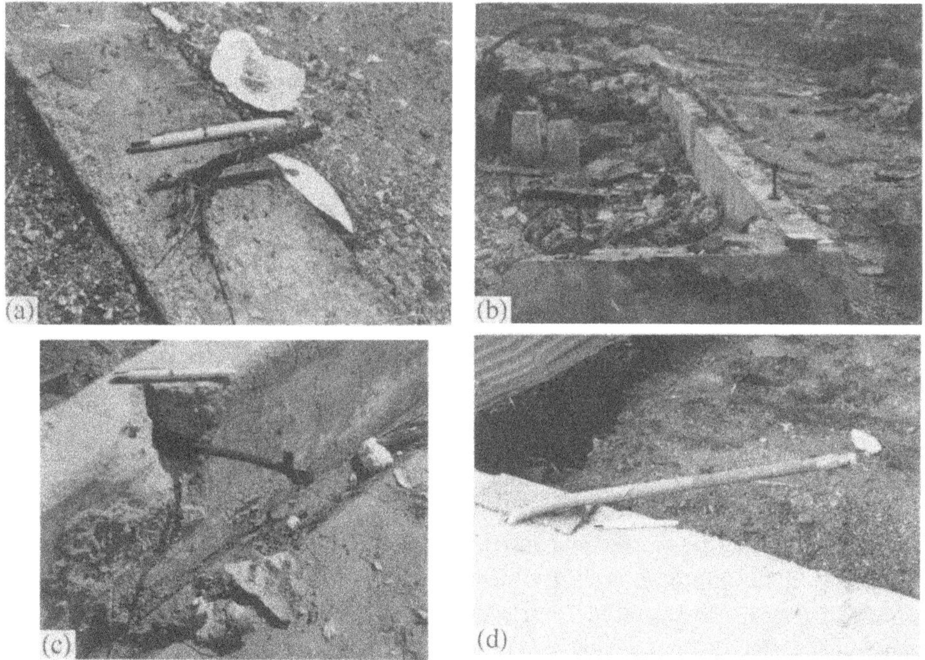

Figure 1

Iron bolts and water pipes on the concrete foundations of wooden houses in heavily damaged areas, as indicators of the advancing direction of tsunami. (a) A bent-down iron bolt on the concrete foundation of a carried-away wooden house, (b) a protruding iron bolt on the foundation of a burned wooden house, (c) an iron bolt turned down by a heavy construction machine during the clearance of a damaged house, and (d) a bent-down iron water pipe on the foundation of a swept-away wooden house.

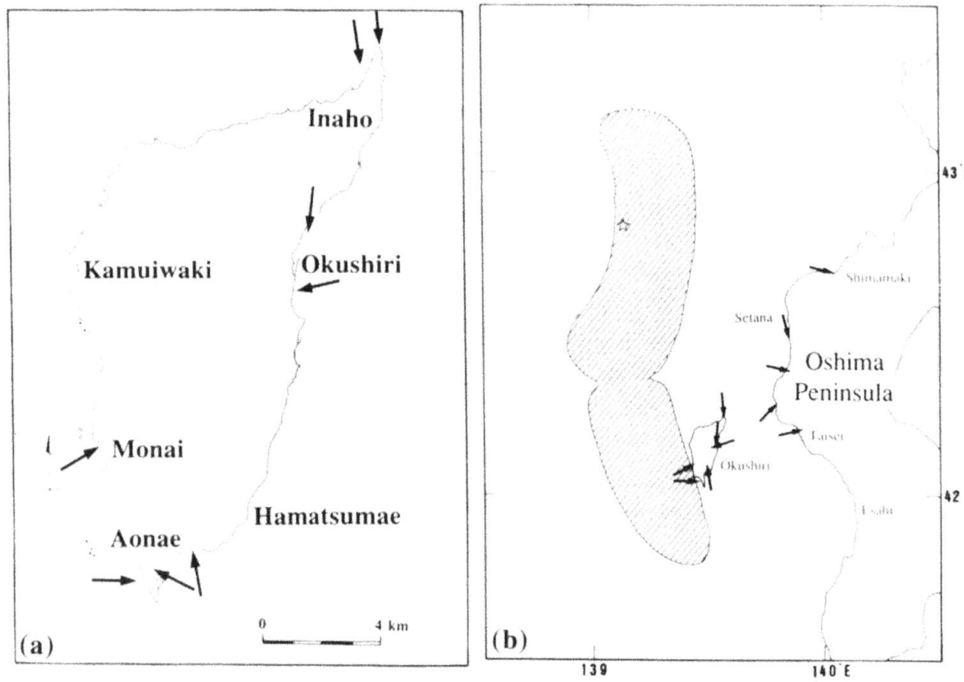

Figure 2

The advancing direction of tsunami as determined from bent-down iron bolts and pipes and overturned trees (a) on Okushiri Island and (b) on Okushiri Island and along the northwestern coast of the Oshima Peninsula. Arrows show representative advancing directions near the coast. Shaded region and the star in (b) indicate, respectively, the aftershock area and the epicenter of the Southwest Hokkaido Earthquake after NAKANISHI and KIKUCHI (1993).

their bent-down direction can barely be altered by later tsunami movement. Moreover, several bolts or in some cases more than ten bolts are used in building a house and many data can be collected in heavily damaged areas. Thus the bolts became ideal indicators of the movement direction of tsunami on land. Frequently the bent-down bolts and concrete foundation are the only remains of carried-away houses, and the bolts are the last message from those houses. We have measured the bent-down direction of more than 800 bolts, all that can be measured, on Okushiri Island and along the southwest coast of Hokkaido.

Aonae at the southern end of Okushiri Island (Figure 2a) is by far the hardest hit area in terms of casualties. Parts of central Aonae at low elevations escaped tsunami disaster, but much of the area with houses remaining was burnt down completely by earthquake-caused fire (Figure 3; the shaded area in Figure 4). The bolts from the remains of burned wooden houses were protruding upright (Figure 1b), and the boundary between the burnt area and the area swept away by tsunami can be identified by inspecting only those bolts. Also, some houses, which were heavily damaged but not carried away by tsunami, were quickly cleared away with heavy construction equipment. At first sight, the foundations of these cleared houses cannot easily be distinguished from those swept away by tsunami. However, the bolts on the foundations of those "cleared" houses are mostly bent in irregular orientations, most easily identifiable. In the photograph in Figure 1c, the edge of the foundation was broken off and dropped down at the foot of the concrete foundation, indicating that the bolt was bent at the time of the clearance of the house. Otherwise the tsunami would have carried the edge piece much further away. A thin layer of dust was sometimes left atop the foundation with bent bolts, and the bending of the bolts cannot be due to tsunami since there is virtually no chance for the dust to stay after being swept away by the tsunami. Thus the bolts can reveal whether a house was carried away or was severely damaged but was not carried away by tsunami.

To supplement data on the bent bolts, we also measured the bent-down directions of iron water pipes (Figure 1d). At places where there is mostly insignificant return flow of tsunami judging from topography (i.e., at a nearly flat place), we have used fell-down directions of big trees (Figure 5a) and grass (Figure 5b).

An Overview of the Advancing Direction of Tsunami

Figure 2 shows the advancing direction of tsunami in heavily damaged areas on Okushiri Island and along the southwest coast of Hokkaido, together with the epicenter (star) and the aftershock area (shaded region) of earthquake. The representative data on bent-down bolts and fell-down trees near shore are displayed in the figures to show the advancing direction of tsunami as close as possible to the shore.

In the Shimamaki and Setana districts on the northwestern part of the Oshima Peninsula, southwest Hokkaido, tsunami originated from the WNW and NNW ~ WNW directions, respectively, almost directly from the epicentral region of the earthquake (Figure 2b). Monai and southern Aonae on the southwestern part of Okushiri Island were struck by tsunami from WSW and W directions, respectively, within 5 to 10 minutes after the earthquake, and the tsunami seems to have originated at the southern portion of the focal region immediately to the west of the areas. The runup height was generally very high and exceeded ten meters at many places along the coast from Monai to Aonae (TSUJI *et al.*, 1994a; MATSU-TOMI and SHUTO, 1994). This is probably due to large moment release at the southern end of the focal region during the earthquake (NAKANISHI and KIKUCHI, 1993; KIKUCHI, 1994; KUGE *et al.*, 1994).

Inaho district, also heavily damaged, at the northern end of Okushiri Island was battered by tsunami from N to NNW directions (Figure 2). The aftershock area is bent in the middle, in the northwestern direction from Okushiri Island (Figure 2b; NAKANISHI and KIKUCHI, 1993). The pattern of the aftershocks in the middle bent region is complex and the moment release there is small (KIKUCHI, 1993; KUGE *et*

Figure 3
Aerial photograph of Aonae district at the southern end of Okushiri Island, taken on the next day of the earthquake (courtesy of Kokusai Kogyo Co., Ltd.). Note smoke emerging from the burnt area due to the fire that started just after the earthquake.

al., 1994). Quite interestingly, tsunami causing the most severe damage at Inaho did not emanate from the focal region immediately to the northwest, but came from the N nearly directly from the epicentral region. This is probably due to the small moment release in the bent region of aftershocks. The runup height of tsunami was generally lower than 10 m along the northwestern coast of the Okushiri Island (TSUJI *et al.*, 1994a; MATSUTOMI and SHUTO, 1994).

The advancing direction of tsunami on the east coast of Okushiri Island is complex, as a result of reflection and interactions of tsunami waves (Figure 2a).

Tsunami Disasters in Aonae District, Okushiri Island

Severe tsunami damage at Aonae district at the southern end of Okushiri Island has become a symbol of tsunami disasters due to the Southwest Hokkaido

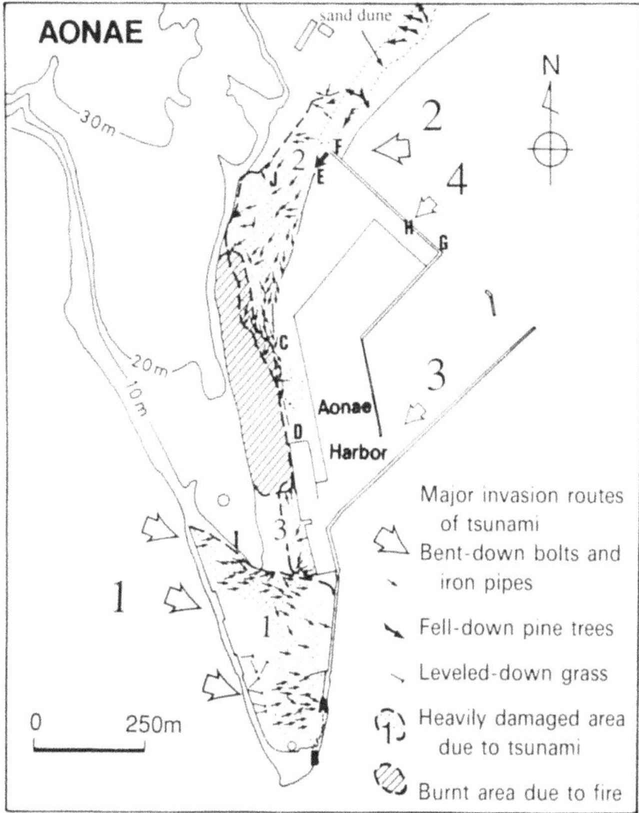

Figure 4

Advancing direction of tsunami at Aonae district, Okushiri Island, as determined from bent-down iron bolts and water pipes, toppled pine trees, and leveled-down grass. Each small arrow indicates the vector mean of the orientations of the bent-down bolts and pipes on the foundation of each carried-away house. Thick arrows on the sand dune at northern Aonae indicate the vector means of the orientation of fallen pine trees. A thick arrow near point E indicates the mean orientation of bent-down poles of a guardrail on the road side. Big open arrows indicate major invasion routes of tsunami. The stippled and shaded zones show, respectively, wiped-out area by tsunami and burnt-down area by fire.

earthquake (see an aerial photograph in Figure 3, taken on the next day of the earthquake). In particular, the southern end of Aonae was devastated completely by tsunami that overflowed the sea wall at the shore, 4.5 m high from the level of sea water. Aonae subsided by about 0.5 m during the earthquake (TSUTSUMI *et al.*, 1993), so that the sea wall was about 4 m high when tsunami struck southern Aonae. All 82 wooden houses were carried away (Figure 6a), and a broken public toilet with a reinforced concrete wall was the only remaining structure there (Figure 6b). 263 bent bolts on the foundations of 58 swept-away houses indicate the easterly movement of tsunami as shown by small arrows in Figure 4. Each arrow indicates the vector mean of the orientations of bent-down bolts (mostly 5 to 10) at each

Figure 5

Other indicators of the advancing direction of tsunami: (a) uprooted pine trees due to tsunami on the sand dune in northern Aonae, Okushiri Island, and (b) leveled-down grass by tsunami and tiny grooves formed by tsunami water flow at southern Aonae.

foundation using the method of CURRAY (1956). The orientation of turned-over grass is also shown in the figure as a thin arrow since the southern part of Aonae is flat and there was no signficant return flow. It took just 5 to 10 minutes after the earthquake for the tsunami to strike southern Aonae. This, together with the advancing direction of tsunami from the west, strongly suggests that the tsunami derived directly from the southern part of the earthquake focal region, immediately to the west of Aonae (invasion route 1 in Figure 4).

Damage to the sea wall, to the tide embankments and to the breakwater was relatively minor at Aonae, in marked contrast with the complete devastation of wooden houses. The tide embankment, 4.5 m high from the sea-water level before the earthquake, collapsed at the distance of about 80 m at the southeastern corner of the Aonae Cape (from A to B in Figure 4; Figure 7a). The tide

Figure 6

(a) Completely devastated southern Aonae due to tsunami that came from the west and went over the sea wall 4.5 m high from the sea-water level before the earthquake, and (b) a damaged public toilet building with reinforced concrete wall that survived tsunami at the southeastern part of the Aonae Cape.

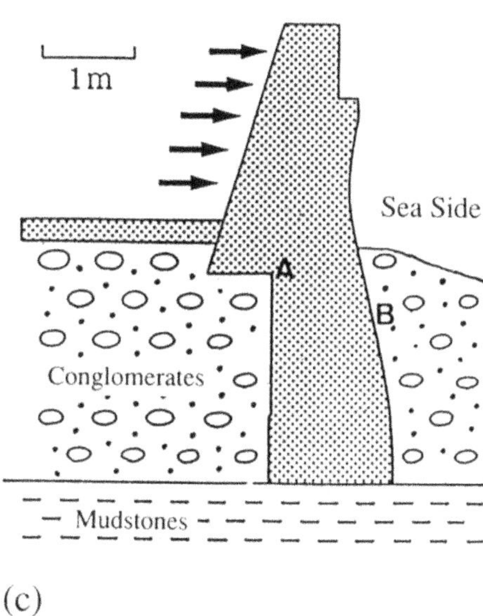

Figure 7

Collapsed tide embankment at the southeastern corner of the Aonae Cape. (a) Turned-over tide embankment, (b) the broken portion of the tide embankment at the northern end of the collapsed part, and (c) a schematic cross section showing the tide embankment and its surrounding conglomerates and mudstones at the basement. Note that mudstone can be seen on the bottom surface of overturned tide embankment in (a).

embankment here was constructed on top of the Miocene mudstones (8 to 13 Ma old) and was buried in unconsolidated Holocene conglomerates (see a schematic profile in Figure 7c). The mudstones and conglomerates are weak and even breakable by hand, and the tide embankment caved in nearly intact after being hit by tsunami from the land side (Figure 7a). The northern end of the tide embankment was broken just underneath the ground level for the width of about 1.5 m (Figure 7b). The broken portion corresponds to AB in Figure 7c, and evidently a

Figure 8
Northern and central Aonae after the disaster (refer to Figure 4 for the locations of the photographs).
(a) A bent-down guardrail on the roadside near point E, (b) burned-down area in central Aonae due to
fire caused by the earthquake, (c) highly eroded corner zone between the sand dune and the tide
embankment at the northern end of the Aonae Harbor (point F), and (d) a fishing boat thrown on top
of a fire engine by tsunami that invaded from the entrance for ships to the Aonae Harbor.

fracture causing the failure initiated at the corner near A due to the stress
concentration. The northern block of about 10 m in width was mostly carried out
to sea for a distance of about 20 m, and this distance decreases towards the south
(Figure 7a). Thus the collapse of the tide embankment must have initiated from the
northern part. The conglomerates near the fell-down embankment were eroded by
1 to 2 m deep for an area of about 150 m², implying stronger water flow compared
to other areas with much less erosion (for a detailed description of the erosion and
deposition due to tsunami, see SATO *et al.*, 1994). Most likely, many houses and
even people must have been carried out into the Japan Sea from there.

Part of tsunami that pounded the tide embankment between A and the Aonae
Harbor (Figure 4) must have turned its direction northward towards the harbor to
carry fishing boats there to the north, some striking two small concrete buildings
near point D in Figure 4. Also, some people who became trapped by tsunami at
southern Aonae were recovered at the northern part of the Aonae Harbor,
according to the Okushiri Fire Department.

The northern part of Aonae district also suffered almost as much damage as did the southern part (stippled region 2 in Figure 4). The tsunami swept away nearly 100 houses, claiming 35 lives. The orientations of bent bolts at the foundation of carried-away houses (small arrows), those of toppled pine trees on sand dune (thick arrows; see also Figure 5a), and those of the bent-down poles of a guard rail (heavy arrow near point E; see also Figure 8a) indicate that tsunami advanced toward the corner F of the sand dune (about 8 m high from the sea-water level) and the tide embankment at the northern margin of the Aonae Harbor (invasion route 2 in Figure 4). The tsunami devastated northern Aonae, moving towards the SW–SSW directions along the old tide embankment between E and C, gradually changing its direction to S to SSE towards the harbor (Figure 4). The shaded region in the figure escaped severe tsunami damage, but burnt down completely due to a fire which originated in two places just after the earthquake (Figure 8b). There is a small overlapped region between the burned region and the damaged area due to tsunami (Figure 4). The second strike of the tsunami, which came in about 17 minutes after the earthquake, caused serious damage to northern Aonae (TSUJI *et al.*, 1994a).

The corner of a tide embankment and shore (point F in Figure 4) is probably the most dangerous blind spot recognized in the tsunami disasters due to the Southwest Hokkaido earthquake. The width of the wave front becomes narrow and the tsunami wave can be amplified substantially as it approaches such a corner. Thus for the prevention of tsunami disasters, such corners must be fortified, for example, with tetrapods and with higher tide embankment towards the shore. Whereas at many harbors, breakwater and tide embankment seem to have been constructed lower towards the shore. At the northern part of the Aonae Harbor, the height of tide embankment decreases from 4.5 m at the northeastern corner to 2.7 m at the shore (from the sea–water level before the earthquake). Thus enormous amounts of water invaded from this corner, causing severe damage to northern Aonae. Sand dune originally extended to the edge of the tide embankment, however tsunami eroded 2,000 to 4,000 m³ of sand, leaving a pond at the corner (Figure 8c; for details see SATO *et al.*, 1994).

Orientation data of bent bolts on concrete foundations in the stippled zone 3 in Figure 4 strongly suggest that the tsunami entered from the entrance from the sea to the harbor and washed houses and fishing boats away to southern Aonae (note scattered houses and boats at southern Aonae in Figure 3). However, damage due to tsunami from this invasion route 3 was considerably smaller than at the above two areas. The photograph in Figure 8d, taken near the southern end of the Aonae Harbor, shows a fishing boat from the harbor crushing over a fire engine due to the tsunami from this route. The rectangular region immediately to the west of the stippled zone is elevated 4 m from the sea-water level and escaped serious tsunami damage. There is a small cliff, about 1 m high, at the western margin of the stippled zone 3. Part of tsunami changed its direction to the north, after impacting this cliff, to transfer three houses to the north at the northern end of the zone.

Figure 9
Broken part of tide embankment at the northeastern end of the Aonae Harbor (point H in Figure 4). (a) The broken top portion of the tide embankment, and (b) the broken block in the left center which was carried in the SW direction for a distance of about 60 m.

The entrance for ships to the harbor is sizable, and it is desirable to improve its design in the future. For instance, breakwater can be constructed, or tetrapods may be installed at the outside of the entrance to stop tsunami without obstructing the entrance. Also, residential houses and important facilities should not be built at low elevations within a harbor immediately behind the entrance.

The routes 2 and 3 in Figure 4 are thus the main invasion routes of tsunami from the eastern side, and no massive tsunami surged over the tide embankment and breakwater (4.5 m high from the sea-water level before the earthquake). However, some damage has been caused by a tsunami wave which inundated the northern tide embankment of the Aonae Harbor (invasion route 4 in Figure 4). The very top portion of the bank at point H was broken and transported to SW for a distance of 60 m (big open arrow in Figure 4; Figures 9a and 9b). The broken surface was very fresh, indicating that the embankment did not break at a weak discontinuity. SHUTO (1993) therefore suspects that a heavy object such as a fishing boat struck the broken embankment. However, the top portion next to this broken part was also fractured (Figure 9a), but no scratch marks, suggestive of a collision with a heavy object, were left on the seaward side of the bank. In any case, a large tsunami must have hit at least this broken part of the embankment and went over it. Moreover, three houses near point C were swept away into SSW direction (Figure 4), which is different from the overall trend of water movement towards the harbor there. A road here crosses the old tide embankment between E and C, and tsunami seems to have invaded from the sea side through this opening in the road. Since point C is located at an extension of tsunami movement at the northeastern end of the Aonae Harbor (H in Figure 4), damage to the three houses seems to have been caused by tsunami that went over the tide embankment around H and G. The north side of the reinforced concrete wall of a small Fishermen's Association building near point D was broken southwards. Tsunami from route 2 did not reach this far, and therefore the damage to the wall must also have been due to tsunami that went over the tide embankment at the northeastern end of the Aonae Harbor. Thus damage from route 4 was minor.

The Aonae district was nearly completely devastated by tsunami and fire, with the casualty figure rising to total of 107 (2 being due to fire) except for the area on the marine terraces higher than 10 m, the small rectangular area on the southwestern corner of low land, and the northern end low land behind the sand dune (Figure 4). At a glance, coastal structures may seem to have been ineffective in reducing the tsunami damage. Indeed, the southwestern sea wall was ineffective preventing the tsunami disaster. In the northeastern part, however, tide embankment and breakwater no doubt reduced the damaged area substantially, even though the corner between the tide embankment and sand dune was a blind spot in the prevention of tsunami disaster. Without those coastal structures the tsunami damage would have been extensively more disastrous, probably devastating all houses on the low land including the burnt area.

Tsunami Disasters in Other Heavily Damaged Areas

Monai District, Okushiri Island

Monai is located at the southwest coast of Okushiri Island, very close to the southern part of the focal region (Figure 2). The maximum runup height of 31.5 m

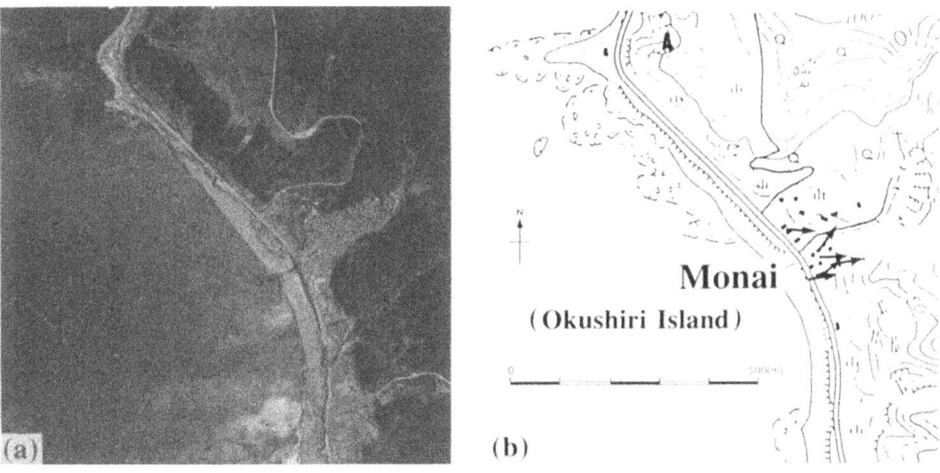

Figure 10
(a) Aerial photograph of the Monai district on the southwestern coast of Okushiri Island (courtesy of Kokusai Kogyo Co., Ltd.), and (b) the topographic map of the same area published by the Geographical Institute of Japan. The highest tsunami was recorded along a small valley at point A. Arrows in (b) show the vector mean of the orientations of bent-down bolts and pipes on the foundations of lost houses.

Figure 11

Scenes of the Monai district after being struck by a giant tsunami. (a) A coast south of Monai village where tsunami uprooted land plants over electric cables several meters above the ground, (b) devastated Monai village, (c) fragments of wooden houses scattered along a stream near Monai village, and (d) the roof of a two-storied house at Monai after being hit by tsunami whose true height was 7–8 m. Note that plants on the slope were leveled down completely by tsunami in (a) and (b).

was recorded along a little valley several hundred meters north of the small village of Monai (TSUJI *et al.*, 1994a; A in Figures 10a and 10b). The runup height on the mountain slope at the entrance to this valley was 23 m and it decreased gradually to about 15 m at Monai village. The tsunami runup height exceeded 10 m at many locations from Monai to Aonae. Tsunami heaved land plants over electric power lines on the road side, and grass hanging from electric lines several meters high (Figure 11a) were broadcast on TV news many times after the earthquake. The southwestern part of Okushiri Island sustained the most gigantic tsunami hit on the coast.

Except for a house on a little hill, 15 m high above the sea-water level, all of the other 7 houses in Monai were destroyed and completely carried away, leaving debris of shattered wooden houses along a nearby little stream (Figures 11b,c). Orientations of bent bolts on the foundations indicate that tsunami advanced from W direction directly from the southern part of the focal region (Figure 10b, Figure 2b).

Hamatsumae District, Okushiri Island

Hamatsumae, located northeast of Aonae, also suffered from tragic damage as tsunami swept away 36 wooden houses which resulted in 32 deaths (Figure 12a). A house at a slightly higher elevation was the only house that escaped the disaster. The orientation data on the bent bolts indicate that tsunami advanced toward N to NNW directions (Figure 12b). A vivid photograph of Hamatsumae (AS-AHIGRAPH, July 25, 1993 issue, pp. 22–23. Asahi News Paper Co. Ltd., Tokyo, Japan), taken immediately after the earthquake, clearly shows that the second floors

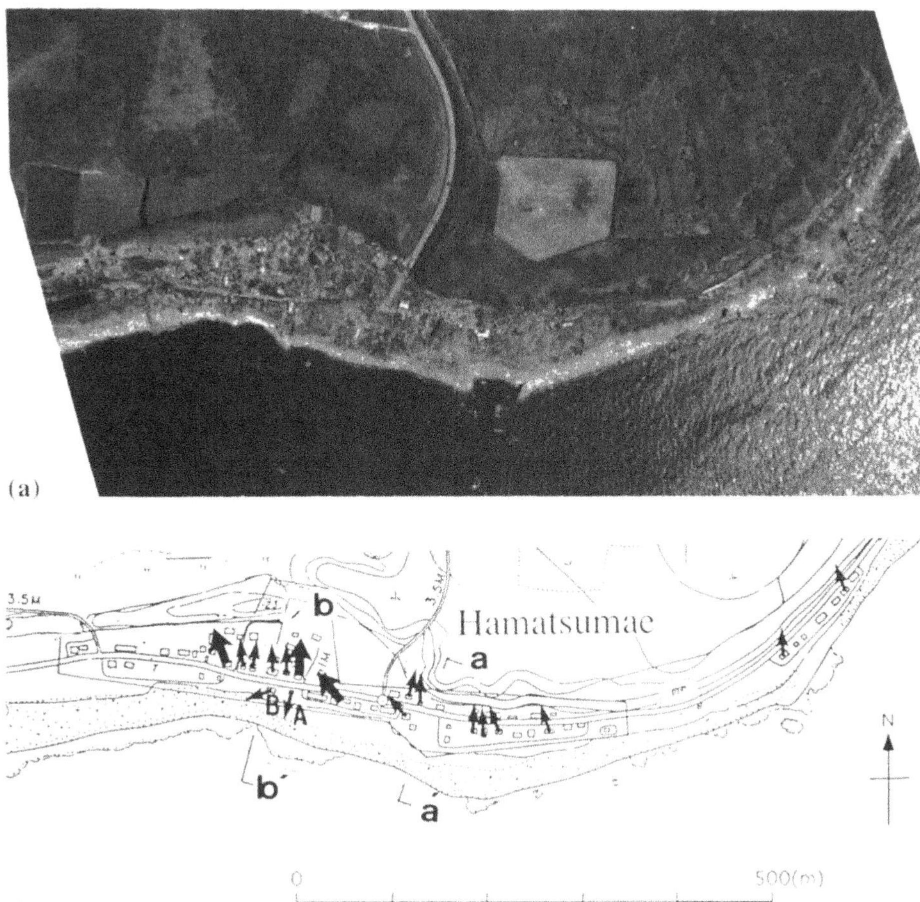

Figure 12

Hamatsumae on the southeast coast of Okushiri Island devastated by tsunami. (a) Aerial photograph (courtesy of Kokusai Kogyo Co., Ltd.), and (b) topographical map of approximately the same area after the Geographical Institute of Japan. Arrows in (b) show mean orientations of bent-down bolts and pipes on the foundations of lost houses, and heavy arrows indicate the displaced orientations of three two-storied houses.

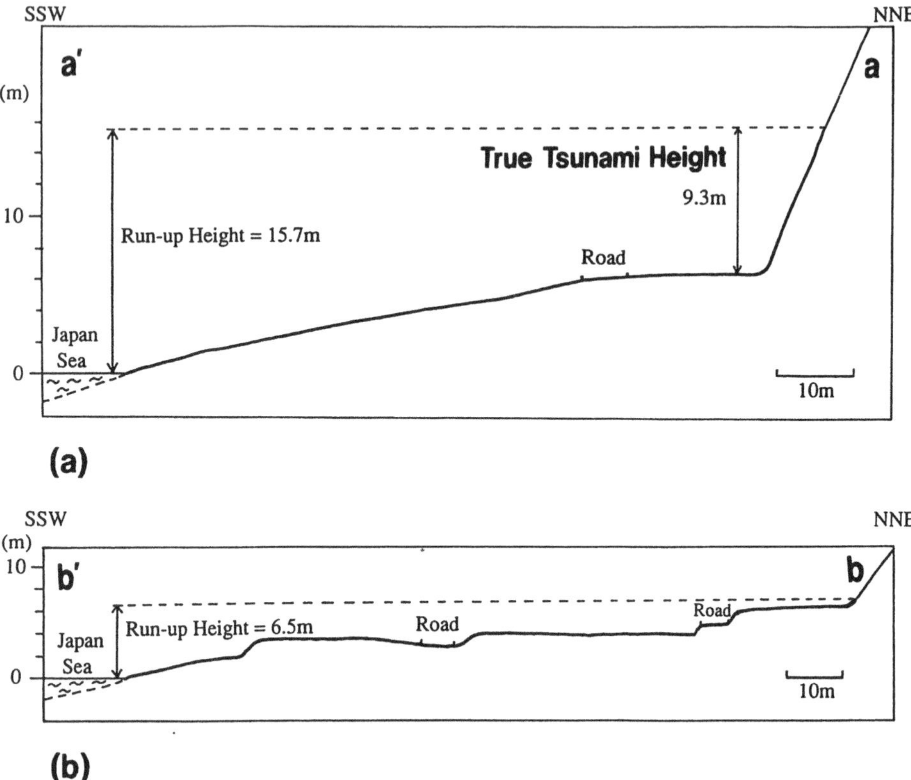

Figure 13

Geomorphological cross sections, cutting aa' and bb' in the previous figure, drawn parallel to the advancing direction of tsunami.

with the roofs of two-storied houses, remained nearly intact even though the first floors were broken apart and the houses were carried away. The original locations of three of those second floors were identified by the testimony of local residents and by a member of the Okushiri Fire Station, and heavy arrows in Figure 12b indicate the orientation from the original locations of the houses towards their stopped positions. These orientations are consistent with the data on the bent bolts.

Tsunami swept away Hamatsumae and struck the steep slope of a marine terrace nearly perpendicularly, leaving clear marks on the slope. The runup height was as high as 16 m for an interval of about 30 to 40 m at the eastern part of Hamatsumae (Figure 12a). It decreased rapidly northeastwards towards Matsue and westwards towards the western part of Hamatsumae. The runup height at the central and western parts of Hamatsumae was 6 to 9 m and fluctuated by as much as 10 m within short distances. The edge of tsunami runup on land was thus quite irregular. We recognized two main factors as the cause of the disaster in this district; first, there were no coastal structures such as tide embankments, and second the slope from the sea to the residential area was very gentle and the tsunami could run up easily (Figure 13). In addition, TSUJI *et al.* (1994a) analyzed that the unusually high runup height at eastern Hamatsumae was due to the focusing effect of submarine topography on tsunami.

Three bent-down bolts on the foundation at point A in Figure 12b indicate seaward motion, and the houses must have been detached from their foundations by the return flow of the tsunami that hit the terrace slope. Moreover, three bolts on the foundation at point B exhibit northward (landward) motion, but four bolts and a water pipe on the same foundation display southward (seaward) motion of tsunami. Presumably, the advancing tsunami destroyed and detached part of the house, and the remaining part of the house was carried away by the return flow. An arrow near point B points out the vector mean of these bimodal data and does not indicate physically meaningful orientation. Of all the foundations of lost houses and buildings we have inspected at damaged areas, those are the only cases that manifest the detachment direction of a house in the direction of the return flow. Thus the advancing tsunami is substantially more powerful than its return flow.

The southeastern coast of Okushiri Island from Aonae to Hamatsumae forms a gentle bay. The advancing directions of tsunami in these districts (Figures 2a and 12b) exhibit a fan-like arrangement, thus the tsunami must have changed its advancing direction into a fan reflecting the submarine geomorphology of the bay.

Inaho District, Okushiri Island

The village of Inaho on the northwestern end of Okushiri Island also suffered severe tsunami damage with about 35 lost houses and 11 casualties (Figure 14). The orientations of bent-down bolts and pipes at the Inaho Cape indicate that the advancing direction of the tsunami was from N to NNW (Figure 15). The damage at Inaho brought out the danger of tsunami advancing nearly parallel to the coastline. Such a tsunami proceeds for a long distance along the shallow sea near the coast, with its speed slowed and its height increased substantially because the shallower the sea, the slower the speed of tsunami. The tsunami wave which hit Inaho was not so large (see next section), but it did cause severe damage to the Inaho Primary School located about 6 m in elevation from the sea-water level (Figure 16a; location A in Figure 15). The coastline and a road change their direction from SW to W near the Inaho Fishing Port (Figure 15), cutting into the passage of tsunami nearly parallel to the coast. The massive tsunami overran the coast road, about 3 to 4 m high from the sea water, and completely wiped out all houses in the village. Thus the southwestern part of Inaho was located just at the passage of tsunami advancing nearly parallel to the coast.

The orientation of bent bolts on the foundations of lost houses reveal two flow patterns; one to the S to SW indicating the direct motion of tsunami from the sea, and the other to the SW to W probably indicating the motion of tsunami that goes

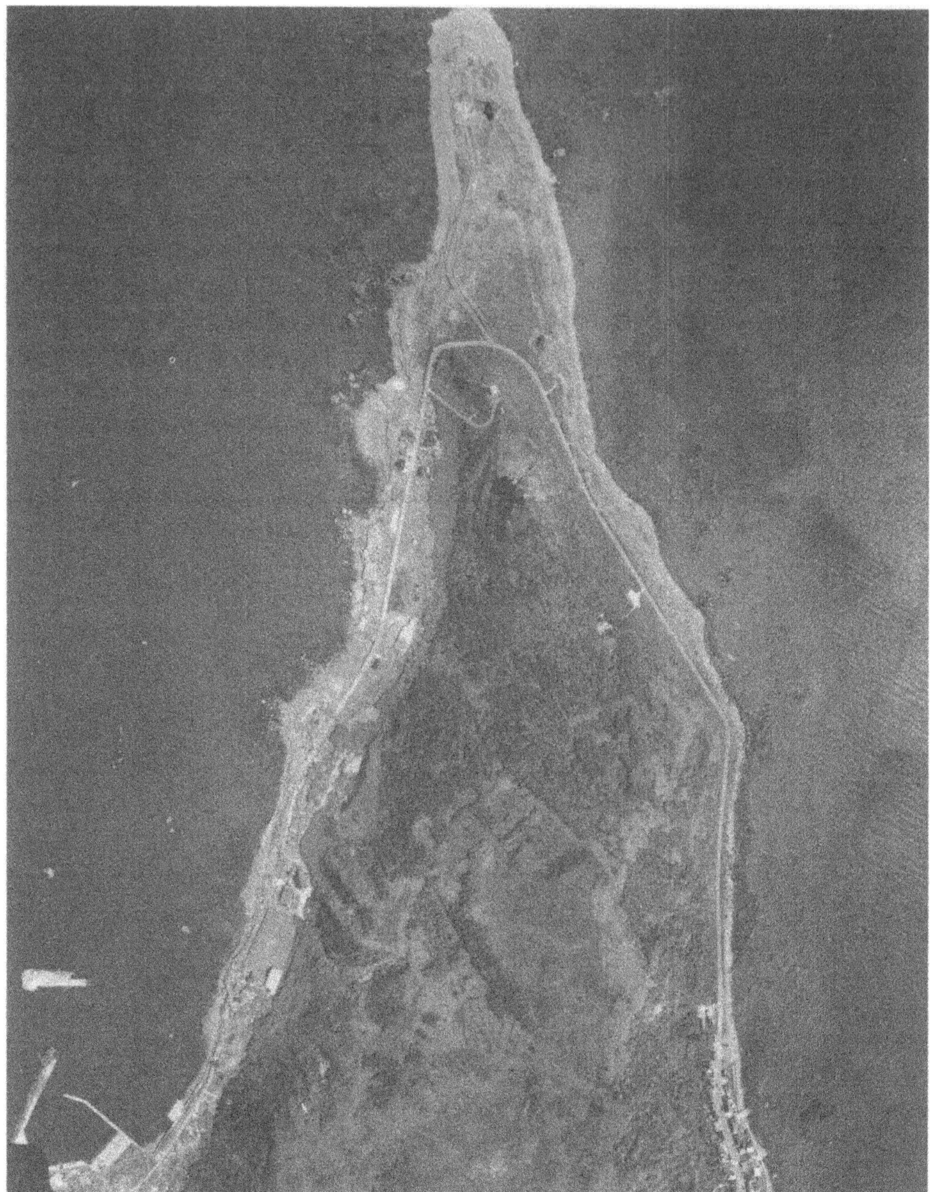

Figure 14
Aerial photograph of the Inaho Cape (upper center), Inaho district (lower left) and Kantahama district (lower right) at the northern end of Okushiri Island (courtesy of Kokusai Kogyo Co., Ltd.).

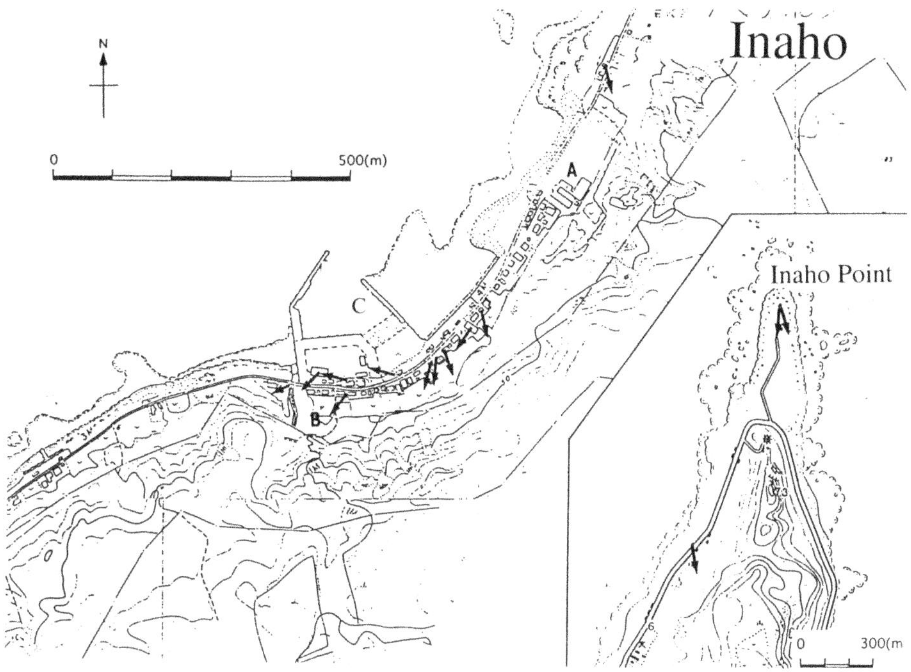

Figure 15
The advancing direction of tsunami determined from bent-down bolts and pipes at the Inaho district.
Topographical map from the Geographical Institute of Japan.

around the foot of the mountain behind the village. The maximum runup height at Inaho was 11 m at point B in Figure 15. This location was immediately behind the entrance to the port and its nearby point C, where the breakwater was broken down for a width of about 10 m.

True Tsunami Height and its Indicators

Definition of True Tsunami Height

The maximum runup height of 31.5 m in Monai was broadcast repeatedly after the earthquake and most ordinary people, even those of local government offices, had the impression that a giant tsunami, higher than 10 to 20 m, attacked almost everywhere in the damaged areas on Okushiri Island and even in part of the southwestern coast of Hokkaido. Moreover, the runup height as measured from the sea-water level was easily confused with the height as measured from the ground to intensify the biased view of tsunami height. We also strongly felt that testimonies on tsunami height were not reliable since splashes of tsunami as it struck houses

Figure 16
(a) Damaged building of Inaho Primary School, (b) a part of the pine forest on the sand dune
at northern Aonae which survived tsunami, (c) branches and small twigs of a pine tree, and (d)
nearly intact second floor of destroyed two-storied house at Nonamae, just southwestern of Inaho.
Note in (c) that tiny twigs survived tsunami even though a bigger branch was broken at the left
center.

and structures must have given a false impression of the real height of tsunami,
excluding its splashes. Indeed, 1–2 m high tsunami often partially damaged even
the second floor of a house. The inundation depth sometimes can be identified by
markers left on inundated houses, such as horizontal stains on white walls. This
depth, however, is determined by the drainage conditions of the surrounding areas
and does not necessarily imply that a tsunami of the height of the inundation depth
impacted the structures with a powerful momentum.

From the standpoint of tsunami disaster prevention, therefore, one must determine the height of advancing tsunami, excluding splashes which cannot destroy the
entire structures completely, as measured from the ground. We shall call this height
herein "true tsunami height" and distinguish it from "runup height," i.e., height of
the point tsunami reached, as measured from the mean level of sea water. True
tsunami height near shore is of particular significance in designing coastal structures
to prevent future tsunami disasters.

Indicators of True Tsunami Height

Now the question is determining what to use for estimating the true tsunami height objectively. After searching for suitable indicators, we happened to notice that twigs of trees are an ideal indicator of the true tsunami height. In the northern part of Aonae, an 8-m high sand dune (Figure 4) is covered with a pine forest planted as a wind barrier. Tsunami went over this sand dune and leveled more than one thousand pine trees (Figure 5a). However, some pine trees remained standing even though their lower branches and twigs were severed (Figure 16b). Some twigs were connected to the trunk with tiny stems, often less than a few millimeters in diameter (Figure 16c). Upon closer inspection of the pine trees, we recognized that below a certain height no such twigs were left, and that above a certain height some twigs were left even though some bigger branches were broken, probably due to the splashes of tsunami. Figure 17 displays the upper bound of no remaining twigs and the lower bound of at least some remaining twigs, both measured from the ground. The data clearly show that the true tsunami height was around 2 m above the sand dune, while the runup height of tsunami was 9.5 m.

The clue to determining the true tsunami height is to use fragile objects that definitely break upon being struck by tsunami. Namely, the height below which all or most of those fragile objects were broken, gives the true tsunami height at that point. This is in contrast with using stiff objects such as bolts in determining the advancing direction of tsunami. Windowpanes turned out to be an even more

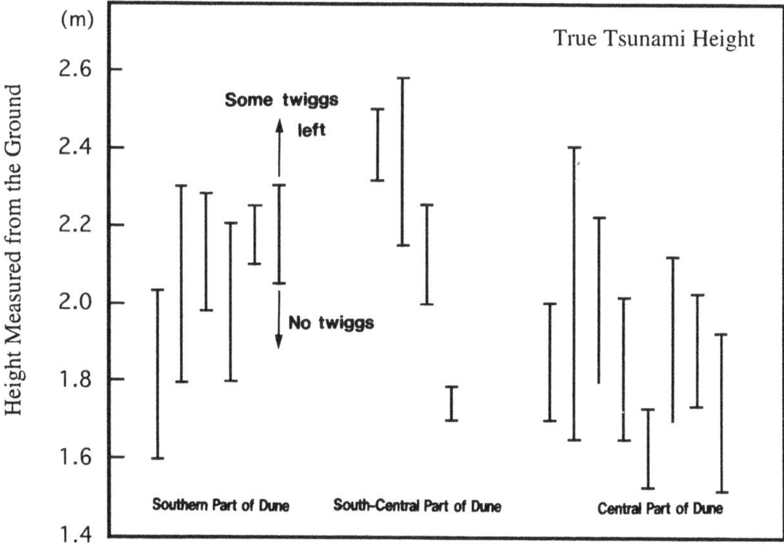

Figure 17

Estimation of the true tsunami height from twigs of pine trees that did not fall after being hit by tsunami. Vertical bars tie the height above which some twigs are left and the height below which no twigs remain, both measured from the ground. Data are for three survived groups of pine trees.

suitable object than twigs of trees for the present purpose. For instance, the lower row of windowpanes of Inaho Primary School was mostly broken, but the upper row remained mostly unbroken (Figure 16a). Thus the true tsunami height there was between the lower and upper heights of the lower windowpanes (1.5–2.4 m from the ground).

When all houses were carried away, the estimation of the true tsunami height was very difficult. In such cases we used a topographic cross section drawn parallel to the advancing direction of tsunami and estimated the true tsunami height by plotting the point the tsunami reached on the cross section. Topography sometimes changes abruptly from a flat surface to a steep slope such as the foot of a mountain. Consequently the height of the point the tsunami reached on the slope, as measured from the ground, should approximately give the true tsunami height near the foot of the slope. Figure 13a shows such a cross section at the eastern part of Hamatsumae, and the point the tsunami reached on the slope of a marine terrace is about 9 m above the ground. In general, tsunami tends to runup somewhat further when it hits a steep slope. Observations in Monai suggest that the amount of this runup is of the order of 0.5–1 m (see SHIMAMOTO *et al.*, in preparation). In view of this running-up effect, the true tsunami height at eastern Hamatsumae is found to be around 8 m.

True Tsunami Height and Characteristics of Tsunami Damage

True Tsunami Height in Heavily Damaged Areas

We have determined the true tsunami height in many places at Okushiri Island and in the southwestern part of Hokkaido, using mostly the above methods (SHIMAMOTO *et al.*, in preparation). We focus here only on heavily damaged areas.

The hardest place to estimate the true tsunami height is probably southern Aonae, because all residential houses were swept away. There is no slope in the Aonae Cape and even the runup height of tsunami has not been determined. The only material evidence here is the public toilet building with a reinforced concrete wall near the southeastern end of the Cape (Figure 6b; just west of B in Figure 4). Tsunami entered from the entrance of the toilet on the west (from the left side of the photograph) and broke walls on both sides outward, but it did not penetrate the ceiling from inside. Several fishermen's floats were trapped on a little bolt protruding from the concrete roof, indicating that tsunami reached at least this roof, 2.9 m above the ground. Moreover, synthetic roofing plate, glued to the concrete ceiling with cement, was peeled off for an area of about 3 m², as can be seen in Figure 6b. Only splashes of tsunami would be unable to peel off the roofing plate, so that the true tsunami height here must have exceeded the height of the roof. On the other hand, the thin peeled-off plate, which easily can be carried away by strong water flow, remains on the roof. This implies that the tsunami height did not overly exceed the height of the roof. We thus infer that the true tsunami height at the southern end of the Aonae Cape was of the order of 3 to 4 m.

The only other material evidence for the tsunami height in southern Aonae is the clear marks of tsunami left on a grassy slope of Marine terrace at the northern margin of southern Aonae (just south of I in Figure 4). The point the tsunami reached on the slope was 4.2 m high from the road at the foot of the slope. Subtracting the running-up distance of tsunami on the slope, the true tsunami height is also estimated here as 3 to 4 m.

Thus, the true tsunami height in southern Aonae was probably of the order of 3 to 4 m. The elevation of ground from the sea-water level is from 2 to 4 m, increasing northward, and hence the tsunami height as measured from the sea-water level is around 6 to 8 m in southern Aonae. Tsunami therefore was not necessarily gigantic, but it completely obliterated southern Aonae.

The true tsunami height on top of the sand dune in northern Aonae was estimated as about 2 m (Figure 17), and this is consistent with the estimate from the topographical cross section across the dune (2.3–2.8 m; see SHIMAMOTO *et al.*, 1994). The two-storied building of the Farmers Association at point J in Figure 4 was damaged, but was not carried away. Damage to the windowpane and wall indicates that the true tsunami height was less than 1.5 m at this point. The true tsunami height might have reached 7–8 m very locally at the corner of the tide embankment and sand dune where the massive tsunami invaded northern Aonae, judging from the topography of the corner area. However, the true tsunami height in northern Aonae as a whole was around 2 m or even less. And yet, the tsunami carried away nearly 100 houses there.

The true tsunami height as high as 7–8 m was recognized only at Monai, on the eastern part of Hamatsumae, and just on the corner of tide embankment and sand dune on northern Aonae. The true tsunami height was around 2 to 4 m at the central and western parts of Hamatsumae and at Inaho.

Characteristics of Tsunami Damage

To facilitate future assessment of tsunami disasters, we shall summarize the relationship between the true tsunami height and the degree of damage to the buildings. There were several concrete buildings, all two-storied, in heavily damaged areas, therefore most data are for Japanese-style wooden houses. We restrict our arguments to those houses that were bolted to concrete foundations because non-bolted, mostly old houses can float themselves and can be easily carried away by tsunami even less than 1 m in true height. It is out of question to build non-bolted wooden houses from the standpoint of the prevention of tsunami disasters.

The true tsunami height is estimated fairly accurately at around 2 m in northern Aonae (Figure 17), and this can be used as a standard of the degree of damage caused by tsunami of that height. Tsunami of that height is strong enough to detach nearly all bolted houses completely and demolish the first floor. However, the upper level of a two-storied house immediately behind the sand dune was carried away, but remained nearly intact, according to testimony by a nearby resident. Figure 16d is an example of such cases from Nonamae, just southwest of Inaho, where the true tsunami height was 2–3 m. The upper level part in the photograph appears to be a little damaged one-storied house, but there is no entrance to it.

This peculiar mode of damage is due to the structures of Japanese wooden houses which have a strong framework of rafters on the floor of the second floor and on the roof. Those frameworks are connected with the rather weak walls of the first and second floors. If the true tsunami height is smaller than the height of the upper-level floor (normally 3–4 m high from the ground), the tsunami may demolish the first floor, but the second floor can remain nearly intact. If a two-storied house is hit by tsunami higher than 4–5 m in true height, the floor and wall of the upstairs will be destroyed, but the roof can remain nearly intact. Figure 11d is such an example from Monai where the true tsunami height was estimated at 7–8 m.

Thus, the nearly intact upstairs are a useful diagnostic feature of a tsunami typically of 2–3 m in true height when one examines vivid photogrpahs taken soon after the earthquake. The Aonae Harbor, due to the invasion route 3 in Figure 4, northern Aonae, and the central and western portions of Hamatsumae are typical of such cases.

When the true tsunami height is less than 1 m, tsunami can cause substantial damage to houses, but most bolted houses were not carried away and stayed repairable. Tsunami of 1 to 1.5 m in true height is transitional between the one causing catastrophic disaster and the one causing serious, but not disastrous damage. Thus, a practical goal for the future assessment of a tsunami disaster is to keep the true tsunami height below 1 m.

Conclusions and Lessons for Future Assessment of Tsunami Disasters

Based on our field work, we have recognized the following cases for tsunami disasters (see SHIMAMOTO *et al.* (in preparation) for cases that are not described in this paper). Multiple factors may apply to a region.
(1) Unusually high tsunami struck a village or town (Monai, and eastern Hamatsumae).
(2) There were breakwaters, tide embankments and/or sea walls, but high tsunami went over them (southern Aonae, Inaho).
(3) Tsunami hit a village or a town where there were no coastal structures to prevent tsunami (Monai, Hamatsumae, etc.).

(4) Tsunami was amplified at the corner of the tide embankment and sand dune and hit a town (northern Aonae).

(5) Tsunami advancing nearly parallel or at a low angle to the coast was amplified and hit a town or a village (Inaho, Baikatsu north of Setana, northern part of the Setana Harbor).

(6) Tsunami invaded through the entrance for ships to a harbor or a fishing port and caused damage to houses immediately behind the entrance (the Aonae Harbor, the Kamuiwaki Fishing Port of Okushiri Island).

(7) Tsunami advancing parallel or at an angle to the coast hit the mouth of a river at the coast and river banks (Yachi on the central east coast of Okushiri Island, Chihase in Shimamaki village).

(8) Tsunami surged up the gentle slope where boats are pulled up for inspection and caused damage to houses immediately behind the slope (Kamuiwaki Fishing Port, Sanbon-sugi just north of the Setana Harbor).

(9) Tsunami did not go over the tide embankment, but invaded from passage ways through the embankment and caused damage to houses just behind the passage ways (Yachi, Sanbon-sugi).

Thus, only a few restricted places were struck by unusually high tsunami which may be impossibly difficult to stop with coastal structures. We have the impression that with proper coastal structures and the proper planning of residential areas near the coast, damage caused by tsunami due to the Southwest Hokkaido earthquake could have been reduced substantially.

Among above cases, (4) to (9) are more or less due to blind spots in the prevention of tsunami disasters, and counteractions must be undertaken against them in tsunami-prone areas. Whenever tide embankments or breakwaters are constructed nearly normal to the coast, the danger of tsunami amplification at their corners always arises. Most harbors and ports in Japan have been designed to countermeasure storm surge, and tide embankments and breakwaters have been constructed high on the front side towards the sea, but low towards the coast. As a result, areas near harbors or ports have become weak as seen in cases (4) and (5) above. The tsunami caused by the Southwest Hokkaido earthquake also clearly brought out the danger of the entrances for ships to a harbor and to a fishing port. Either way some manner of breakwater with tetrapods must be constructed on the outside of the entrance, or residential houses should not be built at low sites immediately behind the entrance.

This paper has not described cases from (7) to (9), since the damage in each case was minor. However, their overall damage is considerable. In particular, the potential danger of (7) should not be overlooked because the areas near the mouth of a river are often densely populated areas and also natural topography there often forms gentle slopes from the river. If tsunami hit the mouth of a river nearly parallel to the coast, much of the tsunami would runup along the river. But if it hits the mouth of a river at an angle to the coast, tsunami will invade from the mouth and deluge the river bank, even if the ocean coast is fortified with a sea wall or tide embankment. As with the entrance to a harbor or a port from the sea, the mouth of a river is another fairly large entrance for the tsunami. We have seen a few cases in southwest Hokkaido where the sea wall terminates at the mouth of river and does not extend along the river bank. In such cases, areas near the mouth of a river are not protected from the tsunami.

We have thus clarified in detail how damage was caused in each case, by knowing the advancing direction of the tsunami on land and the true tsunami height. We also have conducted *in situ* strength tests on a roadside guardrail bent by tsunami at northern Aonae (Figure 8a) and estimated the bending moment exerted on the poles of the guardrail by the tsunami. Assuming a steady flow of tsunami, we have shown that the particle velocity of tsunami reached 10 to 18 m/s (TSUTSUMI *et al.*, 1994). Such a high tsunami velocity is probably due to the fact that the Japan Sea, about 2,000 m deep near the focal region of the Southwest Hokkaido earthquake, becomes abruptly very shallow towards Okushiri Island and tsunami could not be slowed down appreciably even at the coast.

Those results will be fundamental data in designing proper coastal structures and in developing coastal city or town plans to prevent tsunami disasters. If the results reported herein and in the paper by TSUTSUMI *et al.* (1994) can be predicted quantitatively, the analysis will be a powerful tool for the precise assessment of tsunami disasters in other areas. It would be quite interesting to perform simulations demonstrating the extent to which the disasters due to the Southwest Hokkaido earthquake tsunami can be prevented by constructing various coastal structures. A criterion for preventing catastrophic disasters is to keep the true tsunami height below 1 m in populated areas, as revealed by the present study. Moreover, a region of particular interest is the Suruga Bay in the Tokai area where the Tokai earthquake is expected to occur (e.g., MOGI, 1985). This area may be similar to Okushiri Island in the sense that the area is close to the focal region and the deep sea near the expected focal region becomes very abruptly shallow towards the coast. A detailed assessment of tsunami disasters in the Tokai region is imperative.

Acknowledgements

We would like to express our condolences to the 230 people who perished due to the Southwest Hokkaido earthquake. We also wish to thank Kokusai Kogyo Co., Ltd. for permitting us to use their aerial photographs taken immediately after the earthquake. Critical review by K. Satake and three anonymous reviewers who improved the manuscript is greatly acknowledged.

REFERENCES

ABE, K. (1994), *An Island Effect on the Propagation of Tsunami due to the Southwest Hokkaido Earthquake* (in Japanese), Earth Monthly, Suppl. 7, 185–191.

CURRAY, J. R. (1956), *The Analysis of Two-dimensional Orientation Data*, J. Geol. 64, 117–131.

GOTO, A., TAKAHASHI, H., UTSUGI, M., ONO, S., NISHIDA, Y., OHSHIMA, H., KASAHARA, M., TAKENAKA, H., and SAITA, T. (1994), *Tsunami due to the Southwest Hokkaido Earthquake: From Otaru to Reibun Island* (in Japanese), Earth Monthly, Suppl. 7, 153–158.

KIKUCHI, M. (1994), *The Nicaraguan Tsunamigenic Earthquake as Recorded in Wide-band Seismometers* (in Japanese), Earth Monthly 16, 116–122.

KUGE, K., KIKUCHI, M., and ZHANG, J. (1994), *Complex Source Processes of the Southwest Hokkaido Earthquake (July 12, 1993) as Revealed from Far-field Body and Surface Waves* (in Japanese), Earth Monthly, Suppl. 7, 21–28.

IMAMURA, F., TAKAHASHI, T., and TAKAHASHI T. (1994), *Is the Earthquake Fault of the Southwest Hokkaido Earthquake Dipping to West, or to the East?: An Inference from Tsunami Data* (in Japanese), Earth Monthly, Suppl. 7, 179–184.

MATSUTOMI, H., and SHUTO, N. (1994), *The Southwest Hokkaido Earthquake Tsunami near Okushiri Island and its Velocity on Land* (in Japanese), Earth Monthly, Suppl. 7, 132–138.

MOGI, K., *Earthquake Prediction* (Academic Press 1985), 355 pp.

NAKANISHI, I., and KIKUCHI, M. (1993), *Characteristics of earthquake and induced groundmotion* (in Japanese). In *Special Issue on the 1993 Southwest Hokkaido Earthquake*, Japan Society for Earthquake Engineering Promotion, NEWS 133, 1–5.

SATO, H., SHIMAMOTO, T., TSUTSUMI A., and KAWAMOTO, E. (1994), *Onshore Tsunami Deposits Caused by the 1993 Southwest Hokkaido Earthquake and the 1983 Japan Sea Earthquake*, Pure and Appl. Geophys., this issue.

SATAKE, K., and TANIOKA, Y. (1994), *Tsunami Generation of the 1993 Hokkaido Nansei-oki Earthquake*, Pure and Appl. Geophys., this issue.

SHIMAMOTO, T., TSUTSUMI, A., KAWAMOTO, E., MIYAWAKI, M., and SATO, H. (1994), *Tsunami Disasters due to the Southwest Hokkaido Earthquake and Lessons for Tsunami Disaster Prevention* (in Japanese), Earth Monthly, Suppl. 7, 219–231.

SHIMAMOTO, T., TSUTSUMI, A., KAWAMOTO, E., MIYAWAKI, M., and SATO, H. (in preparation), *Advancing Direction and True Tsunami Height of Southwest Hokkaido Tsunami and their Implications for the Prevention of Future Tsunami Disasters* (in Japanese), to be submitted to Bull. Earthquake Res. Institute.

SHUTO, N. (1993), *Damage due to tsunami* (in Japanese). In *Special Issue on the 1993 Southwest Hokkaido Earthquake*, Japan Society for Earthquake Engineering Promotion NEWS 133, 21–24.

TAKAHASHI, T., ORTIZ, M., TAKAHASHI, T., and SHUTO, N. (1994), *Initial Profile of the Hokkaido Nansei-oki Earthquake Tsunami for a Better Simulation* (in Japanese), Abstracts for Japan Earth and Planetary Science Joint Meeting, p. 264.

TANIOKA, Y., SATAKE, K., and RUFF, L. (in review), *Total Analysis of the 1993 Hokkaido Nansei-oki Earthquake Using Seismic Wave, Tsunami, and Geodetic Data*, Geophys. Res. Lett.

TSUJI, Y., KATO, K., ARAI, K., HAN, S., and YAMANAKA, Y. (1994a), *Characteristics of Tsunami due to the Southwest Hokkaido Earthquake* (in Japanese), Earth Monthly, Suppl. 7, 110–122.

TSUJI, Y., KATO, K., ARAI, K., ARAI, K., and UEDA, K. (1994b), *Run-up Height Distribution of Tsunami due to the Southwest Hokkaido Earthquake along Coast of Southwest Japan* (in Japanese), Earth Monthly, Suppl. 7, 110–122.

TSUTSUMI, A., SHIMAMOTO, T., MIYAWAKI, M., SATO, H., and KAWAMOTO, E. (1993), *Subsidance of Okushiri Island during the 1993 Southwest Hokkaido Earthquake* (abstract in Japanese), Abstract for the Seismological Society of Japan Meeting, 2, 62.

TSUTSUMI, A., SHIMAMOTO, T., KAWAMOTO, E., and LOGAN, J. M. (1994), *Estimation of Near-shore Velocity of the Southwest Hokkaido Earthquake Tsunami* (in Japanese), Earth Monthly, Suppl. 7, 166–172.

(Received September 28, 1994, revised March 30, 1995, accepted April 8, 1995)

PAGEOPH, Vol. 144, Nos. 3/4 (1995)

0033-4553/95/040693-25$1.50 + 0.20/0

Onshore Tsunami Deposits Caused by the 1993 Southwest Hokkaido and 1983 Japan Sea Earthquakes

HIROSHI SATO,[1] TOSHIHIKO SHIMAMOTO,[1] AKITO TSUTSUMI[1] and
EIKO KAWAMOTO[1]

Abstract—Onshore tsunami deposits resulting from the 1993 Southwest Hokkaido and 1983 Japan Sea earthquakes were described to evaluate the feasibility of tsunami deposits for inferring paleoseismic events along submarine faults. Tsunami deposits were divided into three types, based on their composition and aerial distribution: (A) deposits consisting only of floating materials, (B) locally distributed siliclastic deposits, and (C) widespread siliclastic deposits. The most widely distributed tsunami deposits consist of the first two types. Type C deposits are mostly limited to areas where the higher tsunami runup was observed. The scale of tsunami represented by vertical tsunami runup is an important factor controlling the volume of tsunami deposits. The thickest deposits, about 10 cm, occur behind coastal dunes. To produce thick siliclastic tsunami deposits, a suitable source area, such as sand bar or dune, must be available in addition to sufficient vertical tsunami runup. Estimation of the amounts of erosion and deposition indicates that tsunami deposits were derived from both onshore and shoreface regions. The composition and grain size of the tsunami deposits strongly reflect the nature of the sedimentary materials of their source area. Sedimentary structures of the tsunami deposits suggest both low and high flow régimes. Consequently, it seems very difficult to identify tsunami deposits based only on grain size distribution or sedimentary structure of a single site in ancient successions.

Key words: Onshore tsunami deposits, 1993 Southwest Hokkaido earthquake, 1983 Japan Sea earthquake, eastern coast of the Japan Sea, paleoseismicity.

1. Introduction

The determination of paleoseismic events along active faults is of fundamental significance in the long-term risk assessment of seismic hazard, including that due to tsunamis. Paleoseismic events have been detected by trenching across inland active faults (e.g., SIEH, 1978) and by acoustic profiling and piston coring for shallow submarine active faults (e.g., OKAMURA *et al.*, 1993). The analysis of tsunami deposits has received considerable attention recently in determining paleoseismic events along deep submarine faults (ATWATER, 1987, 1992; KATO, 1987; MINOURA and NAKAYA, 1991; ATWATER and MOORE, 1992) and paleo-tsunami

---

[1] Earthquake Research Institute, University of Tokyo, Yayoi, Bunkyo, Tokyo, 113, Japan.

events generated by submarine land slides (DAWSON *et al.*, 1988; LONG *et al.*, 1989; MOORE and MOORE, 1988; MOORE *et al.*, 1989).

The principle of the last method is simple: if tsunami deposits are recognized within a succession with known ages, the timing of a paleo-tsunami event can be decided from the ages of sediments immediately above and below the tsunami deposits. The main uncertainty of this method still lies in the identification of tsunami deposits. Typical tsunami deposits were identified as a thin sand sheet, showing a decreased thickness toward the sea (e.g., ATWATER, 1992; CLAGUE and BOBROWSKY, 1994), graded bedding (KON'NO *et al.*, 1961; NISHIMURA and MIYAJI, 1995) and erosional contact with underlying sediments (MINOURA and NAKATA, 1994). However, the grain size of tsunami deposits varies from mud (KON'NO *et al.*, 1961) to boulder (KATO, 1987; NAKATA and KAWANA, 1993) and tsunami deposits occurring in a variety of sedimentary structure. Despite fine documentation of modern tsunami deposits due to the 1960 Chile earthquake (KON'NO *et al.*, 1961; WRIGHT and MELLA, 1963), descriptions of more modern

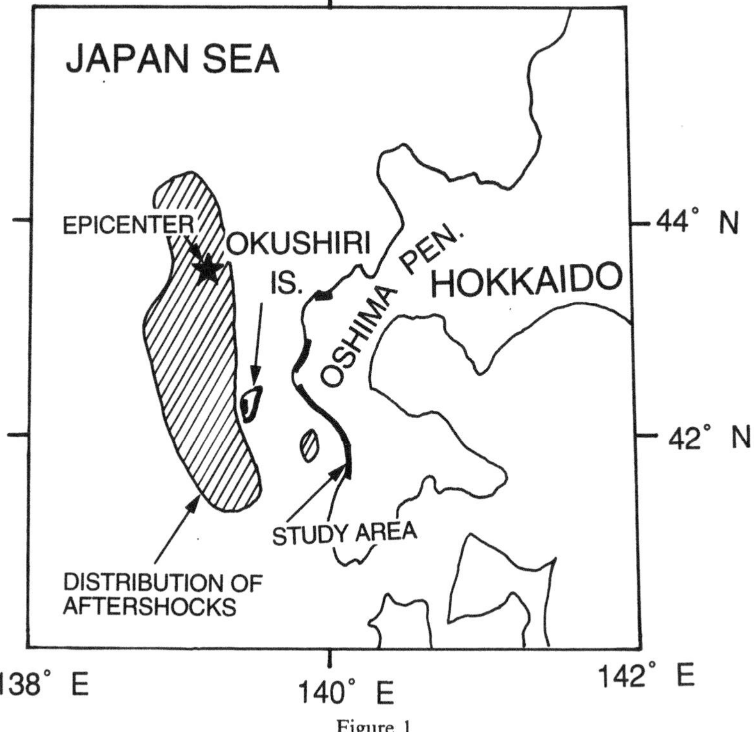

Figure 1

Location of the study area. The epicenter and distribution of aftershocks of the 1993 Southwest Hokkaido earthquake are after NAKANISHI and KIKUCHI (1993).

tsunami deposits are needed to evaluate the feasibility of tsunami deposits for inferring paleo-tsunami events.

The 1993 Southwest Hokkaido earthquake (M 7.8, July 12) generated extensive tsunami disasters along the shoreline of Okushiri Island and in southwestern Hokkaido (Figure 1). We have attempted to describe tsunami deposits (their composition, thickness, spatial distribution and sources) with a view toward a better understanding of the processes forming tsunami deposits. In particular, we tried to understand the conditions that produce thick tsunami deposits. Our field survey was carried out through three periods; July 15–21 (immediately after the earthquake), July 31 to August 15, and December 10–18, 1993.

The tsunami due to the 1993 Southwest Hokkaido earthquake surged mainly onto rocky coastal areas. However, the tsunami caused by the 1983 Japan Sea earthquake (M 7.7, May 26) largely swept ashore across sand beaches. We compared deposits from these two tsunami events to evaluate the effect of the substratum and topography on tsunami deposits.

2. Tsunami Deposits Caused by the 1993 Southwest Hokkaido Earthquake

2.1. Classification of Tsunami Deposits

The 1993 Southwest Hokkaido earthquake on July 12 (at 13:17 GMT, M 7.8) struck near Okushiri Island off the coast of southwest Hokkaido, Japan. The epicenter location and distribution of aftershocks are shown in Figure 1. The earthquake and subsequent tsunami caused approximately 250 deaths. Most of the fatalities were attributed to the tsunami. Tsunami vertical runup measurements varied between 15 to 30 m over the southern part of Okushiri Island, with several 10-m values on the northern portion of the island. Along the west coast of Hokkaido, no survey values exceeded 10 m (HOKKAIDO TSUNAMI SURVEY GROUP, 1993).

The tsunami due to the 1993 Southwest Hokkaido earthquake deposited sediments in most of the area traversed by the tsunami. We describe only onshore tsunami deposits in this paper. Onshore tsunami deposits were identified in the area where the tsunami swept over normal tidal zone, such as on roads, rice paddies, farm fields, grass, and the floors of houses.

Tsunami deposits were classified based on the density of their constituent clasts and their spatial distribution. Particles forming tsunami deposits were divided into three types: particles light enough to float, bioclasts, and siliclastic material. Floating materials include wood, plants, artificial products, and light bioclasts such as sea urchins. Bioclasts, such as shells, coral, and foraminifera, have a density intermediate between siliclastic and floating materials. Siliclastic fragments include gravel, sand and mud.

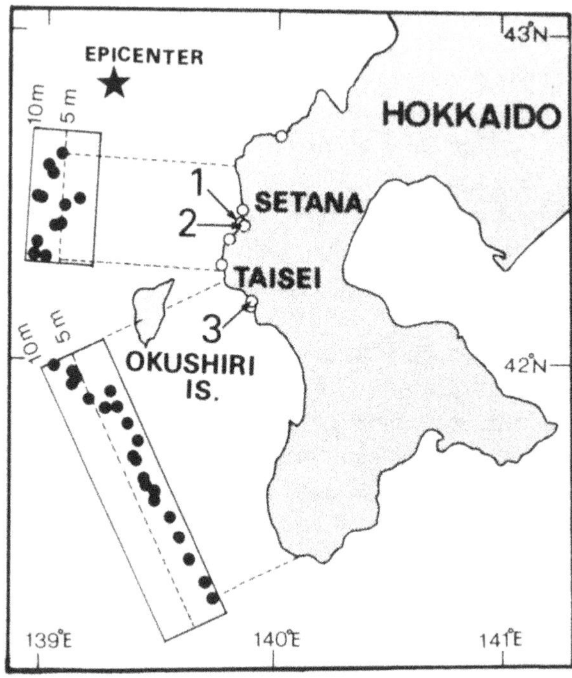

Figure 2

Distribution of tsunami deposits and vertical tsunami runup caused by the 1993 Southwest Hokkaido earthquake on Oshima peninsula, Hokkaido. White circles: tsunami deposits consisting only of floating materials and locally distributed sand or gravel. Vertical tsunami runup measurements are after the HOKKAIDO TSUNAMI SURVEY GROUP (1993).

Deposits consisting of floating materials were found in most areas traversed by the tsunami. In contrast, deposits consisting of siliclastic material were more restricted in their distribution. Spatial distribution of tsunami deposits is important information for evaluating the preservation potential of tsunami deposits. To classify the spatial distribution of tsunami deposits, we further subdivided siliclastic tsunami deposits into two types: small pockets that extend only a few tens of meters, and more widespread sheets.

Based on these criteria, we classified the tsunami deposits generated by the 1993 Southwest Hokkaido earthquake into three types; (A) deposits consisting of only floating materials, (B) locally distributed siliclastic deposits, and (C) more widespread siliclastic deposits. Most of the flooded area was occupied by type A and B deposits, and type C was rarely encountered (Figures 2 and 3).

In this paper, the terms describing the thickness of tsunami deposits are different from those used in sedimentology (INGRAM, 1954). Since the maximum thickness of reported onshore tsunami deposits is about 1 m (BOURGEOIS and REINHART, 1989), we use the term "thick" for tsunami deposits more than 10-cm thick, and "thin" for deposits less than 3-cm thick.

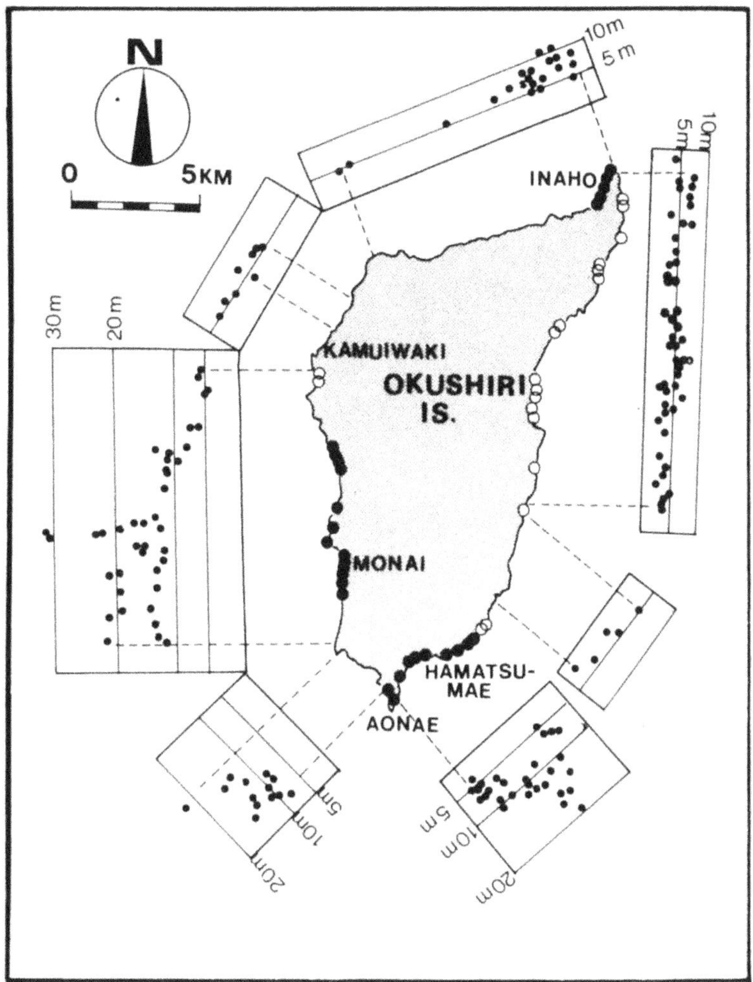

Figure 3

Distribution of tsunami deposits and vertical tsunami runup caused by the 1993 Southwest Hokkaido earthquake on Okushiri Island. White circles: tsunami deposits consisting only of floating materials and locally distributed sand or gravel. Filled circles indicate widespread siliclastic tsunami deposits. Vertical tsunami runup measurements are after the HOKKAIDO TSUNAMI SURVEY GROUP (1993).

2.2. *Tsunami Deposits Consisting of Only Floating Materials (Type A)*

Floating materials largely consisted of grass, wood, and artificial materials. There was a small amount of sea weed and sea urchin. This type of tsunami deposit shows the most extensive distribution and is characterized by the lack of siliclastic particles. Similar tsunami deposits were reported by KON'NO *et al.* (1961). Floating materials rafted by the tsunami tended to deposit near the front of the submerged area.

A typical example of Type A deposits was observed on the western coast of Hokkaido (Site 1 of Figure 2), where the tsunami traversed a road located on the flank of sand dune. Judging from the distribution of bent and brown colored grasses by the salt water, we estimate that about 30 cm of water covered the road. Although this site is located next to an appropriate source of sediment for tsunami deposits, we found no distinctive lithic or siliclastic tsunami deposits. This was true in a rice paddy that was located 130 m onshore (Site 3 of Figure 2).

A similar example was observed along the river about 1.8 km inland from the river mouth (Site 2 of Figure 2). The tsunami went upstream, and spilled over grass-covered river banks that were about 2 m higher than the river surface. Near the front of the area submerged by the tsunami, plants accumulated (Figure 4a). Although the submerged area was a grass-covered slope, there were no lithic tsunami deposits. This implies that the tsunami did not transport sand or gravel to the observed site.

2.3. Locally Distributed Siliclastic Deposits (Type B)

Tsunami deposits consisting of locally distributed sand or gravel were observed along the shoreline of Okushiri Island and several places along the western coast of Hokkaido. At Site 3 of Figure 2, 2–3 cm of poorly sorted sandy mud was washed onto an asphalt-paved road. Judging from its color and grain size, the sediment could have been derived from an adjacent farm field located inland. The primary current lineations were formed on the soil surface of the farm field by outgoing flow (Figure 4c). It suggests that the siliclastic tsunami deposits at this site traveled only a few tens of meters from their source area.

A 2–3-cm thick mud layer within a few tens of meters in length was sporadically distributed along the eastern coast of Okushiri Island (Figure 3). On the air photographs taken the day after the tsunami, a veneer of brown-colored sediment can be identified on asphalt-paved roads, at stream mouths, and near farm fields. These sediments are presumed to be derived from soil located nearby by mainly outgoing flow (backwash).

Further evidence for onshore erosion and sedimentation was observed at Port Kamuiwaki (Figure 4b). Here, several-cm-thick gravel patches several meters in length were deposited on an asphalt-paved road. The deposits are very poorly

---

Figure 4

Tsunami deposits consisting of only floating materials (type A) and locally distributed siliclastic materials (type B). a. Tsunami deposits consisting of floating materials from alongside the Ushiro-shiribeshi River, Setana, Southwest Hokkaido on July 16, 1993. Floating materials consisting of land plants which are concentrated at the edge of flood area (arrow). b. Locally distributed siliclastic deposits in Port Kamuiwaki, Okushiri Island on July 18. Gravel was derived from the unpaved part on the right. c. Primary current lineations on a field on the Hirahama Coast, Taisei Town, Hokkaido on July 16. The current direction was the same as the outgoing flow (arrow) of the tsunami.

sorted and vary from fine-grained sand to large cobbles. The distribution pattern of this layer suggests that the deposits were derived from adjacent unpaved gravely ground.

Figure 5
Widespread tsunami deposits distributed along the southwestern coast of Okushiri Island. Photographs were taken on July 18, 1993. a. Tsunami deposits consisting of granules on the asphalt-paved road in Monai. Land plants were suspended on the electric wires by the tsunami. b. Widespread tsunami deposits consisting of fine to medium grained sand. c. Sand layer deposited in a hollow between rocks. The arrow shows the direction of tsunami runup.

To summarize, the locally distributed siliclastic deposits are typically less than a few centimeters in thickness and extend over several dozen centimeters to ca. 30 m in length. As deposits were commonly derived from the nearest source of erodable sediment, they include sediment of mud to gravel size, reflecting the source material.

2.4. *Widespread Siliclastic Deposits (Type C)*

Widespread siliclastic deposits are geologically the most important, due to their greater potential for preservation in ancient successions. This type of tsunami deposit occurred at several sites along the shoreline of Okushiri Island (Figure 3). At these sites tsunami runup heights of over 10 m were measured by the HOKKAIDO TSUNAMI SURVEY GROUP (1993), the FIELD SURVEY TEAM of TOHOKU UNIV. and AKITA UNIV. (1993), and KATO and TSUJI (1993).

Along the southwestern coast of Okushiri Island the tsunami reached a runup level of 10–30 m and revealed its maximum vertical runup at Monai (Figure 3). As the coast is a rocky or gravel beach, the observed tsunami deposits were thin in spite of the large vertical runup. A 2–3-cm thick layer of coarse-grained sand and granule was distributed on the asphalt-paved road in the southern part of Monai (Figure 5a). These sediments were estimated to be derived from the adjacent beach. Several mm thick well-sorted fine to medium grained sand layers were also deposited on the road to the south of this site (Figure 5b). Beside the road, several centimeter thick sand patches were sporadically distributed on the inland side of large stones (Figure 5c). Some cobbles that were derived from the gravel beach accumulated in a roadside ditch.

A veneer of deposits was recognized on the asphalt-paved, coast-parallel road in Hamatsumae and Inaho (Okushiri Island). These deposits stretched at least 50 m along the shoreline. The thickness of the sand layer varied perpendicular to the shoreline direction, and had a tendency to thin toward the seaward edge of the road from a few cm to zero.

2.5. *Deposition and Erosion by Tsunami in Aonae, Okushiri Island*

Widespread siliclastic deposits also occurred in Aonae, at the southern tip of Okushiri Island. Here, the direction of the tsunami surge can be reconstructed from the damage effecting artificial constructions, such as iron bolts that connected wooden houses to their concrete foundations, and water and gas pipes (SHI-MAMOTO *et al.*, 1995; Figure 6). In particular, bent iron bolts were good direction indicators because of their abundance. Their orientation reflects peak runup conditions. Weeds were so sensitive to currents that their orientation only reflects the final directions of the tsunami, usually the offshore-directed return flow.

The tsunami that reached Aonae divided into four routes (Figure 6). The most destructive tsunami (Route 1) surged from WNW to ESE and crested over the

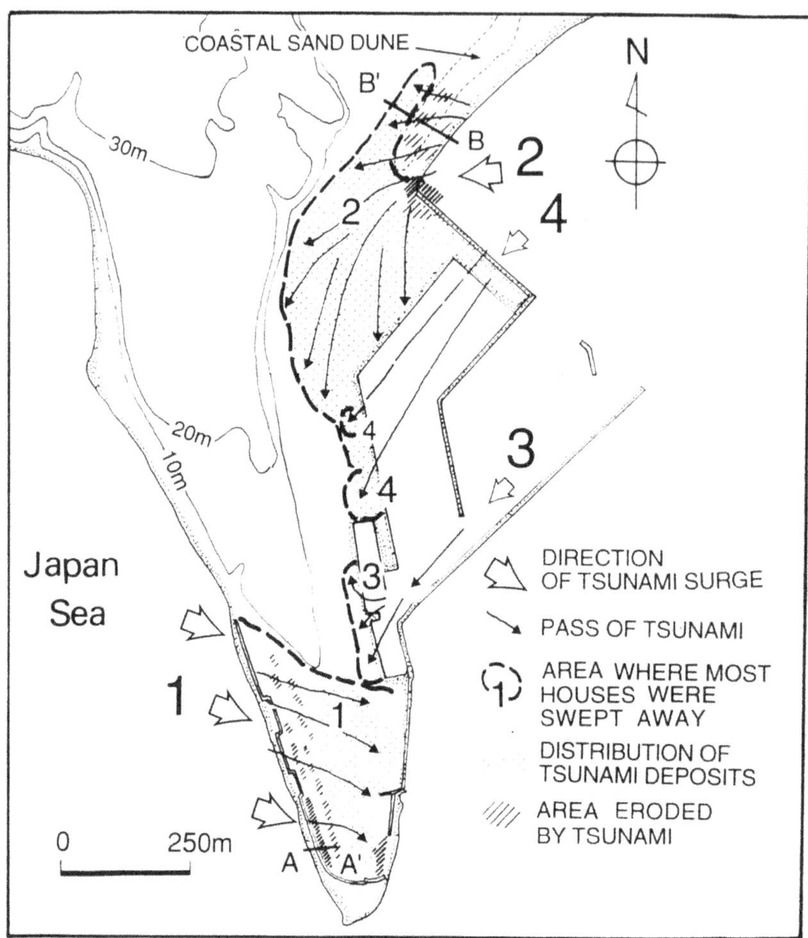

Figure 6

The direction and path of the tsunami surge in Aonae, Okushiri Island. The direction of the main tsunami surge is shown by the numbers 1 to 4. The area surrounded by the thick broken lines shows the area damaged by the tsunami surge labeled with the same number. A-A' and B-B' are locations of cross sections shown in Figure 7.

western shore protection wall. It swept away all the houses and destroyed part of the eastern breakwater before returning to sea. This tsunami surge left deposits over a wide area. Another important tsunami surged in the ENE direction into the northeastern part of Aonae (Route 2; Figure 6). This tsunami swept over the coastal dune and breakwater, removed many houses, and scattered a sand layer throughout the residential area. In contrast, the tsunami that followed Routes 3 and 4 did not produce significant deposits.

The Route 1 tsunami (Figure 6) washed over the western shore protection wall of Aonae, which is 4 meters high, and crossed a Holocene coastal terrace that is 2

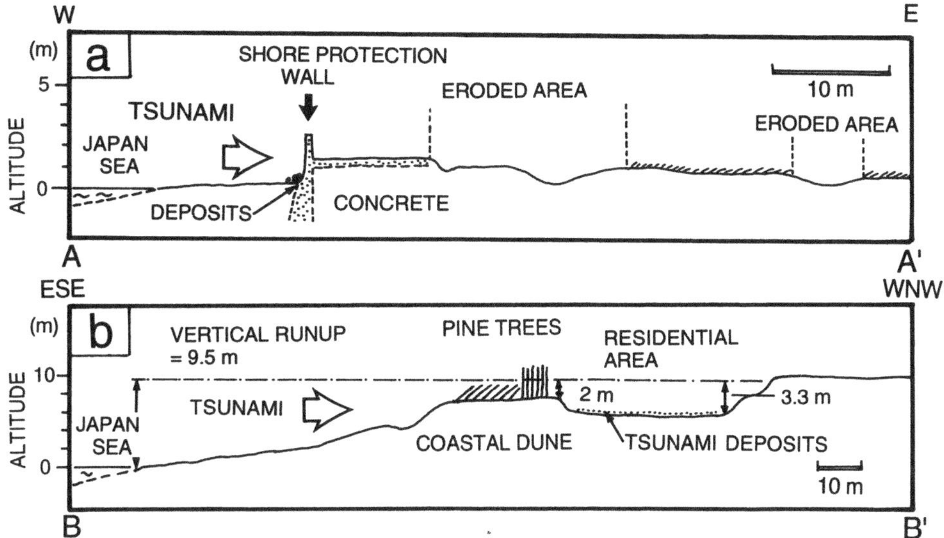

Figure 7

Topographic cross sections showing areas of erosion and deposition. a. Erosion by tsunamis in the southwestern coast of Aonae. Tsunamis swashed over the shore protection wall and produced two swaths of erosional grooves trending parallel to the beach. b. Profile across the coastal dune of the northeastern part of Aonae. The tsunami swashed over the coastal dune and scattered sands in the residential area. The location of each cross section (A-A', B-B') is shown in Figure 6.

to 4 m above mean sea level (Figure 7a). The damage to houses suggests that the coastal terrace was covered by 3–4 m of water (SHIMAMOTO *et al.*, 1995). The tsunami dumped patches of gravel, several tens of cm thick and about 2 m wide, at the seaward base of the shore protection wall. Fifty to one hundred meters further onshore, the tsunami eroded grooves several meters in diameter and trending parallel to the shoreline (shaded area on Figure 6). In the southern portion of Aonae, the erosional grooves occurred in two swatches oriented parallel to the shoreline (Figure 6). Figure 7a shows a cross section of this location. The erosional grooves were located about 20 m and 40 m onshore from the shore protection wall (Figures 7a, 8). Their maximum depth was 40 cm. To the north of cross-section line A-A' (Figure 6), the grooves were shallower and their distribution was more disorganized, reflecting the irregular topography. The erosional groove, immediately behind the protection wall, was formed by water sweeping over the wall. The distance from the shore protection wall to the onshore erosional groove strongly reflects the velocity of the tsunami that swept over the shore protection wall.

To the east of the erosional area, tsunami deposits were widely distributed (dotted area in Figure 6). Figure 9a shows southern Aonae 5 days after the tsunami hit. The road in the center of this photograph (Figure 9b) was completely covered with gravel averaging 7–10 cm in thickness.

Figure 8

Erosion by tsunami in the southern part of Aonae, Okushiri Island. Photographs were taken on December 16, 1993. a. An erosional groove to the east of the tide protection wall. The tsunami scraped glass and formed a groove about 40 cm in depth. b. Erosional depressions 5 m in diameter.

The organized style to the erosion in southern Aonae allows us to estimate the amount of erosion. The erosional area was clearly identified by scraped grasses. The volume of small topographic features was measured with a laser aided surveying instrument. Compared with the topographic map before the tsunami invasion, the total amount of erosion was estimated to be $700-1100 \, \text{m}^3$, whereas tsunami deposits covered an area of about $1300 \, \text{m}^2$. If we assume that the density of the

Figure 9

Deposition by tsunami in the southern part of Aonae, Okushiri Island. a. Destroyed residential area on July 17, five days after the tsunami hit. This photograph is a view toward the north. On the left, a destroyed light house can be recognized. b. Same area as Figure 9a on December 17, five months after the tsunami hit. The asphalt-paved road can be seen after removal of the tsunami debris. The building in the right upper corner is the same in both a and b.

eroded and deposited material was the same, the eroded sedimentary material can produce a 5–8-cm thick deposit, nearly coinciding with the observed thickness. This implies that most of the tsunami deposits were derived from the adjacent source area and transported less than 200 m.

Another remarkable area of tsunami erosion in southern Aonae was at the side of the back part of buildings (Figure 10a) where turbulent currents could easily form. Another area eroded by the tsunami was around the collapsed part of the eastern breakwater (Figures 10b,c). About 50 m of the breakwater collapsed, with large

Figure 10

Erosion by tsunamis in southern Aonae, Okushiri Island. a. The side of a concrete construction was deeply eroded. Photographed on August 7, 1993. b–c. Erosion by tsunamis at the collapsed part of the eastern breakwater. The photographs continue from b to c. A large amount of gravel was eroded and formed a concave depression due to the convergence of the outgoing flow. Photographed on December 16, 1993.

amounts of the outgoing tsunami passing through the new opening. Due to the concentration of the outgoing flow, about 1500 m^3 of gravel from the Holocene coastal terrace gravel was eroded, resulting in a 0.5–1.5-m deep bowl-shaped depression (Figures 10b,c).

Figure 11

Erosion by tsunami in northern Aonae, Okushiri Island. a. The coastal dune and tide embankment in April 1992, before the tsunami hit. b. The fence erected to prevent erosion of sand by winds in November 1992. Photographs a and b were provided by the Fisheries Department of the Hokkaido Development Agency. c. Hollow formed by tsunami erosion on December 13, 5 months after the tsunami hit. The scenery is the same as in Figures 10a and b. The hollow was later filled by rainwater and the resultant pond is frozen over in this photograph.

The Route 2 tsunami that affected the northeastern part of Aonae resulted when sea water crested over the join between an 8-m high coastal sand dune and a tidal breakwater (Figure 6). As the tsunami approached the corner between these topographic features, its velocity and height were amplified due to the reduced space. Although the coastal sand dune continued to the tidal embankment before the tsunami invasion (Figures 11a,b), the tsunami swashed away a large amount of sand between them and formed a bowl-shaped depression about 1 m deep (Figure 11c). The Fishery Department of Hokkaido had surveyed the topography of this location in March of 1991 and we resurveyed the area in December 1993. These surveys indicate that the amount of erosion by the tsunami was 2000–4000 m³.

The deposits scattered by the Route 2 tsunamis behind the coastal dune consisted of well-sorted fine to medium grained sand. The deposits were 5 cm in thickness at a location 30 m landward from the coastal dune (Figure 12a) and locally exceeded ten cm. The thickness and distribution of this sand layer were very irregular due to the complex currents resulting from constructions such as house foundations and walls. The average thickness of the sand layer is estimated to be about 3–4 cm. The amount of sand eroded from beach and coastal dunes (see above) could produce a 2 to 3.5-cm thick layer over the area covered by the tsunami deposits (110,000 m²). Therefore, most of the tsunami deposits were again derived from the nearest source area without transportation over a long distance.

The tsunami deposits contain several kinds of sedimentary structures. Low angle cross stratification was observed in the northeastern part of Aonae (Figure 12a). Well-sorted medium to fine grained sand showed planar, low angle cross stratification with a 5° dip toward the seaward forming antidunes. This sedimentary facies was very similar to upper flow régime plane beds produced on the foreshore (CLIFTON, 1976). In the southern part of Aonae, current ripples and primary current lineations were common on the surface of tsunami deposits (Figure 12b). The current directions reconstructed from these sedimentary structures are in agreement with directions determined from bent grass, implying that these directions represent the outgoing flow (backwash) of the tsunami.

In short, comparison of the amounts of erosion and deposition by the tsunami in Aonae suggests that the tsunami deposits were derived from the nearest source area. The grain size of tsunami deposits directly reflects source area characteristics. For example, tsunami deposits in northeastern Aonae, derived from coastal dune or beach sand, were well sorted sands, whereas tsunami deposits in southern Aonae,

---

Figure 12

Tsunami deposits in Aonae, Okushiri Island. a. Medium grained sand showing low angle cross stratification deposited on the floor of a bathroom in northeastern Aonae. Photographed on July 21, 1993. b. Fine grained tsunami deposits showing current ripples in southern Aonae. Photographed on July 18, 1993.

derived from soil and gravel, were very poorly sorted pebbly sands. Sedimentary structures in the tsunami deposits vary, and represent a wide range of flow régimes.

3. *Tsunami Deposits Caused by the 1983 Japan Sea Earthquake*

The tsunami generated by the Japan Sea earthquake (M 7.7) on May 26 of 1983 attacked the Japan Sea coast of northern Honshu and Hokkaido repeatedly. The coastal area of southern Aomori and northern Akita prefectures suffered the most serious damage by the tsunami. Since the tsunami surged ashore mainly across sand beaches, rather than rocky coastlines as with the 1993 Hokkaido earthquake,

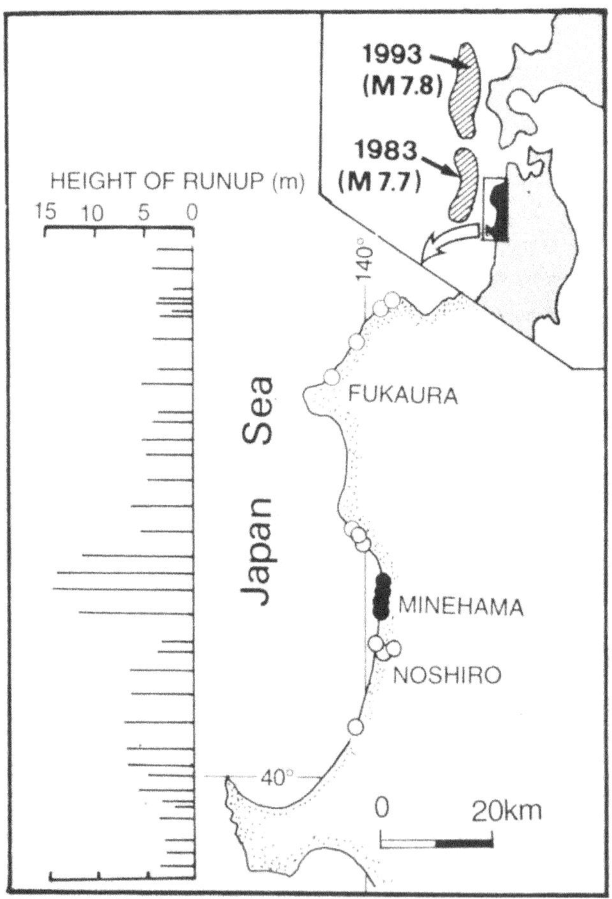

Figure 13

Distribution of tsunami deposits and vertical tsunami runup caused by the 1983 Japan Sea earthquake. White circles: tsunami deposits consisting of floated material and locally distributed sand or gravel. Filled circles indicate widespread lithic tsunami deposits. The runup height is after SHUTO (1984).

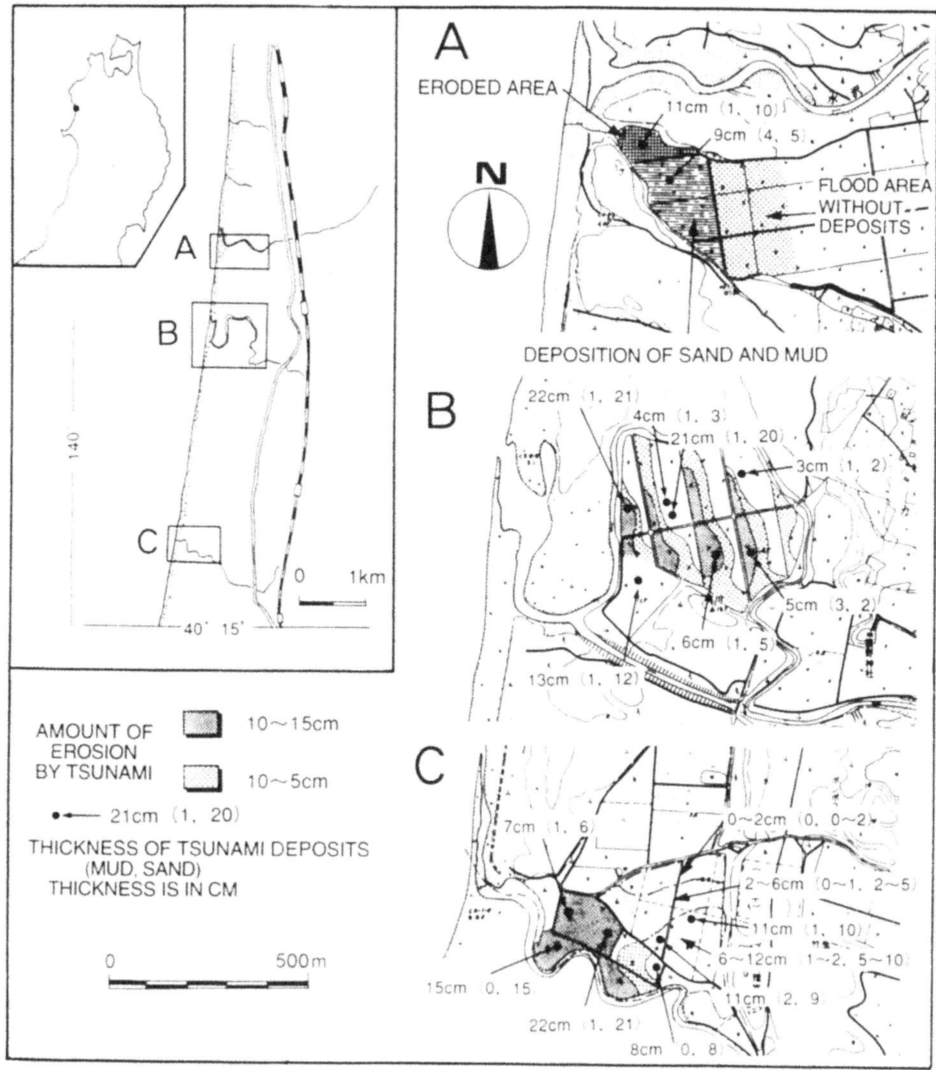

Figure 14

Erosion and deposition by tsunamis generated by the 1983 Japan Sea earthquake in Minehama, northern Akita Prefecture. Based on Minehama Village (1984).

comparison of the tsunami deposits provides further information regarding the effects of substratum on the type of deposit. As most of the tsunami deposits associated with the 1983 Japan Sea earthquake were removed by recovery construction, data on the tsunami deposits was largely gathered from interviews, photographs, documents accumulated in local government offices, and field survey.

We divided deposits into two groups; (1) those consisting of locally distributed sediments and floating materials, and (2) widespread siliclastic deposits. Tsunami

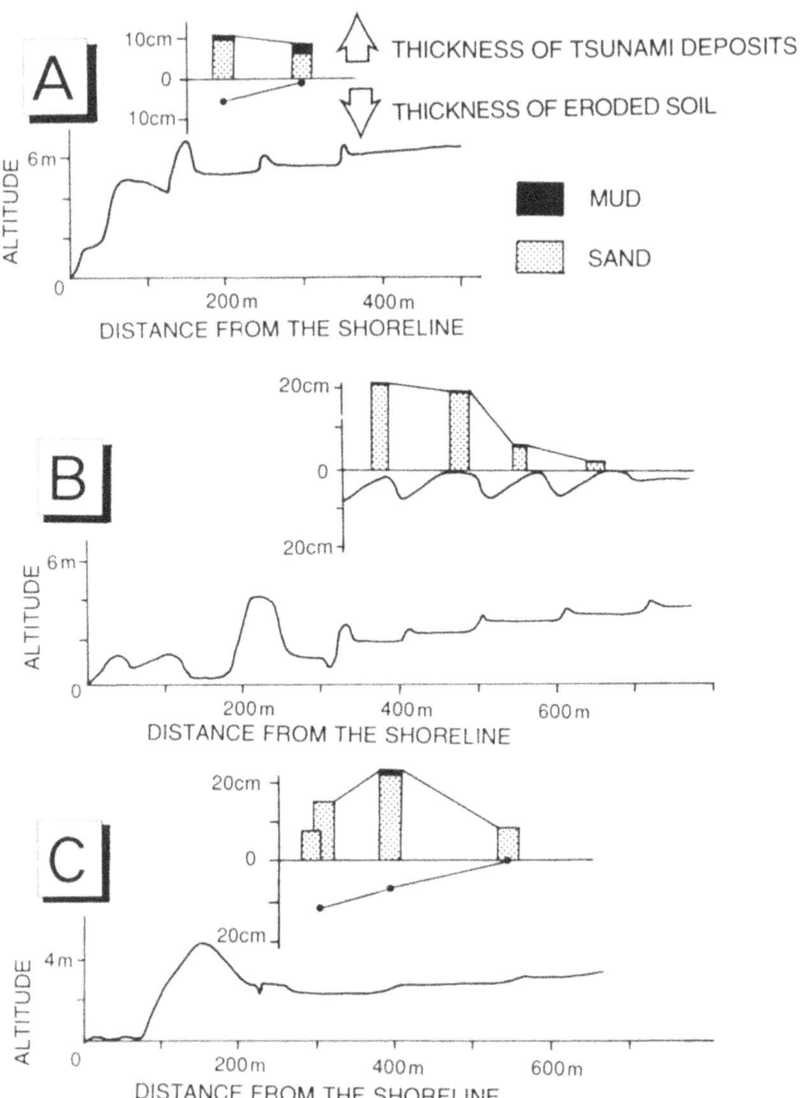

Figure 15

Relationship between the distance from the shoreline, the amount of erosion and the thickness of tsunami deposits generated by the 1983 Japan Sea earthquake.

deposits consisting only of floating materials could not be recognized due to the second hand way of our approach (Figure 13). Most of the flooded area was covered by scattered patches of sand or gravel and floating material (Figure 13). Using the photographs taken just after the incursion of the tsunami (FUKAURA TOWN, 1983; HACHIMORI TOWN, 1984), the thickness of deposits was estimated to be less than 20 cm.

More widespread tsunami deposits of sand were reported in Minehama Village, where the highest tsunami runup of 15 m above m.s.l. (mean sea level) was observed (Figure 13) (SHUTO, 1984). This tsunami washed over a coastal dune and crossed rice paddies located near river mouths. Deposition of sand and floating materials and erosion of soil by the tsunami caused serious damage to the rice paddies (MINEHAMA VILLAGE, 1984). A survey of this damage was performed by the local government in order to plan rehabilitation projects (MINEHAMA VILLAGE, 1986).

Rice paddies in the three districts (Figure 14) A, B, and C were most seriously damaged when the tsunami swept ashore and incoming and outgoing waves produced both erosion and deposition. Deposits in these three districts consist of a sand layer capped by a thin mud layer (Figures 14 and 15). The tsunami transported sediment at least 650 m inland from the shoreline and the maximum thickness of deposits was 22 cm. The thickness of deposits exhibits a gradual decrease in land, except for district C (Figure 15). There, the maximum thickness occurred 400 m from the shoreline (Figure 15). These rice paddies are located directly behind the sand beach and coastal dune, suggesting that this was the source. The upper mud layer may be derived from the rice paddies or the surrounding land surface. This inference is also supported by the observed erosion of the rice paddies by tsunami.

We can estimate the amount of erosion using the thickness distribution of soil after the tsunami incursions (MINEHAMA VILLAGE, 1986), if we assume that the thickness of soil was homogeneous through each rice paddy (15 cm). All three districts suffered soil erosion. The amount of erosion was greatest in the rice paddies located just inland of the river mouth, suggesting a concentration of tsunami currents at these locations. The amount of erosion was also larger than normal across topographic highs, such as the low ridges, 30–50 cm in height, between rice paddies, as observed in district B (Figure 15). The soil erosion by tsunamis found in Minehama Village accords well with the existence of rip-up clasts within tsunami deposits caused by the 1960 Chile earthquake (BOURGEOIS and REINHART, 1989).

4. Controls on the Nature of Tsunami Deposits

What is the derivation of tsunami deposits? Most of the deposits resulting from the two tsunami events we studied consist of subaerially-derived sediments; marine biogenic materials were rare in tsunami deposits. In Aonae on the Okushiri Islands, the volume of deposits was nearly the same as the amount of onshore erosion, implying that most of the sediment was derived from this onshore erosional area.

Before discussing the relationship between the distribution and thickness of tsunami deposits and vertical tsunami runup, we must emphasize the role played by the density of the constituent clasts of the type of deposit. The behavior of low density particles, such as biogenic materials, within currents is quite different from that of high density particles. As the tsunami deposits produced by the 1993 and 1983 earthquakes contained little biogenic material, we mention only siliclastic deposits in the following discussion.

The widespread siliclastic deposits generated by the 1993 and 1983 earthquakes only occurred in areas where the higher vertical tsunami runup was measured (Figures 2, 3 and 13). Therefore, the scale of tsunami represented by vertical tsunami runup is an important factor controlling the thickness + distribution of tsunami deposits. Due to artificial constructions along the shoreline of Northern Japan, it is difficult to determine the lower limit of runup height to generate widespread tsunami deposits. The recent two tsunamis (Figures 2, 3 and 13) suggest that 10 m of vertical tsunami runup is high enough to produce widespread tsunami deposits.

Source supply was also shown to be an important control on the volume and thickness of tsunami deposits. In Monai, southwestern Okushiri Island, although 20–30 m of vertical tsunami runup was measured (HOKKAIDO TSUNAMI SURVEY GROUP, 1993), the tsunami produced a deposit less than 3-cm thick. In contrast, at Minehama, deposits 20-cm thick were formed by tsunami with 15 m of vertical runup. The main reason for the difference was the sediment source. At Minehama, a sand beach and coastal dune were possible source areas, whereas at Monai the rock and gravel beach could not supply sufficient material to produce thick tsunami deposits.

The tsunami from the 1993 Southeast Hokkaido and 1983 Japan Sea earthquakes only produced widespread tsunami deposits greater than 10-cm thick when the tsunami washed over the coastal dunes, such as in the northeastern part of Aonae, Okushiri Island and Minehama Village. The measured vertical tsunami runup in both areas was less than 15 m. Similarly in Pto. Saavedra (1960 Chile earthquake), tsunami deposits more than 30-cm thick were formed by erosion of a sand bar (WRIGHT and MELLA, 1963). A common feature throughout these examples is that thick tsunami deposits are only produced in areas with a suitable onshore source (except for situations of unusually large runup). The presence of unvegetated coastal sand bars provided the best source terrain.

The maximum thickness of tsunami deposits due to the 1983 and 1993 earthquakes was less than 22 cm (MINEHAMA VILLAGE, 1986). The tsunami from the 1960 Chile Earthquake (M 8.5) produced deposits more than 1-m thick (BOURGEOIS and REINHART, 1989), although the general thickness was less than 30 cm (WRIGHT and MELLA, 1963). The 1896 Sanriku Coast tsunami also produced deposits thick enough to bury dead bodies (YOSHIMURA, 1984) and deposits were probably more than 1 m in thickness. As the earthquakes responsible for the two recent tsunamis described in this paper were of the same magnitude, further comparative studies will be needed to evaluate the effect of earthquake magnitude on the thickness of tsunami deposits.

As mentioned above, the composition, grain size distribution, facies and sorting and thickness of tsunami deposits vary widely with the nature of the source material. Consequently, it seems very difficult to identify tsunami deposits based only on grain size distribution or sedimentary structure of a single site in ancient successions. Additional information such as the lateral change of deposits and occurrence of marine fauna in an otherwise subaerial succession is important for the identification of tsunami deposits, except for the identification problem with storm surge deposits (DAWSON *et al.*, 1991; DAWSON, 1994).

In short, due to the strong dependence of tsunami deposits on their source terrain, it seems difficult to indicate the magnitude or frequency of paleoseismic events from the thickness and/or abundance of onshore tsunami deposits. In particular, deposits from vertical runup will be extremely localized and thin and rarely preserved in geological successions.

5. Summary

The results of our survey of onshore tsunami deposits due to the 1993 Southwest Hokkaido and 1983 Japan Sea earthquakes are as follows:

1. Comparison of the amounts of erosion and deposition suggests that the subaerial siliclastic tsunami deposits were derived from subaerial source areas such as beaches and coastal sand dunes.
2. Thick and widespread siliclastic tsunami deposits occurred where higher vertical tsunami runup was observed.
3. To produce thick siliclastic tsunami deposits, a suitable source area, such as a sand bar or dune, must be available in addition to sufficient vertical tsunami runup.
4. Tsunami deposits contain sedimentary structures formed under both low and high flow régimes, their composition and grain size being controlled by the nature of the sedimentary materials in the source area.

Acknowledgments

We wish to thank Dr. Geoff Orton (McMaster University) whose comments helped to improve our manuscript. We are grateful to Prof. A. Dawson (Coventry University) and an anonymous reviewer for their critical reviews and improvement of our manuscript. With regards to our survey of tsunami deposits caused by the 1993 Southwest Hokkaido earthquake, we wish to thank the Fisheries Department of the Hokkaido Development Agency for providing the photographs and surveying data, and Kokusai Kogyoh Co., Ltd., for permission to use air photographs. For our survey of tsunami deposits from the 1983 Japan Sea earthquake, we appreciate Prof. Takeshi Ohguchi (Akita University) for his helpful advice.

REFERENCES

ATWATER, B. F. (1987), *Evidence for the Great Holocene Earthquakes along the Outer Coast of Washington State*, Science 236, 942–944.

ATWATER, B. R. (1992), *Geologic Evidence for Earthquakes during the Past 2000 Years along the Copalis River, Southern Coastal Washington*, J. Geophys. Res. 97, 1901–1919.

ATWATER, B. F., and MOORE, A. L. (1992), *A Tsunami about 1000 Years Ago in Puget Sound, Washington*, Science 258, 1614–1617.

BOURGEOIS, J., and REINHART, M. A. (1989), *Onshore Erosion and Deposition by the 1960 Tsunami at the Rio Lingue Estuary, South-Central Chile*, EOS 70, 1331.

CLAGUE, J. J., and BOBROWSKY, P. T. (1994), *Tsunami Deposits beneath Tidal Marshes on Vancouver Island, British Columbia*, Geol. Soc. Am. Bull. 106, 1293–1303.

CLIFTON, H. E. (1976), *Wave-formed sedimentary structures — a conceptual model*, In *Beach and Nearshore Sedimentation* (eds. DAVIS, R. A. Jr., and ETHINGTON, R. L.), Spec. Publ. Soc. Econ. Paleontol. Miner. 24, 126–148.

DAWSON, A. G. (1994), *Geomorphological Effects of Tsunami Runup and Backwash*, Geomorphology 10, 83–94.

DAWSON, A. G., LONG, D., and SMITH, D. E. (1988), *The Storegga Slides: Evidence from Eastern Scotland for a Possible Tsunami*, Marine Geology 82, 271–276.

DAWSON, A. G., FOSTER, I. D L., SHI, S., SMITH, D. E., and LONG, D. (1991), *The Identification of Tsunami Deposits in Coastal Sediment Sequenees*, Sci. Tsunami Hazards 9, 73–82.

FIELD SURVEY TEAM of TOHOKU UNIV. and AKITA UNIV. — MATSUMORI. H., IKEDA, H., SHUTO, N., ORITZ, M., TAKAHASHI, T., KAWAMATA, S., NOJI, M., KABUTOYAMA, H., ITO, T., IMAMURA, F., YAMAKI, S., and MATSUYAMA, M. (1993). *Field Survey of the 1993 Hokkaido Nansei-Oki Earthquake*. Abstr. 1993 Seismological Soc. Jpn. 2, 51 (in Japanese).

FUKAURA TOWN, *Document of the Hazard of Mid-Japan Sea Earthquake* (Fukaura Town Office, Akita 1983) (in Japanese).

HACHIMORI TOWN, *Document of the 1983 Japan Sea Earthquake and Tsunamis* (Hachimori Town Office, Akita 1984) (in Japanese).

HOKKAIDO TSUNAMI SURVEY GROUP (1993), *Tsunami Devastates Japanese Coastal Region*, EOS 74, 417–432.

INGRAM, R. L. (1954), *Terminology for the Thickness of Stratification and Parting Units in Sedimentary Rocks*, Bull. Geol. Soc. Am. 65, 935–938.

KATO, K., and TSUJI, Y. (1993), *Numerical Calculation of the Tsunami of the 1993 Hokkaido-Nansei-Oki Earthquake with Comparison of Surveyed Height*, Abstr. 1993 Seismological Soc. Jpn. 2, 54 (in Japanese).

KATO, Y. (1987), *Run-up Height of Yaeyama Seismic Tsunami (1977)*, J. Seismol. Soc. Jpn. 40, 377–381 (in Japanese with English abstract).

KON'NO, E., ed. (1961), *Geological Observations of the Sanriku Coastal Region Damaged by Tsunami due to the Chile Earthquake in 1960*, Contrib. Inst. Geol. Paleontol. Tohoku Univ. 52, 1–45 (in Japanese with English abstract).

LONG, D., SMITH, D. E., and DAWSON, A. G. (1989), *A Holocene Tsunami Deposit in Eastern Scotland*, J. Quat. Sci. 4, 61–66.

MINEHAMA VILLAGE, *Japan Sea Earthquake — Attacking Tsunamis* (Minehama Village Office, Akita 1984) (in Japanese).

MINEHAMA VILLAGE, *Report on the Damage of Rice Paddy Caused by Japan Sea Earthquake* (Minehama Village Office, Akita 1986) (in Japanese).

MINOURA, K., and NAKAYA, S. (1991), *Traces of Tsunami Preserved Inter-tidal Lacustrine and Mash Deposits: Some Examples from Northeast Japan*, J. Geol. 99, 265–287.

MINOURA, K., and NAKATA, T. (1994), *Discovery of an Ancient Tsunami Deposit in Coastal Sequences of Southwest Japan: Verification of a Large Historic Tsunami*, Island Arc 3, 66–72.

MOORE, G. W., and MOORE, J. G. (1988), *Large-scale bed forms in boulder gravel produced by giant waves in Hawaii*. In *Sedimentologic Consequerces of Convulsive Geologic Events* (CLIFTON, H. E., ed.), Geol. Soc. Am. Spec. Pap. 229, 101–110.

MOORE, J. G., CLAUGUE, D. A., HOLCOM, R. T., LIPMAN, P. W., NORMARK, W. R., and TOR-
REWSAN, M. E. (1989), *Prodigious Submarine Landslides on Hawaiian Ridge*, J. Geophys. Res. *94*,
17465–17484.

NAKANISHI, I., and KIKUCHI, M. (1993), *Features of the 1993 Southwest Hokkaido Earthquakes*, Japan
Soc. for Earth. Engin. Promotion News *133*, 1–5 (in Japanese).

NAKATA, T., and KAWANA, T. (1993), *Historical and prehistorical large tsunamis in the southern
Ryukyus, Japan*. In *Tsunami '93*, Proc. IUGG/IOC Int. Tsunami Symp. Wakayama Japan, August
23–27, 1993, pp. 297–307.

NISHIMURA, Y., and MIYAJI, N. (1995), *Spacial Distribution and Lithofacies of Tsunami Deposits
Associated with the 1993 Southwest Hokkaido Earthquake and the 1640 Hokkaido Komagatake
Eruption*, Pure and Appl. Geophys., this issue.

OKAMURA, M., SHIMAZAKI, K., NAKATA, T., CHIDA, N., MIYATAKE, T., MAEMOKU, H., TSUTSUMI,
H., NAKAMURA, T., YAMAGUCHI, C., and OGAWA, M. (1993), *Submarine Active Faults in the
Northwestern Part of Beppu Bay, Japan — On a New Technique for Submarine Active Fault Survey —*,
Mem. Geol. Soc. Japan *40*, 65–74 (in Japanese with English abstracts).

SIEH, K. E. (1978), *Prehistoric Large Earthquakes Produced by Slip on the San Andreas Fault at Pallet
Creek, California*, J. Geophys. Res. *83*, 3907–3939.

SHIMAMOTO, T., TSUTSUMI, A., KAWAMOTO, E., MIYAWAKI, M., and SATO, H. (1995), *Tsunami
Disasters Caused by the 1993 Southwest Hokkaido Earthquake and Their Implication for the Prevention
of Future Tsunami Disasters*, Pure and Appl. Geophys., this issue.

SHUTO, N. (1984), *Trace Height of Tsunami of the Japan Sea Earthquake of 1983*, Tsunami Engin.
Technical Rept. Tohoku Univ. *1*, 88–267 (in Japanese).

WRIGHT, C., and MELLA, A. (1963), *Modifications to the Soil Pattern of South-central Chile Resulting
from Seismic and Associated Phenomena during the Period May to August 1960*, Bull. Seismol. Soc.
Am. *53*, 1367–1402.

YOSHIMURA, A., *Large-scale Tsunamis on the Sanriku Coast* (Chuou-Koron, Tokyo 1984).

(Received September 1, 1994, revised February 8, 1995, accepted February 27, 1995)

PAGEOPH, Vol. 144, Nos. 3/4 (1995)

0033–4553/95/040719–15$1.50 + 0.20/0
© 1995 Birkhäuser Verlag, Basel

Tsunami Deposits from the 1993 Southwest Hokkaido Earthquake and the 1640 Hokkaido Komagatake Eruption, Northern Japan

YUICHI NISHIMURA[1] and NAOMICHI MIYAJI[2]

Abstract —The southwest Hokkaido tsunami of July 12th, 1993, left continuous onshore sand deposits along the west coast of Oshima Peninsula, Hokkaido, northern Japan. We investigated spatial distribution and lithofacies of the new tsunami deposits for its identification of ancient tsunami deposits. An eyewitness account and bent plants helped our interpretation of the onshore tsunami behavior. We regard the following properties as typical of the coastal tsunami sand deposits: (1) The deposits cover the surface almost continuously on gentle topography. (2) Deposit thicknesses and mean grain sizes decrease with distance from the sea. (3) Deposit thicknesses and lithofacies vary greatly across local surface undulation. (4) Graded bedding reflecting tsunami runup and backwash is present in thick deposits. (5) The deposits are widely distributed along the coast and extend inland several tens of meters to 100 m. We examined a candidate for the paleo-tsunami deposits associated with the 1640 Komagatake eruption, and confirmed that the similar patterns are typical of ancient tsunami deposits.

Key words: Tsunami deposit, distribution, lithofacies, 1993 Hokkaido tsunami, 1640 Komagatake eruption.

Introduction

New tsunami sand deposits provide excellent opportunities to study processes of tsunami runup and depositional features for characterizing paleo-tsunami deposits (KON'NO, 1961; WRIGHT and MELLA, 1963; MINOURA and NAKAYA, 1991; SHI *et al.*, 1993; SATO *et al.*, 1994). The massive tsunami waves generated by the July 12th, 1993, submarine earthquake (NAKANISHI *et al.*, 1993) provided an unusual opportunity to study new tsunami deposits and how these deposits were distributed along the populated southern coast of Hokkaido. In this paper, we describe spatial distribution and lithofacies of tsunami deposits caused by the 1993 Hokkaido earthquake. The objective is to construct criteria to identify paleo-tsunami deposits. We then compare the geological features of new tsunami deposits to a candidate for the paleo-tsunami deposits from the 1640 Komagatake tsunami.

---

[1] Usu Volcano Observatory of Hokkaido University, 59 Sohbetsu, Hokkaido, 052-01, Japan.
[2] Hokkaido National Agricultural Experiment Station, 1 Hitsujigaoka, Toyohira, Sapporo, 062, Japan.

Field Observations

We conducted a field survey of the 1993 southwest Hokkaido earthquake tsunami deposit along the west coast of the Oshima Peninsula two to four days after the tsunami occurred. The observation area extends from Honme to Taisei (Fig. 1). The maximum height of the tsunami was 5.9 m above sea level at Honme, 7.0 m at Enoshima, 6.4 m at Setana, and 5.7 m at Taisei (TSUJI *et al.*, 1994). Since the purpose of our survey was to view the tsunami deposits on the land slope, we sought sites that faced sandy beaches without intervening artificial structures that might have changed tsunami directions or intensities.

Through the survey, we acquired general aspects regarding the onshore tsunami deposits: (1) Surged areas are easily identified by affected surface plants. (2) Steep slopes at the beach are sometimes eroded. (3) Sand or gravel is deposited at the foot

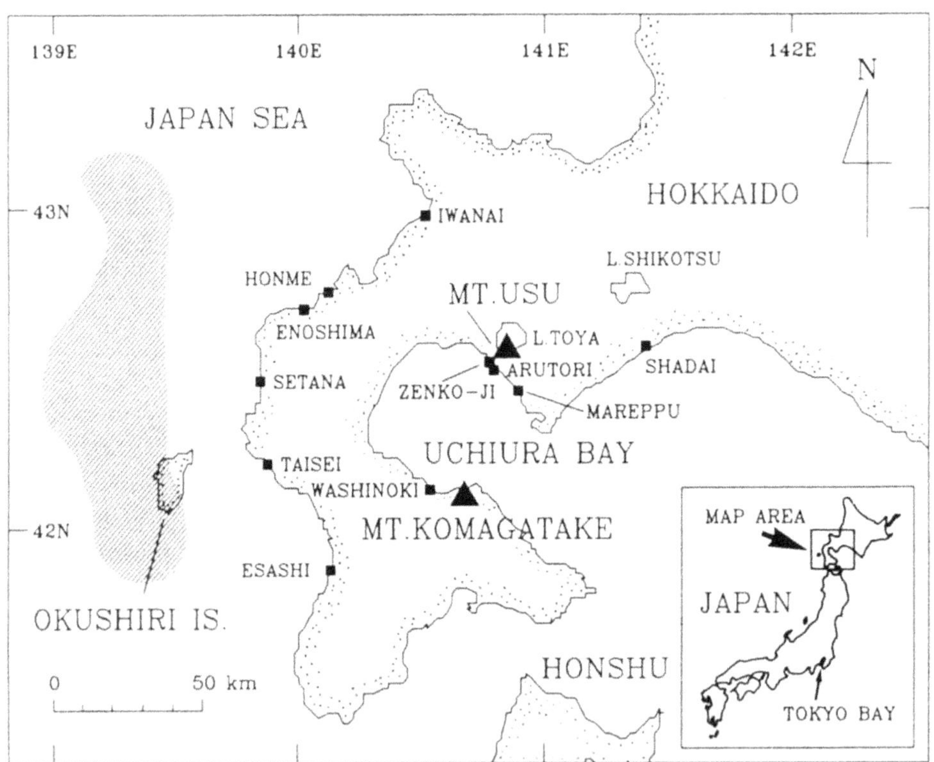

Figure 1

Map showing Oshima Peninsula and field observation areas for the 1993 southwest Hokkaido earthquake tsunami (from Taisei to Honme) and the 1640 Komagatake tsunami (around Uchiura Bay). Shaded area is the aftershock distribution of the 1993 event determined by Hokkaido University (NAKANISHI *et al.*, 1993).

of steep slopes, on gentle slopes, and on flat fields. (4) Thickness of the sand deposits varies with distance from the sea and with surface topography.

At each observation site, we made the following observations. First, we measured the orientations of plants knocked over by the waves. Second, we measured the thickness and lithofacies of sandy deposits carried up by the tsunami waves, and collected samples for grain size analysis. Finally, we surveyed local topography of the site by means of an electro-optical distance-meter (EDM). The heights were measured from sea level at the observation time and converted to the heights from the mean sea level of Tokyo Bay, using the tide gage record at Esashi and Iwanai. In constructing maps of the observation sites, we referred to aerial photographs taken before and after the tsunami.

Eyewitness Account, Bent Plants, and Land Deposits

The relationship between the actual tsunami runup behavior reported by an eyewitness and the resulting deposits on land was examined. Since the tsunami arrived in the middle of the night, there was only one person from whom we could obtain detailed descriptions of the tsunami behavior. The site is Honme, where the tsunami wave did not surge directly because it was located inside a small north-facing bay (Fig. 1).

According to the eyewitness, the first tsunami wave arrived 7 to 8 minutes after the earthquake. The second wave came 2 or 3 minutes after the first wave. It was the maximum wave and it reached the foot of the road, 5.2 m above sea level (Fig. 2). The wave soon receded to P2 and maintained the water level there for a few minutes, and then backwashed. The third and the later waves reached approximately the level of P3.

The stand lines of dry branches and grasses are the most obvious tsunami deposits at this site. Such deposits are commonly observed at beaches where tsunamis have invaded (KON'NO, 1961; SATO et al., 1994). The materials were carried by the tsunami waves and positioned at the uppermost level of the beach reached by each wave, and were distributed almost parallel to the coast. Considering the eyewitness description, the landward debris traces the maintained water level (P2). In this case the maximum runup height was about 1 m higher than the debris border. The seaward line of debris (P3) was deposited by the following repeated waves.

We observed small amounts of sand distributed on the leaves and around the stumps in the area seaward from P2. We believe the deposits were produced by the tsunami. The thickness of the deposit decreases with distance from the sea: about 1 cm at the beach side, 5 mm around P3, and zero to 3 mm between P2 and P3. No sand was deposited between P2 and P1.

Bent grasses served as good indicators of tsunami runup and backwash directions. At the maximum height of the tsunami waves (P1), the grasses were bent

Figure 2
Photograph showing two lines of debris (P2 and P3) composed of dried branches and grasses associated with the 1993 southwest Hokkaido tsunami at Honme. The maximum wave reached to about P1.

landward. On the other hand, grasses on the submerged area were bent seaward, indicating that they were generated in the final backwash process. These features are probably common at all of the observation sites.

Sand Deposits on Gentle Slope

Thickness distribution and lithofacies of the sand deposit on a typical gentle slope were examined. Figure 3 (top) shows the surface topography at Taisei. Judging from the bent grasses and the debris distribution, the tsunami waves crossed the road and invaded the upper fallow paddy field. The maximum runup height there is estimated to be 6.4 m. No evidence of surface erosion was detected.

The surface of the slope was continuously covered with sand deposits less than 3 cm thick. We infer that the sand was carried from the beach, because composition and grain size distribution of the tsunami deposits were similar to those of the beach sand.

At topographical break points, we took sand samples for grain size analysis. Analyses were made by the conventional sieving method using half phi sieve intervals. Each sample was sieved for 10 minutes on a vibro screen shaker. Grain size parameters such as mean and sorting were calculated following INMAN (1952).

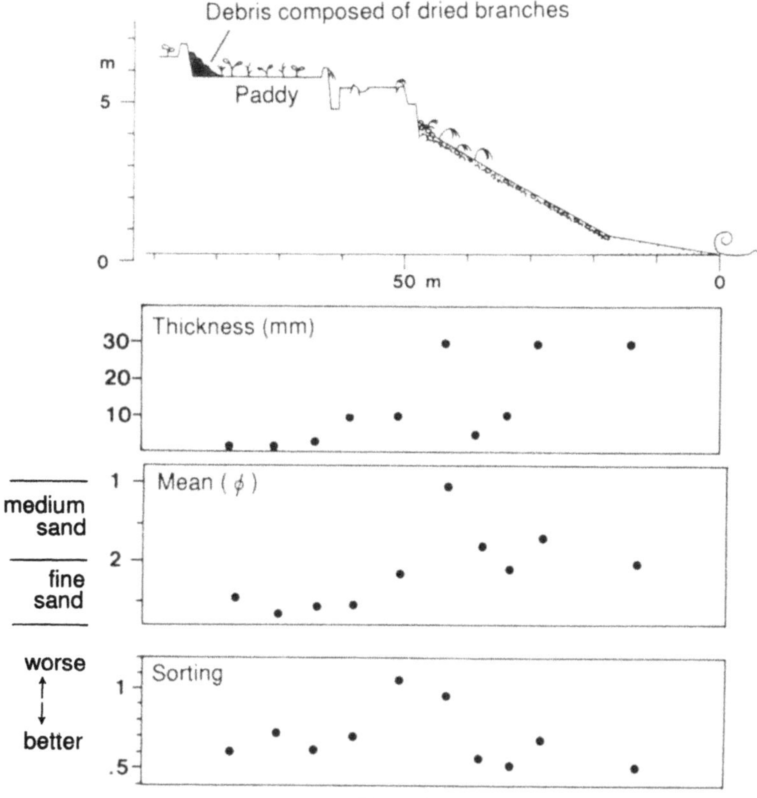

Figure 3

Profile of the beach slope at Taisei, and results of the grain size analyses for the tsunami sand deposits.

Thickness and lithofacies of the sand deposits vary, associated with topographical breaks such as a small bank (Fig. 3). At the sea side foot of the bank, the sand was deposited to thicknesses of several mm to 3 cm, and mean grain sizes are relatively large. At the base of the bank, poor sorting was observed. On the other hand, on the flat surface beyond the bank, the sand layer is thinner, the grain sizes are finer and the sorting is better.

Traces on Steep Slope

The tsunami eroded some of the steep slopes (e.g., Fig. 4). The maximum runup height overflowed the top of this slope, probably about 6.4 m, Wash-up materials, mainly composed of concrete blocks, were distributed at the base of the slope. Some freshly formed gullies with widths of several cm are also evident on the surface of the slope.

Figure 4
Photograph showing a steep slope eroded by the tsunami, at Enoshima. Wash-up concrete blocks are
located at the base of the slope.

The thickness of the sand containing gravels at the foot of the slope is about
12 cm. Graded bedding structures are shown in these thick tsunami deposits. In this
case, the sediment is composed at most of the four layers. In the three lower layers,
many concrete block fragments of pebble which had constructed the original slope
were found. Plants in the sediments are bent seaward. The sequence suggests that
erosion and deposition processes repeated at least three times at this site. On the
other hand, the uppermost layer is composed of the same sands as the original
surface of the slope, indicating that this layer was deposited in the final backwash
waves of the tsunami episode.

Thick Tsunami Deposits on a Flat Field

Graded bedding is also present in thick tsunami deposits in a paddy field at
Enoshima (Figs. 5–6). The maximum tsunami height here was 3.7 m. The direction
of the flattened plants indicates the final backwash direction of the tsunami.
Crescent marks directed toward the sea were found. Those were probably produced
by fast backwash waves of the tsunami. The crescent marks were crossed by current

Figure 5

Photograph looking southwest across a section of fallow paddy field covered by the thick tsunami deposit at Enoshima.

ripple, indicating that the water stagnated before draining back to the sea via the adjacent channel.

Thin sandy silt layer or crust covered the surface of the tsunami deposits on a part of the paddy fields. The origin of the silt is probably footpaths separating the rice fields, since some walls of the footpaths were eroded. Similar sandy silt deposits were observed on the surface of paddy fields at Sanriku coasts, Japan, after the tsunami generated by the 1960 Chile earthquake (KON'NO, 1961).

The thicknesses of the sand deposits at Enoshima decrease inland from 15 cm at point 3 through 8 cm at point 2, to 3 cm at point 1 (Fig. 7). Graded bedding is present at all three points. Bent plants in the sediments indicate the direction of sand movement during deposition. Plants in the lower and middle layers were bent landward, indicating that the sands were carried by the advancing tsunami waves. On the other hand, plants on the surface and in the uppermost layer were bent seaward, indicating that they were caused by wave backwash. These observations demonstrate that tsunami deposits may preserve the original runup and backwash process.

On the basis of our observations, we regard the following properties as typical of the 1993 Hokkaido tsunami deposits. (1) The deposits cover the surface almost

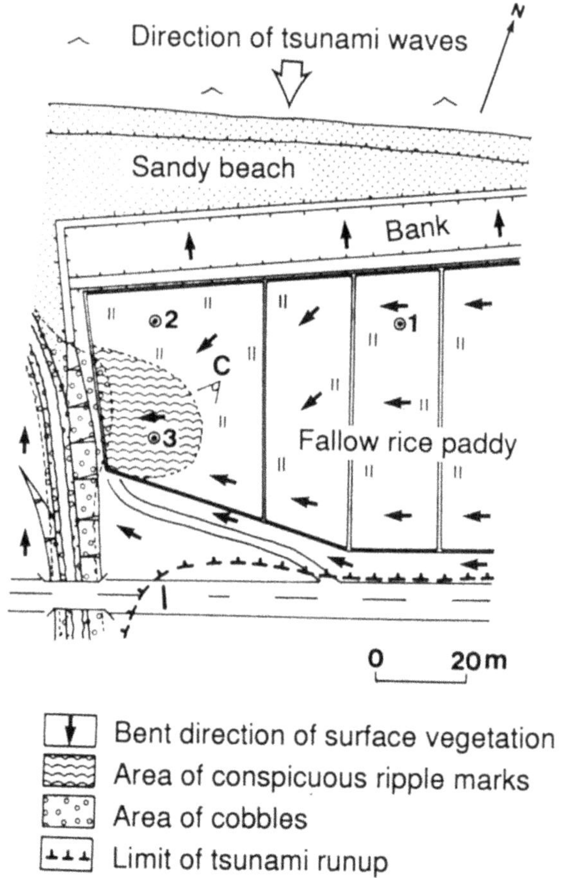

Figure 6

Map showing the backwash directions of the tsunami waves as recorded by plants flattened in the fallow rice field shown partially in Figure 5. Numbered points denote sites of columnar sections in Figure 7. C indicates camera site and direction for Figure 5.

continuously on a simple gentle slope. (2) Deposit thicknesses and mean grain sizes decrease with distance from the sea. (3) Deposit thickness and lithofacies vary greatly across local surface undulation. (4) Graded bedding reflecting tsunami runup and backwash is present in thick deposits. (5) The deposits are widely distributed along the coast and extend inland several tens of meters to 100 m.

The 1640 Komagatake Tsunami Deposits

In this section, we show an ancient sand deposit satisfying the above criteria for tsunami deposit. At Arutori (Fig. 1), a continuous sand layer overlies a volcanic ash soil (Fig. 8). The layer thins with distance from the sea. The sand layer immediately

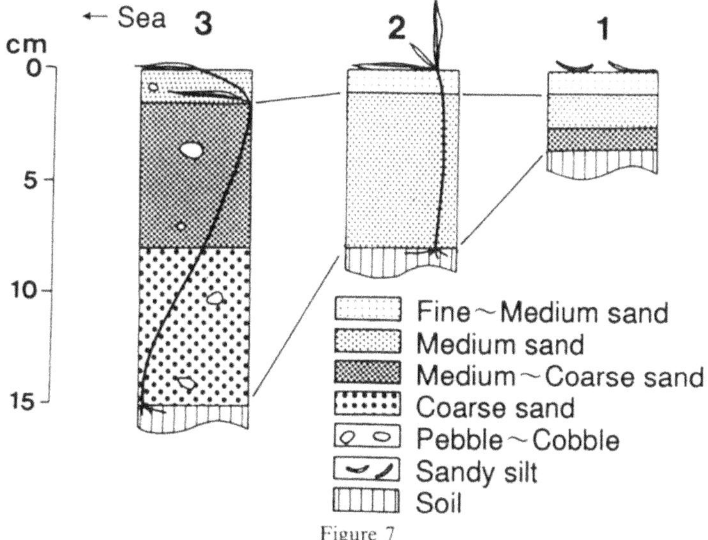

Figure 7

Columnar sections of the thick tsunami deposits in Figures 5–6.

Figure 8

Photograph showing a vertical section of the 1640 Komagatake outcrop at Arutori. The uppermost part of the sand layer contains some of the 1640 Komagatake ash (Ko-d). Us-b is the ash layer associated with a series of eruptions of Mt. Usu in 1663.

underlies the 1640 Komagatake ash layer (Ko-d). The sand layer also underlies the 1663 pumice fall from Usu volcano (Us-b). Ko-d was identified by comparing the mineral composition, shape, and refractive indices of volcanic glasses to an established ash catalogue (MIYAHARA, personal communication). These key tephra layers can be traced around most of Uchiura Bay (KATSUI *et al.*, 1975).

We infer that the sand records the tsunami caused by the 1640 Komagatake eruption, because this eruption produced both a large tsunami and a large amount of tephra (Ko-d). On July 31, 1640, Mt. Komagatake broke into eruption after more than 1000 years of dormancy. A summit portion of the volcano collapsed by phreatic eruptions and a large-scale debris avalanche was generated. Part of the avalanche entered the sea and caused a tsunami. This tsunami is known to have killed more than 700 people around Uchiura Bay (KATSUI *et al.*, 1975). The maximum height of the deposit traced at Arutori is about 7.5 m above present sea level. This is consistent with a documentary description of an 8.5-m height for the 1640 tsunami at Zenko-ji (TSUJI, 1989).

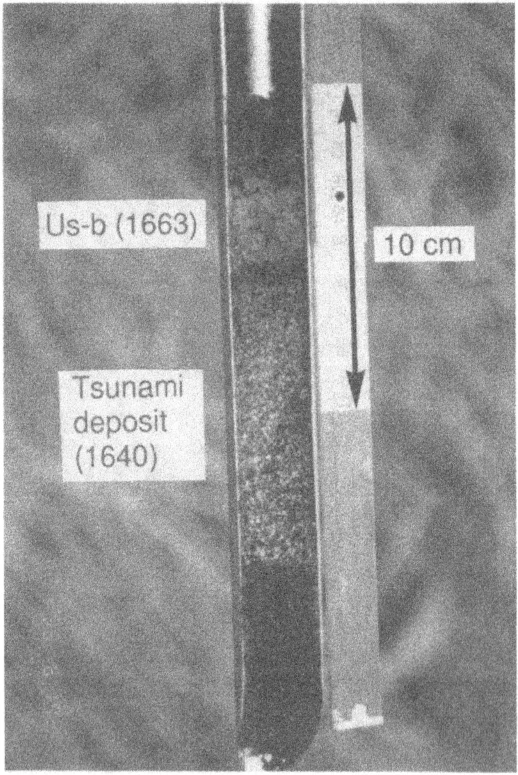

Figure 9
Photograph showing a core from a depth of about 1 m below the present grassland surface at Mareppu.

Where outcrops are inadequate, we studied shallow underground stratigraphy by means of a 3-cm-diameter soil auger. Part of a core from about 1 m depth beneath the present grassland surface at Mareppu contains a sand layer about 10 cm thick with graded bedding (Fig. 9). As these sand deposits are overlaid directly with Ko-d and/or indirectly with Us-b, the sand deposits also originated from the 1640 Komagatake tsunami. The maximum height traced the tsunami deposit is 2.9 m above the present sea level at Shadai. Based on a simple computer simulation, the tsunami height at Shadai was estimated to be about the same as around Uchiura Bay: that is, about 5 to 10 m (NISHIMURA and SATAKE, 1993).

At Shadai, we took some 40 core samples and examined the thickness pattern and lithofacies of the sand deposits. Generally, the thickness of the sand layer at Shadai decreases with distance from the sea (Fig. 10). The layer is especially thick at the seaward foot of the dunes (Fig. 11). The mean grain size and the sorting of the deposits vary across the undulating surface. In general the deposits become finer and better sorted in a landward direction.

Two possible depositional processes could have resulted in the above thickness patterns. (1) The tsunami carried the beach sand and deposited them seaside of the dune wall while invading the field. (2) The tsunami eroded the seaward wall of the

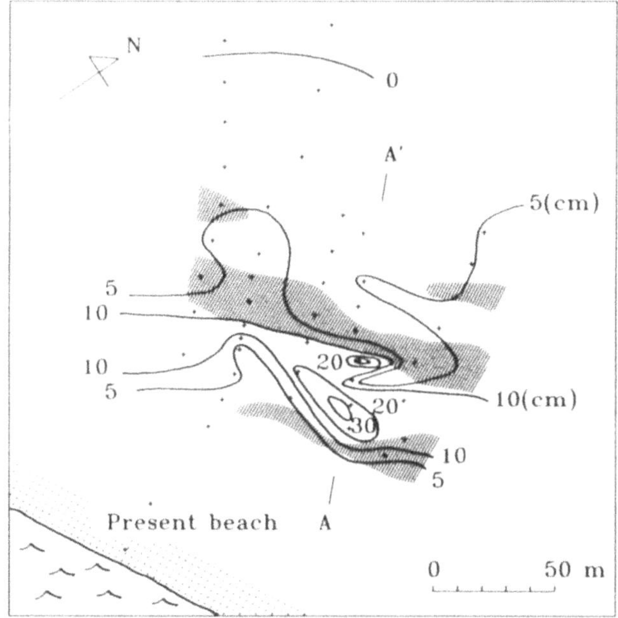

Figure 10

Contour maps for the thickness of the 1640 tsunami deposits at Shadai, revealed by soil cores. Crosses are the sampling points. Shaded areas are old dunes. A–A' shows the cross section referred to in Figure 11.

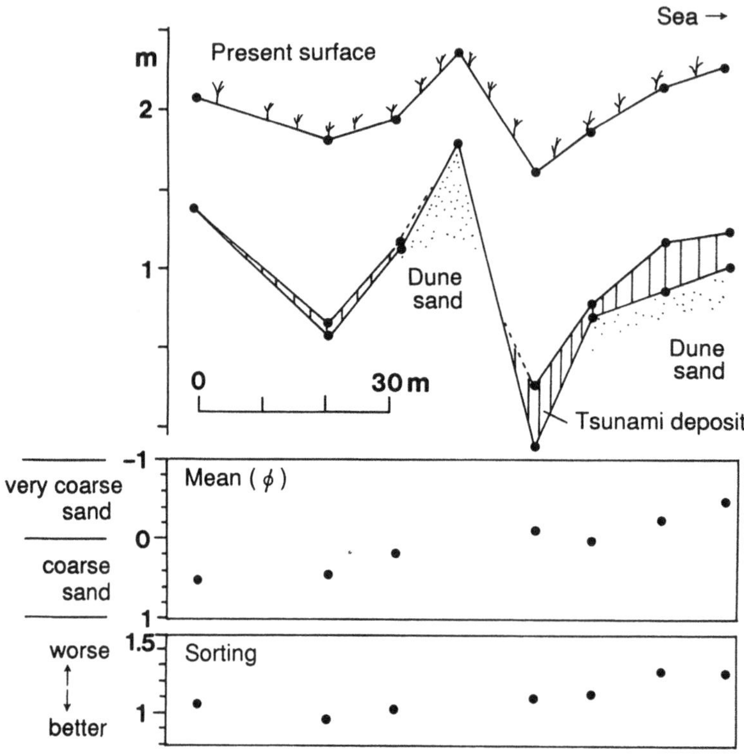

Figure 11

Vertical cross section of A–A' line in Figure 10, and results of the grain size analyses of the underground sand deposits (Fig. 3).

dunes and deposited them at the foot. We consider the first explanation to be suitable because the composition of the accumulating sand is completely different from that of the dune. The dunes contain significant amounts of green rock fragments, but the accumulated sand includes little. Sorting of the dune sand is also considerably better than that of the accumulated sand.

At four different sites around Uchiura Bay (Arutori, Washinoki, Mareppu and Shadai), we found sand layers manifesting similar patterns of grain size and sorting as the deposits from the 1993 southwest Hokkaido tsunami. The maximum heights we could trace up the deposits at each site were 7.3, 5.8, 4.5 and 2.9 m, respectively. Those heights represent the minimum heights for tsunami runup.

Discussion

Investigations to identify prehistoric tsunami deposits have been conducted to reveal the original paleo-seismic events (ATWATER, 1987; DAWSON et al., 1988; ABE *et al.*, 1990; MINOURA and NAKAYA, 1991; ATWATER and YAMAGUCHI, 1991). For the ancient tsunami deposits, the following features have been reported. (1) Tsunami deposits consisting of sand include fossils of marine origin (MOORE and MOORE, 1984; DAWSON *et al.*, 1993). (2) The thickness of the deposits decreases with distance from the sea (MOORE and MOORE, 1984; ATWATER and MOORE, 1992). (3) Likewise, fining and better sorting are also seen (MINOURA and NAKAYA, 1991; ATWATER and MOORE, 1992; CLAGUE and BOBROWSKY, 1994).

Ancient tsunami deposits cannot be identified solely from the sedimentary faces and/or grain size distribution of the samples (SATO *et al.*, 1994). However, with additional information concerning the spatial distribution of deposits and the variation of the lithofacies with respect to onshore topography, tsunami deposits might be identified with some confidence. Even thin (less than 10 cm) sand layers preserve their original geological features where an environment conducive to preservation has been present (e.g., ATWATER, 1987). The 1640 Komagatake deposits may have been aided by burial by volcanic ash soon after the tsunami occurred.

Coastal sand layers can also result from marine transgression, from eolian deposition, and from river floods. However, geological features for such sand layers are different from those of tsunami deposits. For a deposit caused by transgression, the upper limit of the spatial distribution should be constant and the thickness of the deposit should not depend on the distance from the sea. Dune sand shows good sorting and does not show grading structure. Flood deposits have poor sorting and are usually distributed within the small area around the mouth of a river.

It is, of course, necessary to be careful when estimating the tsunami runup height from the upper limit of the deposits. In the case of the 1993 Hokkaido tsunami, the maximum runup height is about 1 m higher than the deposits. The same observation was reported by SHI *et al.* (1993) for the 1992 Flores event.

SATO *et al.* (1994) established that small (less than 10 m) tsunami waves caused by the 1993 Hokkaido earthquake widely failed to produce any tsunami deposits composed of sand or gravel. Their results differ from ours. It is probably because the observation sites were different. We think that wave frequency of tsunami is one of the important factors which determines whether tsunami leave deposits or not. The 1960 Chile earthquake generated a long-period tsunami along the Sanriku coast, Japan, and produced a large amount of sand deposits, although the tsunami heights were about 5 m or less (KON'NO, 1961).

Conclusion

The field survey of the tsunami deposits from the 1993 southwest Hokkaido tsunami shows that tsunami waves can carry beach sand inland even where the wave height is less than 10 m. We regard the following properties as typical of the

1993 Hokkaido tsunami deposits. (1) The deposits cover the surface almost continuously on a simple gentle slope. (2) Deposit thicknesses and mean grain sizes decrease with distance from the sea. (3) Deposit thicknesses and lithofacies vary greatly across local surface undulation. (4) Graded bedding reflecting tsunami runup and backwash process is present in thick deposits. (5) The deposits are widely distributed along the coast and extend inland several tens of meters to 100 m.

Consequently, using information pertaining to the spatial distribution of the deposits and the variation of the lithofacies with respect to onshore topography, ancient tsunami deposits might be identical. We applied the above criteria in our identification of the 1640 Komagatake tsunami deposits.

Acknowledgements

We thank Kunihiko Endo for valuable discussions, Naoya Miyahara for identifying the 1640 Komagatake tephra, and the Hokkaido Development Bureau for providing the tide gage data at Esashi and Iwanai. We also express gratitude to Brian F. Atwater, an anonymous reviewer and David K. Yamaguchi for their critical comments on earlier versions of the manuscript.

REFERENCES

ABE, H., SUGENO, Y., and CHIGAMA, A. (1990), *Estimation of the Height of the Sanriku Jogan 11 Earthquake-tsunami (A. D. 869) in the Sendai Plain*, Zishin *43*, 513–525 (in Japanese with English abstract).

ATWATER, B. F. (1987), *Evidence for Great Holocene Earthquakes along the Outer Coast of Washington State*, Science *236*, 942–944.

ATWATER, B. F., and YAMAGUCHI, D. K. (1991), *Sudden, Probably Coseismic Submergence of Holocene Trees and Grass in Coastal Washington State*, Geology *19*, 706–709.

ATWATER, B. F., and MOORE, A. L. (1992), *A Tsunami about 1000 Years Ago in Puget Sound, Washington*, Science *258*, 1614–1617.

CLAGUE, J. J., and BOBROWSKY, P. T. (1994), *Evidence for a Large Earthquake and Tsunami 100–400 Years Ago on Western Vancouver Island, British Columbia*, Quat. Res. *41*, 176–184.

DAWSON, A. G., LONG, D., and SMITH, D. E. (1988), *The Storegga Slides: Evidence from Eastern Scotland for a Possible Tsunami*, Mar. Geol. *82*, 271–276.

DAWSON, A. G., LONG, D., SMITH, D. E., SHI, S., and FOSTER, I. D. L., *Tsunamis in the Norwegian Sea and North Sea caused by the Storegga submarine landslide. In Tsunamis in the World* (ed. Tinti, S.) (Kluwer Academic Publishers 1993) pp. 31–42.

INMAN, D. L. (1952), *Measures for Describing the Size Distribution of Sediments*, J. Sed. Petrol. *22*, 125–145.

KATSUI, Y., YOKOYAMA, I., FUJITA, T., and EHARA, Y. (1975), *Komagatake, its Volcanic Geology, History of Eruption, Present State of Activity and Prevention of Disaster*, Committee for Prevention of Disasters of Hokkaido, Sapporo, 194 pp. (in Japanese).

KON'NO, E., ed. (1961), *Geological Observations of the Sanriku Coastal Region Damaged by Tsunami due to the Chile Earthquake in 1960*, Contrib. Inst. Geol. Paleontol. Tohoku Univ. *52*, 1–45 (in Japanese with English abstract).

MINOURA, K., and NAKAYA, S. (1991), *Traces of Tsunami Preserved in Inter-tidal Lacustrine and Marsh Deposits: Some Examples from Northeast Japan*, J. Geol. *99*, 265–287.

MIYAJI, N., and NISHIMURA, Y. (1994), *Tsunami Deposits Associated with the 1993 Southwest Hokkaido Earthquake, Japan*, Bull. Nat. Dis. Sci. Data Center, Hokkaido *9*, 25–48 (in Japanese with English abstract).

MOORE, J. G., and MOORE, G. W. (1984), *Deposit from a Giant Wave on the Island of Lanai, Hawaii*, Science *226*, 1312–1315.

NAKANISHI, I., KODAIRA, S., KOBAYASHI, R., KASAHARA, M., and KIKUCHI, M. (1993), *The 1993 Japan Sea Earthquake: Quake and Tsunamis Devastate Small Town*, EOS, Trans. Am. Geophys. Union *74*, 377–379.

NISHIMURA, Y., and SATAKE, K. (1993), *Numerical Computations of Tsunamis from the Past and Future Eruptions of Komagatake Volcano, Hokkaido, Japan*, Proc. of IUGG/IOC International Tsunami Symposium, Wakayama, Japan, 573–583.

SATO, H., SHIMAMOTO, T., TSUTSUMI, A., KAWAMOTO, E., and MIYAWAKI, M. (1994), *Onshore Tsunami Sediments Caused by the 1993 Southwest Hokkaido Earthquake and the 1983 Japan Sea Earthquake*, Katsudanso Kenkyu *12*, 1–23 (in Japanese).

SHI, S., DAWSON, A. G., and SMITH, D. E. (1993), *Geomorphological Impact of the Flores Tsunami of 12th December, 1992*, Proc. of IUGG/IOC International Tsunami Symposium, Wakayama, Japan, 689–696.

TSUJI, Y. (1989), *Tsunami from the Eruption of Hokkaido Komagatake on July 13, 1640*, Program and Abstracts of Seismological Society of Japan *1*, 336 (in Japanese).

TSUJI, Y., KATO, K., ARAI, K., HAN, S., and YAMANAKA, Y. (1994), *Characteristics of the Southwest Hokkaido Tsunami*, Kaiyo Monthly *7*, 110–122 (in Japanese).

WRIGHT, C., and MELLA, A. (1963), *Modifications to the Soil Pattern of South-central Chile Resulting from Seismic and Associated Phenomena During the Period May to August 1960*, Bull. Seismol. Soc. Am. *53*, 1367–1402.

(Received September 23, 1994, revised March 14, 1995, accepted March 15, 1995)

PAGEOPH, Vol. 144, Nos. 3/4 (1995)

0033-4553/95/040735-11$1.50 + 0.20/0
© 1995 Birkhäuser Verlag, Basel

Modeling of the Runup Heights of the Hokkaido-Nansei-Oki Tsunami of 12 July 1993

KATSUYUKI ABE[1]

Abstract — The Hokkaido-Nansei-Oki earthquake (M_w 7.7) of July 12, 1993, is one of the largest tsunamigenic events in the Sea of Japan. The tsunami magnitude M_t is determined to be 8.1 from the maximum amplitudes of the tsunami recorded on tide gauges. This value is larger than M_w by 0.4 units. It is suggested that the tsunami potential of the Nansei-Oki earthquake is large for M_w. A number of tsunami runup data are accumulated for a total range of about 1000 km along the coast, and the data are averaged to obtain the local mean heights H_n for 23 segments in intervals of about 40 km each. The geographic variation of H_n is approximately explained in terms of the empirical relationship proposed by ABE (1989, 1993). The height prediction from the available earthquake magnitudes ranges from 5.0–8.4 m, which brackets the observed maximum of H_n, 7.7 m, at Okushiri Island.

Key words: Tsunami magnitude, runup, tsunami warning.

1. Introduction

A large earthquake (M_w 7.7) occurred on July 12, 1993, off the southwestern coast of Hokkaido in the Sea of Japan. This strong shock and resulting tsunami caused 230 deaths, 197 of which occurred in Okushiri Island. Of these 197, 169 persons perished from the tsunami and 28 from the landslide. The hypocenter parameters given by the Japan Meteorological Agency are the following: the origin time, 22:17:11.7(JST = UT + 9); location 42.780N, 139.183E; depth, 35.1 km; and $M(JMA)$, 7.8. This earthquake is officially called the "Hokkaido-Nansei-Oki Earthquake" (earthquake off the coast southwest of Hokkaido), consequently we abbreviate it to the "Nansei-Oki earthquake." This earthquake is the largest tsunamigenic event in the Sea of Japan since the Akita-Oki earthquake (M_w 7.9) of May 26, 1983. These two earthquakes represent the back-arc thrust along the assumed plate boundary between the North American and the Eurasian plates, and both set off devastating tsunamis (NAKANISHI *et al.*, 1993). In this paper, we first

---

[1] Earthquake Research Institute, The University of Tokyo, Yayoi 1-1-1, Bunkyo-ku, Tokyo, Japan
113.

determine the tsunami magnitude of the Nansei-Oki tsunami in order to quantify the physical size of the tsunami. Secondly, we compare observed runup heights with the prediction from earthquake magnitudes by using the method of ABE (1989, 1993), and assess how the method is useful for tsunami warning purposes to elaborate on the feasibility of conducting this type of study in real time.

2. Tsunami Magnitude, M_t

The tsunami magnitude scale M_t is unique among the many scales in seismology because it is based on instrumental tsunami-wave amplitude; its level being adjusted to agree with moment magnitude M_w of an earthquake. Values of M_t have been assigned to large events in the Pacific (ABE, 1979) and tsunamis around Japan (ABE, 1981, 1985, 1988). Now the physical size of a tsunami can be discussed in terms of M_t.

The following formula is used for determining M_t for local events:

$$M_t = \log \dot{H} + \log R + D \tag{1}$$

where H is the maximum amplitude or the maximum crest-to-trough amplitude of tsunami waves in meters measured by tide gauges, R is the distance in km from the earthquake epicenter to the tide station along the shortest oceanic path, $D = 5.80$ when using amplitudes and $D = 5.55$ with crest-to-trough amplitudes.

Considerable data of tide gauges are available for the Nansei-Oki tsunami. Figure 1 shows the variation of the maximum amplitude with the propagation

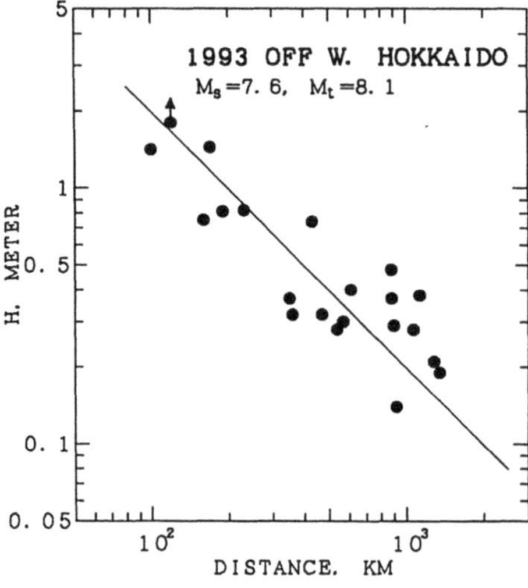

Figure 1

Maximum tsunami-wave amplitude vs. propagation distance for the 1993 Nansei-Oki tsunami. The straight line represents the relation for $M_t = 8.1$. The arrow indicates scale-out.

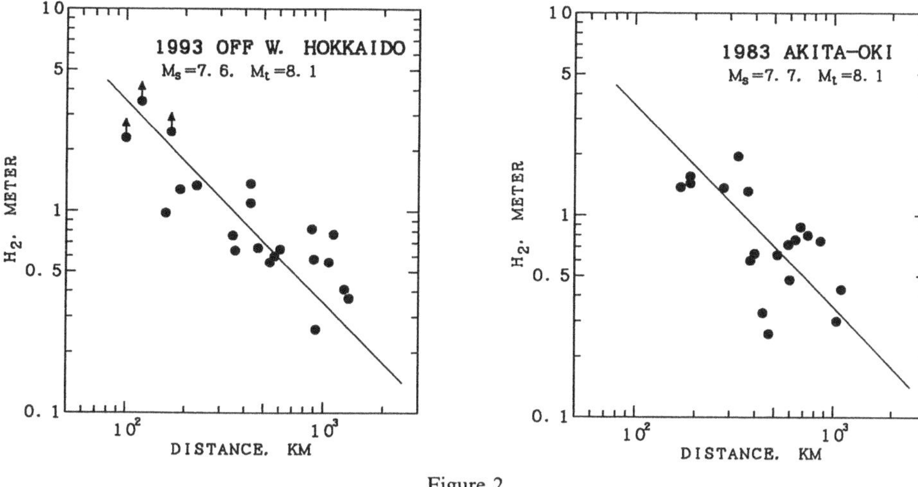

Figure 2

Maximum crest-to-trough amplitude vs. propagation distance for the Nansei-Oki and the Akita-Oki tsunamis. Straight lines represent the relation for $M_t = 8.1$. Arrows indicate scale-out.

distance. The symbol depicting an arrow nearby (station Esashi in Hokkaido) is scaled out. The average of the single-station magnitudes is obtained as 8.09 ± 0.20 from 21 observations. The straight line in the figure shows the least-squares fit for $M_t = 8.1$. Figure 2 shows the observed crest-trough amplitudes (H_2) for the Nansei-Oki and the 1983 Akita-Oki tsunamis. The average of M_t for the Nansei-Oki tsunami is estimated to be 8.09 ± 0.22, which is consistent with M_t from the maximum amplitude. The value of M_t for the Akita-Oki tsunami is also 8.1 (ABE, 1985), indicating that the physical sizes of the Nansei-Oki and the Akita-Oki tsunamis are the same. The tsunami potential energy E_t is estimated from M_t (ABE, 1988) through

$$\log E_t(\text{erg}) = 2M_t + 4.3. \tag{2}$$

Giving $M_t = 8.1$, we obtain $E_t = 3.2 \times 10^{20}$ erg.

The moment magnitude M_w of the Nansei-Oki earthquake has been estimated to be 7.7 (e.g., NAKANISHI et al., 1993). This value is smaller than the value of M_t by 0.4 units. However, it is known that the M_t values for the Japan Sea events tend to be larger than M_w. Figure 3 displays a comparison between M_t and M_w of tsunamigenic earthquakes around Japan for which the seismic moments have been determined from seismic wave data. The M_t values for the Pacific events (solid circles) agree very well with the M_w values over a wide magnitude range. On the other hand, the M_t values for the Japan Sea events (open circles) are on the average 0.2 units larger than the M_w values. This important feature can be reasonably explained in terms of the effective difference both in the source depth of the earthquake and the geometry of the fault motion (ABE, 1985). For the Nansei-Oki

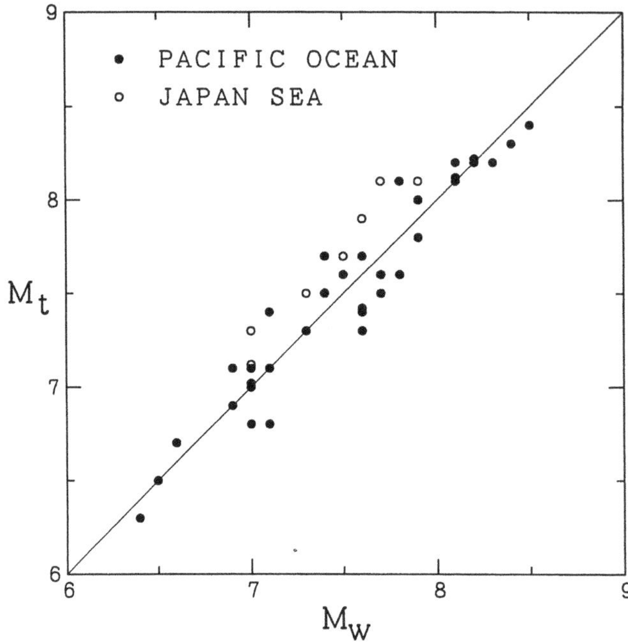

Figure 3

M_t vs. M_w for events around Japan. Solid circles denote Pacific events and open circles denote Japan Sea events. Open circles with the largest M_t correspond to the Nansei-Oki and the Akita-Oki events. Straight lines show $M_t = M_w$.

event, M_t minus 0.4 is equivalent to M_w, the suggestion being that the maximum amplitude of tsunami waves from the Nansei-Oki earthquake is 2.5 times as large as that from earthquakes along the Pacific coast with a comparable M_w.

3. Tsunami Runup Heights

Measured Data

The results of field surveys of tsunami runup heights were reported by many research groups (SHUTO, 1994; HOKKAIDO TSUNAMI SURVEY TEAM, 1993). These data are accumulated for a total range of about 1000 km along the coast where major tsunamis were experienced. The Japan Sea coast from Wakkanai in the northernmost location in Hokkaido to Niigata Prefecture in Honshu is divided into 23 segments approximately 40 km long. The numbering of each segment and the aftershock area are shown in Figure 4. In each segment n, all the observed height values are averaged logarithmically to produce a single value H_n and the standard deviation for $\log H_n$. The number of height samples depends on the topographic

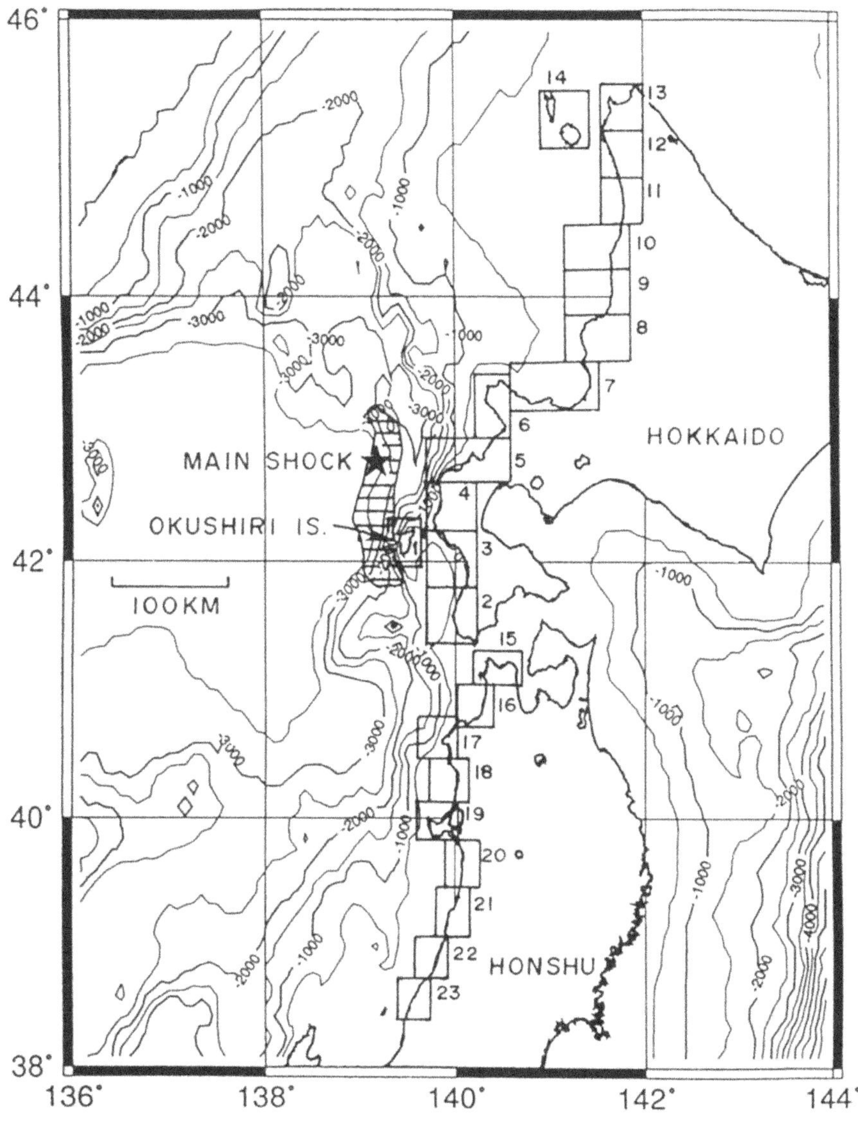

Figure 4

Map showing identification number of each segment along the Japan Sea coast for use in averaging runup data. The main shock epicenter and the aftershock area are also shown. Bathymetry is shown in meters.

features as well as the interest of the survey teams, and as such, the sampling density of data is inhomogeneous. A summary of data for each segment is given in Table 1. The total number of samples is 325.

The maximum runup, $H_{max} = 31.7$ m, was found at one point deep in a small valley on Okushiri Island (segment number 1). This height was found to be

Table 1

Summary of observed heights of the 1993 Hokkaido-Nansei-Oki tsunami ($M_t 8.1$)

Segment number n	Number of samples N	Average height H_n (m)	Log std. dev. SD_n	Maximum height H_{max} (m)	Minimum height H_{min} (m)
1	164	7.74	0.246	31.7	2.3
2	5	1.50	0.088	1.8	1.2
3	16	3.27	0.224	8.5	1.7
4	14	6.51	0.155	9.2	2.5
5	22	4.39	0.142	8.8	2.7
6	11	2.98	0.131	4.9	1.9
7	5	0.81	0.156	1.6	0.6
8	4	1.29	0.123	1.9	1.0
9	4	1.04	0.044	1.2	0.9
10	5	1.16	0.019	1.2	1.1
11	3	0.84	0.134	1.1	0.6
12	2	1.16	0.188	1.6	0.9
13	3	0.58	0.111	0.8	0.5
14	6	1.26	0.149	1.8	0.8
15	4	1.23	0.264	1.9	0.5
16	3	0.46	0.351	1.0	0.2
17	4	1.58	0.156	2.2	1.0
18	9	2.31	0.122	3.5	1.6
19	13	1.46	0.083	2.0	1.1
20	6	1.60	0.103	2.1	1.2
21	11	1.49	0.080	2.0	1.2
22	5	1.06	0.153	1.8	0.7
23	6	2.00	0.187	2.8	0.9
ALL	325	4.01	–	31.7	0.2

exceptionally large, and the average of 17 samples in a range of about 10 km around that location is reduced to 13.07 m (the maximum 31.7 m and the minimum 5.8 m). The maximum value of the local mean H_n is 7.74 m in Okushiri Island.

Regional Maximum

The maximum value of H_n, denoted by $H_{n,max}$, is 7.74 m in Okushiri Island. For a stretch of about 300 km near the epicenter (segment numbers 1–7), the average height $\langle H \rangle$ (logarithmic average of H_n) is 3.89 m with the mean standard deviation 0.163 in the logarithmic scale. The ratio of $H_{n,max}$ to $\langle H \rangle$ is 1.99. This ratio indicates that the Nansei-Oki tsunami manifested normal runup in Okushiri Island as compared with the case of average tsunamis for which $H_{n,max}/\langle H \rangle$ is about 2 (KAJIURA, 1983).

Given an earthquake magnitude M, the maximum of the local-mean heights ($H_{n,max}$) and the maximum runup (H_{max}) can be predicted by

$$\log H_{n,\max} = 0.5M - 3.30 + C \tag{3}$$

$$H_{\max} = 2H_{n,\max} \tag{4}$$

where $C = 0.0$ for a fore-arc earthquake, and $C = 0.2$ for a back-arc earthquake (ABE, 1989). The moment magnitude of the Nansei-Oki earthquake is estimated to be 7.7. Given $M = 7.7$ and $C = 0.2$, $H_{n,\max}$ is calculated to be 5.6 m, which is slightly smaller than the observed height 7.7 m. The surface-wave magnitude M_s is 7.6. Giving $M = 7.6$, we obtain $H_{n,\max} = 5.0$ m.

Since the local magnitude of the Japan Meteorological Agency, $M(JMA)$, is determined soon after a strong earthquake, $M(JMA)$ is important in Japan for tsunami warning purposes where rapid evaluation of tsunami potential is required. Although $M(JMA)$ was regressed against M_s, it is known that there is a systematic deviation between $M(JMA)$ and M_s, thus we must convert it to utilize (3). The following relation is suggested by ABE (1989)

$$M_s = (M(JMA) - 1.44)/0.79. \tag{5}$$

Giving $M(JMA) = 7.8$, we obtain 8.4 m, which is slightly larger than the observation 7.7 m. Observation is made that the height prediction from the available magnitudes, 5.0–8.4 m, brackets the observed height. On the other hand, the observed maximum 31.7 m is higher than the estimated $H_{\max}$ through (4). However, the average height around the singular point, 13 m, can be reasonably predicted from the estimated $H_{\max}$. It is emphasized that a quick determination of earthquake magnitude as well as hypocenter leads to rapid evaluation of the tsunami potential.

Eleven large events in the vicinity of Japan in this century sustained a recorded maximum runup height larger than 4 m. Table 2 lists the values of various

Table 2

Maximum runup heights of major tsunamis with $H_{\max} > 4\,m$

Year	Mo	Dy	Region	M_t	M_w	M_s	$M(JMA)$	$H_{\max}$ (m)	$H_{n,\max}$ (m)	H_r (m)
1923	9	1	Kanto	8.0	7.9	8.2	7.9	12.0	3.9	4.5
1933	3	3	Sanriku-Oki	8.3	8.4	8.5	8.1	28.7	8.0	7.9
1944	12	7	Tonankai	8.1	8.1	8.0	7.9	10.0	5.0	5.6
1946	12	21	Nankaido	8.1	8.1	8.2	8.0	6.5	4.4	5.6
1952	3	4	Tokachi-Oki	8.2	8.1	8.3	8.2	6.5	4.0	5.6
1964	6	16	Niigata	7.9	7.6	7.5	7.5	6.4	3.5	5.0
1968	4	1	Miyazaki-Oki	7.7	7.4	7.6	7.5	4.6	1.9	2.5
1968	5	16	Tokachi-Oki	8.2	8.2	8.1	7.9	6.8	4.5	6.3
1973	6	17	Nemuro-Oki	8.1	7.8	7.7	7.4	4.5	3.0	4.0
1983	5	26	Akita-Oki	8.1	7.9	7.7	7.7	13.8	7.5	7.1
1993	7	12	Hokkaido	8.1	7.7	7.6	7.8	31.7	7.7	5.6

$H_{\max}$ = Observed maximum height. $H_{n,\max}$ = Observed maximum of local-mean height. H_r = Maximum local-mean height estimated from moment magnitude.

magnitudes, observed maximum heights ($H_{\max}$), observed maximum of local-mean heights ($H_{n,\max}$), and the value $H_{n,\max}$ estimated from M_w. The 1964 Niigata, the 1983 Akita-Oki and the 1993 Nansei-Oki events took place in the Sea of Japan while the others occurred in the Pacific Ocean. The average ratio K of the estimated height to the observed $H_{n,\max}$ over all 11 tsunamis is 1.16 if the height is estimated from M_w. The average 1.16 indicates an overestimate by 16 percent on the average. The ratio K for M_s and $M(JMA)$ is 1.18 and 1.35, respectively. The average ratio to the observed maximum, $H_{\max}$, over all the tsunamis is 1.10 for M_w, 1.13 for M_s, and 1.29 for $M(JMA)$. For practical purposes, the overestimate is more favorable than an underestimate, and the amount of the present overestimate may be tolerable.

Distribution of Runup Heights

KAJIURA (1983) noticed that the mean value $\langle SD_n \rangle$ of SD_n (standard deviation of $\log H_n$) for four tsunamis is close to 0.16; they are the 1896 Sanriku, 1933 Sanriku, 1960 Chile and 1968 Tokachi-Oki tsunamis. The mean value given in Table 1 results as 0.148. This indicates that the height ratio of the single observation H to the local mean H_n lies mostly between $0.5 < H/H_n < 2.0$ (less than $2\langle SD_n \rangle$) in each segment.

The decay of tsunami height with distance is examined by plotting the overall distribution of average heights with respect to their locations. In Figure 5, solid circles represent the observed local-mean height H_n in each segment interval, and error bars denote the standard deviation. The segments shown in the figure are ordered from south to north. The maximum value of the local-mean height is 7.74 m in Okushiri Island (segment number 1), as mentioned before. The relationship between H_n and distance is roughly, inversely linear. This observation is expected from the formula of height prediction proposed by ABE (1989)

$$\log H_n = M - \log R - 5.55 + C \tag{6}$$

where H_n is tsunami height in meters, M is earthquake magnitude, R is distance in km, $C = 0.0$ for a fore-arc earthquake, and $C = 0.2$ for a back-arc earthquake. When using $M(JMA)$, we must convert $M(JMA)$ to M by (5). In Figure 5, cross symbols represent the estimated heights for $J(JMA) = 7.8$. Since the Nansei-Oki earthquake took place in the Sea of Japan, the constant C is taken to be 0.2. It is discovered that the predicted values simulate well the observed heights and the distance-decay pattern over the wide stretch of the coast some 1000 km, particularly near the center of tsunami attack (segment numbers 1, 3, 4–6).

When comparing the observed and predicted heights, a relatively large negative deviation is evident in the northern branch. The location of this area (segment numbers 7–9) is along the central Hokkaido coast facing a wide continental shelf and is more or less in the shadow zone of large Shakotan Peninsula. In the southern

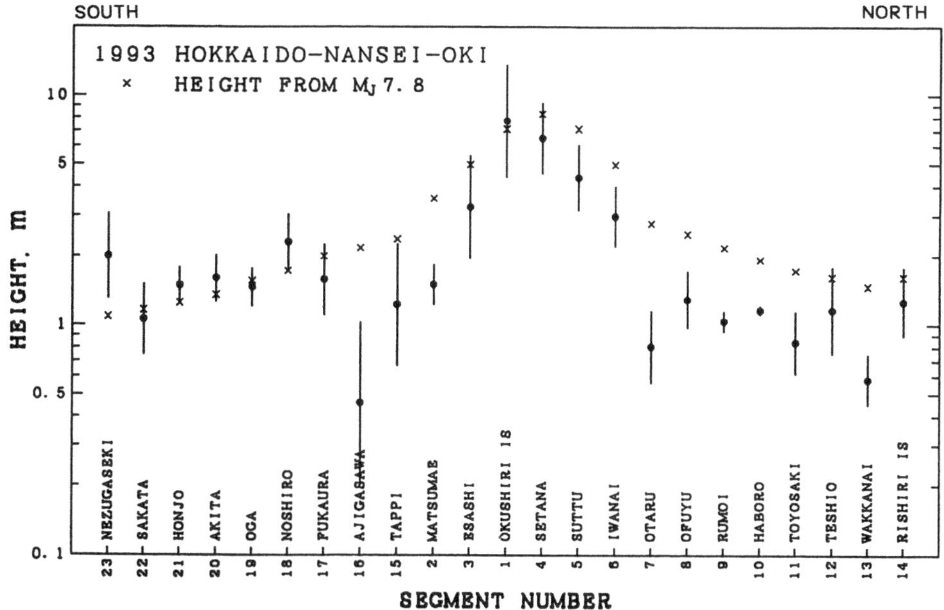

Figure 5

Distribution of observed and estimated heights for the Nansei-Oki tsunami. Solid circles represent local-mean height H_n in segment intervals of about 40 km (see Fig. 4). Error bars represent standard deviation. Crosses represent heights estimated from $M(JMA)$.

branch, a notable negative deviation is seen along the southern Hokkaido and the northern Honshu coasts (segment numbers 2, 15, 16). This area may be strongly affected by the shallow topography of Okushiri Spur which extends from Okushiri Island to the south.

The overall fitness can be assessed in terms of the average ratio K of estimated heights to observed heights (AIDA, 1978; ABE, 1989). The index K for the entire area is 1.42 and its logarithmic standard deviation k is 0.193 if utilizing $M(JMA)$. This indicates that the estimated heights are on average 42 percent larger than the observed heights. The values of K for various magnitudes are summarized in Table

Table 3

Comparison of observed and estimated local-mean heights

Year	Region	M_u	M_v	M (JMA)	K for M_u	K for M_v	K M(JMA)	k
1993	Nansei-Oki	7.7	7.6	7.8	0.63	0.50	1.42	1.56
1983	Akita-Oki	7.9	7.7	7.7	1.14	0.72	1.21	1.28

K = Average ratio of estimated heights to observed local-mean heights (H_n), showing how many times the estimated heights are on the average as large as the observed heights. k = Standard deviation in terms of linear scale.

3 and compared with the case of the 1983 Akita-Oki tsunami (ABE, 1989). The prediction from M_w and M_s is more or less underestimated. The overall fit to provide $K = 1.0$ is calculated to be $M = 7.9$. This value is consistent with $M_t = 8.1$ as mentioned, but slightly larger, by 0.2–0.3, than the values of M_w and M_s. This difference is partly related to the underestimate of the prediction.

4. Conclusion

The 1993 Nansei-Oki earthquake in the Sea of Japan set off a damaging tsunami. The tsunami magnitude M_t is determined to be 8.1 from the maximum amplitudes of the tsunami recorded on tide gauges. This value is larger by 0.4 than the moment magnitude 7.7, indicating that the tsunami height of the Nansei-Oki event is on average 2.5 times as large as that of Pacific events with a comparable M_w. The empirical relation between the earthquake magnitude and the mean runup height predicts the maximum of the local-mean height and the distance-decay pattern of the local-mean height. The reasonable agreement between prediction and observation suggests again that the relation proposed by ABE (1989) provides a practical and robust method for near-field tsunami warning purposes.

Acknowledgments

Assistance by Miss Ikuko Kato throughout the computational work is gratefully acknowledged.

REFERENCES

ABE, K. (1979), *Size of Great Earthquakes of 1837–1974 Inferred from Tsunami Data*, J. Geophys. Res. *84*, 1561–1568.

ABE, K. (1981), *Physical Size of Tsunamigenic Earthquakes of the Northwestern Pacific*, Phys. Earth Planet. Inter. *27*, 194–205.

ABE, K. (1985), *Quantification of Major Earthquake Tsunamis of the Japan Sea*, Phys. Earth Planet. Inter. *38*, 214–223.

ABE, K. (1988), *Tsunami Magnitude and the Quantification of Earthquake Tsunamis around Japan*, Bull. Earthquake Res. Inst. Tokyo Univ. *63*, 289–303 (in Japanese with English abstract)

ABE, K. (1989), *Estimate of Tsunami Heights from Magnitudes of Earthquake and Tsunami*, Bull. Earthquake Res. Inst. Tokyo Univ. *64*, 51–69 (in Japanese with English abstract).

ABE, K. (1993), *Estimate of Tsunami Heights from Earthquake Magnitudes*, Proceedings of the IUGG/ IOC International Tsunami Symposium TSUNAMI '93, Wakayama, Japan, Aug. 23–27, pp. 495–507.

AIDA, I. (1978), *Reliability of a Tsunami Source Model Derived from Fault Parameters*, J. Phys. Earth *26*, 57–73.

HOKKAIDO TSUNAMI SURVEY TEAM (1993), *Tsunami Devastates Japanese Coastal Region*, EOS Trans. AGU *74*, 417.

KAJIURA, K., *Some statistics related to observed tsunami heights along the coast of Japan.* In *Tsunamis — Their Science and Engineering* (eds. Iida, K., and Iwasaki, T.) (Terra Sci. Publ., Tokyo 1983) pp. 131–145.

NAKANISHI, I., KODAIRA, S., KOBAYASHI, R., KASAHARA, M., and KIKUCHI, M. (1993), *The 1993 Japan Sea Earthquake*, EOS Trans. AGU *74*, p. 34.

SHUTO, N. (1994), *Tsunami Runup Heights for the Hokkaido-Nansei-Oki Earthquake*, Tsunami Engineering Tech. Rep., Disaster Control Research Center, Tohoku Univ., 120 pp.

(Received July 31, 1994, revised September 27, 1994, accepted September 28, 1994)

PAGEOPH, Vol. 144, Nos. 3/4 (1995)

0033-4553/95/040747-21$1.50 + 0.20/0

Source Models for the 1993 Hokkaido Nansei-Oki Earthquake Tsunami

TOMOYUKI TAKAHASHI,[1] TAKEYUKI TAKAHASHI,[1] NOBUO SHUTO,[1] FUMIHIKO IMAMURA,[2] and MODESTO ORTIZ[3]

Abstract—A source model for the 1993 Hokkaido Nansei-Oki tsunami must satisfy certain conditions. Such conditions are presented in this paper, and two methods are used to determine the best source model for this event. A trial-and-error method selects DCRC-17a as the best among 24 different models. This model has three fault planes dipping westward. To reproduce well the tide gauge records at two locations, an inversion analysis is used to modify the dislocation of DCRC-17a.

Key words: Tsunami numerical simulation, inversion analysis, fault model, Japan Sea.

Introduction

An earthquake of $M_w = 7.8$ occurred off the southwest coast of Hokkaido, Japan at 22:17 on July 12, 1993. A targe tsunami was generated and impacted the coastlines surrounding the Japan Sea. The highest runup of 31.7 m was found near Monai on the island of Okushiri (HOKKAIDO TSUNAMI SURVEY TEAM, 1993). Severe damages from the tsunami were reported from Japan as well as Russia and Korea.

The aim of the present paper is to determine the initial profile of the tsunami by taking seismic, geodetic and tsunami data into consideration. After a brief review of the position of the fault, several fault models derived from the seismic data are summarized. Once the conditions to be satisfied by the fault model are clearly stated, numerical simulations of the tsunami are carried out.

Two methods are selected to estimate the best model of the initial tsunami profile. One is a trial-and-error method which uses forward numerical simulations

---

[1] Disaster Control Research Center, Faculty of Engineering, Tohoku University, Sendai 980-77, Japan.

[2] School of Civil Engineering, Water Resources Engineering, Asian Institute of Technology, G.P.O. Box 2574, Bangkok 10501, Thailand.

[3] Departamento de Oceanografia Fisica, Centro de Investigacion Cientifica y de Educacion Superior de Ensenada (CICESE), Apartado Postal 2732, Ensenada, Baja California 22800, Mexico.

in an iterative scheme to match computed and measured runup distributions. A numerical simulation of the tsunami is performed, assuming an initial profile. A computed runup distribution is compared to a measured runup. The initial profile for the next numerical simulation is then modified, based upon this comparison. This procedure is repeated until there is reasonable agreement between the computed and measured runup distributions. In this iterative scheme, the dislocation is assumed to be uniform throughout the fault plane.

The second method is an inversion method (SATAKE, 1989), which is applied to study the heterogeneity of the fault by using tide gauge records. The dislocation in a fault plane is assumed not to be unique but spatially variable. A fault is divided into segments which have common fault parameters with the exception of the dislocation. The dislocation of each segment is determined by inverting tsunami records at two tide gauge stations on the coasts of Hokkaido, Esashi and Iwanai.

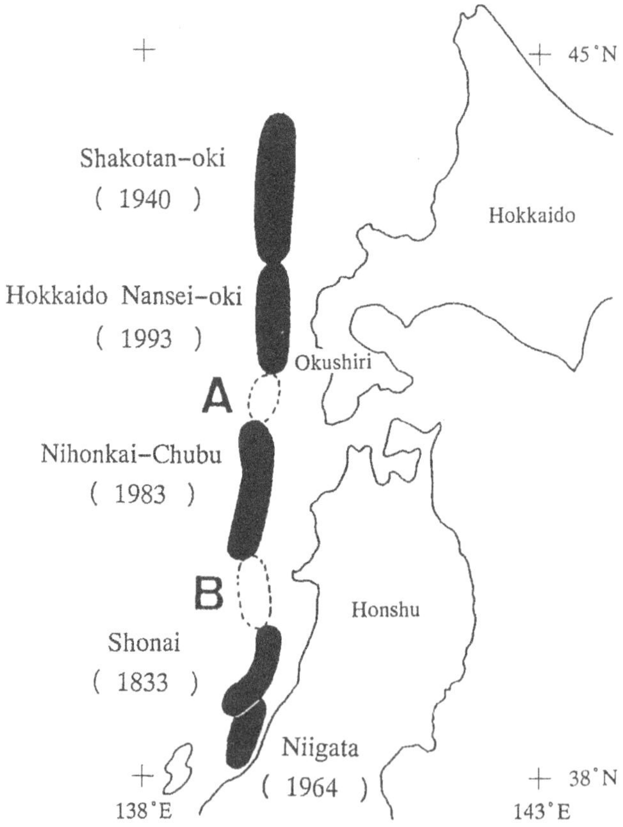

Figure 1
Tsunami sources in the Japan Sea during the last 100 years from OTAKE (1993).

Position of the Tsunami Source

Residents in coastal regions of the Japan Sea were surprised ten years ago when the 1983 Nihonkai-Chubu earthquake generated a large tsunami. Many believed that tsunamis could not be generated in the Japan Sea. After a careful revision, experts acknowledged that a large tsunami could occur approximately once every 100 years. However, only ten years later in 1993, yet another large tsunami was generated in the Japan Sea.

Figure 1 (OHTAKE, 1993) shows source locations of tsunamis which have occurred over the past 100 yeras. They include the 1940 Shakotan-Oki tsunami at the northern end, the 1993 Hokkaido Nansei-Oki tsunami, the 1983 Nihonkai-Chubu tsunami, the 1833 Shonai tsunami and the 1964 Niigata tsunami. The two blank areas denoted by dotted lines, A and B, remain between the source areas of the above events. It is believed that all of these tsunami sources exist along the boundary of the Eurasian and North American plates, which is the back-arc subduction zone of the Japan Trench in the Pacific Ocean. This boundary is now recognized to be one of the most active seismic areas in the world.

Fault Model Proposed from the Seismic Analysis

Table 1 summarizes fault models and their parameters (the rigidity constant of 3.0×10^{11} g/cm/s^2 is assumed), proposed by different researchers and institutions. Differences exist as as a result of the seismic data used in the analysis and in the procedure of the analysis. For example, the Harvard University CMT solution indicates a single fault dipping eastward with a small angle of inclination, while the USGS model presents a fault dipping westward with a small angle of inclination. As was pointed out by NAKANISHI and KOBAYASHI (1993) and TANIOKA *et al.*

Table 1

Fault parameters of the proposed models

Fault model	M_0 ($\times 10^{27}$)	Depth	Strike (°)	Dip (°)	Slip (°)	Length	Width	Dislocation
Harvard	5.60	15 km	1	24	84	100 km	40 km	4.70 m
USGS	0.93	18 km	1	60	67	100 km	40 km	0.80 m
ERI	4.20	10 km	9	35	97	100 km	40 km	3.50 m
Hok. Univ.	6.30	30 km	12	49	102	100 km	40 km	5.30 m
JMA	6.30	37 km	3	41	72	100 km	40 km	5.30 m
Kikuchi								
North	1.92	8 km	198	34	80	50 km	40 km	3.20 m
Central	0.54	8 km	187	65	111	30 km	40 km	1.50 m
South	2.30	8 km	192	79	102	30 km	40 km	6.40 m
Total	4.76							

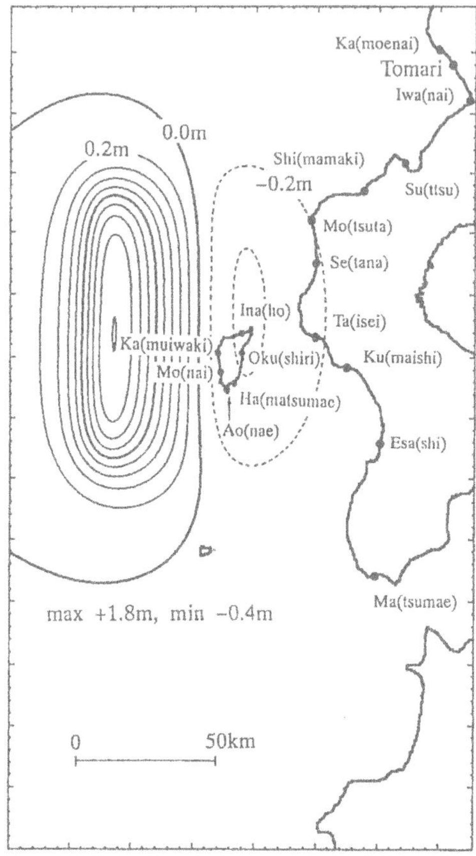

Figure 2
Tsunami source model: Harvard University CMT solution.

(1993), the Harvard CMT solution, obtained by using seismic surface waves, gives an integrated view of the fault motion, while the body wave analysis used by the USGS is potentially sensitive to the initial rupture process. This difference is reflected in the two models. A multiple fault solution obtained by KIKUCHI (1993) supports the above-mentioned discussion. The northern fault in Kikuchi's solution has the same mechanism as the USGS solution and his southern fault has the same mechanism as the Harvard CMT solution.

Figures 2 through 7 show the initial tsunami profiles computed using the MANSINHA-SMYLIE method (1971) for different fault models. In these figures, solid lines show the upheaval of the water surface and dotted lines illustrate the subsidence.

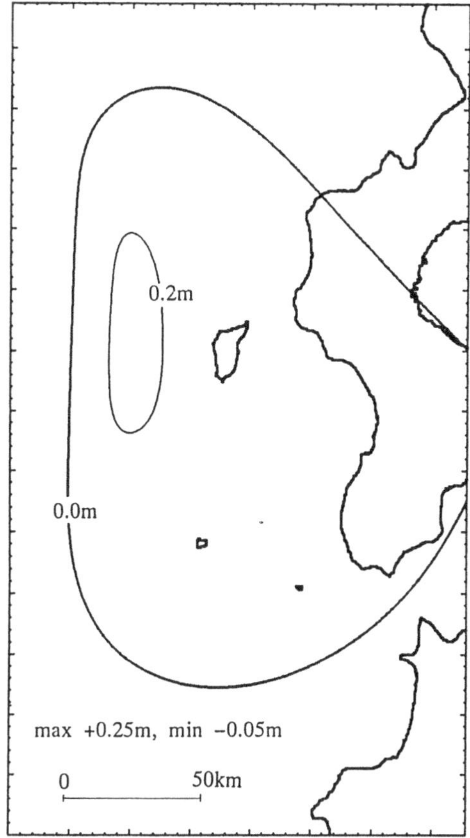

Figure 3
Tsunami source model: USGS.

Conditions to be Satisfied

A model of the 1993 Hokkaido Nansei-Oki earthquake tsunami should satisfy the following conditions. For details, refer to SHUTO and MATSUTOMI (1994).

(1) The fault parameters of the model, its location and mechanism, should be consistent with the horizontal and vertical distribution of aftershocks as shown in Figure 8 (RESEARCH GROUP for AFTERSHOCKS of the July 12, 1993 Hokkaido Nansei-Oki earthquake, 1993).

(2) The vertical displacment of the ground, calculated using the fault parameters above, should be consistent with the measured subsidence on Okushiri Island shown in Figure 9 (KUMAKI *et al.*, 1993).

(3) The simulated tsunami should explain the witnessed arrival time and the time shown by drowned watches as summarized in Table 2 as well as tide gauge records.

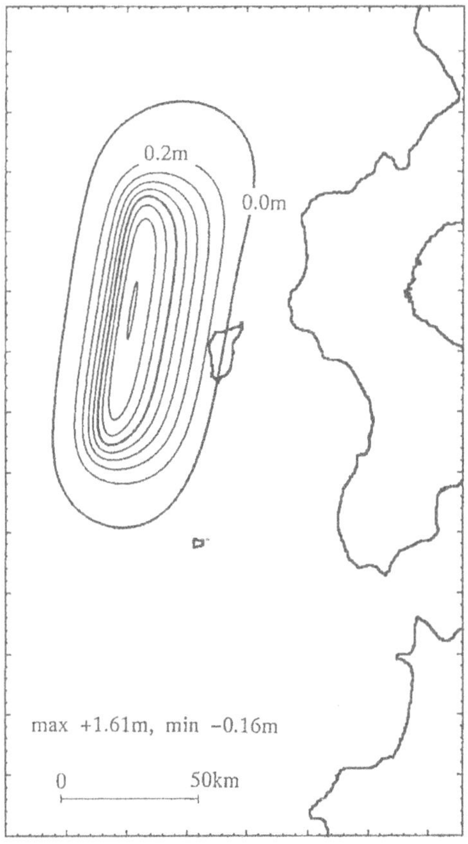

Figure 4
Tsunami source model: ERI, University of Tokyo.

(4) The simulated tsunami should explain the measured runup distribution (e.g., Hokkaido TSUNAMI SURVEY TEAM, 1993), in particular, the high runup height of 20 meters at Hamatsumae on the east coast of Okushiri Island which is sheltered by the Aonae Point from the tsunami source. In the neighborhood of Hamatsumae there is no local topography which cannot be simulated by using spatial grids of 50 meters.

(5) The elevation time histories (at least their beginning parts) obtained at the Esashi and Iwanai tide gauge stations on the coast of Hokkaido should be reproduced in numerical simulations. In addition, the simulation should explain the current velocity records obtained at Tomari in the neighborhood of Iwanai.

(6) The highest runup of 31.7 m, measured near Monai on Okushiri Island, needs no reproducing, because it occurred as a result of local topography. In order to simulate this runup, spatial grids smaller than 5 m are required in the numerical simulation. Therefore, this local value is excluded in the comparison.

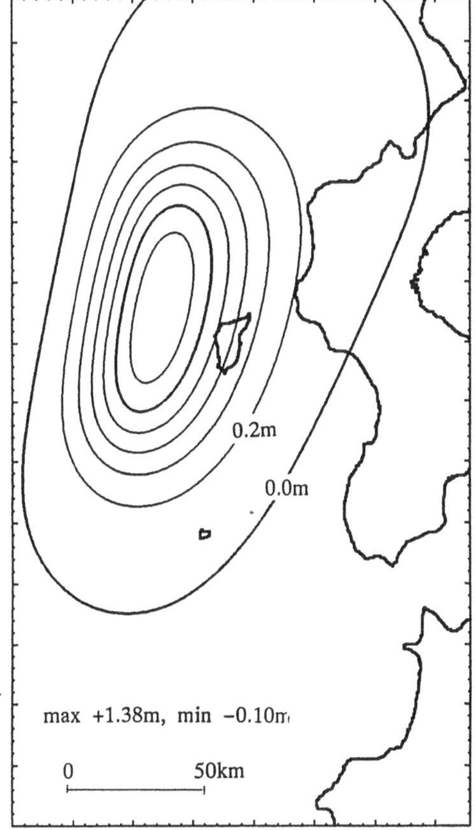

Figure 5
Tsunami source model: Hokkaido University.

Trial-and-error Method

The First Stage: Comparison and Modification

The linear long wave theory is used as the governing equation in the numerical simulation. The domain is represented using a coarse grid 450 meters wide and with a perfect reflection condition at the land boundary and a transmissive condition at the ocean boundary. The area covered extends from 138.30° E to 140.33° E and from 40.31° N to 43.18° N.

For a selected fault model, the initial profile, the runup distribution and the arrival time are computed and are compared qualitatively with the measured values. The fault model is then modified to adjust any disagreement. When an overall agreement is obtained, the computed and measured values are compared quantita-

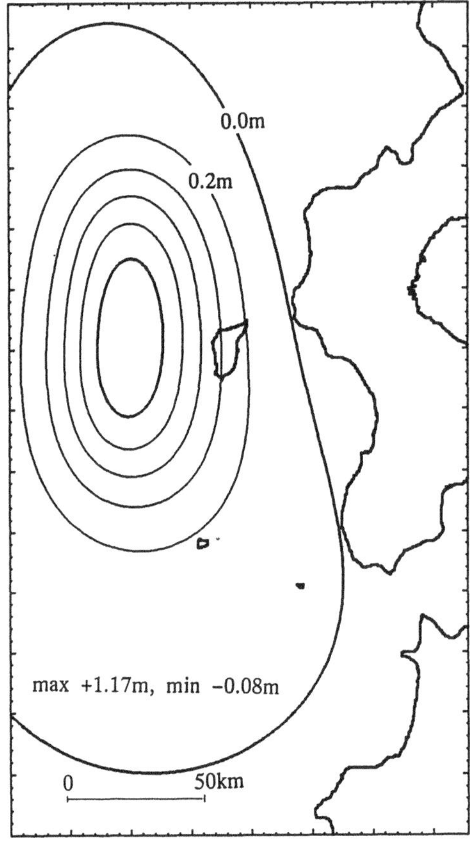

Figure 6
Tsunami source model: Japan Meteorological Agency.

tively, for example, in terms of the Aida criteria K and k (AIDA, 1978), or by examining the agreement at special points such as Hamatsumae. Further modifications are then made as needed.

The simulation, comparison and modification start with the Harvard CMT solution. Figure 2 shows the initial tsunami profile (= vertical ground displacement) computed for this fault model. Although Okushiri is contained within the area of subsidence, the subsidence pattern differs from the measured one shown in Figure 9. The computed maximum subsidence of 0.4 m is smaller than the measured value of 0.8 m. In Figure 10, the computed runup heights for the Harvard CMT solution (dotted lines) are compared with the measured heights (crosses) along the coast of Hokkaido in the upper figure and along the coast of Okushiri in the lower figure. For the extended name of the locations, refer to Figure 2. In general, the Harvard CMT solution gives smaller values.

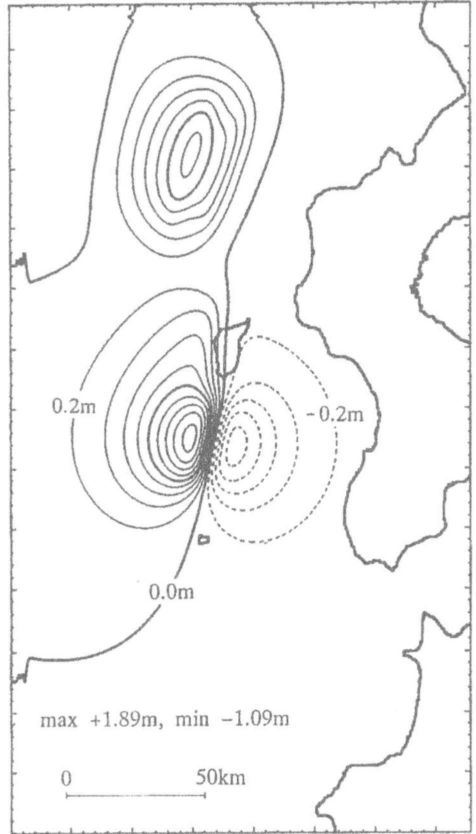

Figure 7
Tsunami source model: Kikuchi.

Initial Profiles of DCRC-17a

After repeating the procedure described above for 24 models, the authors arrived at the DCRC-17a model, which provides satisfactory results for observations along the coast of Okushiri. However, the authors acknowledge that further modification may be required to satisfy the tsunami observations along the coast of Hokkaido.

The DCRC-17a model consists of three fault planes, as summarized in Table 3. The southern and central faults have the same dip, slip and width. Conceptually, these two faults can be thought of as being one fault with different dislocations. We assume that the three faults dip westward with large angles of inclination, which is consistent with the vertical distribution of aftershocks. The total earthquake moment of 6.62×10^{27} dyne-cm is 30% larger than that of the Harvard CMT

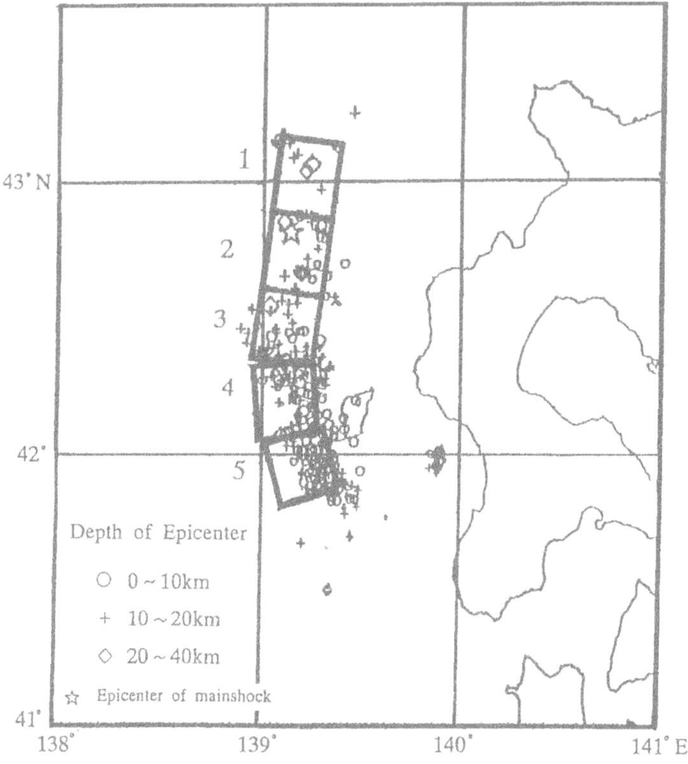

Figure 8
Distribution of aftershocks and divided segments for inversion analysis.

solution. Another interpretation proposed by IMAMURA *et al.* (1994) indicates the possibility that the north fault dips westward and the south fault dips eastward.

The computed initial tsunami profile is shown in Figure 11. The subsidence of 0.8 m at the southern end of Okushiri and the pattern of subsidence are very close to the measured subsidence shown in Figure 9. The position and strike direction of the southern fault were adjusted to satisfy the arrival time of the tsunami at Aonae, which was impacted 4 to 5 minutes after the earthquake. The strike direction is also necessary to explain the tsunami observations in Hamatsumae, an area sheltered by Aonae Point. The Okushiri Spur, a wide shallow region in the sea south of Okushiri Island, refracts the tsunami and concentrates it to Hamatsumae.

The introduction of the central fault is required to reproduce the tsunami heights measured along the coastal cliff between Kamuiwaki and Inaho, on the northwestern coast of Okushiri Island. If the central and southern faults are assumed to form one homogeneous fault, the disagreement between the measured and computed values becomes large along this portion of the cliff coast.

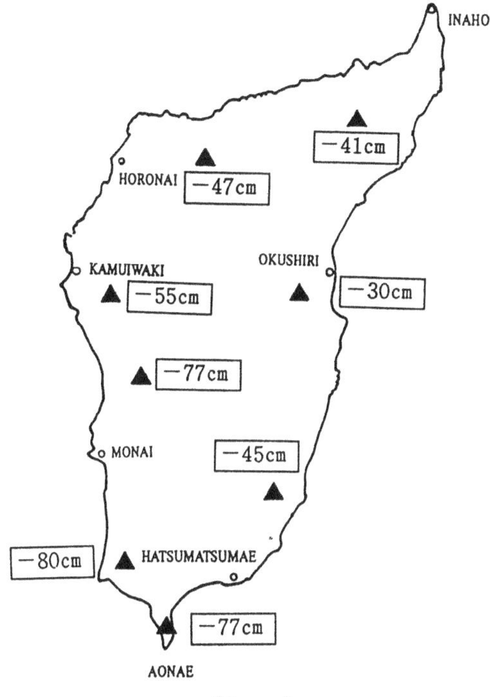

Figure 9
Subsidence on Okushiri Island.

The Second Stage: The Tsunami Computed with DCRC-17a

To see more agreement between computed results and observations, another numerical simulation was carried out for the DCRC-17a model. In this simulation, the domain is represented using fine spatial grids of 50 meters (near and on land), 150 meters (near Okushiri) and 450 meters (in deep sea). The shallow-water theory is applied as the governing equation (GOTO and OGAWA, 1982) for the domains of grids 150 m wide with a reflection condition and 50 m wide with an inundation condition. Figures 10 and 12 show the runup heights (solid lines) computed with spatial grids of 450 m and 50 m.

On Okushiri, the agreement is $K = 1.05$ and $k = 1.47$ in terms of the Aida criteria (if the highest runup of 31.7 m near Monai is excluded by the reason explained above). These values for the Aida criteria indicate good agreement between the observations and computed results.

On Hokkaido, the agreement is poorer than that for Okushiri. Along the concave coast at the southern part from Matsumae to Kumaishi, the computed runups are larger than the measured runups. Along the coast from Taisei to Shimamaki which projects toward the tsunami source, the computed runups are of

Table 2

Arrival times of the tsunami

Okushiri Island

Place	Observed	Harvard	DCRC-17a
South Aonae	$4 \sim 5$	$8(3 \sim 4)$	$6(1 \sim 2)$
North Aonae	$20 \sim 21$	$12(-9 \sim -8)$	$10(-11 \sim -10)$
Hamatsumae	5	12(7)	10(5)
Monai	4	7(3)	6(2)
Hoyaishikawa	6	6(0)	5(-1)

Unit; min

( ); computed minus observed

Hokkaido Island

Place	Observed	Harvard	DCRC-17a
Kamui Cap.	$5 \sim 6$	$23(17 \sim 18)$	$18(21 \sim 13)$
Numamae Cap.	$25 \sim 35$	$21(-14 \sim -4)$	$16(-19 \sim -9)$
Kawashiro Cap.	10	20(10)	16(6)
Sannai	10	21(11)	16(6)
Ryuujin Cap.	$5 \sim 10$	$23(13 \sim 18)$	$20(10 \sim 15)$
Iwanai	15[28]	30(15, 2)	23(8, -5)
Enoshima	5	15(10)	12(7)
Sukki	3	9(6)	7(4)
Shimauta	5	10(5)	8(3)
Setana	5	13(8)	11(6)
Hutoro	5	12(7)	10(5)
Ota	5	13(8)	12(7)
Miyano	5	19(14)	18(13)
Hirahama	5	19(14)	17(12)
Esashi	[21]	20(-1)	18(-3)

Unit; min

( ); Computed minus observed

[ ]; Tide record

Table 3

Fault parameters of DCRC-17a

Fault	$M_0(\times 10^{27})$	Depth	Strike(°)	Dip(°)	Slip(°)	Length	Width	Dislocation
North	3.85	10 km	188	35	80	90 km	25 km	5.71 m
Central	0.56	5 km	175	60	105	30 km	25 km	2.50 m
South	2.21	5 km	163	60	105	24.5 km	25 km	12.00 m
Total	6.62							

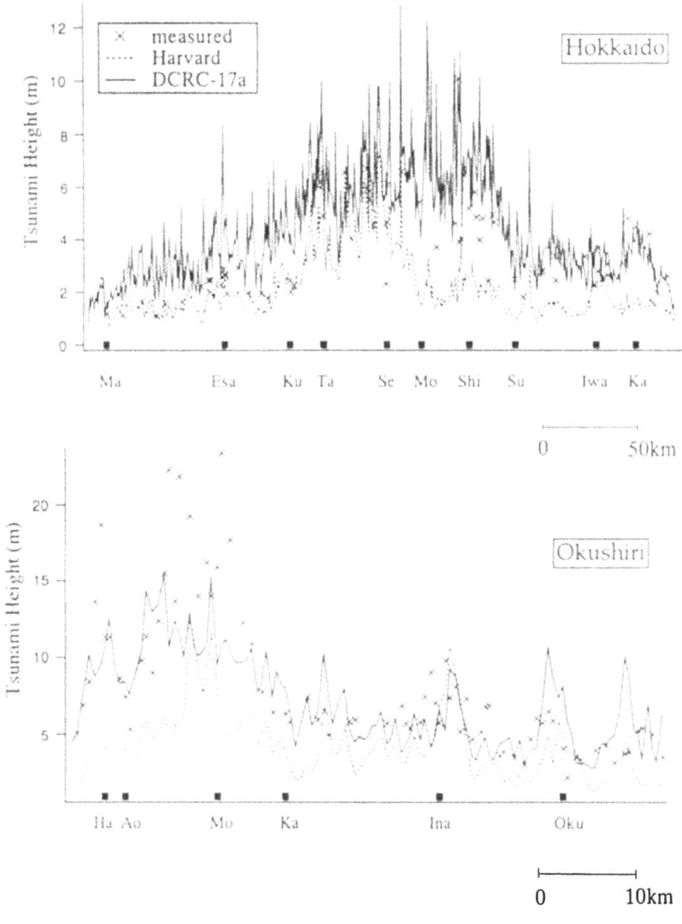

Figure 10
Comparison of the measured and computed runups.

the same magnitude as the measured ones. However, the computed arrival is late compared with that witnessed, as shown in Table 2. Along the coast north of Suttsu, there is reasonable agreement.

Figures 13 and 14 compare the computed waveforms, using the DCRC-17a model (solid lines) and the tide gauge records (dotted lines). The Harvard CMT solution results (broken lines) are included for reference. The tide gauge records are adjusted to account for the effect of hydraulic filtering (SATAKE *et al.*, 1988) and the water depth, using the Green formula (DEAN and DALRYMPLE, 1984).

At Esashi (Figure 13), the initial small fall of water level is reproduced by the two models. The Harvard CMT solution manifests better agreement with the tide record. The recording stopped about 50 minutes after the earthquake because the

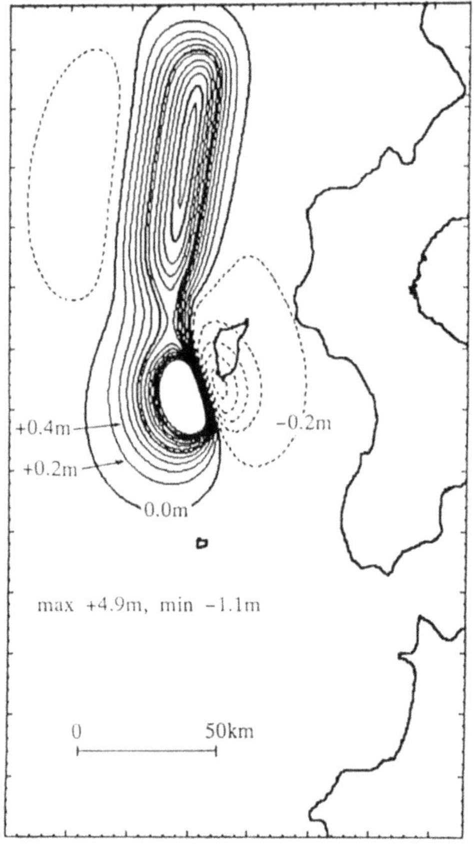

Figure 11
Initial tsunami profile of DCRC-17a.

writing pen broke the paper when the surface of flooded water lowered. At Iwanai (Figure 14), the Harvard CMT solution provides better results, again. There is a 5-minute difference between the computed waveform for DCRC-17a and the tide record.

To our embarrassment, the current velocity recorded at Tomari, in the neighborhood of Iwanai, indicates a faster arrival time (= a first crest) of the tsunami, 20 minutes after the earthquake, as shown by the dotted lines in Figure 15. This arrival time coincides well with the computed waveform for DCRC-17a, but with nonnegligible differences in the shape and magnitude of velocity time histories. The authors consider only early velocity records of the tsunami for the present comparison, since the current velocity is affected more by the local topography than the water level (NAGANO *et al.*, 1991). For a better comparison, more detailed topographies are needed near the current velocity measurement station.

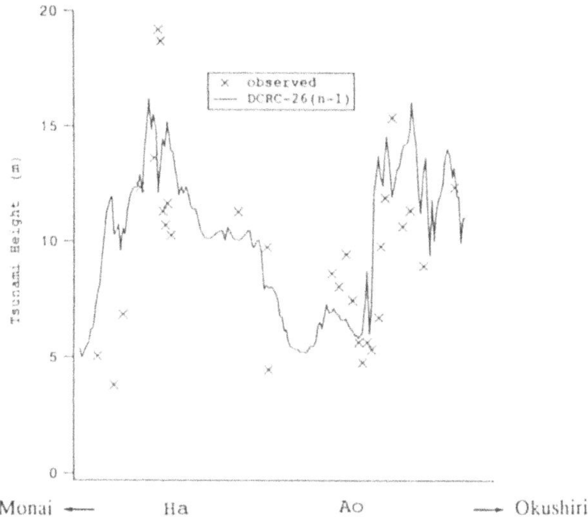

Figure 12
Comparison of the measured and computed runups on the south part of Okushiri Island.

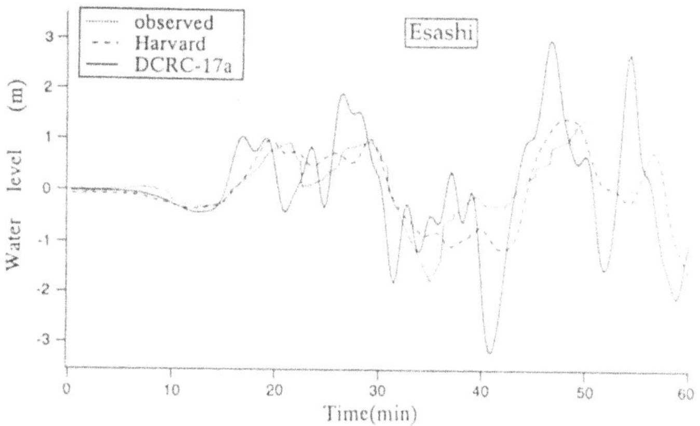

Figure 13
Tide record at Esashi and the computed waveform using DCRC-17a.

Inversion Analysis

Since a tsunami waveform is dependent on the fault parameters, the fault parameters can be estimated, in turn, from the observed tsunami waveform. AIDA (1972) developed the inversion method to estimate the initial tsunami profile from tide gauge records. In the 1980s, seismologists discovered that the fault motion of large earthquakes was not uniform but had heterogeneities described as asperities

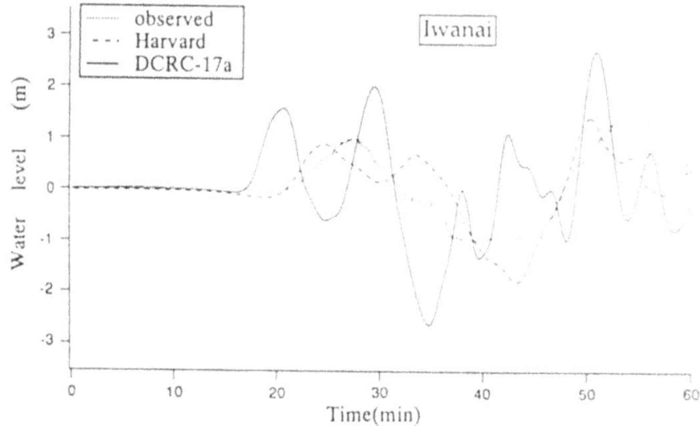

Figure 14
Tide record at Iwanai and the computed waveform using DCRC-17a.

or barriers. SATAKE (1993) introduced the joint inversion method, in which the dislocations of subfaults are determined by inversion of geodetic and tsunami data. The present authors use SATAKE's method with no geodetic data (1989), by allowing the unknowns to be only the dislocations in segments of the faults.

Application to the Present Tsunami

With the exception of the dislocation, the mechanism and size of the faults are assumed the same as those of the DCRC-17a model. The faults are divided into 5 segments as shown in Figure 8. Segments 1 to 3 have the same fault parameters as the north fault of DCRC-17a, segment 4 has those of the central fault and segment 5 has those of the south fault. The dislocations are left unknown. A unit dislocation is assumed for each individual fault segment. The vertical displacement of the ground is then computed with the Mansinha and Smylie method, the numerical simulation of the tsunami is carried out and waveforms are computed at tide gauge stations. The time history at the tide gauge station is the Green function in the inversion method.

For inversion, the first 1300 second recorded at Esashi and the first 1700 second recorded at Iwanai (both of which are denoted by vertical dotted lines in Figures 16 and 17) were used, because these parts are considered less contaminated by the local topography effect. The computed waveforms agree well with the records in the initial part of the time histories as shown in the same figures.

Table 4 summarizes the dislocations obtained through inversion as well as the other fault parameters. The most remarkable difference from the original DCRC-17a is found in the dislocation of segment 1 at the northernmost end of the entire fault. The dislocation estimated by the inversion analysis is zero, to account for the

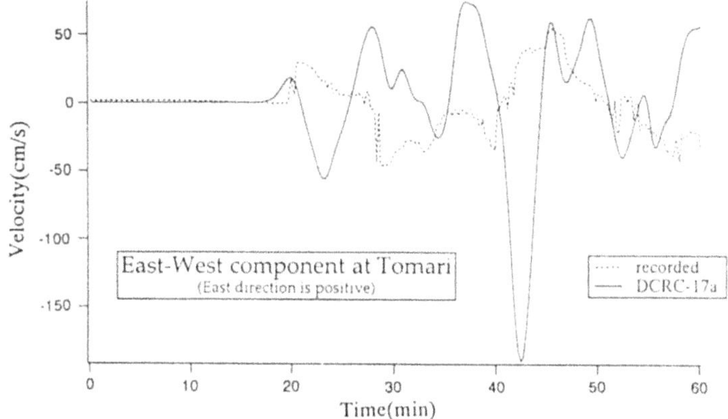

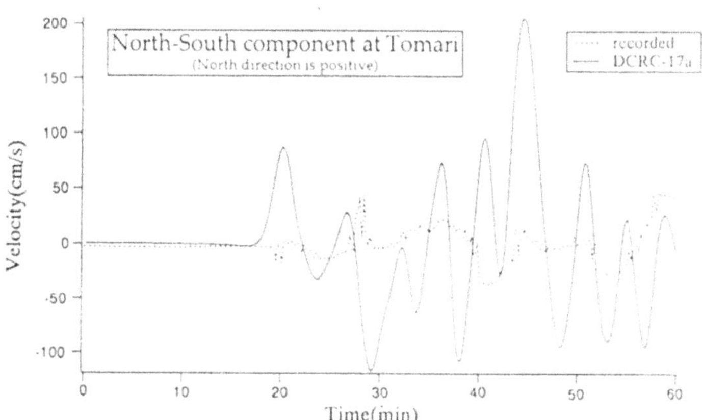

Figure 15

Current velocity measured at Tomari and the computed waveform using DCRC-17a. (a) East-West component of velocity. (b) North-South component of velocity.

Table 4

Fault parameters of the inversion solution

Segment	$M_0(\times 10^{27})$	Depth	Strike(°)	Dip(°)	Slip(°)	Length	Width	Dislocation
1	0.00	10 km	188	35	80	30 km	25 km	0.00 m
2	0.06	10 km	188	35	80	30 km	25 km	0.10 m
3	0.32	10 km	188	35	80	30 km	25 km	1.70 m
4	2.31	5 km	175	60	105	30 km	25 km	11.98 m
5	0.23	5 km	163	60	105	24.5 km	25 km	1.21 m
Total	2.92							

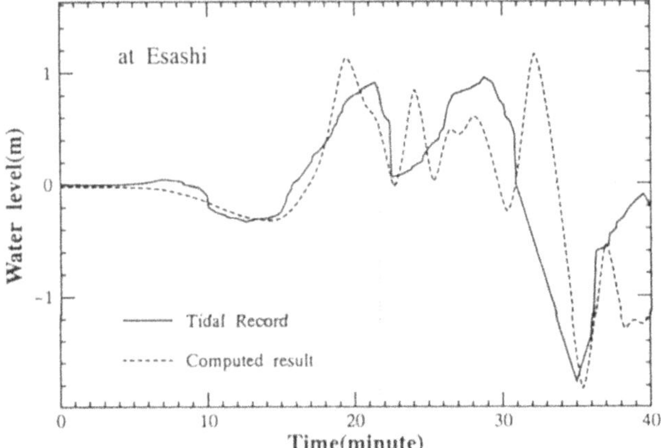

Figure 16

Tide record at Esashi and the computed waveform using the inversion solution.

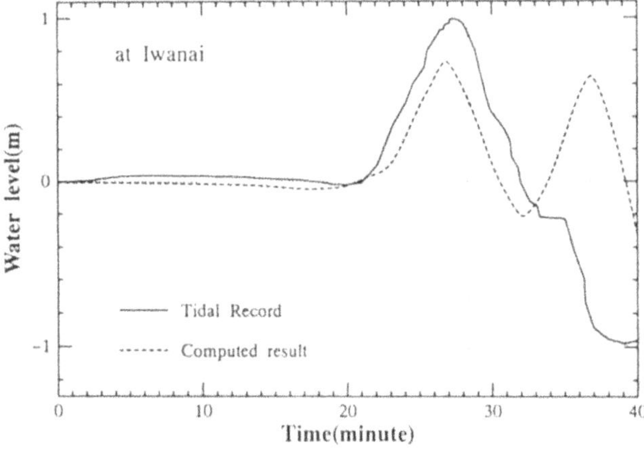

Figure 17

Tide record at Iwanai and the computed waveform using the inversion solution.

late arrival time of the tsunami at Iwanai. With these fault parameters, the initial tsunami profile is computed as shown in Figure 18. The horizontal extension of the initial tsunami profile is smaller than that of DCRC-17a.

Computed and observed runup distributions along the coasts of Hokkaido and Okushiri Island are compared in Figure 19. On Hokkaido, agreement is favorable along the coast from Matsumae to Taisei, with Esashi at the center. Agreement is sound but less so in the neighboring place. On Okushiri, agreement is poor except along the east coast.

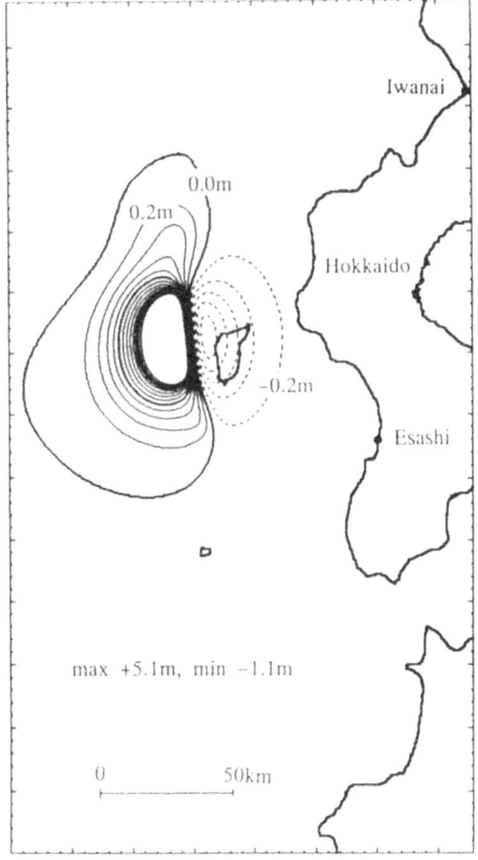

Figure 18
Initial tsunami profile modified by the inversion analysis.

Conclusions

The DCRC-17a model is proposed as the best fault model to reproduce well the tsunami runup distribution measured on Okushiri. It consists of three faults. The parameters of the south fault are determined in a manner such that high runups at Hamatsumae in the sheltered area are reproduced. The introduction of the central fault is required to explain the runup heights along the northwestern coast of Okushiri. The fault parameters are given in Table 3, and the initial tsunami profile is shown in Figure 11.

A defect of the DCRC-17a is that the computed arrival time is faster than that recorded by the tide gauge at Iwanai. In order to improve this defect, an inversion analysis is carried out. The modified fault parameters are given in Table 4, and the corresponding initial tsunami profile is shown in Figure 18.

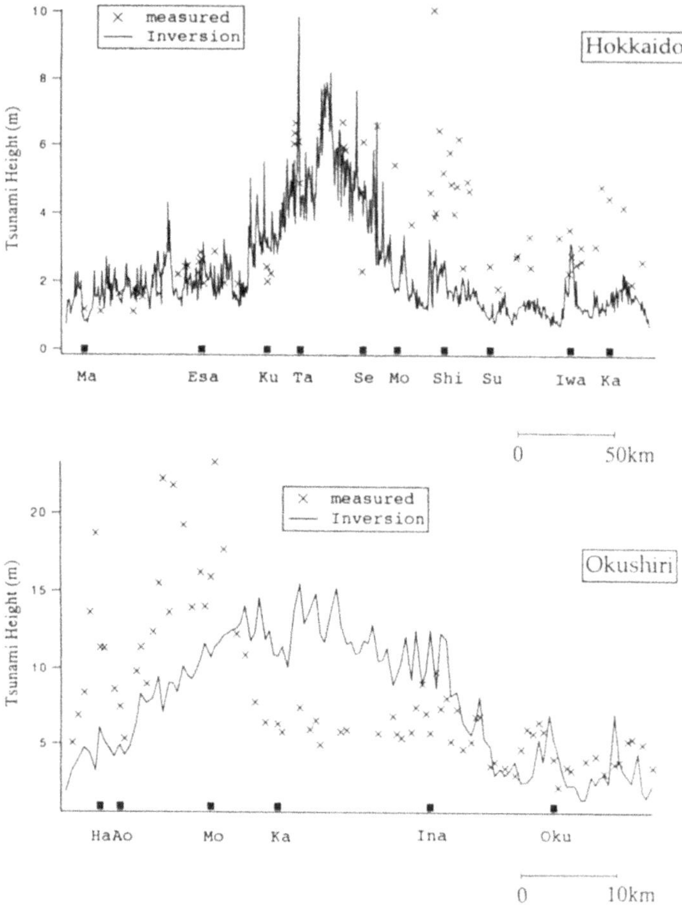

Figure 19
Comparison of the measured runup and the computed runup using the inversion solution.

There is, however, a riddle which we must solve. The current velocity record at Tomari, in the proximity of the Iwanai tide gauge station shows an early arrival time of the tsunami, which agrees well with the computed tsunami of the original DCRC-17a. Future study will focus on the modification of the north fault of DCRC-17a.

Acknowledgements

The publication of the present study is financially supported by the Ogawa Commemorative Fund.

REFERENCES

AIDA, I. (1972), *Numerical Estimation of a Tsunami Source*, Zisin 2 (25), 343–352.

AIDA, I. (1978), *Reliability of a Tsunami Source Model Derived from Fault Parameters*, J. Phys. Earth. 26, 55–73.

DEAN, R. G., and DALRYMPLE, R. A., *Water Wave Mechanics for Engineers and Scientists* (World Scientific, London 1984).

GOTO, C., and OGAWA, Y., *Numerical Method of Tsunami Simulation with the Leap-frog Scheme* (Dept. of Civil Eng., Fac. of Eng., Tohoku University, Sendai 1982).

HOKKAIDO TSUNAMI SURVEY TEAM (1993), *Tsunami Devastates Japanese Coastal Region*, EOS 74 (34), 417–420.

IMAMURA, F., TAKAHASHI, TOMO, and TAKAHASHI, Take., *Fault of the 1993 Hokkaido-Nansei-Oki earthquake dipping toward east of west?—Interpretation from tsunami data*. In *Monthly Ocean* (Kaiyou Shuppanpp, 1994) pp. 179–184.

KIKUCHI, M. (1993), *Source Process of the Hokkaido-Nansei-Oki Earthquake of July 12, 1993 Inferred from Teleseismic Body Wave Inversion*, Abst. of SSJ Meeting 2, 28.

KUMAKI, Y. et al. (1993), *Vetrical Seismic Crustal Movement of the 1993 Hokkaido-Nansei-Oki Earthquake Based on Coastal Landform Changes in Okushiri Island, West of Hokkaido, Japan*, Abst. of SSJ Meeting 2, 63.

MANSINHA, L., and SMYLIE, D. E. (1971), *The Displacement Field of Inclined Faults*, Bull. Seismol, Soc. Am. 61, 1433–1440.

NAGANO, O., IMAMURA, F., and SHUTO, N. (1991), *A Numerical Model for Far-field Tsunamis and its Application to Predict Damages Done to Aquaculture*, Natural Hazards 4, 235–255.

NAKANISHI, I., and KOBAYASHI, R. (1993), *Source Mechanism of the 1993 Southwest Hokkaido Earthquake*, Abst. of SSJ 2, 29.

OHTAKE, M. (1993), *Time-space Distribution of Large Earthquakes along Eastern Margin of Japan Sea*, Abst. of SSJ Meeting 2, 37.

RESEARCH GROUP FOR AFTERSHOCKS of the July 12, 1993 (1993), *Hokkaido-Nansei-Oki Earthquake, Geometry of the Aftershocks of the July 12, 1993 Hokkaido-Nansei-Oki Earthquake*, Abst. of SSJ Meeting 2, 15.

SATAKE, K., OKADA, M., and ABE, Ku. (1988), *Tide Gauge Response to Tsunamis: Measurements at 40 Tide Gauge Stations in Japan*, J. Marine Res. 40, 557–571.

SATAKE, K. (1989), *Inversion of Tsunami Waveforms for the Estimation of Heterogenous Fault Motion of Large Submarine Earthquakes: The 1968 Tokachi-Oki and the 1983 Japan Sea Earthquakes*, J. Geophys. Res. 94, 5627–5636.

SATAKE, K. (1993), *Depth Distribution of Coseismic Slip along the Nankai Trough, Japan, from Joint Inversion of Geodetic and Tsunami Data*, J. Geophys. Res. 98, 4553–4565.

SHUTO, N., and MATSUTOMI, H. (1994), *Field Survey of the 1993 Hokkaido Nansei-Oki Earthquake Tsunami*, Pure Appl. Geophys. (to appear).

TANIOKA, Y., SATAKE, K., and RUFF, L. (1993), *Mechanism of the 1993 SW Hokkaido Earthquake*, Abst. of SSJ Meeting 2, 30.

(Received August 15, 1994, revised March 1, 1995, accepted March 6, 1995)

PAGEOPH, Vol. 144, Nos. 3/4 (1995)

Finite Element Modeling of the July 12, 1993 Hokkaido Nansei-Oki Tsunami

EDWARD P. MYERS[1] and ANTÓNIO M. BAPTISTA[1]

Abstract—A fault plane model and a finite element hydrodynamic model are applied to the simulation of the Hokkaido Nansei-Oki tsunami of July 12, 1993. The joint performance of the models is assessed based on the overall ability to reproduce observed tsunami waveforms and to preserve mass and energy during tsunami propagation. While a number of observed characteristics of the waveforms are satisfactorily reproduced (in particular, amplitudes and arrival times at tidal gauges relatively close to the source, and general patterns of energy concentration), others are only marginally so (notably, wave periods at the same gauges, and wave heights along Okushiri); differences between observations and simulations are traceable to both the fault plane and the hydrodynamic models. Nonnegligible losses of energy occur throughout the simulated tsunami propagation. These losses seem to be due to a combination of factors, including numerical damping and possible deficiencies of the shallow water equations in preserving energy.

Key words: Hokkaido Nansei-Oki, tsunami, finite element modeling, energy conservation.

1. Introduction

An earthquake occurred west of the islands of Hokkaido and Okushiri, in the Sea of Japan, at 13 h 17 m (UTC) on July 12, 1993. The epicenter was located at 42.3°N and 139.4°E and registered a magnitude of $M_S = 7.8$. The earthquake was felt in coastal regions throughout the Sea of Japan, and the induced tsunami strongly impacted Okushiri and Hokkaido. Most of the damage and loss of life occurred on Okushiri, particularly along the south and southwestern coastlines where runups were often in the range of 15 to 30 meters. Runups up to 10 meters were also observed along the central western coast of Hokkaido.

Several aspects of this event make it a good test for numerical models of tsunami generation and propagation. First, available field data are relatively abundant, including (a) a reasonably detailed bathymetry, (b) observations of coastal subsidence/uplift along the coast of Okushiri, (c) combined records of tides

[1] Center for Coastal and Land-Margin Research and Department of Environmental Science and Engineering, Oregon Graduate Institute of Science and Technology, P.O. Box 91000, Portland, Oregon 97291-1000, U.S.A.

and tsunami waveforms at several tidal gauges throughout the Sea of Japan, and (d) observations of tsunami runup at numerous locations along the coasts of Okushiri and Hokkaido, and, to a lesser extent, along the coasts of Korea and Russia. Second, the source mechanism is relatively well constrained, and geodetic data on coastal uplift/subsidence on Okushiri permit comparisons with results of bottom deformation models.

This paper describes and assesses the joint application of a fault plane model and a numerical hydrodynamic model to the *a posteriori* simulation and interpretation of the observed tsunami. Waveforms simulated under alternative assumptions of source mechanisms are compared against recorded waveforms, and extensive checks of mass and energy conservation are conducted.

2. Methods

The methods used in this study consist primarily of (a) a conventional fault plane model, (b) an hydrodynamic model, and (c) a set of conservation measures for mass and energy.

The fault plane model is the same proposed by OKADA (1985), which assumes a simple source configuration and an isotropic homogeneous half-space. This model has been extensively used in the generation of initial conditions for tsunami simulations, often extended to include (as in our case) a linear superposition of two or more sources. The formulation is well-known and will not be revisited here.

Applied numerical modeling of tsunami propagation has, in the past, been primarily performed using finite difference schemes, with regular or quasi-regular grids. The hydrodynamic model used in this study is based on the finite element solution of the depth-averaged shallow water equations, and allows for flexible variation of the grid resolution throughout the domain. The model was originally developed by LUETTICH et al. (1991), primarily for the study of tidal circulation (GRENIER et al., 1994) and storm surges (WESTERINK et al., 1992). For tsunami simulations, we introduce two modifications: the continuity equation is extended to account for the bottom deformation, and support for transmissive boundary conditions is added. The grids necessary to support the application of the model were built using ACE/gredit (TURNER and BAPTISTA, 1991), a flexible semi-automatic finite element grid generator.

Section 2.1 reviews the hydrodynamic model and describes our modifications. Associated measures of mass and energy conservation are described in Section 2.2.

2.1. Numerical Model

2.1.1. Modified continuity equation. To represent the tsunami generation, a time and spatially dependent bottom deformation is introduced in the depth-averaged continuity equation. Rather conventionally, this is accomplished by incorporating the bottom deformation into the kinematic boundary condition, and leads to the following modified equation

$$\frac{\partial \eta}{\partial t} - \frac{\partial \gamma}{\partial t} + \frac{\partial}{\partial x} uH + \frac{\partial}{\partial y} vH = 0, \tag{1}$$

where γ is the bottom deformation (positive for uplift), η is the free surface elevation, u and v are the depth-averaged velocities, $H = h + \eta - \gamma$ is the total water depth, and h is the water depth relative to a reference level.

The time interval over which the bottom deformation is imposed should be consistent with the rise time of the earthquake, which, following GELLER (1976), can be approximated as

$$\Upsilon_s = \frac{\bar{D}}{\dot{D}} \approx \frac{\mu \bar{D}}{\beta \Delta \sigma}, \tag{2}$$

where Υ_s is the theoretical rise time, $\bar{D}$ the average dislocation, $\dot{D}$ the dislocation velocity, μ the average rigidity, β the shear wave velocity, and $\Delta\sigma$ the mean stress drop.

2.1.2. Governing equations and numerical formulation. The adopted finite element model is based on the shallow water equations, with the momentum equations written in nonconservative form, and the continuity equation written in a generalized wave form. The use of wave continuity rather than primitive continuity equations was introduced by LYNCH and GRAY (1979) and later generalized by KINMARK and GRAY (1984); the approach is effective in eliminating the spurious $2\Delta x$ oscillations often associated with early finite element solutions of coastal flow simulations.

To review the derivation of the generalized wave continuity equation (GWCE), we let L represent the primitive continuity equation written as before

$$L \equiv \frac{\partial \eta}{\partial t} - \frac{\partial \gamma}{\partial t} + \frac{\partial}{\partial x} uH + \frac{\partial}{\partial y} vH = 0 \tag{3}$$

and let M represent the nonconservative form of the momentum equations subject to the Boussinesq, hydrostatic, and incompressibility assumptions

$$M \equiv \frac{\partial}{\partial t} \mathbf{v} + \mathbf{v} \cdot \nabla \mathbf{v} + \mathbf{f} \times \mathbf{v} + \tau \mathbf{v} + \nabla \left[\frac{p_s}{\rho_0} + g(\eta - \alpha \Psi) \right]$$

$$+ \frac{E_h}{H} \left[\frac{\partial^2}{\partial x^2} \mathbf{v} H + \frac{\partial^2}{\partial y^2} \mathbf{v} H \right] + \frac{\tau_{sx}}{\rho_0 H} = 0, \tag{4}$$

where $\mathbf{v}$ is the depth-averaged velocity (u, v), $\mathbf{f}$ is the Coriolis vector, $\tau = c_f((u^2 + v^2)^{1/2}/H)$, c_f is the bottom friction coefficient, p_s is the atmospheric pressure at the free surface, ρ_0 is the reference density of water, g is the acceleration due to gravity, α is the effective earth elasticity factor, Ψ is the Newtonian equilibrium tide potential, E_h is the horizontal eddy diffusion coefficient, and τ_{sx} is the applied free surface stress.

If we represent the conservative form of the momentum equations as

$$M^c = (H)(M) + (\mathbf{v})(L) \tag{5}$$

the continuity wave equation, W, is constructed as

$$W \equiv \frac{\partial L}{\partial t} + \tau L - \nabla \cdot M^c = 0, \tag{6}$$

where τ is the same as the friction factor used in the momentum equations. The GWCE is an extension of Equation (6) where τ is replaced by a generic weighting factor G

$$W^G \equiv \frac{\partial L}{\partial t} + GL - \nabla \cdot M^c = 0 \tag{7}$$

or

$$
\begin{aligned}
W^G \equiv{} & \frac{\partial^2 \eta}{\partial t^2} + G \frac{\partial \eta}{\partial t} - \frac{\partial^2 \gamma}{\partial t^2} - G \frac{\partial \gamma}{\partial t} \\[1em]
& + \frac{\partial}{\partial x} \left\{ u \frac{\partial \eta}{\partial t} - uH \frac{\partial u}{\partial x} - vH \frac{\partial u}{\partial y} + fvH - H \frac{\partial}{\partial x} \left[\frac{p_s}{\rho_0} + g(\eta - \alpha \Psi) \right] \right\} \\[1em]
& + \frac{\partial}{\partial x} \left\{ -E_h \frac{\partial^2 \eta}{\partial x\, \partial t} + \frac{\tau_{sx}}{\rho_0} - (\tau - G)uH \right\} \\[1em]
& + \frac{\partial}{\partial y} \left\{ v \frac{\partial \eta}{\partial t} - uH \frac{\partial v}{\partial x} - vH \frac{\partial v}{\partial y} - fuH - H \frac{\partial}{\partial y} \left[\frac{p_s}{\rho_0} + g(\eta - \alpha \Psi) \right] \right\} \\[1em]
& + \frac{\partial}{\partial y} \left\{ -E_h \frac{\partial^2 \eta}{\partial y\, \partial t} + \frac{\tau_{sy}}{\rho_0} - (\tau - G)vH \right\} = 0.
\end{aligned}
\tag{8}
$$

The advective terms in Equation (8) are formulated in nonconservative form in order to be consistent with the nonconservative advective terms in Equation (4) (KOLAR *et al.*, 1994). The larger the value of G, the more primitive the GWCE will be. Thus, if G is too large, spurious oscillations may arise. However, if G is too small, the solutions will likely be plagued with mass balance errors. A balance must therefore be achieved for an optimal G.

Equations (4) and (8) are solved with a Galerkin finite element method. The solution involves three stages (see WESTERINK *et al.*, 1992 for details): first, symmetrical weak weighted residual statements are developed for the GWCE and primitive momentum equations; second, the equations are time-discretized, with either two or three time-level schemes applied selectively to different terms within each equation; finally, the finite element method is implemented, by expanding the variables over linear triangular elements, developing discrete equations on an elemental level, assembling global systems of equations, and enforcing boundary conditions.

2.1.3. Boundary conditions. The original model allows for the specification at the boundaries of either elevations (enforced in the discrete GWCE equation) or normal velocities (enforced in the discrete momentum equations). We have added the ability to specify transmissive boundary conditions, enforced in the discrete GWCE equation.

Transmissive boundary conditions are imposed by first backtracking from the boundary node in the direction of the incoming wave and then interpolating the elevation from the previous time step at that spatial location. The incident angle of the incoming wave is approximated as,

$$\theta = \operatorname{atan}\left(\frac{v}{u}\right). \tag{9}$$

The wave is backtracked a distance,

$$S = \Delta t \sqrt{gH} \tag{10}$$

in the direction prescribed by θ (Figure 1). Once the backtracked positions (X_{int}, Y_{int}) are known, the elevation from the previous time step may be interpolated from the appropriate element. The new elevation at the boundary node may

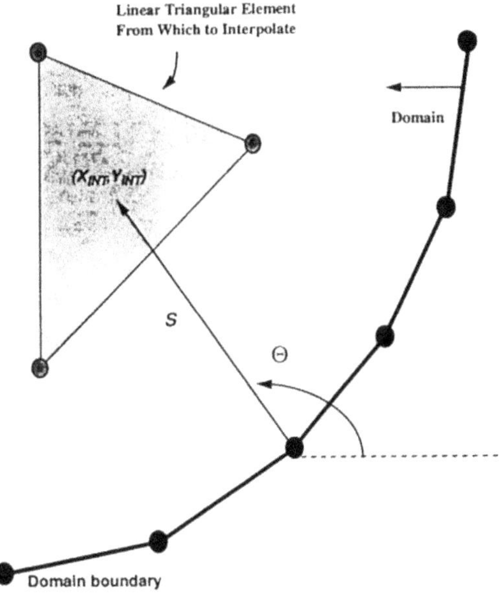

Figure 1
Method of backtracking to the point of interpolation.

then be set equal to this interpolated old elevation, thus allowing the wave to leave the domain of interest. The transmissive boundary condition should be imposed if $\mathbf{v} \cdot \mathbf{n} \geq 0$ and if the velocities are not relatively small compared to what is expected for a particular simulation.

2.2. Conservation Measures

2.2.1. Mass. We adopted, and review here, the algorithm proposed by KOLAR *et al.* (1994) to check mass conservation. The algorithm involves integrating the primitive continuity equation over time and space

$$\int_{t_0}^{t} \int_{\Omega} \left[\frac{\partial \eta}{\partial t} - \frac{\partial \gamma}{\partial t} + \nabla \cdot (\mathbf{v}H) \right] d\Omega \, dt = 0, \tag{11}$$

where Ω is the spatial domain.

Applying the divergence theorem of Gauss to the third term of Equation (11) and integrating over time the first two terms

$$\int_{\Omega} (\eta_t - \eta_{t_0}) \, d\Omega - \int_{\Omega} (\gamma_t - \gamma_{t_0}) \, d\Omega + \int_{t_0}^{t} \left[\int_{\partial \Omega} H\mathbf{v} \cdot \mathbf{n}d(\partial \Omega) \right] dt = 0, \tag{12}$$

where $\partial \Omega$ is the boundary and $\mathbf{n}$ is the normal vector to the boundary.

Because the domain is spatially discretized with linear triangular elements, the three spatial integrals can be evaluated exactly as

$$\int_{\Omega} (\eta_t - \eta_{t_0}) \, d\Omega = \sum_{el} [\bar{\eta}_t - \bar{\eta}_{t_0}]_{el} A_{el} \tag{13}$$

$$\int_{\Omega} (\gamma_t - \gamma_{t_0}) \, d\Omega = \sum_{el} [\bar{\gamma}_t - \bar{\gamma}_{t_0}]_{el} A_{el} \tag{14}$$

$$Q^{\text{net}} \equiv \int_{\partial \Omega} H\mathbf{v} \cdot \mathbf{n}d(\partial \Omega) = \sum_{e_b} \frac{w_e}{6} [2H_1 v_{n1} + H_1 v_{n2} + H_2 v_{n1} + 2H_2 v_{n2}] \tag{15}$$

where A_{el} is the elemental area, w_e is the length of the boundary portion of element e_b, $v_n = \mathbf{v} \cdot \mathbf{n}$, and the indices [1, 2] represent the beginning and ending nodes of each boundary segment. To complete the evaluation of Equation (12), the time integral of Q^{net} is approximated as

$$\int_{t_0}^{t} \left[\int_{\partial \Omega} H\mathbf{v} \cdot \mathbf{n}d(\partial \Omega) \right] dt = \int_{t_0}^{t} Q^{\text{net}} \, dt \approx \sum_{N \text{ timesteps}} \frac{1}{2} [Q_t^{\text{net}} + Q_{t+\Delta t}^{\text{net}}] \, \Delta t. \tag{16}$$

The deviation from zero of the summation of the integrals defined by Equations (13), (14), and (16) provides an effective measure of mass conservation errors.

2.2.2. Energy. The sum of potential and kinetic energy of a tsunami as it propagates through the sea and interacts with the coast should match the energy released by the generating earthquake, except for energy dissipation due to friction.

The equations described in this section are used to check energy conservation in our numerical simulations.

The potential energy E_p of the tsunami at any point in time is defined as

$$E_p(t) = \frac{1}{2}\rho g \int_\Omega \eta^2 \, d\Omega \tag{17}$$

or, for a finite element grid consisting of linear triangular elements

$$E_p(t) = \frac{1}{2}\rho g \sum_{i=1}^{N_{el}} \int_{\Omega_i} \left(\sum_{j=1}^{3} \eta_j \varphi_j\right)^2 d\Omega_i, \tag{18}$$

where N_{el} is the number of elements, φ_j is the shape function in the local coordinate system, and j is the node number in local coordinates.

Neglecting the vertical component of the velocity, w, the kinetic energy E_k is defined as

$$E_k(t) = \frac{1}{2}\rho \int_\Omega \left[\int_0^{(h+\eta-\gamma)} (u^2 + v^2) \, dz\right] d\Omega \tag{19}$$

or, for depth-averaged equations and linear triangular elements:

$$E_k(t) = \frac{1}{2}\rho \sum_{i=1}^{N_{el}} \int_{\Omega_i} \left(\sum_{j=1}^{3} (h+\eta-\gamma)_i \varphi_j\right)\left[\left(\sum_{j=1}^{3} u_j \varphi_j\right)^2 + \left(\sum_{j=1}^{3} v_j \varphi_j\right)^2\right] d\Omega_i. \tag{20}$$

Thus, the total tsunami energy E_t at any point in time is

$$E_t(t) = E_p(t) + E_k(t). \tag{21}$$

Following TOLMAN (1992), the energy loss due to bottom friction, per unit time and per unit bed area, is

$$\dot{E}_f(t) = \tau_b \cdot \mathbf{v} \tag{22}$$

with

$$\tau_b = \frac{1}{2}\rho c_f \mathbf{v}|\mathbf{v}|. \tag{23}$$

Hence, the accumulated energy loss due to friction at any given time is

$$E_f(t) = \int_{t_0}^{t} \left(\int_\Omega \dot{E}_f \, d\Omega\right) dt = \frac{1}{2}\rho \int_{t_0}^{t} \int_\Omega c_f (u^2 + v^2)^{3/2} \, d\Omega \, dt. \tag{24}$$

KAJIURA (1981) developed an empirical relationship between the seismic energy released during the earthquake E_s and the total initial tsunami energy, E_t'

$$\log\left(\frac{E_t'}{E_s}\right) = 0.5 M_w - 7.26, \tag{25}$$

where M_w is the moment magnitude. Since the information which is generally available is in the form of the seismic moment, M_0, then we can compute M_w as

$$M_w = \frac{2(\log M_0 - 16.1)}{3} \tag{26}$$

E_s can also be computed based upon M_0 as

$$E_s = (5 \times 10^{-5})M_0 \tag{27}$$

and, therefore, the initial tsunami energy should be approximately equal to

$$E'_t = \frac{(M_0)^{4/3}}{8.47 \times 10^{16}}, \tag{28}$$

where we used Equations (26) and (27) to substitute for M_w and E_s, respectively.

The instantaneous difference between the empirical (Equation 28) tsunami energy and the energy computed based on results of our finite element model is then

$$\Delta E_t(t) = E'_t - [E_t(t) + E_f(t)], \tag{29}$$

where Equations (21), (24) and (28) define the right-hand side.

3. Model Set-up

3.1. Domain and Grid Discretization

The finite element grid adopted for our simulations has 20,307 nodes and 39,668 triangular elements, with element sizes ranging from 311.8 km² to 7,683 m² (Figures 2a–c). Progressively finer grid resolution is used in regions increasingly closer to Hokkaido and Okushiri, and in the region of the seismic sources. The adopted discretization enables a good representation of initial conditions and of zones of sharp wave height gradients (both of which are important for numerical accuracy), while keeping Courant numbers below 1 (a stability requirement).

Land boundaries are treated as impervious (zero normal velocities imposed) and ocean boundaries are treated as transmissive. Nodal bathymetry is interpolated from a background grid formed by superposition of three databases: ETOPO5, which extends worldwide and has a resolution of 5′; bathymetry for the Sea of Japan at 1′ resolution; and bathymetry in the vicinity of Okushiri Island at 0.1′ resolution.

3.2. Seismic Source Scenarios

Following the July 12, 1993 event, several alternative descriptions of the seismic source were proposed by various institutions. We use eleven such scenarios (Table

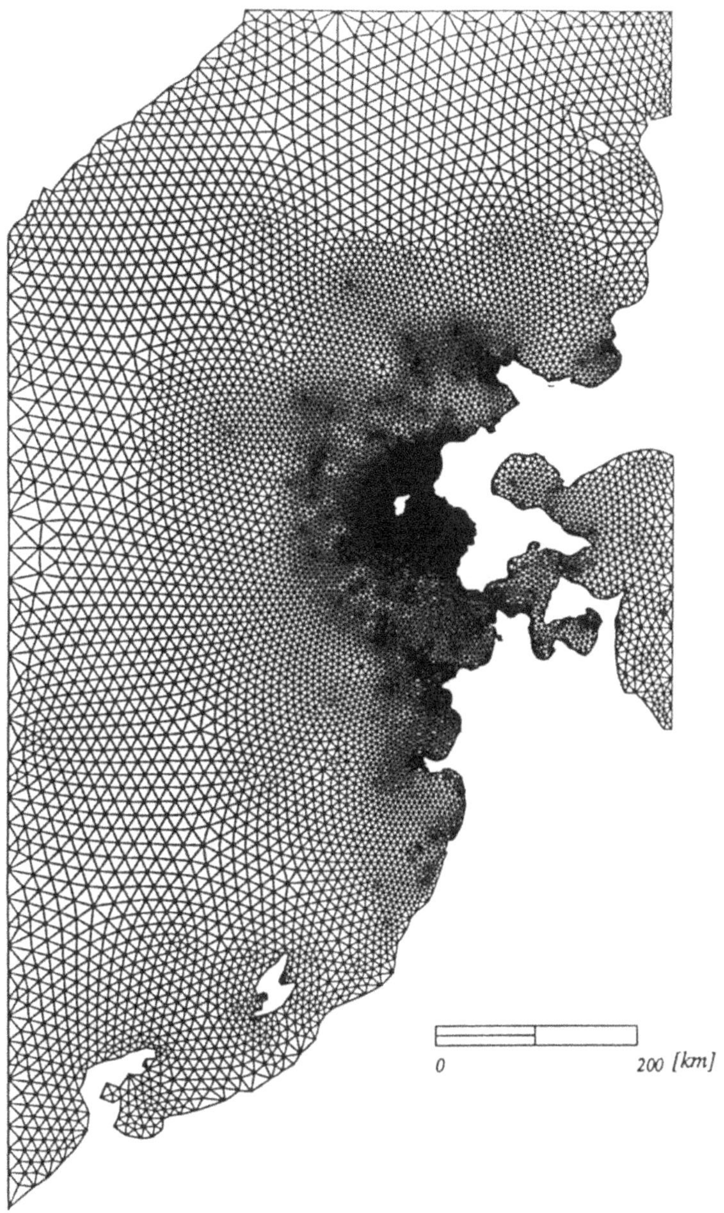

Figure 2(a)
Finite element grid used for Sea of Japan simulations.

1) to generate the initial conditions for the numerical simulations. The first six scenarios consist of two seismic sources, and the remaining five consist of only one source. All scenarios, except those proposed by SATAKE (1994), were developed shortly (hours to days) after the event.

Preliminary geodetic evidence reported by survey teams immediately after the earthquake suggested that the northern tip of the island of Okushiri had been uplifted. A more recent report (KUMAKI *et al.*, 1993) indicates, however, that the entire island may have subsided. Deformations computed with the OKADA (1985) fault plane model, using the parameters of the eleven scenarios of Table 1, are compared in Figure 3 and Tables 2(a–b) with the values reported by KUMAKI *et al.* (1993). All but one scenario show subsidence throughout the island. The

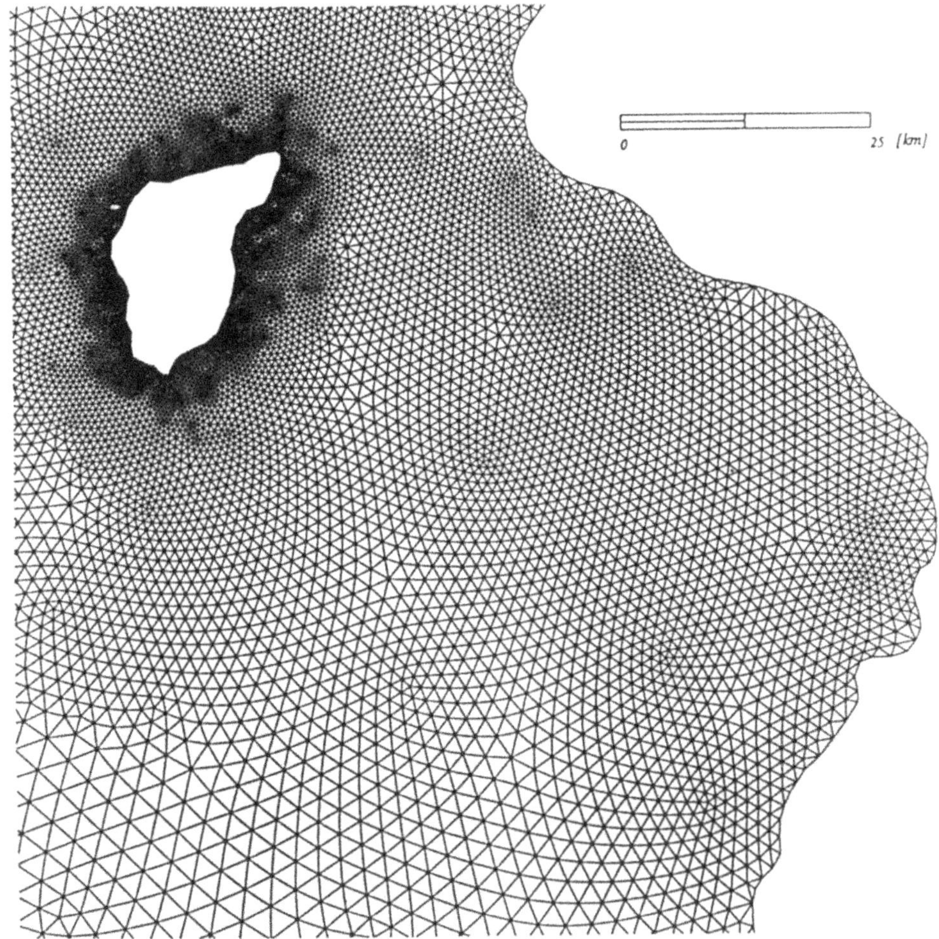

Figure 2(b)

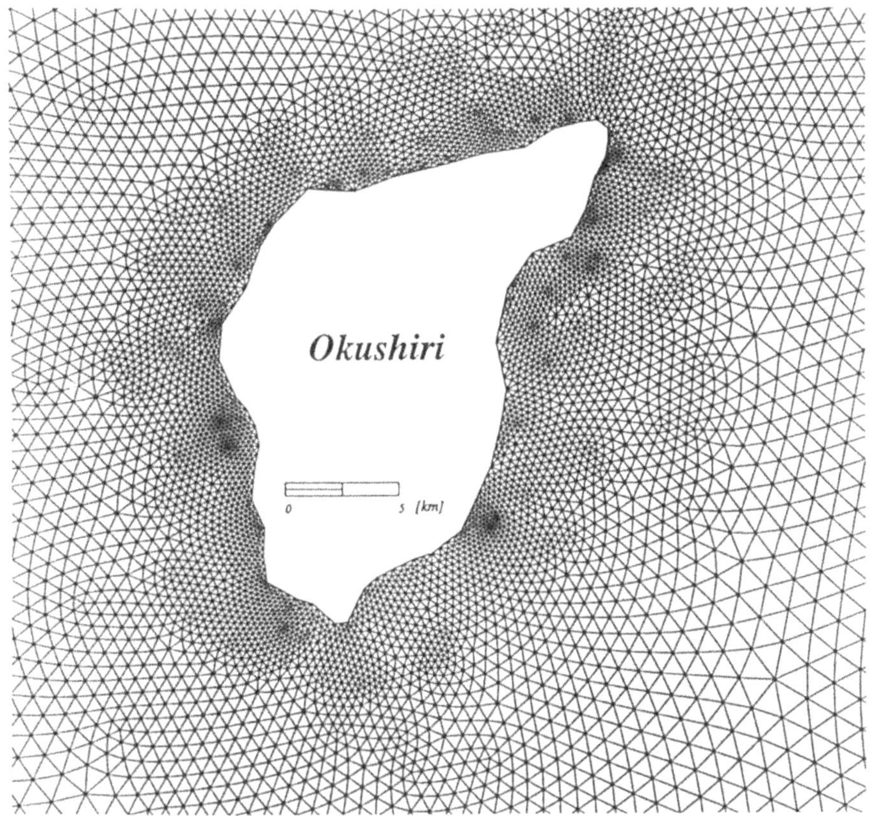

Figures 2(b,c)
Close-ups of the finite element grid near Okushiri.

exception, DCRC2, appears unrealistic, showing strong uplift in parts of the island. The scenarios proposed by SATAKE (1994) fit the observations generally better than the other (more preliminary) scenarios.

3.3. Choice of Parameters for the Hydrodynamic Model

The formulation of the hydrodynamic model contains three primary empirical parameters: the GWCE weighting factor, G; the Manning coefficient, representing bottom friction; and the viscosity coefficient.

The choice of G was based on the notion that this parameter should ensure numerical stability, minimize spurious oscillations, and lead to satisfactory mass conservation. In general, too small a value of G will lead to mass balance errors, while too large a value will lead to significant spurious oscillations and/or numerical instability. To optimize G, we used the EE seismic source scenario as reference, and

Table 1

Seismic parameters for each of the scenarios

Scenario	Length** (km)	Width** (km)	Strike	Dip	Rake	Depth (km)	Slip (m)
EE (SATAKE, 1994)	80.0	40.0	−20.0	30.0	90.0	0.0	3.0
	70.0	40.0	20.0	60.0	90.0	0.0	3.0
EW (SATAKE, 1994)	80.0	40.0	160.0	60.0	90.0	0.0	3.0
	70.0	40.0	20.0	60.0	90.0	0.0	3.0
WW (SATAKE, 1994)	80.0	40.0	160.0	60.0	90.0	0.0	3.0
	70.0	40.0	200.0	30.0	90.0	0.0	3.0
WE (SATAKE, 1994)	80.0	40.0	−20.0	30.0	90.0	0.0	3.0
	70.0	40.0	200.0	30.0	90.0	0.0	3.0
DCRC-1*	55.0	27.5	10.0	60.0	85.0	10.0	3.8
(Tohoku University)	55.1	27.6	320.0	30.0	120.0	5.0	5.6
DCRC-2	55.0	27.5	190.0	30.0	80.0	10.0	3.8
(Tohoku University)	55.1	27.6	140.0	60.0	105.0	5.0	5.6
Harvard	87.8	43.9	1.0	24.0	84.0	15.0	3.7
University of Tokyo	86.6	43.3	9.0	35.0	97.0	10.0	2.8
Japan Meteorological Agency	86.6	43.3	3.0	41.0	72.0	37.0	4.2
Hokkaido University	86.6	43.3	12.0	49.0	102.0	30.0	4.2
USGS	88.0	44.0	1.0	60.0	67.0	18.0	0.6

* Final 7 scenarios compiled by IMAMURA (1993).
** Length and width were computed by the authors based on the energy of the earthquake and assuming that length = 2.0* width.

adopted values of G ranging from 1×10^{-3} to 10. Based on the results of these simulations, we found $G = 0.3$ to provide a good compromise between mass conservation and numerical smoothness and stability. This value was retained for all remaining simulations.

Sensitivity to the Manning coefficient was investigated by comparison of simulations against recorded waveforms at 16 tidal stations on the islands of Hokkaido and Honshu (Figure 4). SATAKE (1994) and co-workers digitized copies of the original paper records of observed elevations at these stations, and filtered the digitized values to remove the effects from astronomical tides. Using the EE source scenario as reference, we made simulations with Manning coefficients of 0.0250, 0.0275, 0.0300, 0.0325 $s \times m^{-1/3}$. Root mean square errors were computed at each station and for each simulation, using the expression:

$$\text{Error} = \left[\frac{1}{N} \sum_{k=1}^{N} \left\{ (\eta_k - \tilde{\eta}_k) - \frac{1}{N} \sum_{p=1}^{N} (\eta_p - \tilde{\eta}_p) \right\}^2 \right]^{1/2}, \qquad (30)$$

where η is the observed tsunami waveform, $\tilde{\eta}$ is the computed tsunami waveform, and N is the number of equally spaced sampling points.

Errors per station and averaged over the domain are displayed in Figures 5(a–b), respectively. The errors tend to be significantly larger for stations close to

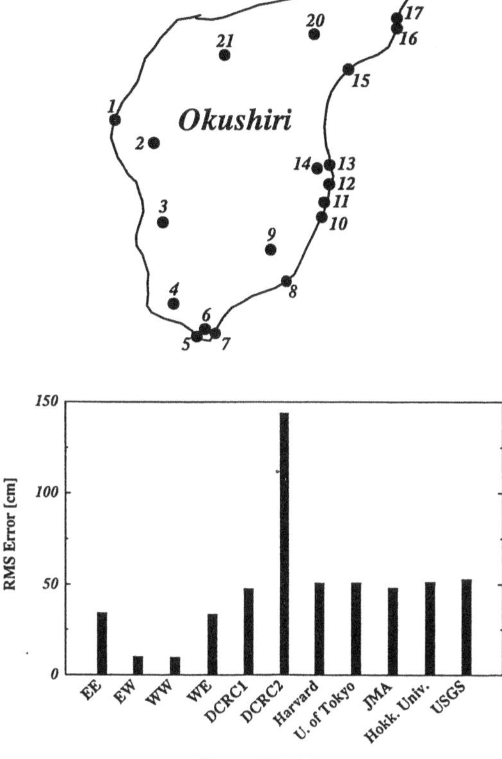

Figure 3(a,b)

(a) Observed subsidence locations (estimated from KUMAKI et al., 1993); (b) Root mean square errors for each seismic source scenario.

the source (e.g., Esashi and Iwanai), a direct consequence of the fact that wave heights are largest at these stations. However, the errors show little sensitivity to the Manning coefficient, which we interpret as an indication that the effect of the choice of this coefficient is secondary relative to other types of errors (in particular, the characteristics of the seismic source). Hereafter we adopt a Manning coefficient of $0.0275 \, \text{s} \times \text{m}^{-1/3}$, which has the lowest average root mean square error and appeared to minimize the error for Iwanai, one of the stations closest to the source. Due to the limitations inherent to this sensitivity analysis, we do not consider this an optimized value.

A similar sensitivity analysis was conducted with regard to the viscosity coefficient, E_h. Error sensitivity to this parameter was marginal (see MYERS, 1994), and we chose to perform the remaining simulations without viscosity (i.e., $E_h = 0 \, \text{m}^2/\text{s}$).

Table 2(a)

Observed and computed subsidence/uplift (in cm) on Okushiri

Location	Observation	EE	EW	WW	WE	DCRC1	DCRC2
1	−80	−25	−65	−67	−26	−9	154
2	−55	−25	−60	−61	−26	−8	155
3	−77	−24	−63	−64	−25	−8	96
4	−80	−24	−66	−66	−25	−8	51
5	−77	−23	−63	−64	−24	−7	37
6	−80	−23	−61	−62	−24	−7	40
7	−75	−23	−60	−61	−24	−7	38
8	−35	−21	−47	−47	−21	−5	50
9	−45	−21	−48	−49	−22	−5	73
10	−40	−19	−40	−41	−19	−4	75
11	−50	−19	−39	−40	−19	−4	85
12	−40	−18	−38	−39	−19	−4	108
13	−45	−18	−38	−38	−19	−4	138
14	−30	−19	−39	−40	−19	−4	140
15	−25	−16	−33	−34	−17	−3	−101
16	−25	−14	−28	−29	−14	−3	−86
17	−20	−14	−28	−29	−14	−3	−85
18	−25	−13	−26	−27	−13	−2	−73
19	−35	−14	−28	−29	−14	−3	−83
20	−41	−17	−36	−36	−18	−4	−76
21	−47	−22	−46	−47	−23	−5	256
Root Mean Square Error (cm)		34.0	9.8	9.5	33.3	47.6	143.7

* Subsidence values are negative, uplift values are positive.

4. Numerical Results

Using the values for the GWCE weighting factor, Manning coefficient, and viscosity coefficient mentioned in the previous section, numerical simulations were made for each of the eleven seismic source scenarios of Table 1. Results for all simulations are compared against observed waveforms at the two stations closest to the seismic source (Iwanai and Esashi) and one remote station (Awashima)—Figures 6 through 8.

Iwanai is likely to be affected first by the waves generated by the northern seismic source. Therefore, the arrival time of the first wave at Iwanai may provide an indication of the correct orientation of the source. Figures 6(a–b) show that scenarios WE, WW, and DCRC2 best match the recorded arrival time of the first waves. Among these scenarios, the WE and WW best reproduce the amplitude of the first wave; we also recall that scenario DCRC2 is unrealistic based on geodetic

Table 2(b)

Observed and computed subsidence/uplift (in cm) on Okushiri

Location	Observation	Harvard	Univ. Tokyo	JMA	Hokk. Univ.	USGS
1	−80	−4	−3	−7	−2	−0.3
2	−55	−3	−3	−6	−2	−0.3
3	−77	−3	−2	−5	−2	−0.3
4	−80	−2	−2	−5	−2	−0.3
5	−77	−2	−2	−4	−3	−0.2
6	−80	−2	−2	−5	−2	−0.2
7	−75	−2	−2	−4	−2	−0.2
8	−35	−2	−2	−5	−2	−0.2
9	−45	−2	−2	−5	−2	−0.2
10	−40	−2	−2	−5	−5	−0.2
11	−50	−2	−2	−5	−2	−0.2
12	−40	−2	−2	−5	−2	−0.2
13	−45	−2	−2	−5	−2	−0.2
14	−30	−2	−2	−5	−2	−0.2
15	−25	−3	−2	−6	−2	−0.3
16	−25	−3	−2	−6	−2	−0.2
17	−20	−3	−2	−6	−2	−0.2
18	−25	−3	−3	−6	−2	−0.2
19	−35	−3	−3	−6	−3	−0.3
20	−41	−3	−3	−6	−2	−0.3
21	. −47	−4	−3	−7	−2	−0.3
Root Mean Square Error (cm)		50.6	50.9	48.2	51.2	52.8

data. Since Esashi is located further south on the island of Hokkaido and will first be impacted by the waves generated from the southern seismic source, Figures 7(a–b) may provide insight on the most likely orientation of the southern source. A close examination of these figures reveals that scenarios EW and WW best reproduce the trough of the first incoming wave.

Combining the information of Figures 6 and 7, it appears that scenario WW best reproduces the observed waveforms at near-source stations. However, Figures 8(a–b) show that none of the eleven scenarios reproduce well the waveforms at Awashima, which holds true for most other remote stations.

The numerical results shown in Figures 6–8 were generated by fully nonlinear simulations. Due to the nonlinear terms, simulations were made using a minimum depth of 5 meters, to avoid wetting and drying of elements and associated instabilities. The minimum depth around Okushiri required a setting of 15 meters due to the large ratios of wave height to depth. This representation of the coastal bathymetry may help explain the difficulty of the model in representing waveforms at remote stations—Figures 8(a–b).

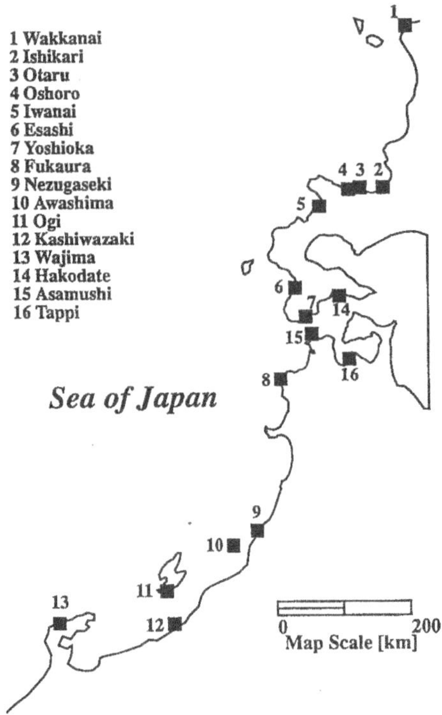

Figure 4
Locations of various tidal stations used in this study.

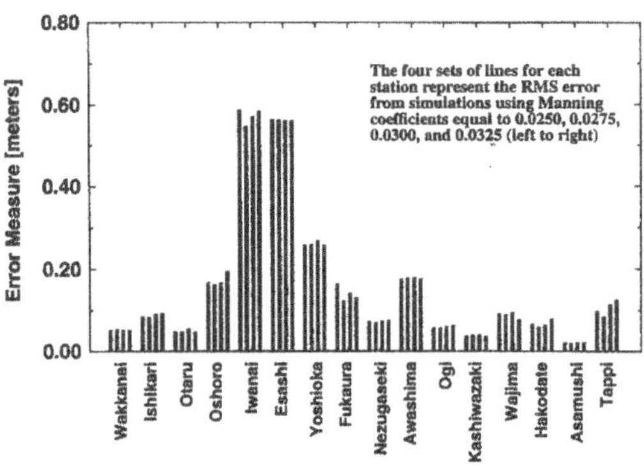

Figure 5(a)
Friction calibration: root mean square errors for each tidal station.

The time histories of mass conservation and of tsunami energy are shown in
Figure 9 (dark lines) for the simulation with the WW source scenario. Mass errors
remain small for approximately 40 to 50 minutes, but increase thereafter up to
about 5% of the volume of the seismic deformation. Possible reasons involve

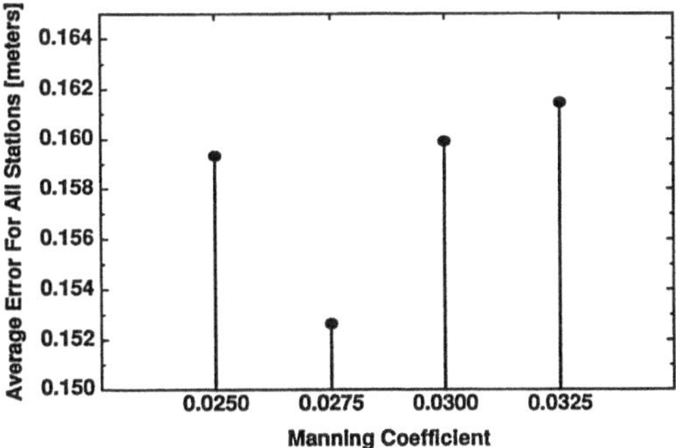

Figure 5(b)
Friction calibration: root mean square errors averaged for all stations.

interaction of the waves with the land boundary and departure of mass through the transmissive boundary. For the Sea of Japan, the waves are unlikely to interact with transmissive boundaries until at least half an hour into the simulation.

Energy conservation is a more significant problem. The tsunami energy corrected for frictional effects (i.e., $E_t + E_f$ in Equation (29)) reaches a peak at the end of the earthquake rise time, which is typically different but of the same order of magnitude of the empirical value (Table 3) given by Equation (28). However, this peak is followed by a significantly prolonged decline of energy, that friction cannot explain (Figure 9). We hypothesize (see discussion in Section 5) that most of the energy decline results from a combination of numerical damping and a possible inability of the shallow water equations to conserve energy (because they neglect pressure gradients associated with vertical acceleration terms). In addition, some energy is also lost through the open boundaries, an effect which is not accounted for in Equation (21).

Table 3

Comparison of theoretical and computed tsunami energy

Scenario	Empirical Energy (E_t') [Joules]	Computed Initial Energy $(E_t + E_f)$ [Joules]
EE	$1.119\,e + 13$	$1.783\,e + 13$
EW	$1.119\,e + 13$	$2.234\,e + 13$
WE	$1.119\,e + 13$	$1.452\,e + 13$
WW	$1.119\,e + 13$	$1.937\,e + 13$
DCRC-1 (Tohoku University)	$1.203\,e + 13$	$1.701\,e + 13$
DCRC-2 (Tohoku University)	$1.203\,e + 13$	$1.701\,e + 13$
Harvard	$1.203\,e + 13$	$1.169\,e + 13$
Hokkaido University	$6.379\,e + 11$	$2.376\,e + 13$
Japan Meteorological Agency	$6.379\,e + 11$	$1.921\,e + 13$
USGS	$1.072\,e + 12$	$4.771\,e + 11$
University of Tokyo	$8.004\,e + 12$	$9.050\,e + 12$

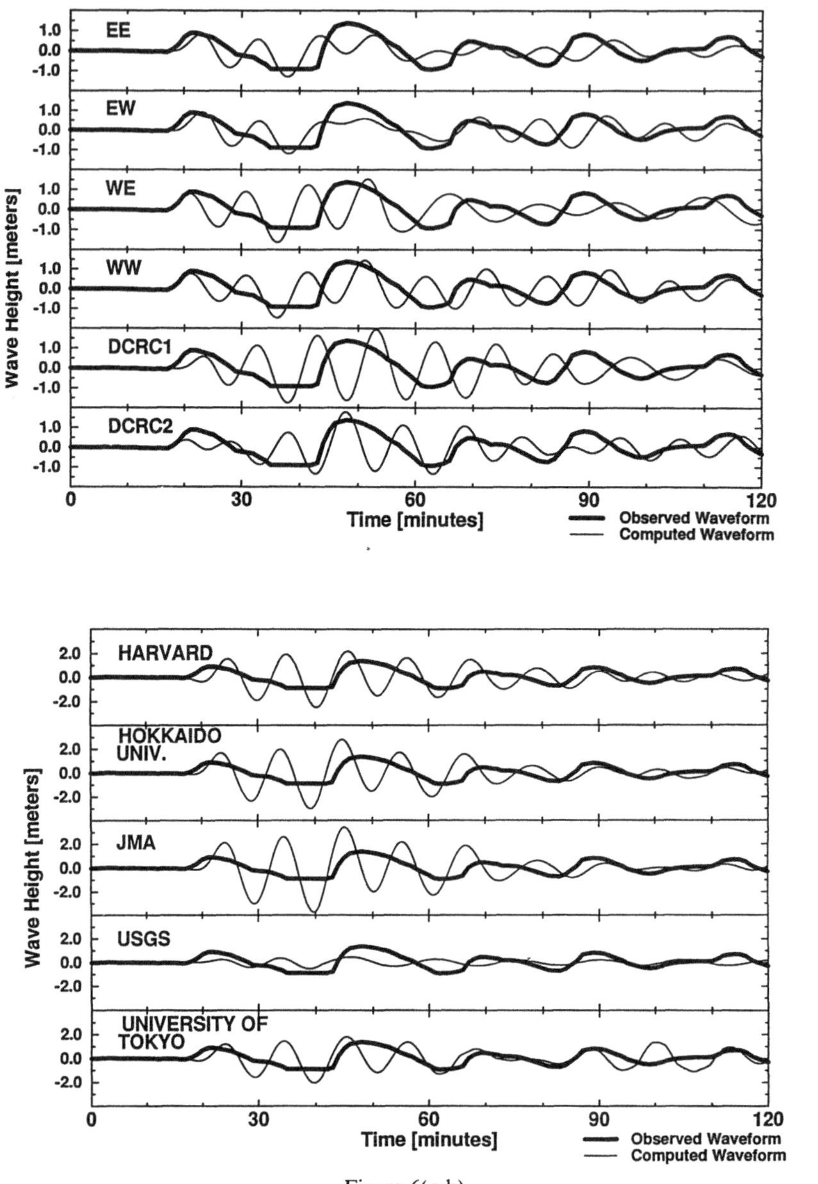

Figure 6(a,b)
Numerical results and observed waveforms at Iwanai.

To examine the effects of the transmissive boundary on both mass and energy variation, the WW scenario was simulated using a grid in which all of the boundaries are artificially closed and made fully reflective. Figure 9 compares mass and energy variation for the WW simulation using both the original grid and this modified grid. Results are shown by the lighter lines in Figure 9. Mass is extremely

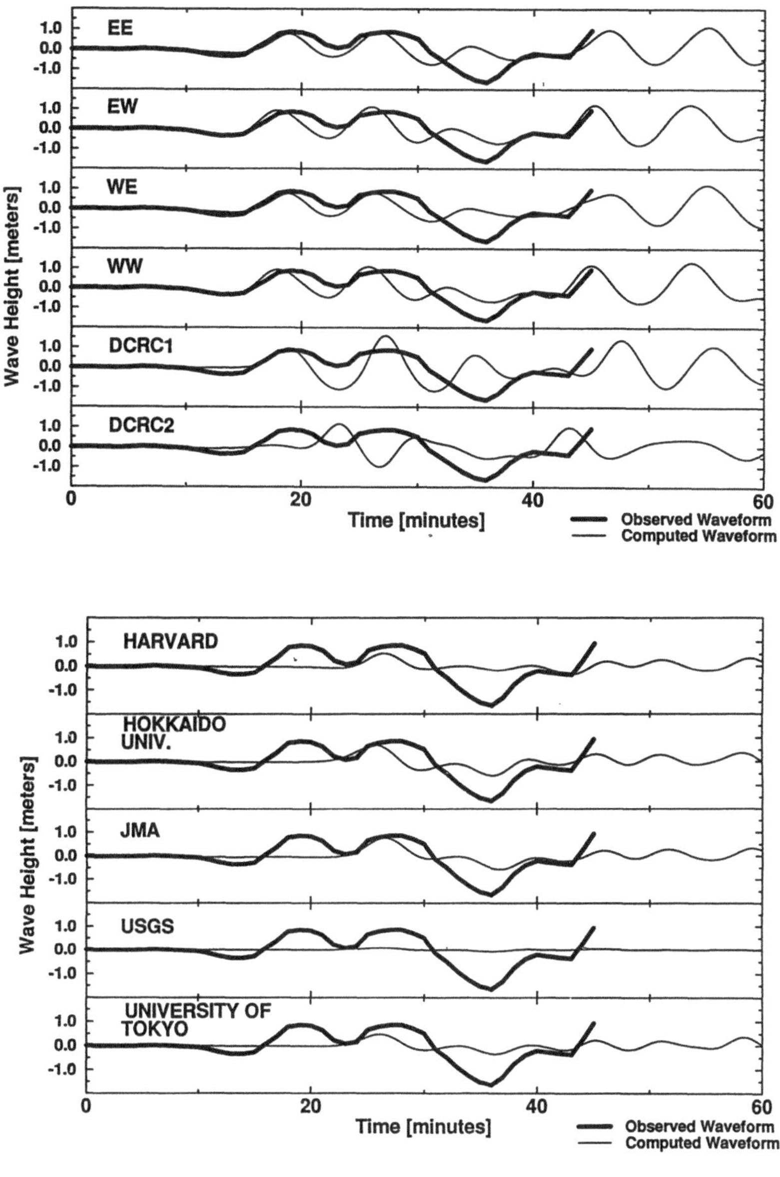

Figure 7(a,b)
Numerical results and observed waveforms at Esashi.

well conserved, which manifests that it is the formulation of the transmissive boundary conditions that does not conserve mass properly. The energy plot, however, shows that while some energy is lost as the wave departs the domain, significant additional energy loss mechanisms are also present. Figure 10 displays

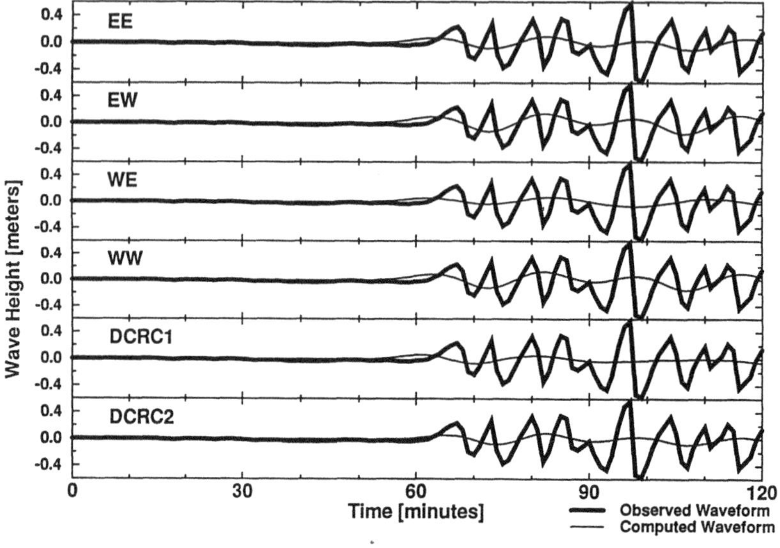

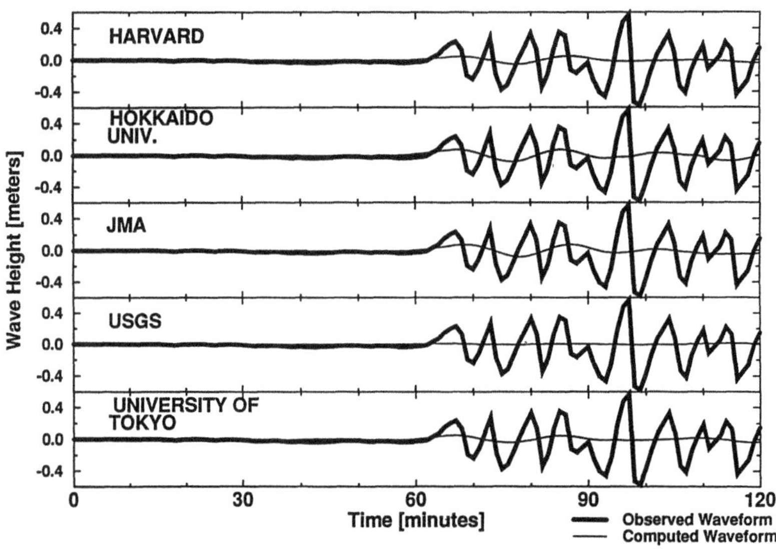

Figure 8(a,b)
Numerical results and observed waveforms at Awashima.

the contributions to the total energy from potential energy, kinetic energy, and energy loss due to friction for the closed boundary case described above.

Figures 11(a–d) compare simulations for the WW scenario against observed waveforms at the 16 stations shown in Figure 4. Results confirm earlier discussions. The model tends to overpredict periods and underpredict amplitudes at remote

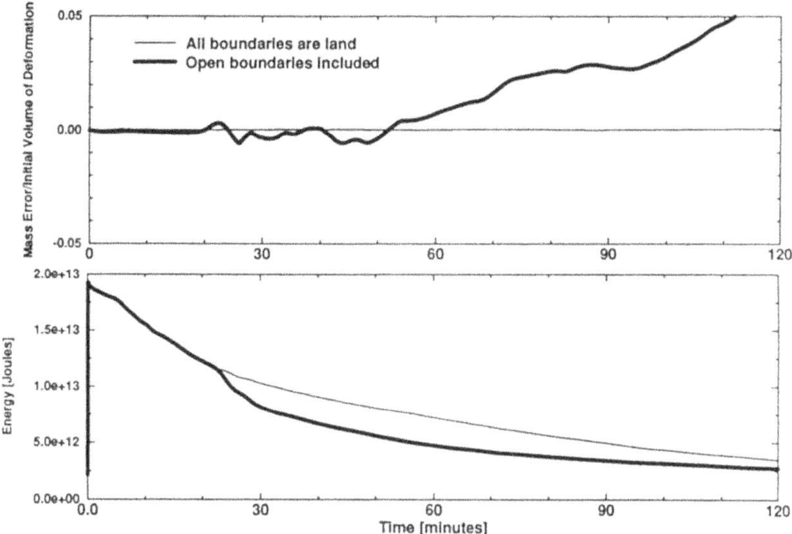

Figure 9
Mass and energy $(E_t + E_f)$ variation for WW source scenario.

stations. Wave amplitudes and arrival times at near-field stations such as Esashi and Iwanai are reasonably reproduced, suggesting that the fault plane model reasonably represents the magnitude of the sea floor deformation.

Simulated patterns of propagation of energy throughout the domain are qualitatively shown in Figures 12(a–b). The darker shades in these figures correspond to

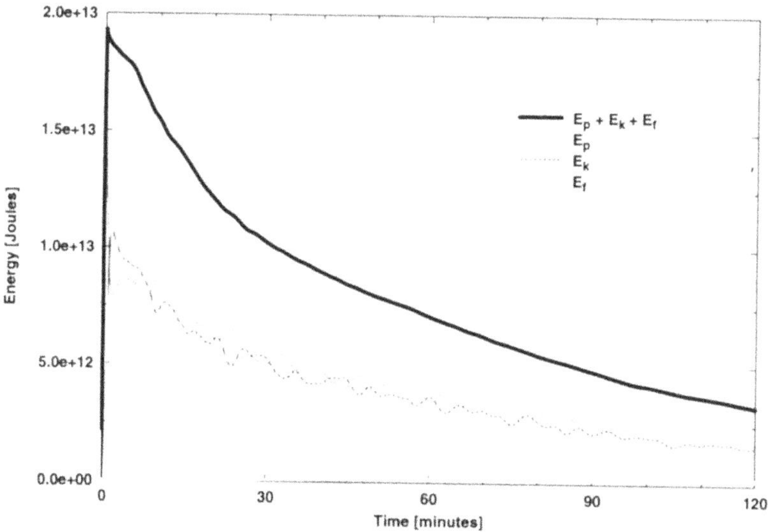

Figure 10
Potential, kinetic, and frictional energy components.

higher energy values. Much of the energy is dissipated once the waves have impacted land boundaries. Figure 13(a) shows simulated isolines of maximum energy throughout the simulation, while Figures 13(b–c) show observed runups along the islands of Okushiri and Hokkaido from post-tsunami survey teams. It can be seen that the larger runup measurements match fairly well the isolines of maximum energy. Okushiri and the northwest coast of Hokkaido sustained the higher energy levels which is consistent with field survey reports noting where the larger runup measurements were observed.

Maximum wave heights along the coastlines of Okushiri and Hokkaido are shown in Figures 14(a–b), for linear and nonlinear simulations. Linear simulations were performed with uncorrected bathymetry, while, as previously discussed, non-linear simulations require a minimum depth (i.e., artificial deepening of near coastal

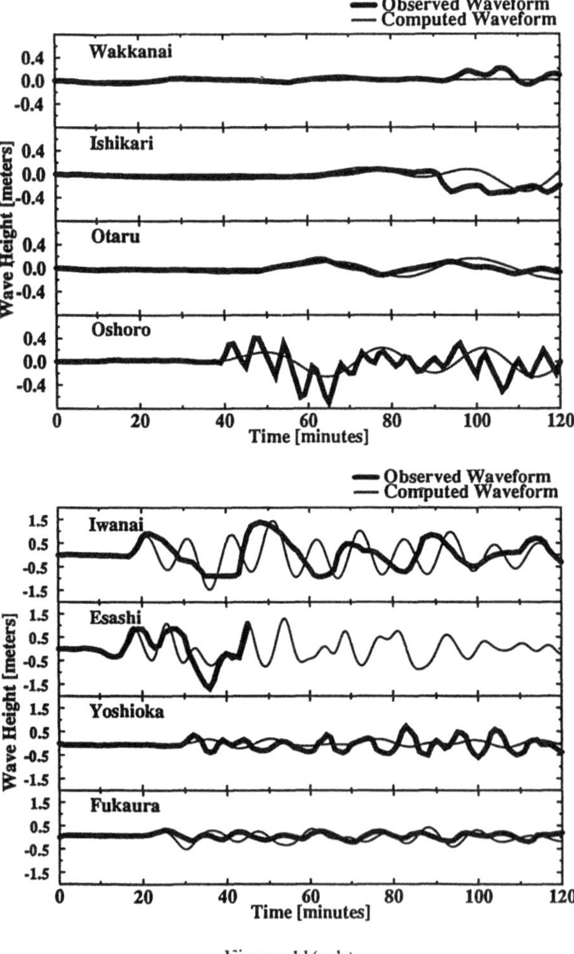

Figure 11(a,b)

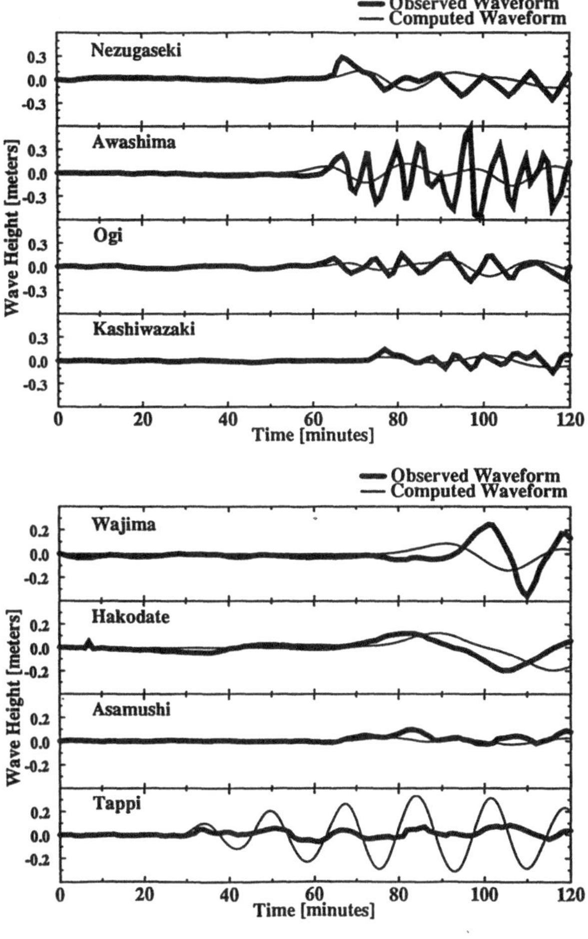

Figure 11(a–d)
Comparisons of source WW results with observed waveforms.

regions) to avoid instabilities. Because of this practical consideration, linear maximum wave heights are considerably larger than their nonlinear counterparts, and significantly better match the observed runups shown in Figures 13(b–c).

5. Analysis and Interpretation

The simulated wave patterns reproduce limited observations. In particular, if we consider the reference numerical simulation (nonlinear, WW scenario, empirical parameters as set in Section 3.3):

Initial Energy Distribution

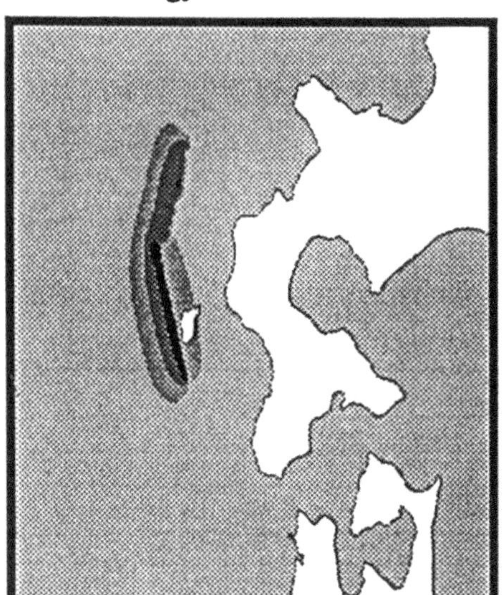

Time = 3 minutes

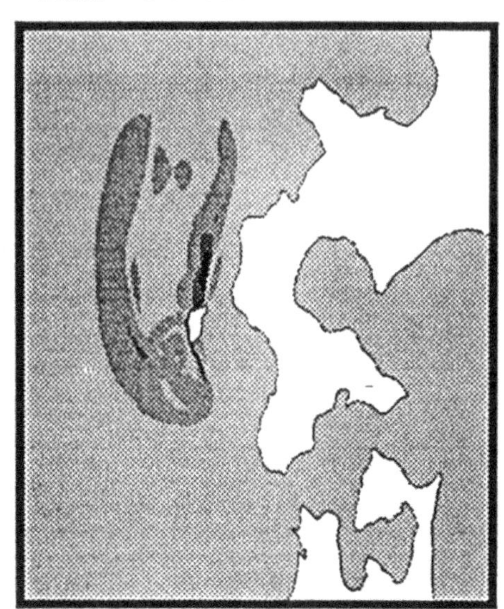

Time = 6 minutes

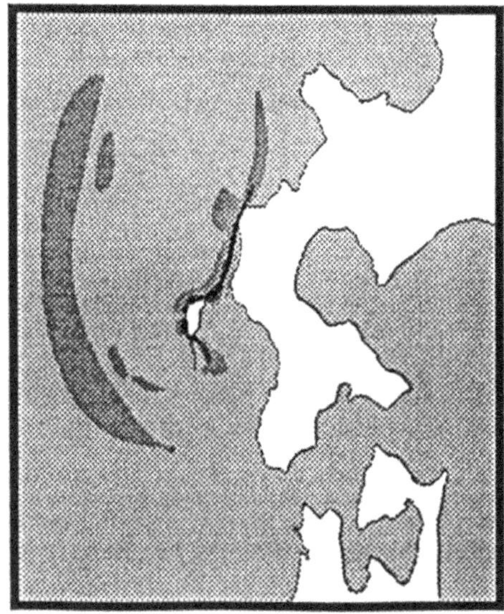

Time = 9 minutes

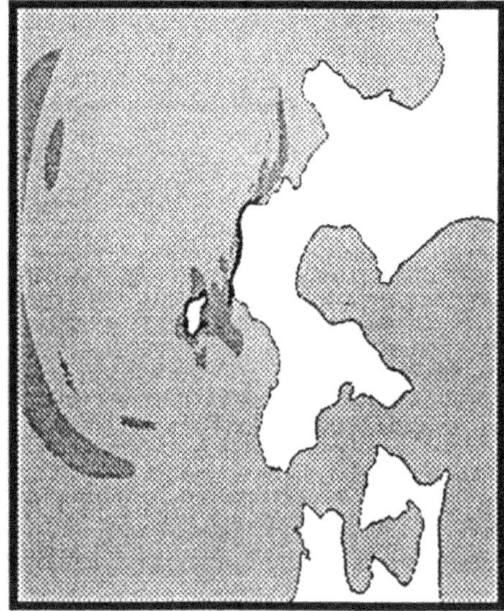

Figure 12(a).

Time = 12 minutes

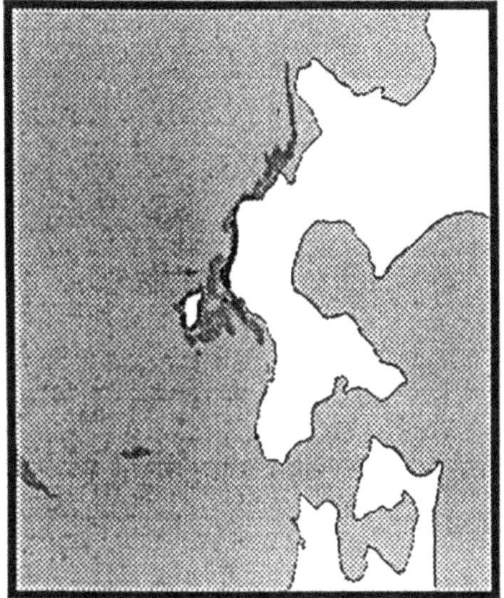

Time = 15 minutes

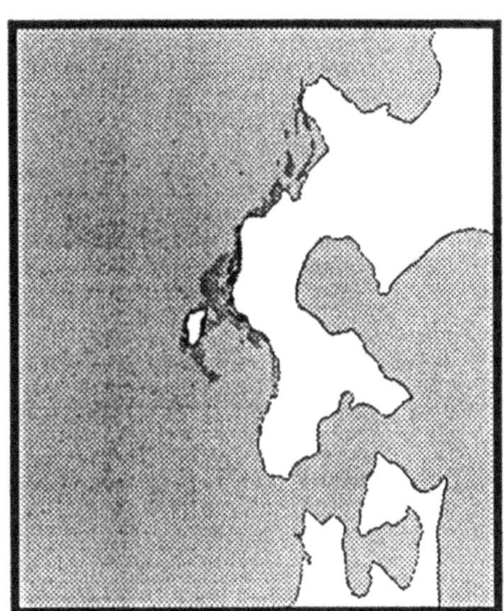

Time = 18 minutes

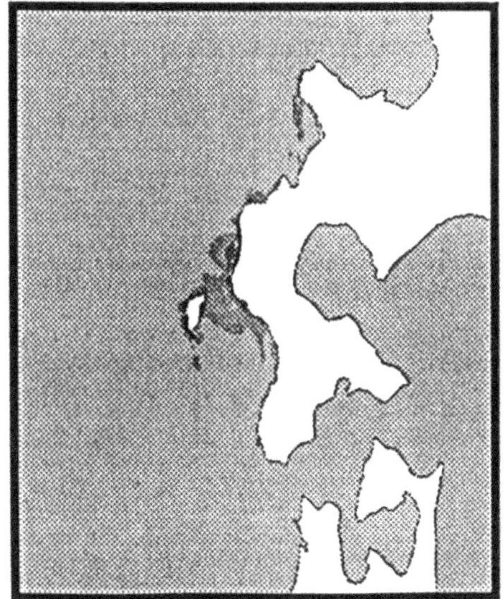

Time = 21 minutes

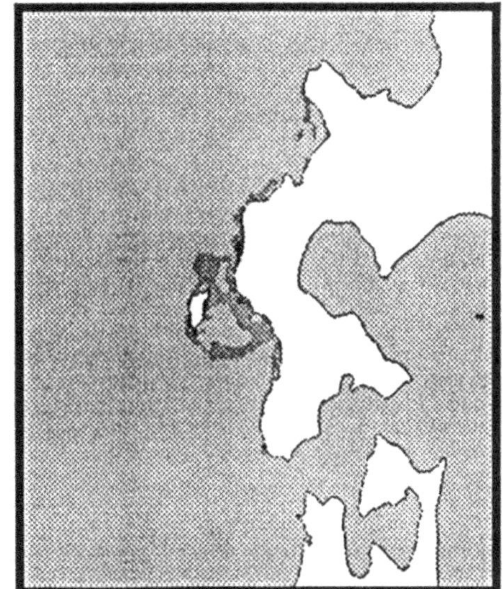

Figure 12(b)

Qualitative propagation of energy $(E_t + E_f)$ over time for WW simulation (darker regions represent higher levels of energy.

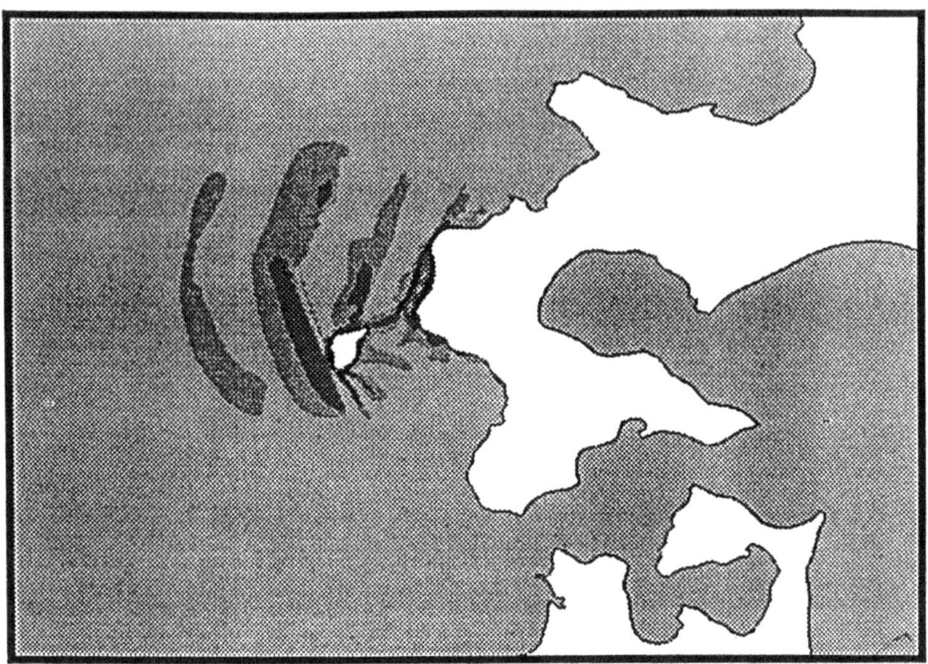

Figure 13(a)

Isolines of maximum energy ($E_t + E_f$) for WW simulation (darker regions represent higher levels of energy).

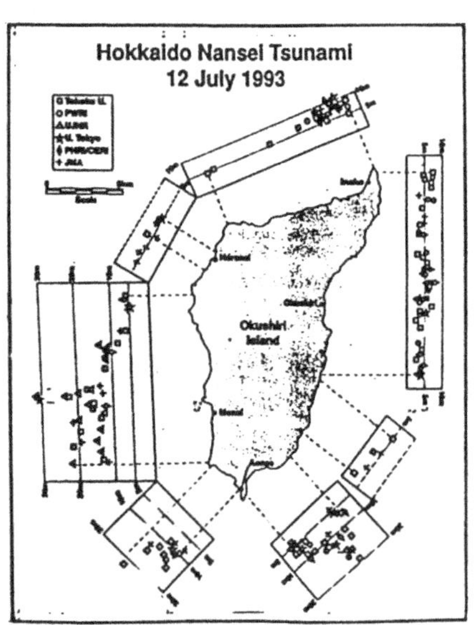

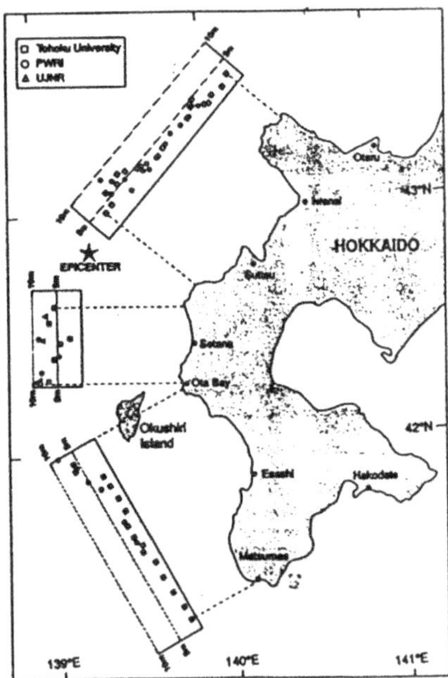

Figure 13(b,c)

Runup measurements (extracted from HOKKAIDO TSUNAMI SURVEY GROUP, 1993).

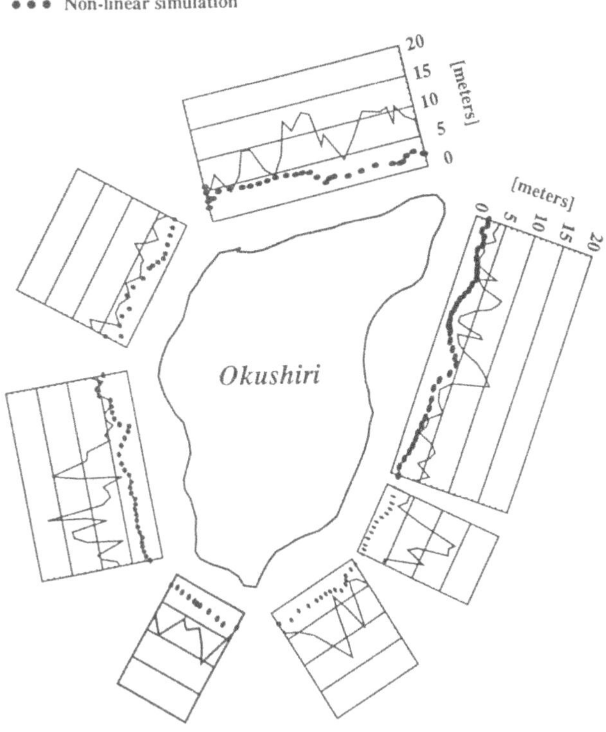

Figure 14(a)
Computed maximum wave heights along Okushiri.

- areas of observed energy concentration along Okushiri and Hokkaido are generally well identified;
- observed maximum amplitudes and arrival times at near-field tidal gauges (specifically, Iwanai and Esashi) are satisfactorily represented;
- significant wave action continues for several days after the earthquake (results not shown).

However:

- observed runups along the coasts of Okushiri and Hokkaido are systematically underpredicted, to a degree that cannot be explained only by the model's inability to represent inundation;
- observed periods at the near-field stations are underestimated;
- observed amplitudes, periods, and arrival times at remote stations are reproduced unevenly;
- observed periods at several remote stations (Oshoro, in northern Hokkaido, and Nezugaseki, Awashima, Ogi, and Kashiwazaki, all on or in the vicinity

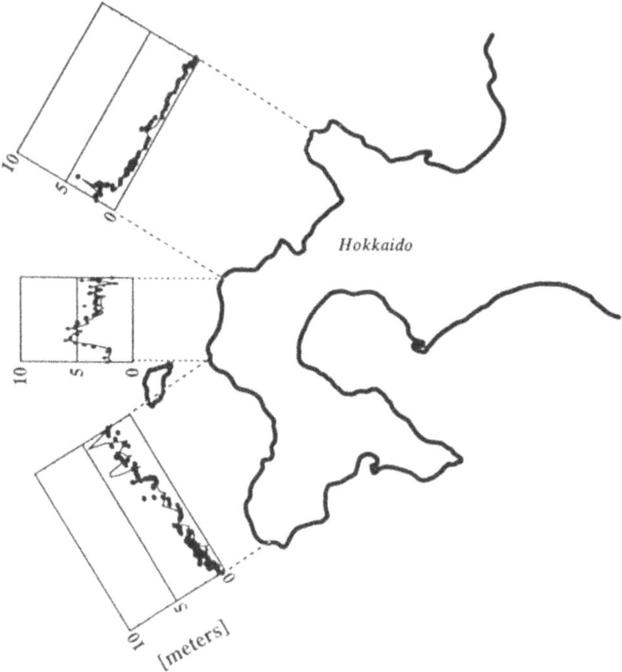

Figure 14(b)
Computed maximum wave heights along Hokkaido.

of Honshu) are significantly overestimated, suggesting that the model may not be correctly representing local effects.

The above differences between observations and simulations may result from a variety of factors. The following interpretation, although at times subjective, puts several of these factors into perspective:

● Fault plane model

The adopted fault plane model (scenario WW) appears to represent some aspects of the source mechanism in a reasonable manner. In support of this argument, we note that (a) there is a rather satisfactory agreement between observed and simulated amplitudes and arrival times at Iwanai and Esashi, and (b) the disagreement between observed and simulated runups at Okushiri and Hokkaido appears to be primarily due to limitations of the hydrodynamic model (see discussion of formulation, below), rather than to the source mechanism. However, the fault plane model (e.g., through a combination of the adopted width and dip angle) may be responsible for the disagreement between observed and simulated periods.

● Hydrodynamic model/formulation

The adopted finite element model has two major limitations: an intrinsic inability to describe inundation, a process that is not included in the model formulation; and a practical inability to remain stable for small water depths (i.e., large η/H) when nonlinearities are accounted for. The latter limitation forced the artificial "deepening" of several coastal regions, which is arguably the leading reason why the nonlinear model systematically and very significantly underpredicts "runups" at Okushiri (where minimum depth was fixed at 15 m); this argument is well supported by the comparison of linear and nonlinear simulations (Figures 14a and 14b), which favor the linear simulations.

● Hydrodynamic model/discretization

The model benefited from the ability, inherent to finite elements, to flexibly place added refinement where needed (e.g., fault area and coast of Okushiri). Furthermore, the use of a versatile grid generator enabled effective grid adjustments throughout the modeling process. While the need to artificially "deepen" certain coastal regions arguably prevented a more full utilization of the available grid flexibility, sensitivity tests conducted with grids finer than the one shown in Figure 2 suggest that the adopted discretization was not a significant source of error.

● Waveform data

The waveform data available for this study have potential inaccuracies, associated with at least three factors: measurement errors (gauges have built-in inertia relative to signals (including tsunamis) with smaller-than-tidal periods); digitization errors, both due to the mechanical reproduction of the original records and to operator error, and processing errors, associated with the removal of tides. None of these factors explains satisfactorily the differences between observations and simulations, though. Indeed, earlier research on the response of Japanese tidal gauges to tsunamis (SATAKE *et al.*, 1988) indicates that Iwanai and Esashi have lag-times of the order of only 1–3 minutes, which appear insufficient to explain why observed periods at these key stations are exceedingly larger than simulated periods. Digitization errors can be roughly estimated by visual examination of paper copies of the tidal gauge records released by the Japanese Meteorological Agency (JMA), and are relatively small. And while we know little about the filtering process, tides are too modest to have a major impact.

Beyond the comparison observations/simulations, our results provide significant insight on this and possibly other models' (in)ability to preserve mass and energy. Transmissive boundaries are clearly responsible for the bulk of the identified mass unbalances (Figure 9).

With regard to energy losses, we hypothesize that they result from a combination of effects, which include:

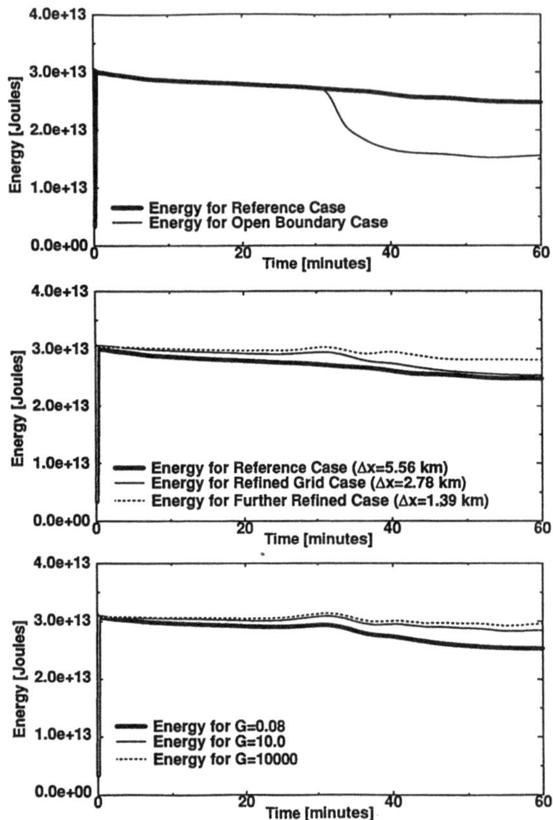

Figure 15(a–c)

Energy (E_t) comparison for reference and test cases: (a) open boundary case, (b) refined grid case, (c) effect of G on energy.

- Numerical damping, which can be controlled in part through additional grid refinement and through the choice of G.
- Interaction of the wave with the land boundary, possibly associated with an inherent inability of the shallow water equations to preserve energy when vertical accelerations cannot be ignored in the pressure balance.

Short of providing a formal justification, we briefly describe the synthetic experiment that motivated this hypothesis. The generation and propagation of a $M_W = 8.8$ tsunami was simulated in a 550 km long, rectangular, frictionless channel, with a nonuniform bottom topography roughly representing the transition from deep ocean to a continental shelf. Six simulations were performed:

- A reference simulation, with a 5.56 km discretization, and all-closed boundaries, showed the energy behavior illustrated in Figure 15 (dark lines).

Energy loss occurs in a fairly steady manner and not as extensively as was observed in the Sea of Japan simulations.

- A second simulation, with a transmissive boundary on the left channel side, and a land boundary on the right channel side, showed that energy was lost as the waves departed through the transmissive boundary soon after 30 minutes into the simulation (Figure 15a, lighter line); this is not a limitation of the model, reflecting rather the fact that our expression for the total tsunami energy does not account for boundary losses.

- A third simulation, with a more refined grid and all-closed boundaries, showed a slightly better energy preservation, but the same rate of decline (Figure 15b, solid light line). A fourth simulation with a further refined grid again displayed better energy preservation (Figure 15b, dotted light line). In these third and fourth simulations, a new feature begins to appear between 30 and 40 minutes. At this time, a slight energy variation becomes more visible with increased refinement. In the fourth simulation, two variations are clearly evident; the timing of each coincides with the arrival at the land boundary of (1) the trough of the wave (~ 30 minutes) and (2) the crest of the wave (~ 40 minutes).

- Finally, two simulations were performed to examine the damping effect of G on energy preservation (Figure 15c). Increasing G from 0.08 (reference simulation) to 10.0 (light line) and 10000.0 (dotted line) leads to better conservation of energy. Thus, as the GWCE becomes more similar to the primitive continuity equation (larger values of G), energy is better preserved. This is consistent with the GWCE introducing numerical damping at high frequencies, to control spatial oscillations.

This experiment shows that numerical damping (through both grid resolution and the value of G) may induce significant energy losses, which justifies part one of our hypothesis. They also suggest, however, that land boundaries disrupt the energy balance. This suggests (but does not prove, further research being needed) part two or our hypothesis, because the hydrostatic approximation is least realistic when vertical accelerations are significant (as during the interaction of the wave with land boundaries).

We note that, if the hydrostatic approximation is indeed unsupportable during tsunami interactions with the coastline, all models based on the shallow water equations should show a similar pattern of energy losses. This appears to be the case with the only other model that we tested (an adaptation of MADER, 1988), but remains to be demonstrated in general.

6. Final Considerations

The difficulties experienced by this and presumably other numerical models to explain, after the fact, certain aspects of the propagation of the reasonably well-constrained and reltively well-monitored Hokkaido Nansei-Oki tsunami, is a sobering reminder of the limitations of the current state-of-the-art, and of its implications for the design of adequate strategies for coastal protection and emergency response.

Enhancements in the current ability to characterize source mechanisms and to correctly describe coastal inundation and associated velocity fields have for some time been identified as critical for a better understanding of tsunami events. Because a number of new approaches are being developed in each of these areas by different research groups, there is a very clear need for the definition of a set of widely accepted benchmarks, based both on actual tsunami events and on laboratory experiments. In this context, the workshop being organized by YEH *et al.* (1995) may prove extremely valuable.

Acknowledgments

We thank J. J. Westerink and R. A. Luettich for providing the base code for the finite element model ADCIRC. Bathymetric data were provided by M. Okada of the Meteorological Research Institute through K. Satake (for the Sea of Japan) and by S. Gusiakov through V. Titov (for the vicinity of Okushiri). Waveform data were provided by JMA, through K. Satake. K. Satake and F. Imamura provided parameters for the seismic source scenarios. Runup data were collected by diverse post-tsunami survey teams, including those from Tohoku University, the Public Works Research Institute, the Port and Harbor Research Institute, the Civil Engineering Research Institute, the United States-Japan Cooperative Program in Natural Resources, the University of Tokyo, JMA, and the Oregon Graduate Institute and the Oregon Department of Geology and Mineral Industries (BAPTISTA *et al.*, 1993). The NOAA/PMEL Tsunami Bulletin Board played a vital role in relaying information and data from this event. This research was partially sponsored by the Oregon Sea Grant and by DOD/AASERT Contract DAALO3-92-G-0065.

REFERENCES

BAPTISTA, A. M., PRIEST, G. R., TANIOKA, Y., and MYERS, E. P. (1993), *A Post-tsunami Survey of the 1993 Hokkaido Tsunami*, EOS Transactions *74* (43), 349.

GELLER, R. J. (1976), *Scaling Relations for Earthquake Source Parameters and Magnitudes*, Bull. Seismol. Soc. Am. *66* (5), 1501–1523.

GRENIER, R. R., LUETTICH, R. A., and WESTERINK, J. J. (1994), *Comparison of 2D and 3D Models for Computing Shallow Water Tides in a Friction-dominated Tidal Embayment*, Proceedings of the 3rd International Conference on Estuarine and Coastal Modeling, American Society of Civil Engineers, 58–70.

HOKKAIDO TSUNAMI SURVEY GROUP (1993), *Tsunami Devastates Japanese Coastal Region*, EOS Transactions, AGU *74* (37), 417.

IMAMURA, F. (1993), *personal communication.*

KAJIURA, K. (1981), *Tsunami Energy in Relation to Parameters of the Earthquake Fault Model*, Bull. Earthq. Res. Inst. *56*, 415–440.

KINMARK, I. P. E., and GRAY, W. G. (1984), *Am Implicit Wave Equation Model for the Shallow Water Equations*, Adv. Water Resources *7*, 2–14.

KOLAR, R. L., WESTERINK, J. J., CANTEKIN, M. E., and BLAIN, C. A. (1994), *Aspects of Nonlinear Simulations Using Shallow Water Models Based on the Wave Continuity Equation*, Comp. Fluids *23* (3), 523–538.

KUMAKI, Y., KISANUKI, J., OHTANI, T., ONO, Y., and KAJIKAWA, S. (1993), *Vertical Seismic Crustal Movement of the 1993 Hokkaido-Nansei-Oki Earthquake Based on Coastal Landform Changes in the Okushiri Island, West of Hokkaido, Japan*, Meeting of the Seismological Society of Japan *A63*, 63 (in Japanese).

LUETTICH, R. A., WESTERINK, J. J., and SCHEFFNER, N. W. (1991), *An Advanced Three-dimensional Circulation Model for Shelves, Coasts, and Estuaries*, Dept. of the Army, U.S. Army Corps of Engineers, Washington, D.C.

LYNCH, D. R., and GRAY, W. G. (1979), *A Wave Equation Model for Finite Element Tidal Computations*, Comp. Fluids *7* (3), 207–228.

MADER, C. L., *Numerical Modeling of Water Waves* (University of California Press, 1988).

MYERS, E. P. (1994), *Numerical Modeling of Tsunamis with Applications to the Sea of Japan and the Pacific Northwest*, M.Sc. Thesis, Department of Environmental Science and Engineering, Oregon Graduate Institute of Science and Technology, Portland, OR, U.S.A.

OKADA, Y. (1985), *Surface Deformation due to Shear and Tensile Faults in a Half-space*, Bull. Seismol. Soc. Am. *75* (4), 1135–1154.

SATAKE, K., OKADA, M., and ABE, K. (1988), *Tide Gauge Response to Tsunamis: Measurements at 40 Tide Gauge Stations in Japan*, J. Mar. Res. *46*, 557–571.

SATAKE, K. (1994), *personal communication.*

TOLMAN, H. L. (1992), *An Evaluation of Expressions for Wave Energy Dissipation due to Bottom Friction in the Presence of Currents*, Coastal Engin. *16*, 165–179.

TURNER, P. J., and BAPTISTA, A. M. (1991), *ACE/Gredit Users Manual: Software for Semi-automatic Generation of Two-dimensional Finite Element Grids*, CCALMR Software Report SDS2(91–2), Oregon Graduate Institute of Science and Technology, Portland, OR, U.S.A.

WESTERINK, J. J., LUETTICH, R. Z., BAPTISTA, A. M., SCHEFFNER, N. W., and FARRAR, P. (1992), *Tide and Storm Surge Predictions Using a Finite Element Model*, ASCE J. Hydraulic Eng. *118* (10), 1373–1390.

YEH, H., LIU, P., and SYNOLAKIS, C. (1995), *Benchmark Problems for the International Workshop on Long-Wave Runup Models* (unpublished).

(Received August 10, 1994, revised February 2, 1995, accepted February 16, 1995)

PAGEOPH, Vol. 145, Nos. 3/4 (1995)

0033–4553/95/040803–19$1.50 + 0.20/0
© 1995 Birkhäuser Verlag, Basel

Tsunami Generation of the 1993 Hokkaido Nansei-Oki Earthquake

KENJI SATAKE[1] and YUICHIRO TANIOKA[1]

Abstract—Heterogeneous fault motion of the 1993 Hokkaido Nansei-Oki earthquake is studied by using seismic, geodetic and tsunami data, and the tsunami generation from the fault model is examined. Seismological analyses indicate that the focal mechanism of the first 10 s, when about a third of the total moment was released, is different from the overall focal mechanism. A joint inversion of geodetic data on Okushiri Island and the tide gauge records in Japan and Korea indicates that the largest slip, about 6 m, occurred in a small area just south of the epicenter. This corresponds to the initial rupture on a fault plane dipping shallowly to the west. The slip on the northernmost subfault, which is dipping to the east, is about 2 m, while the slips on the southern subfaults, which are steeply dipping to the west, are more than 3 m. Tsunami heights around Okushiri Island are calculated from the heterogeneous fault model using different grid sizes. Computation on the smaller grids produces larger tsunami heights that are closer to the observed tsunami runup heights. Tsunami propagation in the nearly closed Japan Sea is examined as the free oscillation of the Japan Sea. The excitation of the free oscillation by this earthquake is smaller than that by the 1964 Niigata or 1983 Japan Sea earthquake.

Key words: Tsunamis, 1993 Hokkaido Nansei-Oki earthquake, Okushiri Island.

1. Introduction

The Hokkaido Nansei-Oki (southwest off Hokkaido) earthquake of July 12, 1993 generated devastating tsunamis that caused significant damage in Japan, particularly on Okushiri Island. The death total (including those still missing) was 230 and the damage totaled 1.2 billion U.S. dollars (ISHIYAMA, 1994). This earthquake was recorded on seismograms worldwide. The tsunamis were recorded on many tide gauges around the Japan Sea, including several in Korea. In addition to these instrumental data, extensive measurements were made on Okushiri Island for tsunami runup heights and crustal deformation. These surveys (e.g., HOKKAIDO TSUNAMI SURVEY GROUP, 1993; KUMAKI *et al.*, 1993; TSUTSUMI *et al.*, 1993; SHUTO and MATSUTOMI, 1995), show that the maximum tsunami runup height was 32 m on Okushiri Island and that the entire island subsided by 5–80 cm.

Along the eastern margin of the Japan Sea (Figure 1), other earthquakes with a similar size have occurred in this century (FUKAO and FURUMCTO, 1975;

---

[1] Department of Geological Sciences, University of Michigan, Ann Arbor, MI 48109-1063, U.S.A.

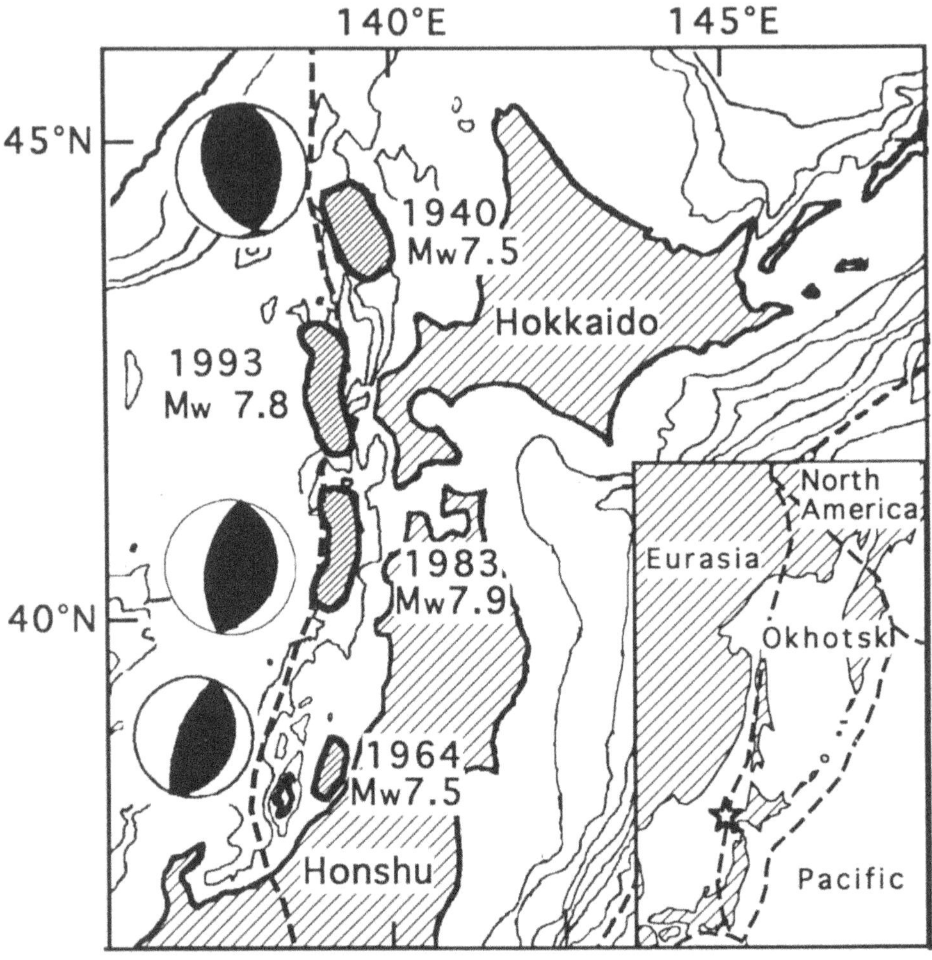

Figure 1

The source regions and focal mechanisms of the earthquakes along the eastern margin of the Japan Sea.
Plate boundaries in the region are shown in the inset map.

SATAKE, 1986). They are the 1940 Shakotan-oki (M_w 7.5), 1964 Niigata (M_w 7.5), and 1983 Japan Sea (M_w 7.9) earthquakes. Because of the semi-periodic occurrence of large earthquakes and geological background, it has been proposed in the last decade (NAKAMURA, 1983; KOBAYASHI, 1983) that the boundary between the North American and Eurasian plates has moved to the eastern margin of the Japan Sea in the last 1–2 Ma. A more recent model introduces another plate, the Okhotsk plate, and argues that the eastern margin of the Japan Sea is a convergent boundary between the Eurasian and Okhotsk plates (SENO and SAKURAI, 1993).

In this paper, we estimate the heterogeneous fault motion of the 1993 Hokkaido Nansei-Oki earthquake and examine the tsunami generation from the fault model.

We first summarize the seismological analyses in Section 2. After we describe the tsunami and geodetic observations in Section 3, we perform inversions of tsunami and geodetic data in Section 4 to estimate the slip, or moment, distribution on the fault. The tsunami heights around Okushiri Island are computed using different grid sizes and are compared with the observed runup heights in Section 5. In Section 6, the tsunami generation is discussed from the viewpoint of free oscillation of the Japan Sea. The excitation coefficients are computed from the fault model and compared with those from other tsunami events in the Japan Sea.

2. Seismological Analyses

2.1 Analyses of Global Data

The Hokkaido Nansei-Oki earthquake occurred at 13 h 17 m 11.9 s on July 12 (GMT). The epicenter is 42.851°N, 139.197°E and the surface wave magnitude (M_s) is 7.6, according to the National Earthquake Information Service. The moment tensor solutions were estimated by Harvard University (DZIEWONSKI *et al.*, 1994) and the USGS (SIPKIN, 1994). The former uses long-period waves whereas the latter uses shorter-period body waves. Both show a reverse fault with the strike approximately in the N–S direction. However, the dip angles were quite different. The Harvard CMT solution shows that the fault plane is either shallowly (with a dip angle of 35°) dipping to the east or steeply (55°) dipping to the west. The USGS solution, on the other hand, shows a shallowly (39°) dipping plane to the west and a steeply (60°) dipping plane to the east. The seismic moment estimates are also different; the Harvard CMT solution gives 4.7×10^{20} Nm, while the USGS solution is 1.1×10^{20} Nm.

Because of these discrepancies, we independently performed both CMT and body wave inversions (TANIOKA *et al.*, 1995). The inversion of teleseismic body waves for Moment Tensor Rate Functions (MTRF inversion) exhibits a similar focal mechanism to that obtained by the USGS; a shallow (24°) dipping plane to the west and steeply (66°) dipping plane to the east. The first motions on global seismic networks are also consistent with this focal mechanism. The moment rate function, or the source time function, shows that about 2×10^{20} Nm of the moment was released within the first 10 s, followed by a continuing moment release up to 1 min. Using this time function as the Green's function, we performed the CMT inversion. The result is still very similar to the Harvard solution; a shallowly (32°) dipping plane to the east and a steeply (63°) dipping plane to the west, and the total seismic moment is 5.5×10^{20} Nm.

The CMT solution represents an average picture of focal mechanism and seismic moment, whereas the MTRF solution from body waves represents an initial rupture. The comparison indicates that about a third of the total seismic moment was released in the first 10 s with a somewhat different focal mechanism (TANIOKA *et al.*, 1993).

2.2 *Analysis of Local Data*

The aftershock distribution has been determined by KASAHARA *et al.* (1994) and shown in Figure 2. The figure shows the epicenters of aftershocks that occurred within 5 months of the main shock. The aftershocks plotted in Figure 2 were all recorded at a station on Okushiri Island and relatively well-located. Nevertheless, the absolute depth may not be well constrained because of poor station distribution and incomplete knowledge of the velocity structure. The dip angle of fault plane is therefore more reliably estimated from focal mechanism studies; the aftershock distribution facilitates our choice of the actual fault plane from the two nodal planes.

The panels on the right in Figure 2 show the cross sections of the aftershock distribution. We will mostly focus on the upper edge of the aftershock clusters. In the northernmost section (A), the distribution shows a planar structure dipping to the east. Along the next two sections (B and C), the clusters, particularly the upper

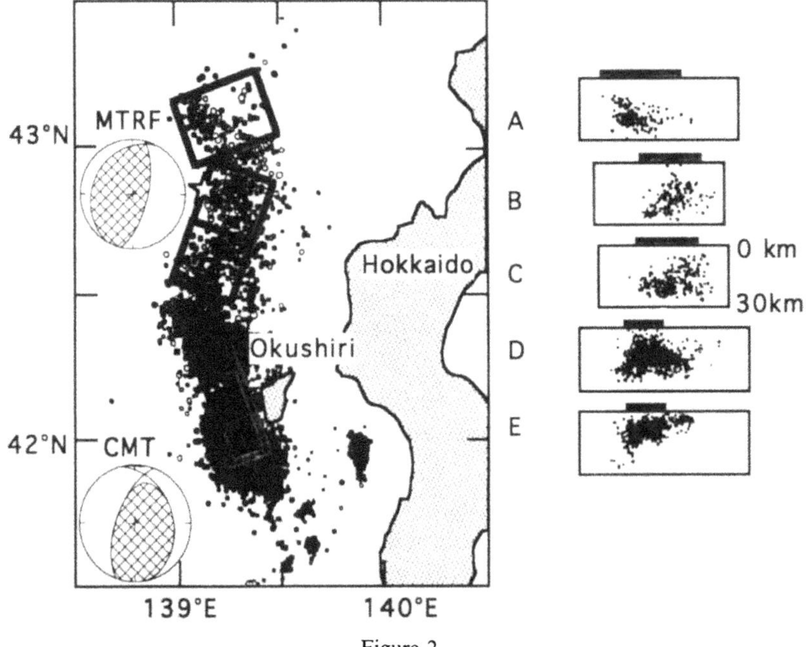

Figure 2

The aftershock distribution determined by KASAHARA *et al.* (1994). Frames in the map show the projections of subfaults that are used for the inversion. Panels in the right show the cross sections perpendicular to the strike of the subfaults. The shaded box indicates the projection of the subfaults. Two focal mechanism solutions are also shown. Star indicates the hypocenter of the main shock.

edge, seem to be dipping to the west, although it is less clear than section A. Along the southern two sections (D and E), it is more difficult to judge the orientation of the fault plane. In particular, along section D, two planes, one shallowly dipping to the east and the other steeply dipping to the west, may be identified.

The two focal mechanism solutions are also shown in Figure 2. As mentioned above, the MTRF solution represents the focal mechanism of the first 10 s of rupture, presumably near the epicenter (in section B). The aftershock distribution suggests that the shallowly (24°) dipping plane to the west is the actual fault plane.

3. Tsunami and Geodetic Observations

3.1 Tsunamis on Tide Gauges

Tsunamis from the Hokkaido Nansei-Oki earthquake were recorded on many tide gauge stations in Japan and Korea. Figure 3 displays the locations of tide

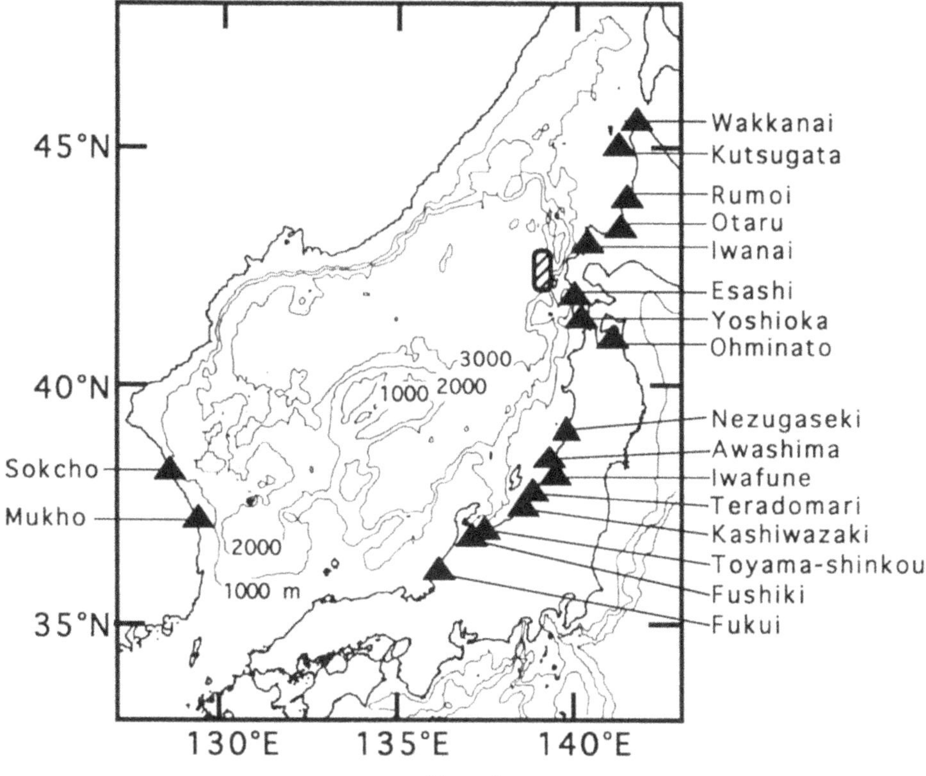

Figure 3

The location of the 1993 tsunami source and the tide gauge stations (solid triangles) used in this study. The map also shows the tsunami computational area and the bathymetry.

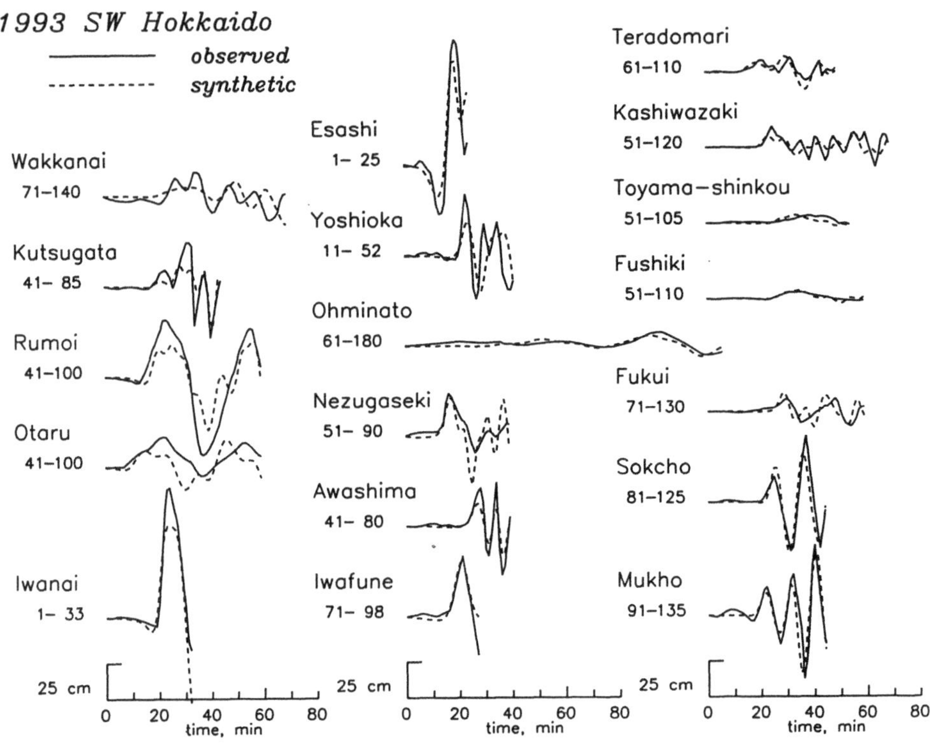

Figure 4

The observed tsunami waveforms (solid curves) compared with the computed ones (dashed curves) from the heterogeneous slip distribution estimated in this study. The time range of the waveform in min from the earthquake origin time is shown below the station name. For example, the Wakkanai record starts at 71 min after the origin time.

gauge stations we used in our analysis. The tsunami waveforms are shown in Figure 4. The closest tide gauge stations to the source area are Iwanai to the north and Esashi to the south. There was no station on Okushiri Island. At Iwanai, the sea level first fell about 12 min after the earthquake (the ground shaking due to the earthquake was recorded on the tide gauge) and a sharp rise started at about 20 min. The amplitude of the first fall was 6 cm and the following rise was about 1 m. The maximum amplitude (140 cm) was registered in the next cycle at about 55 min after the earthquake. At Esashi, the first fall of water level started at 10 min after the earthquake (the ground shaking was also recorded at this station) and the amplitude was about 34 cm. The amplitude of the following peak is 90 cm. After another peak, the gauge went off-scale. Tsunamis are recorded at other stations with smaller amplitudes (Figure 4). The tsunamis arrived at Korean tide gauge stations 100–110 min after the event with amplitudes of more than 1 m.

3.2 Tsunami Runup Data

The tsunami runup heights were extensively measured on Okushiri Island and the Hokkaido and Honshu coasts. On Okushiri Island, the largest tsunami runup was 32 m, recorded in a small valley near Monai. The average runup height is more than 10 m in the southern half of the island, while it is much less than 10 m in the northern half. The detailed survey report can be found in HOKKAIDO TSUNAMI SURVEY GROUP (1993) and SHUTO and MATSUTOMI (1995). We will later compare the calculated tsunami heights with the observed runup heights.

3.3 Geodetic Data

Subsidence of Okushiri Island has been measured by various methods. TSUT-SUMI *et al.* (1993) performed continuous measurements of the water level at several harbors and estimated the water levels, consequently the ground heights, after the earthquake. Comparison of these heights with leveling data before the earthquake revealed the subsidence amounts with errors smaller than 5 cm. KUMAKI *et al.* (1993) made similar estimates on coastal constructions such as breakwaters. They also made GPS measurements. These data are compiled in Figure 5. The amount of subsidence is smaller at the northeastern end of the island (smallest 5 cm, but mostly about 20 cm) and larger at the southwestern end (about 80 cm). The island, as a whole, had tilted to the west.

4. Inversions of Geodetic and Tsunami Data

4.1 Method

Among the various fault parameters such as size and geometry (strike, dip and slip angles), only the slip amount is linearly related to the crustal deformation and consequently tsunami waveforms. We fix the other parameters and estimate the slip distribution on the fault from geodetic and tsunami data. The observational equation is written as (SATAKE, 1993)

$$A_{ij} \cdot x_j = b_i \tag{1}$$

where x_j is the slip amount on subfault j, b_i is the observation (observed subsidence or tsunami waveforms) made at station i, and A_{ij} is the Green's function from subfault j at station i. The Green's function is the calculated crustal deformation or tsunami waveforms from a unit amount (in the present case, 1 m) of slip on each subfault.

For the geodetic Green's functions, we use vertical deformation at each observation point calculated from a finite fault in an isotropic elastic body (e.g., OKADA,

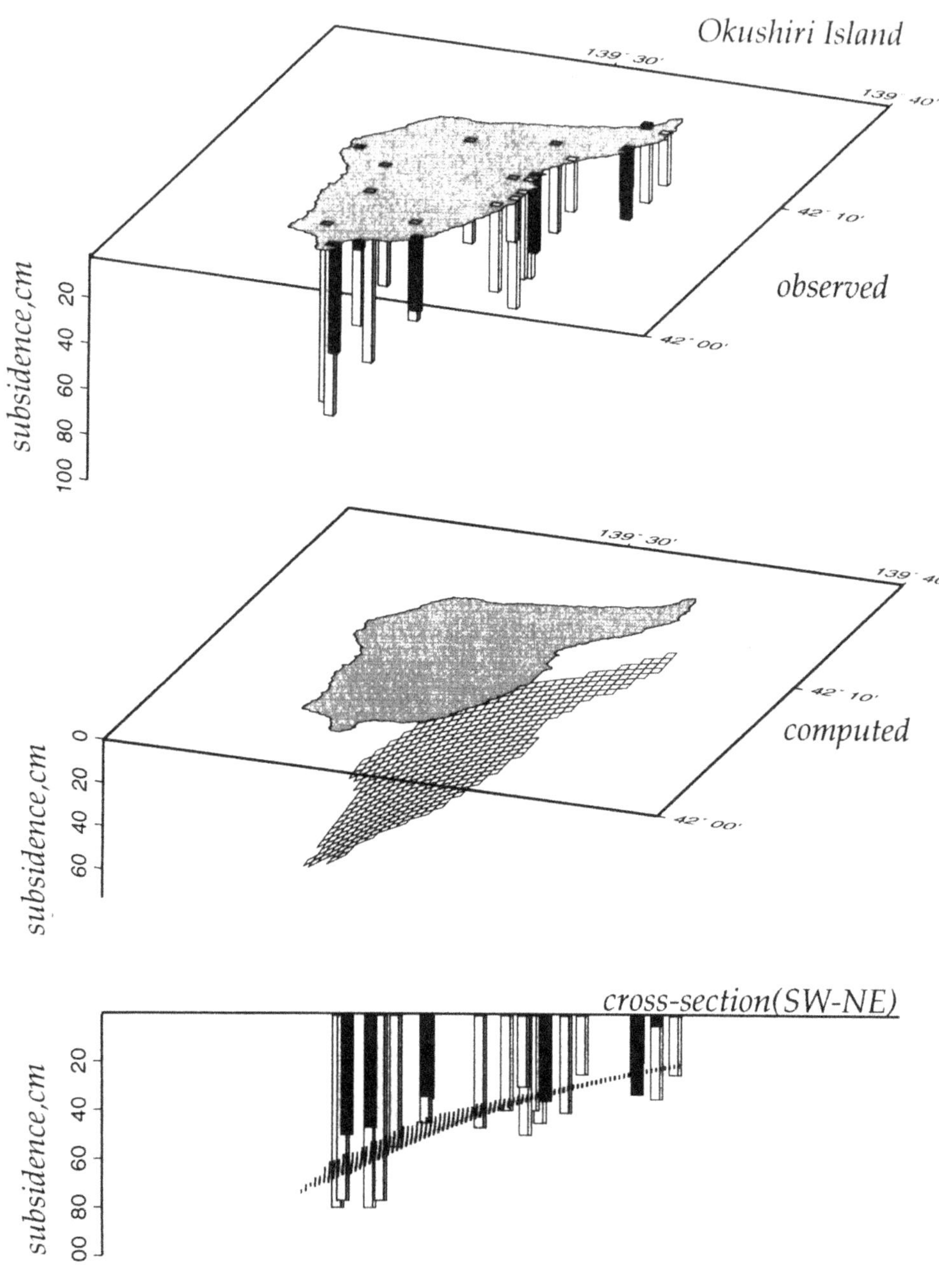

Figure 5

(top) The observed subsidence of Okushiri Island. The dark shaded bars are the data by TSUTSUMI *et al.* (1993) based on the landmark of harbors, the lightly shaded bars are GPS data by KUMAKI *et al.* (1993), and the white bars are the data by KUMAKI *et al.* (1993) based on coastal constructions. *(center)* The computed subsidence from the final slip distribution on the subfaults. *(bottom)* Cross section along a line perpendicular to the strike of subfault E.

1985). For the tsunami Green's functions, we calculated the tsunami waveform at each tide gauge station from bottom deformation due to each subfault. The linear shallow water equation

$$\frac{\partial V}{\partial t} = -g\nabla h \tag{2}$$

and the equation of continuity

$$\frac{\partial h}{\partial t} = -\nabla \cdot (dV) \tag{3}$$

where V is the horizontal velocity of water, g is the gravitational acceleration, h is the water height and d is the water depth, are solved using the finite-difference method. The computational area is shown in Figure 3. The grid size is 1 minute and there are 960×900 grid points. The time step of the computation is 5 s to satisfy the stability condition. For other computational details, see SATAKE (1995).

We divided the source area into five subfaults as shown in Figure 2. The fault size and strike are determined in such a way that the subfaults cover the aftershock area. The size of each subfault is given in Table 1. Top edge of all the subfaults is on the ocean bottom (0 km depth). The dip angles are determined from focal mechanism solutions. The northernmost subfault, A, strikes N20°W and dips 30° to the east. The next two subfaults, B and C, strike N20°E and dip 30° to the west. For the southernmost two subfaults, D and E, we considered both the eastward dipping (30°) fault plane and the westward dipping (60°) fault plane. The strikes are the same, N20°W.

We applied the jackknifing technique (e.g., TICHELAAR and RUFF, 1989) to the error analysis. We inverted the data twenty times, each time randomly dropping 8% of the data. The errors for the slip amount are the standard errors of the twenty inversions multiplied by a scale factor, 3.44 in the present case.

Table 1

Heterogeneous fault model

Subfault	Length km	Width km	Dip	Slip m	Error m	M_0 10^{20} Nm
A	27	40	30°E	2.26	0.06	0.85
B	25	30	30°W	0.52	0.12	0.14
C	25	30	30°W	6.07	0.18	1.59
D	27	30	60°W	3.10	0.13	0.88
E	35	30	60°W	3.79	0.05	1.39
					Total	4.85

4.2 Inversion of Geodetic Data

Since the geodetic data are available only on Okushiri Island, we tried to estimate the slip only on the southernmost subfault, E, which lies next to Okushiri Island. An inversion for the other faults would be very unstable (SATAKE, 1993). The result (Figure 6) demonstrated that slip amount is 9.6 m if the fault plane is dipping to the east, and 4.3 m if it dips to the west. The errors for either case are about 0.1 m, very small, as shown in Figure 6. The result indicates that a larger slip is required on the eastward dipping fault plane in order to explain the subsidence of Okushiri Island.

4.3 Inversion of Tsunami Data

The results of the inversion, using only tsunami data, are displayed in the center column of Figure 6. Both eastward and westward dipping planes of subfaults D and E yield very similar slip amounts, 3–4 m. The dip directions of subfaults D and E affect the estimation of slip amount on subfault C, although its dip direction is fixed to the west. The slip amount is estimated as 7.8 m if subfaults D and E are dipping to the west, and 6 m if they are dipping to the east. The slip amount on subfault A

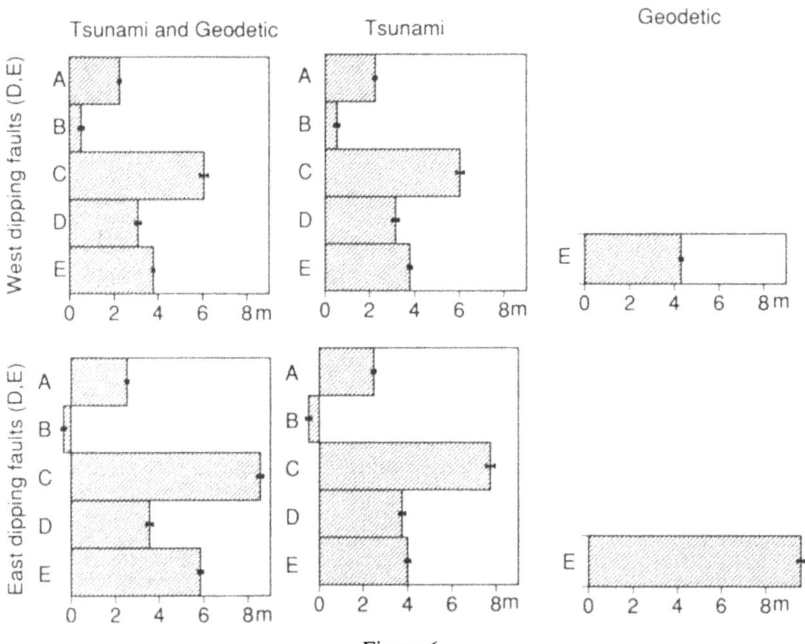

Figure 6

The results of the inversions. The top raw is for the westward dipping subfaults D and E, the bottom raw is for the eastward dipping subfaults D and E. The dip directions of the other subfaults are fixed; subfault A dips to the east and subfaults B and C dip to the west. The left, center, and right columns are for the joint, tsunami and geodetic inversion, respectively. The error bars are also shown.

is about 2 m and it is practically zero for subfault B, whether subfaults D and E are dipping to the east or west. Similar slip distributions for east- and westward dipping planes indicate that it is difficult to distinguish the actual fault plane from tsunami data alone.

4.4 Joint Inversion

We next use geodetic and tsunami data. The result of the joint inversion is shown in the left column of Figure 6. The slip amounts on subfaults A through D are very similar to those from the tsunami inversion. This is not surprising, because the geodetic data on Okushiri Island have poor control on the slip on these subfaults. The slip amount on subfault E is different for the eastward and westward dipping cases. It is 3.8 m for the westward dipping plane, but 5.9 m for the eastward dipping plane. For the westward dipping plane, the geodetic, tsunami, and joint inversions all give similar slip amounts, about 4 m. For the eastward dipping plane, on the other hand, the slip estimated from the geodetic inversion is considerably larger (9.6 m) than that from the tsunami inversion (4.0 m). The joint inversion yields an amount (5.9 m) between these two. This indicates that the westward dipping plane provides a more stable and consistent solution. Furthermore, the variance reduction of the inversion is 81% for the westward dipping faults, whereas it is 71% for the eastward dipping faults. Hence, we prefer the solution on the westward dipping faults and conclude that the fault plane is dipping to the west at subfaults D and E.

4.5 Heterogeneous Fault Model

Table 1 shows the final slip distribution from the joint inversion of tsunami and geodetic data, assuming that subfaults D and E are dipping to the west. Subfault A is dipping east and subfaults B and C are dipping west. The seismic moment on each subfault is calculated by assuming that the rigidity around the fault is 3.5×10^{10} N/m^2. The seismic moment of subfaults B and C, near the epicenter, is 1.7×10^{20} Nm, which is very similar to the moment release in the first 10 s $(2 \times 10^{20}$ Nm) estimated from the MTRF inversion. Therefore the slips on these subfaults occurred in the initial stage of the earthquake. The total seismic moment estimated from the slip distribution is 4.9×10^{20} Nm, which is slightly smaller, but very close to the total moment estimated by our CMT inversion $(5.5 \times 10^{20}$ Nm). This indicates that the slip on the northern (A) and southern (D and E) subfaults occurred after those on subfaults B and C.

The tsunami waveform at each tide gauge station is computed from the final model and shown in Figure 4. The observed and computed tsunami waveforms match very well, both arrival time and amplitudes. The subsidence of Okushiri Island computed from the final slip distribution is shown in Figure 5. The bottom of the figure shows a cross section along the direction N70°E, which is perpendicular to the strike of subfaults D and E. The calculated subsidence delineates the westward tilt of the island with an amount very similar to the observed.

5. Comparison with Runup Heights

We compute tsunami heights around Okushiri Island from the heterogeneous fault model. The computations are now made using nonlinear shallow water equations (for details see SATAKE, 1995). Figure 7 shows the initial condition, or the ocean bottom deformation from the final fault model, as well as the tsunami computational area. We repeated the computation three times using different grid sizes. They are 1 minute (1.9 km × 1.4 km), 20 seconds (0.62 km × 0.46 km) and 6 seconds (0.19 km × 0.14 km). In each case, the tsunami propagation for 20 min after the earthquake is calculated and the maximum tsunami heights around Okushiri Island are recorded. We divide the Okushiri coast into small segments,

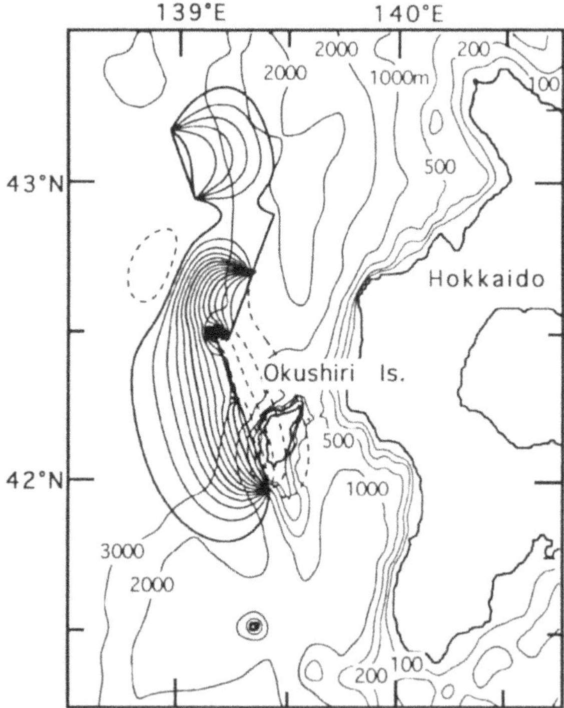

Figure 7

The vertical deformation on the ocean bottom computed from the final slip distribution on the subfaults. The solid curves indicate uplift and the dashed curves indicate subsidence. The contour interval is 20 cm. The nonlinear tsunami computations described in Section 5 are made in the map area.

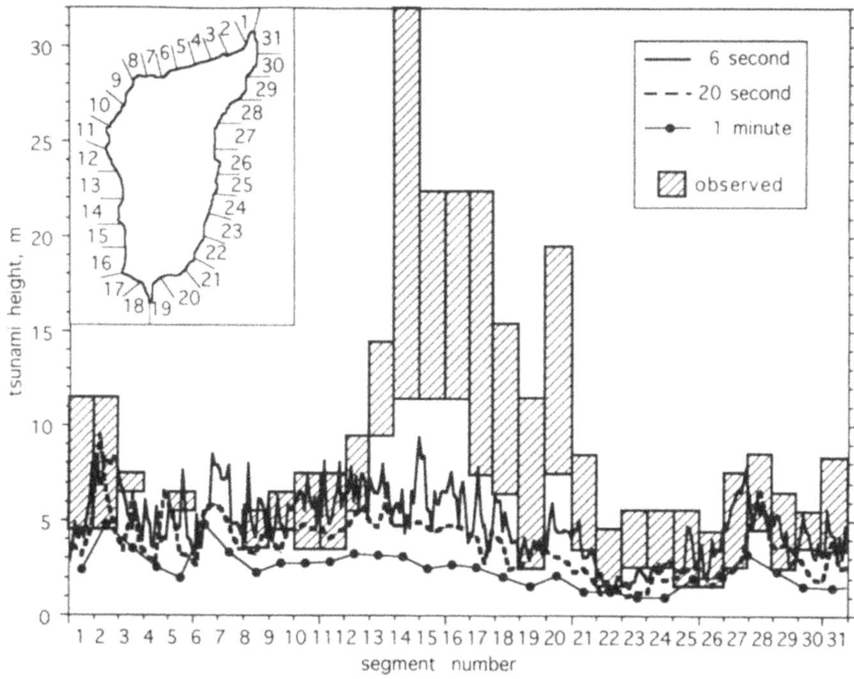

Figure 8
Comparison of the observed tsunami runup heights around Okushiri Island (striped bars) and the computed tsunami heights (curves) using the different grid sizes. The location of each segment is shown in the inset map.

each about 2 km long, as shown in Figure 8. The tsunami heights in each segment are plotted in Figure 8 with the observed runup heights (HOKKAIDO TSUNAMI SURVEY GROUP, 1993).

Figure 8 shows that the computed tsunami heights are smaller than the observed runup heights. The computed heights from the different grid sizes are similar on the northern coast of Okushiri, but differ significantly on the southern coast where the observed runup heights are large. The tsunami heights computed using smaller grid sizes are larger and closer to the observed heights. The computed tsunami heights on the 6 second grids are twice as large as (100% larger than) those on the 1 minute grids, and the tsunami heights on the 20 second grids are as much as 50% larger than those on the 1 minute grids. This is probably due to the fact that smaller grids can represent the coastal topography more accurately than the coarser grid.

The ratio of the observed runup heights to the calculated tsunami heights, or the amplification factor, therefore depends on the grid size. In the present case for Okushiri Island, it is 1–2 for the 6 second grids, 1–3 for the 20 second grids, and 2–4 for 1 minute grids. The result indicates that we can reproduce the observed runup heights when, and only when, we use very small grid size.

6. Free Oscillation of the Japan Sea

The Japan Sea is an almost closed water basin. Tsunamis generated in the Japan Sea therefore continue persistently, because of the multiple reflections on the coasts. SATAKE and SHIMAZAKI (1988a,b) examined tsunamis from the 1964 Niigata and 1983 Japan Sea earthquakes in both wave-theoretical and normal-mode approaches. In the latter approach, they found that the 1983 Japan Sea earthquake excited all the modes equally whereas the 1964 Niigata earthquake mainly excited the regional modes. They interpreted this different excitation as due to the different water depths at the source area; the Niigata earthquake occurred beneath a continental shelf about 100 m deep whereas the water depth at the 1983 epicenter is about 2500 m (SATAKE and SHIMAZAKI, 1988b). The Hokkaido Nansei-Oki earthquake occurred beneath the deep (about 3000 m) part of the Japan Sea, similar to the 1983 event.

The characteristic period of tsunami at the source can be estimated by dividing the source size by the long-wave velocity in the source region. While the source size is similar, about 100 km for all these earthquakes, the long-wave velocity varies from 0.03 km/s for a 100 m deep ocean to 0.17 km/s for a 3000 m depth. Accordingly, the characteristic period of generated tsunamis would be about 10 min for the Hokkaido Nansei-Oki or 1983 Japan Sea events (a source in the deep ocean), but would exceed 50 min for the Niigata earthquake (a source on the continental shelf). In the following, we will examine the difference in the tsunami generation from the viewpoint of excitation of the normal modes.

6.1 Summary of Normal Mode Approach

Here we summarize the normal-mode theory of tsunamis; for more details, see SATAKE and SHIMAZAKI (1987, 1988b). From the linear shallow water equation (2) and the equation of continuity (3), we can derive a wave equation

$$\frac{\partial^2 h}{\partial t^2} = g \nabla \cdot (d \nabla h). \tag{4}$$

Assuming that the water height changes periodically in time, i.e.,

$$h(x, t) = \Phi(x) e^{i\omega t} \tag{5}$$

where Φ describes the spatial distribution of water height and ω is the angular frequency, we can rewrite equation (4) as the following eigenequation,

$$\nabla \cdot (d \nabla \Phi) = \lambda \Phi \quad \text{where} \quad \lambda = -\omega^2/g. \tag{6}$$

The eigenvalue λ is characterized by the eigenfrequency ω. Equation (6) can be numerically solved for actual bathymetry using a supercomputer. SATAKE and SHIMAZAKI (1988b) computed the eigenvalues and eigenfunctions for the lowest 100 modes. The shortest eigenperiod (for mode 100) is about 50 min.

Tsunami waveforms can be written as a superposition of normal modes,

$$h(x,t) = \sum_k C_k \Phi_k(x) e^{i\omega_k t} \tag{7}$$

where C_k is a coefficient, or weight, for each mode. The coefficient can be calculated from the initial water height, h_0, in the source area,

$$C_k = \int_{\text{source}} h_0(x) \Phi_k(x)\, dx \tag{8}$$

for an instantaneous source.

6.2 Excitation of Free Oscillation by the Hokkaido Nansei-Oki Event

We calculated the coefficients of each mode for the 1993 Hokkaido Nansei-Oki earthquake and compared them with those of the 1964 Niigata and 1983 Japan Sea events (SATAKE and SHIMAZAKI, 1988b) in Figure 9. The fault model was proposed by SATAKE and ABE (1982) for the Niigata earthquake, and by SATAKE (1989) for the Japan Sea earthquake. Figure 9 demonstrates that the excitation coefficients for the 1993 event are the smallest among the three. The largest coefficient is about 30 cm for mode 49, whose eigenperiod is 72 min. The coefficients for the other modes are about 10 cm or smaller. This indicates that the tsunami from the 1993 earthquake contained a large component with a period of 72 min. However, the observed tsunami spectra at coastal points such as tide gauge locations are also affected by the value of eigenfunctions at the observational points (see equation (7)).

The variation in excitation coefficients, (for example, the coefficient for mode 4 is 13 cm, but only 3 cm for mode 5), is due to the relationship between the eigenfunction (spatial distribution of water height in each mode) and the location of the source. As shown in Figure 10, the location of the source is near the maximum of eigenfunction for mode 4, whereas it is on the nodal line for mode 5. Therefore, the excitation for mode 5 is considerably smaller than that for mode 4. The eigenfunction of mode 49 has a peak near the source (see Figure 10), which explains why the excitation coefficient of that mode is the largest.

7. Conclusions

The generation characteristics of the 1993 Hokkaido Nansei-Oki earthquakes are examined. The results are summarized as follows.

(1) The aftershock distribution and the joint inversion of geodetic and tsunami data indicate that the fault plane is dipping to the west except at the northern end of the source area where the fault shallowly dips to the east. The fault is shallowly

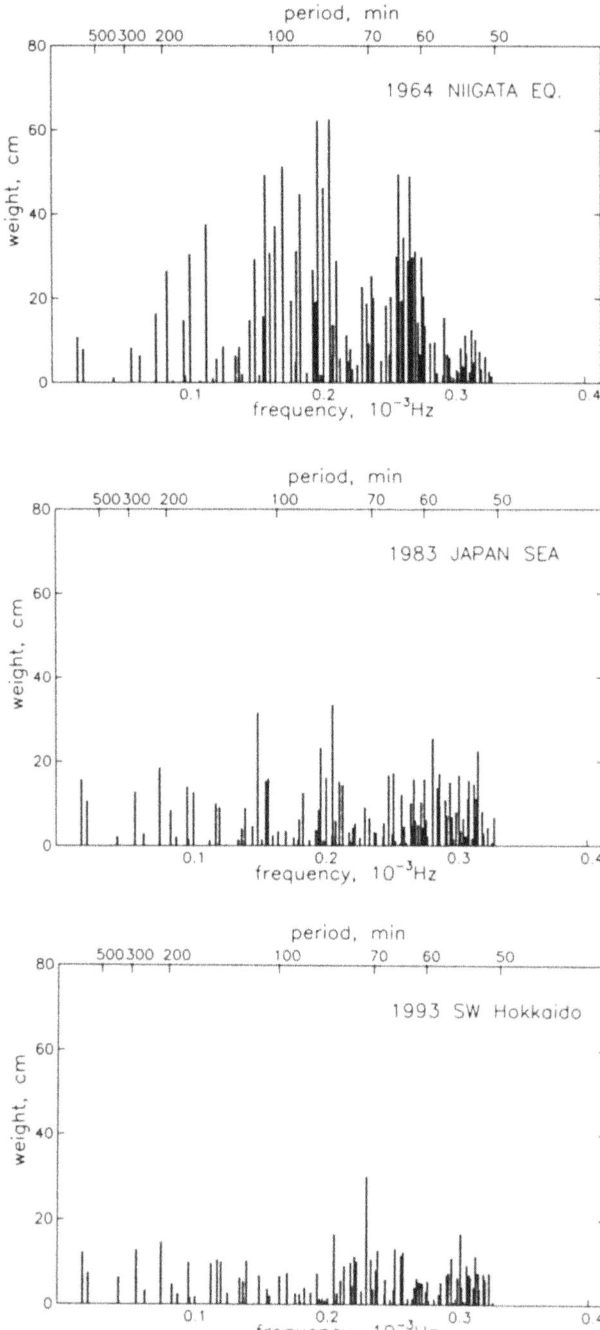

Figure 9
The excitation coefficients for each mode calculated from the 1964 Niigata earthquake *(top)*, the 1983
Japan Sea earthquake *(center)* and the 1993 Hokkaido Nansei-Oki earthquake *(bottom)*.

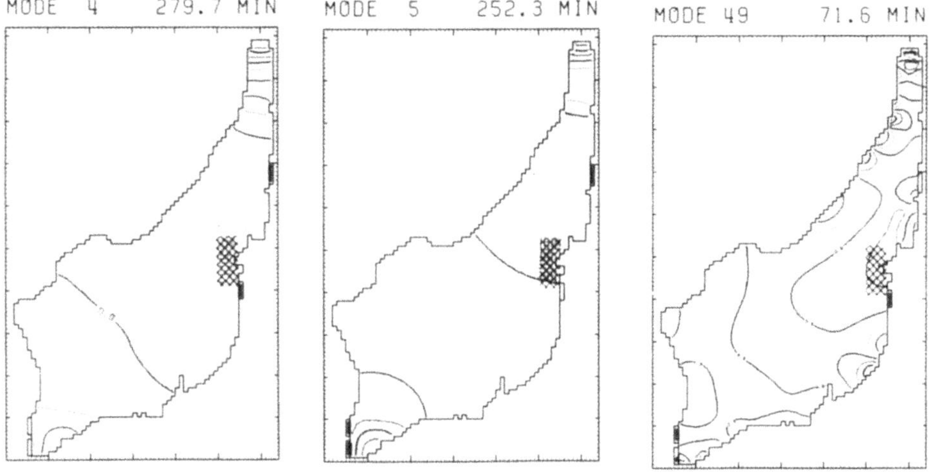

Figure 10

The eigenfunction, or the spatial distribution of water height, for modes 4, 5 and 49. Contour interval is 0.25 for the normalized eigenfunction. Open and shaded areas indicate different signs. The source area of the 1993 Hokkaido earthquake is shown by a hatched rectangle.

dipping to the west near the epicenter, but the dip is steeper on the southern part of the fault.

(2) The joint inversion indicates that the largest slip, about 6 m, occurred just south of the epicenter. This corresponds to the moment release of the first 10 s, as estimated from the seismic wave analysis. The slip on the northern and southern parts of the fault is smaller, 2–4 m.

(3) Computation of tsunamis around Okushiri Island, using different grid sizes, shows that the computed tsunami heights become larger as the smaller grid size is used. This indicates that computation on small grids is necessary to reproduce the observed runup heights.

(4) The normal mode approach indicates that the tsunami excitation from the 1993 Hokkaido earthquake was smaller than the 1964 Niigata and 1983 Japan Sea earthquakes. The excitation was largest for mode 49 with a period of 72 min.

Acknowledgements

Dr. Y. Tsuji at Earthquake Research Institute, University of Tokyo, provided us with the tide gauge records. The bathymetry data were originally compiled by M. Okada, Meteorological Research Institute (the entire Japan Sea) and V. Gusiakov, Novosibirsk Computing Center (detailed data around Okushiri Island). This study would have been impossible to conduct without these data, and we sincerely thank them. We also thank J. Johnson for reading the manuscript and providing valuable comments. This work was supported by the National Science Foundation (EAR 9117800).

REFERENCES

DZIEWONSKI, A. M., EKSTRÖM, G., and SALGANIK, M. P. (1994), *Centroid-moment Tensor Solutions for July–September 1993*, Phys. Earth Planet. Inter. *83*, 165–174.

FUKAO, Y., and FURUMOTO, M. (1975), *Mechanism of Large Earthquakes along the Eastern Margin of the Japan Sea*, Tectonophys. *26*, 247–266.

HOKKAIDO TSUNAMI SURVEY GROUP (1993), *Tsunami Devastates Japanese Coastal Region*, EOS, Trans. Am. Geophys. Union *74*, 417, 432.

ISHIYAMA, Y., *The 1993 Hokkaido Nansei-oki earthquake, tsunami, and their damage.* In *Report for Grant-in-aid (05306012)* (ed. Ishiyama, Y.) (Ministry of Education, Science and Culture 1994) 196 pp. 1–11 (in Japanese).

KASAHARA, M., KODAIRA, S., MOTOYA, Y., TAKANAMI, T., MAEDA, I., OKAYAMA, M., ISHIKAWA, H., ICHIYANAGI, M., YAMAMOTO, A., MATSUMOTO, S. D., TSUMURA, N., OKADA, T., YABE, Y., IIDAKA, T., and HIRATA, N., *Aftershock activity and distribution of the 1993 Hokkaido Nansei-oki earthquake.* In *Report for Grant-in-aid (05306012)* (ed. Ishiyama, Y.) (Ministry of Education, Science and Culture 1994) 196 pp. 13–19 (in Japanese).

KOBAYASHI, Y. (1983), *On the Initiation of Subduction of Plates*, Earth Monthly *5*, 510–514 (in Japanese).

KUMAKI, T., KISANUKI, J., OHTANI, T., ONO, Y., and KAJIKAWA, S. (1993), *Vertical Seismic Crustal Movement of the 1993 Hokkaido-Nansei-oki Earthquake Based on Coastal Landform Changes in the Okushiri Island, West of Hokkaido, Japan*, Prog. Abst. Seism. Soc. Japan *2*, 63.

NAKAMURA, K. (1983), *Possible Nascent Trench Along the Eastern Japan Sea as the Convergent Boundary between Eurasian and North American Plates*, Bull. Earthq. Res. Inst. Univ. of Tokyo *58*, 711–722 (in Japanese).

OKADA, Y. (1985), *Surface Deformation due to Shear and Tensile Faults in a Half-space*, Bull. Seismol. Soc. Am. *75*, 1135–1154.

SATAKE, K. (1986), *Re-examination of the 1940 Shakotan-oki Earthquake and the Fault Parameters of the Earthquakes along the Eastern Margin of the Japan Sea*, Phys. Earth Planet. Inter. *43*, 137–147.

SATAKE, K. (1989), *Inversion of Tsunami Waveforms for the Estimation of Heterogeneous Fault Motion of Large Submarine Earthquakes: 1968 Tokachi-oki and 1983 Japan Sea Earthquakes*, J. Geophys. Res. *94*, 5627–5636.

SATAKE, K. (1993), *The Depth Distribution of Coseismic Slip along the Nankai Trough, Japan, from Joint Inversion of Geodetic and Tsunami Data*, J. Geophys. Res. *98*, 4553–4565.

SATAKE, K. (1995), *Linear and Nonlinear Computations of the 1992 Nicaragua Earthquake Tsunamis*, Pure and Appl. Geophys. *144*, 455–470.

SATAKE, K., and ABE, K. (1983), *A Fault Model for the Niigata, Japan, Earthquake of June 16, 1964*, J. Phys. Earth. *31*, 217–223.

SATAKE, K., and SHIMAZAKI, K. (1987), *Computation of Tsunami Waveforms by a Superposition of Normal Modes*, J. Phys. Earth. *35*, 409–414.

SATAKE, K., and SHIMAZAKI, K. (1988a), *Free Oscillation of the Japan Sea Excited by Earthquakes. I. Observation and Wave-theoretical Approach*, Geophys. J. *93*, 451–456.

SATAKE, K., and SHIMAZAKI, K. (1988b), *Free Oscillation of the Japan Sea Excited by Earthquakes. II. Modal Approach and Synthetic Tsunamis*, Geophys. J. *93*, 457–463.

SENO, T., and SAKURAI, T. (1993), *Can the Okhotsk Plate be Discriminated from the North American Plate?* EOS, Trans. Am. Geophys. Union *74*, Fall Meeting Supplement, 585.

SHUTO, N., and MATSUTOMI, H. (1995), *Field Survey of the 1993 Hokkaido Nansei-oki Earthquake Tsunami*, Pure and Appl. Geophys. *144*, 649–663.

SIPKIN, S. A. (1994), *Rapid Determination of Global Moment-tensor Solutions*, Geophys. Res. Lett. *21*, 1667–1670.

TANIOKA, Y., RUFF, L. J., and SATAKE, K. (1993), *Unusual Rupture Process of the Japan Sea Earthquake*, EOS, Trans. Am. Geophys. Union *74*, 377, 379–380.

TANIOKA, Y., SATAKE, K., and RUFF, L. J. (1995), *Total Analysis of the Hokkaido Nansei-oki Earthquake Using Seismological, Tsunami and Geodetic Data*, Geophys. Res. Lett. *22*, 9–12.

TICHELAAR, B. W., and RUFF, L. J. (1989), *How Good are our Best Models? Jackknifing, Bootstrapping and Earthquake Depth*, EOS, Trans. Am. Geophys. Union *70*, 593, 605–606.

TSUTSUMI, A., SHIMAMOTO, T., MIYAWAKI, M., SATO, H., KAWAMOTO, E., and TANAKA, J. (1993), *Subsidence of the Okushiri Island Caused by the Southwest-off Hokkaido Earthquake*, Prog. Abst. Seismol. Soc. Japan *2*, 62.

(Received August 29, 1994, revised March 6, 1995, accepted March 16, 1995)

PAGEOPH, Vol. 144, Nos. 3/4 (1995)

0033–4553/95/040823–15$1.50 + 0.20/0
© 1995 Birkhäuser Verlag, Basel

Analysis of Seismological and Tsunami Data from the 1993 Guam Earthquake

YUICHIRO TANIOKA,[1] KENJI SATAKE,[1] and LARRY RUFF[1]

Abstract—The fault parameters of the Guam earthquake of August 8, 1993 are estimated from seismological analyses, and the possibility of identifying the actual fault plane from tsunami waveforms is tested. The Centroid Moment Tensor solution of long-period surface waves shows one nodal plane shallowly dipping to the north and the other nodal plane steeply dipping to the south. The seismic moment is 3.5×10^{20} Nm and the corresponding moment magnitude is 7.7. The Moment Tensor Rate Function inversion of P waves also yields a similar focal mechanism and seismic moment. The point source depth is estimated as 40–50 km.

This earthquake generated tsunamis that propagated toward the Japanese coast along the Izu-Bonin-Mariana ridge system. The tsunamis are recorded on ocean bottom pressure gauges and tide gauges. Numerical computation of tsunamis shows that the computed waveforms from the two possible fault planes match well with the observed tsunami waveforms. The numerical computation also shows that the tsunami waveforms at Guam Island, just above the fault, should contain useful information regarding the identification of the actual fault plane. However, the current sampling rate of the tide gauges is so small that the records cannot help the identification.

Key words: Tsunamis, the 1993 Guam earthquake, fault plane.

Introduction

A large earthquake ($M_s = 8.0$) occurred south of Guam Island on August 8, 1993. The National Earthquake Information Center (NEIC) Preliminary Determination of Epicenters (PDE) provides the following estimates: origin time, 08:34:24.93 GMT; epicenter, 12.982°N, 144.801°E; depth 59 km; magnitude, M_s 8.0. The location of the main shock is shown in Figure 1. This is the largest earthquake ever recorded in the Mariana Arc region. This earthquake generated a moderate-size tsunami that was observed on tide gauge stations in Guam and Japan and ocean bottom pressure gauge stations near Japan. A student in Guam, Jody Flores, conducted a survey of local residents and collected descriptive accounts (SIGRIST, 1995).

---

[1] Department of Geological Sciences, University of Michigan, Ann Arbor, MI 48109-1063, U.S.A.

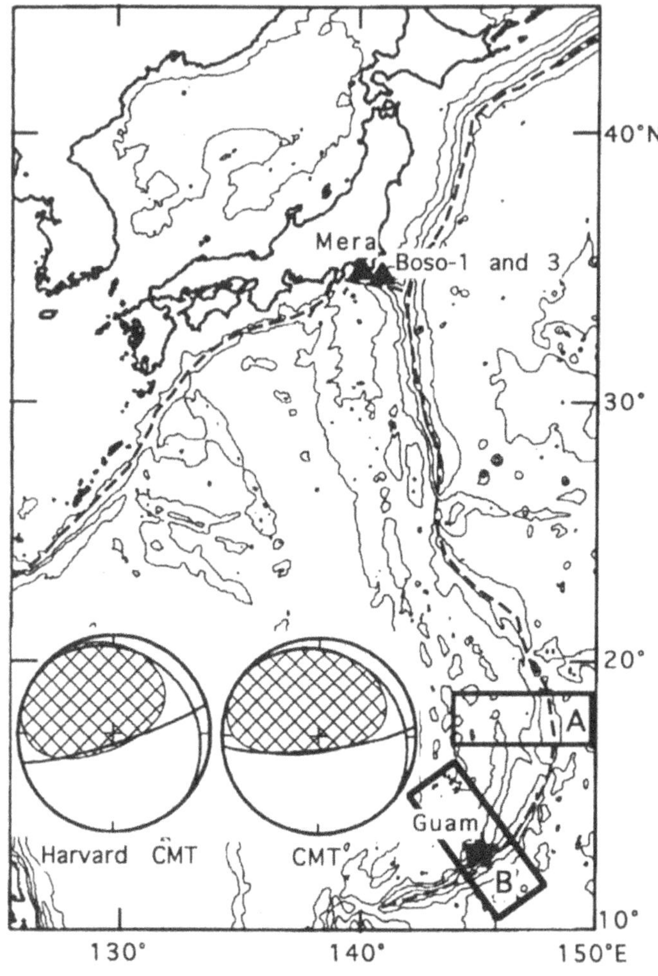

Figure 1

Topography map around the Izu-Bonin-Mariana region with the Harvard and our CMT solutions of the 1993 Guam earthquake. The star shows the epicenter of the Guam earthquake. The triangles show the location of tide gauge and ocean bottom pressure gauge stations. The dashed lines show the plate boundaries in this region. Two boxes represent the locations and limits of the vertical cross section of seismicity in Figure 2.

The largest earthquake previously known in this region occurred on April 5, 1990 ($M_s = 7.5$) with a normal fault mechanism in the outer-rise region (YOSHIDA *et al.*, 1992). The Mariana trench is characterized as a weakly coupled subduction zone where no great thrust event has ever occurred (UYEDA and KANAMORI, 1979). The epicentral data suggest that the Guam earthquake occurred in the subduction zone, not in the outer-rise region. Determination of the source process of this earthquake is very important to understand the characteristics of subduction in the Mariana Arc.

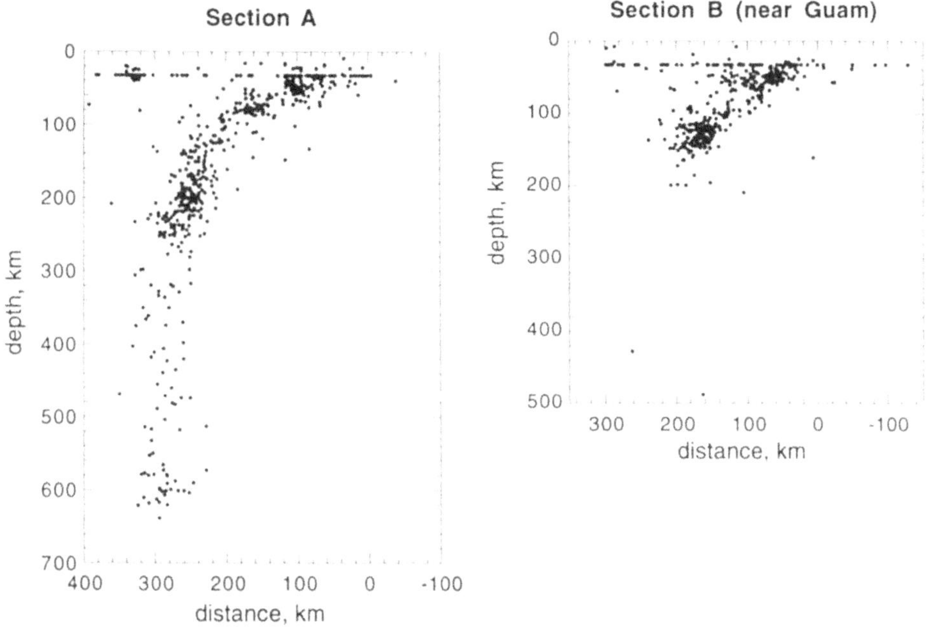

Figure 2

The vertical cross section of the seismicity in sections A and B (see Figure 1 for location). The hypocentral data from 1928 to 1987 are plotted from the NEIC catalog.

In this paper, we use seismic surface waves and body waves to estimate the fault parameters of this earthquake. We further discuss the importance of the identification of the actual fault plane of this earthquake. We then numerically compute tsunami waveforms to test whether we can identify the fault plane from the tsunami waveforms recorded at tide gauge stations. We also examine the bathymetric effect on the tsunami propagation.

Seismological Analyses

a) Seismicity in the Mariana Region

Figure 2 illustrates two vertical cross sections of seismicity in the Mariana region, constructed from the NEIC PDE catalog of hypocentral parameters. Section A (see Figure 1 for the location) shows that the seismic activity is continuously distributed to a depth of 650 km and the distribution below a depth of 200 km is nearly vertical. However, section B, which includes the hypocenter of the 1993 Guam earthquake, shows seismic activity to a depth of only 200 km. Also notice that the apparent dip of the slab suggested from the earthquake distribution

Table 1

Source parameters of the 1993 Guam earthquake

	CMT solution	MTRF solution
strike	310°/84°	238°/67°
dip	17°/78°	24°/67°
rake	135°/78°	82°/94°
M_o	3.5×10^{20} Nm	2.7×10^{20} Nm
M_w	7.7	7.6
depth	50 km (fixed)	40–50 km
duration	—	32 sec

between 50 and 150 km is 40°. This indicates that the geometry of the slab changes between sections A and B. Further, the characteristics of subduction in section B may be different from that in section A, which is defined as a weakly coupled subduction zone (UYEDA and KANAMORI, 1979).

b) Surface Wave Analysis

We performed Centroid Moment Tensor (CMT) inversion (DZIEWONSKI *et al.*, 1981; FUKUSHIMA *et al.*, 1989) using long-period surface waveforms recorded at 12 Incorporated Research Institutions for Seismology (IRIS) network stations (ANMO, CCM, COL, CTAO, HRV, KIP, MAJO, PAB, PAS, RAR, ANTO, and CHTO). The result of this computation is summarized in Table 1. The T axis is -3.2×10^{20} Nm (plunge 55°, azimuth 339°), the P axis is 3.8×10^{20} Nm (plunge 32°, azimuth 184°), and the N axis is -0.6×10^{20} Nm. The seismic moment is estimated as 3.5×10^{20} Nm and the corresponding moment magnitude is 7.7. The focal mechanism has one nodal plane dipping shallowly to the north (strike 310°, dip 17°, rake 135°) and the other steeply dipping to the south (strike 84°, dip 78°, rake 78°) (Figure 1). Figure 1 also shows the focal mechanism from the Harvard CMT which is similar to our estimate. The slip directions from these mechanisms are close to the direction perpendicular to the Mariana trench.

c) Body Wave Analysis

We performed Moment Tensor Rate Function (MTRF) inversion (RUFF and MILLER, 1994) using broadband P waves recorded at 9 stations, 8 IRIS network stations (ANMO, CCM, KONO, OBN, PAS, KIP, ARV, COL) and the University of Michigan station (AAM). We used the first 90 s of the P wave, filtered with a 2 s duration triangle. The MTRFs are sampled at 2 s intervals. We first performed the MTRF inversion to determine the best depth. The average P-wave velocity from the surface to the earthquake source (20–100 km) should be between the crustal P-

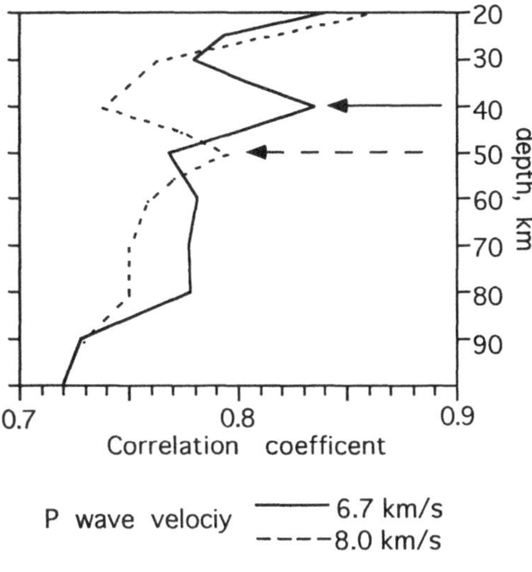

Figure 3
The correlation coefficient between the observed and synthetic seismograms are plotted for different point source depths. The solid line is for P-wave velocity of 6.7 km/s and the dashed line is for P-wave velocity of 8.0 km/s.

wave velocity, 6.7 km/s, and the uppermost mantle P-wave velocity, 8.0 km/s. We therefore performed the inversion using the different velocities, 6.7 km/s and 8.0 km/s. Figure 3 shows the correlation coefficient between the synthetic and observed seismograms as a function of source depth. The best depth is 40 km and 50 km for the assumed P-wave velocity of 6.7 km/s and 8.0 km/s, respectively. The high correlation seen at shallow (20 km) depth is probably an artifact of the inversion for large earthquakes (see CHRISTENSEN and RUFF, 1985). Therefore, the best estimate of point source depth is between 40–50 km. The best double-couple mechanism reduced from the MTRFs for a depth of 40 km with a P-wave velocity of 6.7 km/s is shown in Figure 4 (strike 238°, dip 23°, rake 82°). The strike of the trench in the epicentral region is N55°E (Figure 5). We determined a single source time function from the re-inversion of the seismograms with a fixed double-couple mechanism obtained by the MTRF inversion. Figure 4 shows that the source time history consists of three stages: the initiation (duration 6 s), the first pulse (duration 10 s and seismic moment 1.0×10^{20} Nm), and the second pulse (duration 16 s and moment 1.7×10^{20} Nm). Total seismic moment of 2.7×10^{20} Nm is similar to the result of the CMT inversion, 3.5×10^{20} Nm.

d) Aftershocks

The solid circles in Figure 5 show the aftershock distribution within 2 days, according to PDE. They scatter in the depth range of 20–100 km and do not

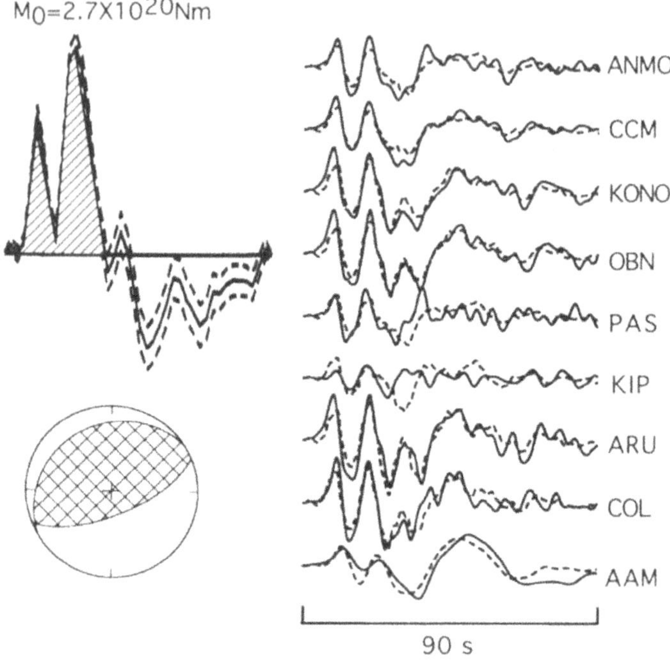

Figure 4

The result of the body wave inversion. The source time function is shown on the upper left. Seismic moment of 3.6×10^{20} Nm was released during the first 32 s (shaded region). The best double-couple mechanism is shown on the lower left. The observed (solid) and synthetic (dashed) seismograms with their station names are shown to the right. The time scale is the same for both the source time function and seismograms.

further identify the actual fault plane. We applied a master event relocation method for arrival time data at common stations for 10 large aftershocks to check the reliability of the epicenters. The result shows that the largest epicentral shift was 24 km. The shifts of the other 9 were less than 7 km. The relocated epicenters were very similar to the original PDE epicenters. As can be seen in Figure 5, the locations of aftershocks extend as far as 150 km along the trench axis.

e) Ambiguity of Fault Plane

Both surface wave and body wave analyses reveal that the best focal mechanism has one nodal plane dipping shallowly to the north and the other dipping steeply to the south. Although the strikes of the nodal plane from both analyses are slightly different, the slip direction is almost perpendicular to the Mariana trench. There are two possible explanations for this earthquake: one is an intraplate rupture due to slab pull; the other is an interplate underthrust event on a fault plane dipping to the north. If the former is the case, then the key feature is the steeply dipping tension axis, and the actual fault plane can be either of the two nodal planes which were

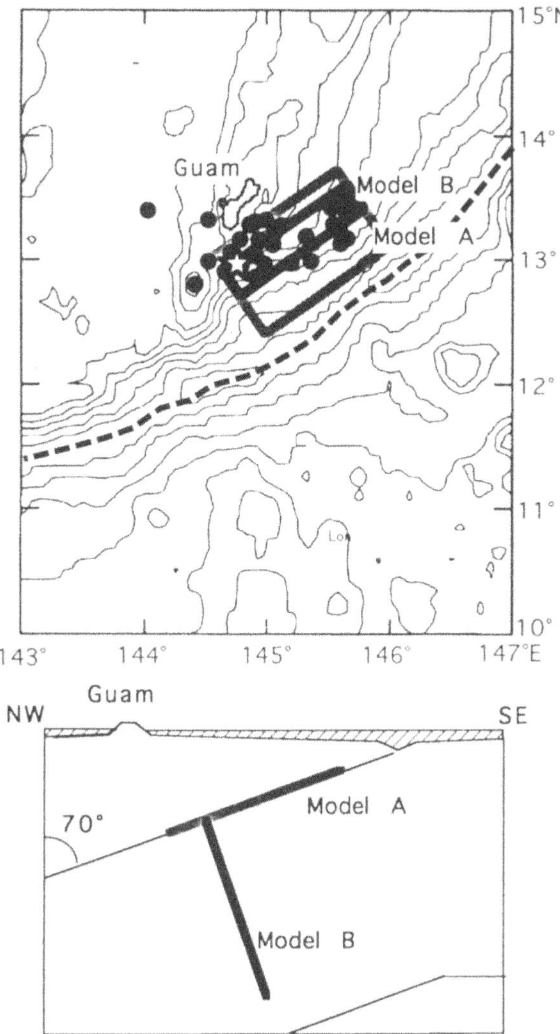

Figure 5

The location of the two fault models. Large and small rectangles show the location of fault models A and B, respectively. The star shows the epicenter of the 1993 Guam earthquake. The solid circles show the epicenters of aftershocks within 2 days, according to PDE. The dashed line represents the trench axis. Bottom part shows the vertical cross section of the two fault models in the northwest-southeast direction.

defined by the surface and body wave analyses. However, if the latter is the case, the actual fault plane has to be the shallow dipping plane. Identification of the actual fault plane is very important to understanding the source process of this earthquake. In the next section, we test the possibility of identifying the actual fault plane from the tsunami waveforms.

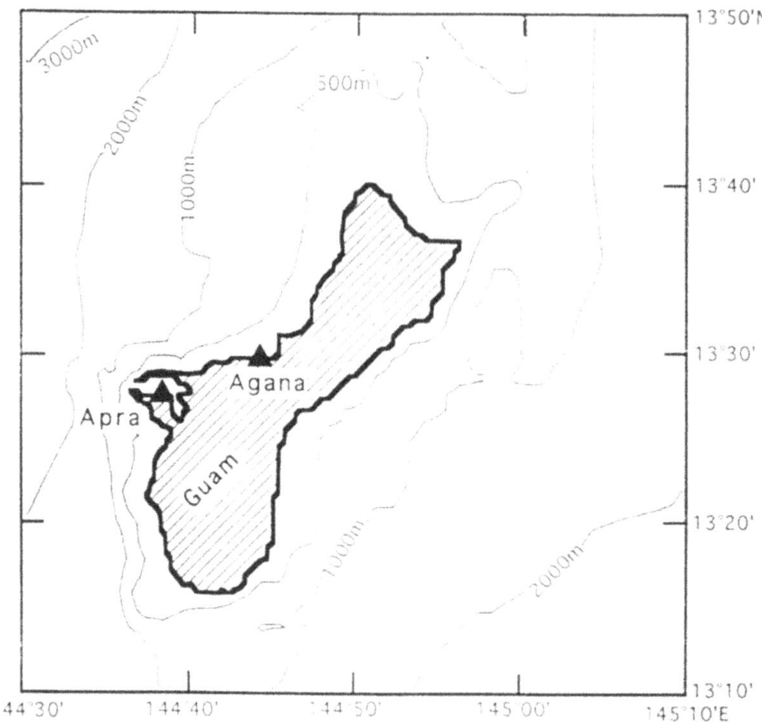

Figure 6
Bathymetry map around Guam Island. The triangles show the location of tide gauge stations, Apra and
Agana.

Tsunami Analysis

a) Tsunami Data

Recently, several pressure gauges have been installed on the deep ocean bottom off the Japanese coast (OKADA, 1991). Records from one tide gauge and two ocean bottom pressure gauge stations near Japan are used as far-field tsunami waveform data. The location and name of these stations are shown in Figure 1. Records from two tide gauge stations in Guam Island are used as near-field tsunami waveform data. The location and name of these stations are shown in Figure 6. These records at Agana and Apra in Guam were sampled at a 15-min and 6-min interval, respectively.

b) Computation of Tsunami

The equation of motion and the equation of continuity for linear long waves are solved by a finite-difference method in the spherical coordinate system (HWANG *et al.*, 1972; KOWALIK and MURTY, 1984; SATAKE, 1995). We computed the far-field tsunami in the area shown in Figure 1. The grid spacing is 2 minutes. A finer grid system of 1 minute was used near Japan to incorporate the small-scale topography near tide gauge and ocean bottom pressure gauge stations. We also computed the near-field tsunami in the area shown in Figure 5 with a grid spacing of 20 seconds. The initial condition of tsunami propagation is an ocean bottom deformation, which is computed using OKADA's (1985) formulas.

c) *Effect of Fault Geometry on the Tsunami Waveform*

Based on the seismological analyses, we tested two different fault models. One model is a steeply-dipping fault to the south (fault model A) shown in Figure 5. The other fault model is a shallowly-dipping fault to the north (fault model B), also shown in Figure 5. The fault length of both models is 130 km and the down-dip width is 85 km. Both fault models cover main shock epicenter and majority of the aftershock area. The average slip is estimated to be 0.8 m from the seismic moment of 3.5×10^{20} Nm obtained from the surface wave analysis, assuming a rigidity of 4×10^{10} N/m^2.

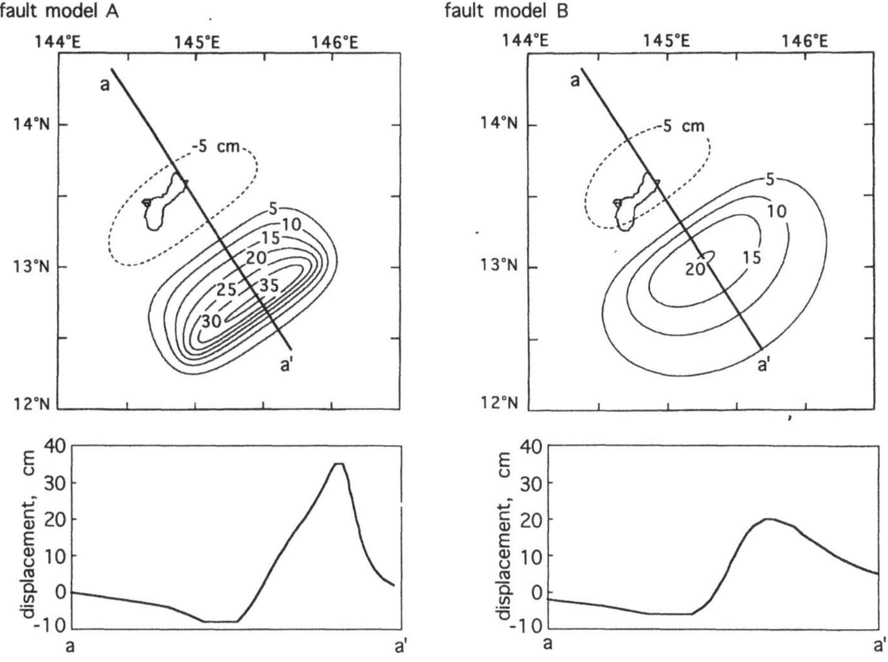

Figure 7

The vertical displacement of fault models A and B. Bottom parts show the cross section of vertical displacements along a-a' line.

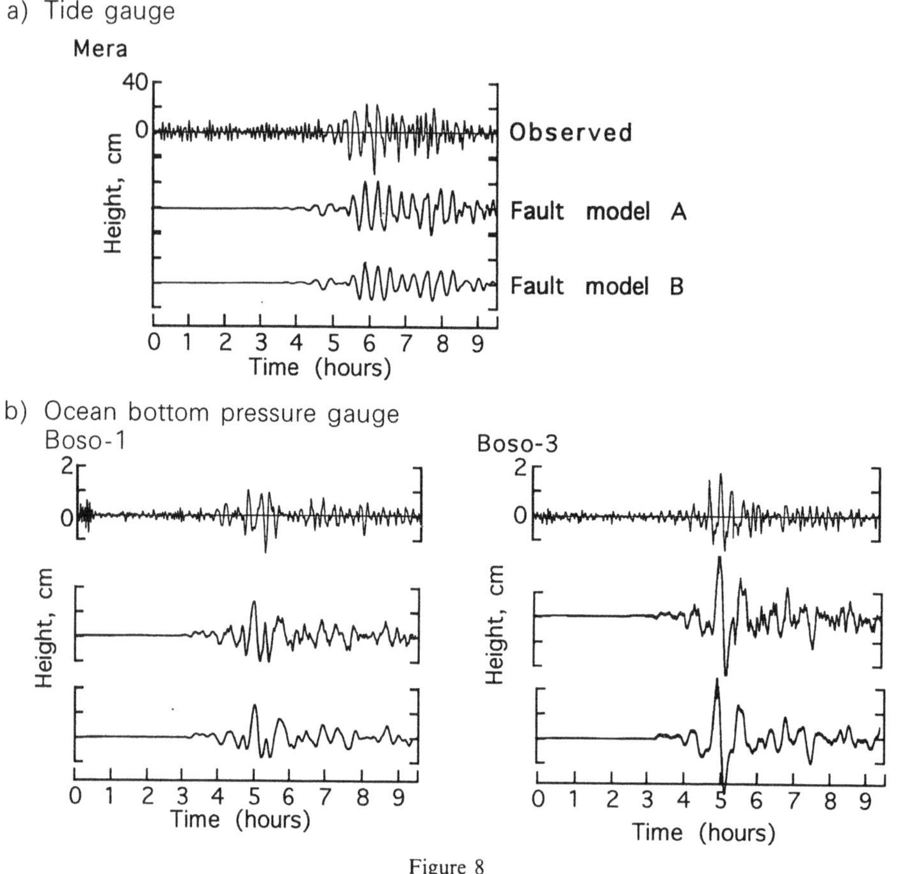

Figure 8

Comparison of the observed and computed tsunami waveforms from fault models A and B at a) the tide gauge at Mera, and b) two ocean bottom pressure gauges (Boso-1 and Boso-3).

The ocean bottom deformation calculated from the two different fault models is shown in Figure 7. It shows that the shape of the vertical deformation from the two fault models is different. The cross section of vertical deformation from fault model A in Figure 7 displays a sharp peak. However, the cross section of vertical deformation from fault B in Figure 7 shows a gentle peak. The deformation pattern for fault model B is much smoother than that for fault model A. In other words, the deformation for fault model A contains shorter wavelength components than that for fault model B. These differences in the vertical deformation would probably cause differences in tsunami waveforms at the nearby tide gauge stations.

The waveform comparisons between fault models A and B are shown in Figures 8 and 9. The computed tsunami waveforms at Mera, Boso-1 and Boso-2 show no significant difference between fault models A and B (Figure 8). This indicates that

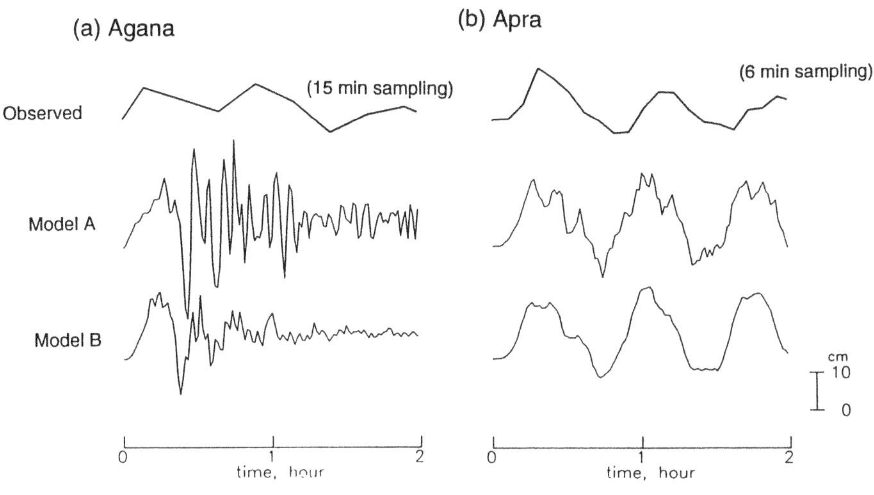

Figure 9

Comparison of the observed and computed tsunami waveforms from fault models A and B on two tide gauges at (a) Agana and (b) Apra in Guam. Computed tsunami are plotted using the sampling rate of 1 sample/min.

the far-field tsunami waveforms are not useful in identifying the actual fault plane for this particular case.

The computed tsunamis (Figure 9a) at Agana, Guam, using fault models A and B, show a small difference in the shape of the first pulse (first 15 min in Figure 9a). This probably corresponds to the difference in the vertical deformation shown in Figure 7. We may be able to determine the actual fault plane from this difference. The computed waveform from fault model A displays high-frequency waves of larger amplitude than the computed waveform from fault model B (Figure 9a). These large, high-frequency waves are probably generated from the sharp peak at the southeast end of the vertical deformation from fault model A (Figure 7). This indicates that the near-field tsunami waveforms are useful for identifying the actual fault plane. Unfortunately, the observed data were sampled at 15 min, causing a loss of information by aliasing (Figure 10a). The period of the high-frequency wave is about 5–6 min; therefore, to identify the actual fault plane, the observed data should be sampled at 2 min or less.

A similar difference in high-frequency waves from two different models is also found for the Apra Harbor tide gauge record, although the amplitude of high-frequency waves is smaller than that in the Agana waveforms (Figure 9b). Unfortunately, as the observed data were sampled at 6 min, the differences were lost by averaging and aliasing. Figure 10b shows the same observed and the computed tsunamis which are resampled at 6 min with 3 min averages. The computed tsunamis in Figure 10b are almost identical. We could not obtain any information concerning the fault geometry from these observations.

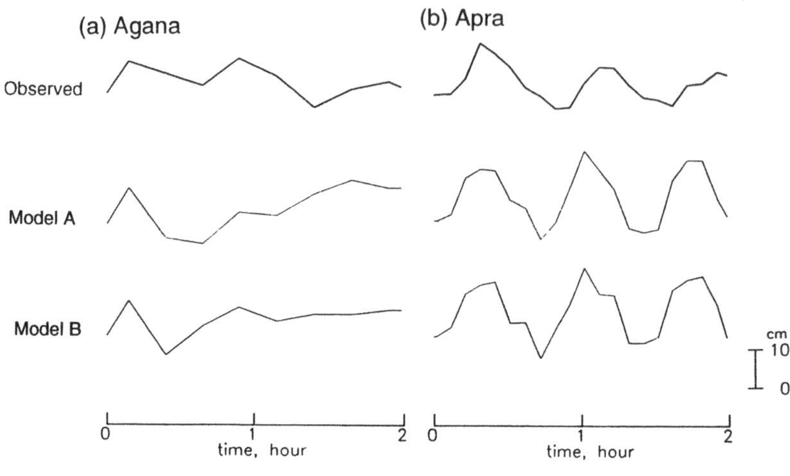

Figure 10

Comparison of the observed and computed tsunami waveforms from fault models A and B on two tide gauges at (a) Agana and (b) Apra. Computed wàveforms are processed in the same way as the observed.

d) Excitation of Tsunami

The observed and computed waveforms fit well at the tide gauge and ocean bottom pressure gauge stations, except at Agana where the observed data are severely undersampled (1 sample/15 min) (Figures 8 and 10). This indicates that both fault models provide a consistent explanation for both seismic and tsunami waves. SCHINDELE *et al.* (1995) categorized this event as a "tsunami-deficient" earthquake, on the basis of a spectral analysis of seismograms. Among the other tsunamigenic earthquakes on the Izu-Bonin-Mariana ridge system, the 1990 Mariana earthquake had a similar feature (SATAKE *et al.,* 1992). But the 1984 Torishima earthquake was different; its tsunami excitation was considerably larger than expected from the seismic wave analysis (SATAKE and KANAMORI, 1991).

e) Tsunami Propagation

Figure 11 exhibits the snapshots of the tsunami propagation at 0, 1, 2, and 3 hours after the origin time for fault model A. As can be seen in Figure 11, the large tsunamis travel slowly along the ridge system (Izu-Bonin-Mariana Ridge), and the tsunami energy is trapped by the ridge system. Figure 8 shows that the largest tsunami arrived at the tide gauge and ocean bottom pressure gauge at about one and a half hours after the first arrival (around 5 hours for Boso-1 and 2, around 6 hours at Mera). This type of wave was also observed in the tsunami caused by the Mariana earthquake of April 5, 1990 ($M_w = 7.3$) and is called a ridge wave (SATAKE *et al.*, 1992).

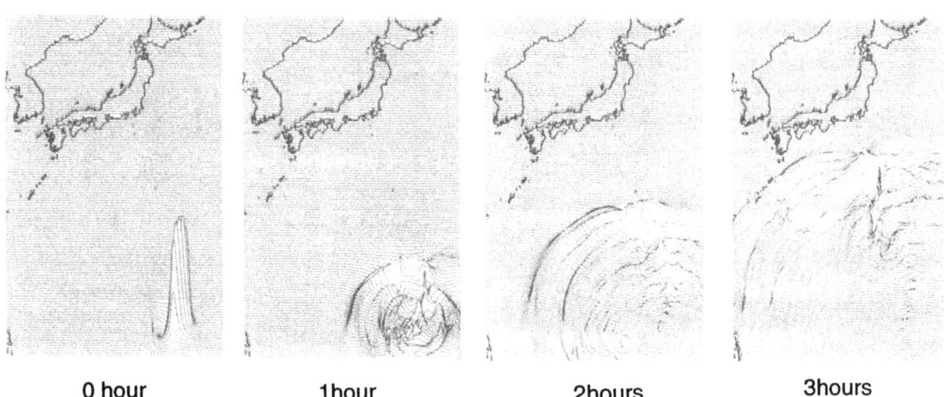

Figure 11
The snapshots of tsunami propagation at 0, 1, 2, 3 hours after the origin time of the earthquake.

f) Effect of Grid Size on the Tsunami Waveform

We compute the tsunami waveforms at the tide gauge station in Apra Harbor, Guam, both on the 1 minute and 20 second grid systems, and compare them with the observation (Figure 12). As can be seen in Figure 12, the computed waveforms are very different on the 1 minute and 20 second grid systems. The latter is closer to the observed waveforms. The period of the computed waveforms on the 1 minute grid system is much shorter than the period of the observed waveforms. This is

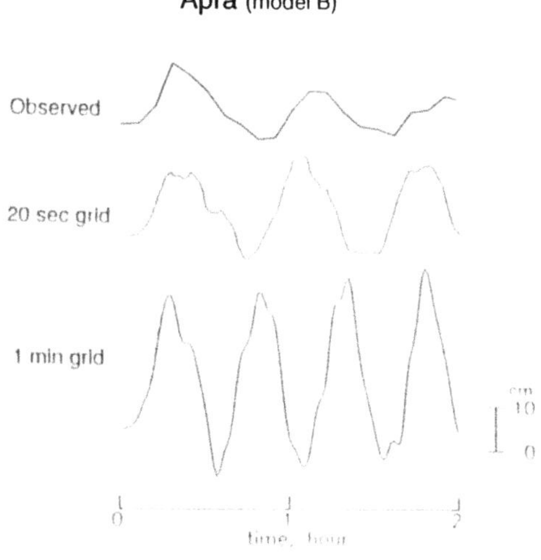

Figure 12
Comparison of the observed and computed tsunami waveforms using two grid systems, 1 minute and 20 seconds, on the tide gauge station at Apra. Fault model B is used for this comparison.

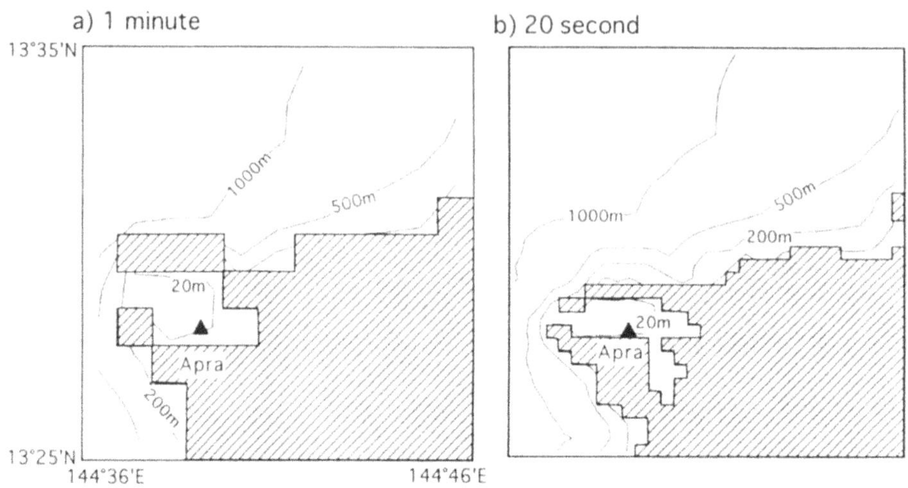

Figure 13
Grided bathymetry around Apra. a) 1-minute grid system, b) 20-second grid system.

because the 1-minute grid system cannot represent the coastal topography around Apra Harbor (Figure 13). The accuracy of the bathymetry is particularly important for computing the tsunami waveforms at tide gauge stations in a bay because the tsunami waveforms are largely controlled by the free oscillations of the bay.

Discussion and Conclusion

We have tried to determine the actual fault plane of the 1993 Guam earthquake. Our analysis of the seismic waves and aftershocks does not allow us to identify the fault plane. Thus we have investigated the possibility of identifying the actual fault plane from tsunami waveforms. The far-field tsunami waveforms at tide gauges or ocean bottom pressure gauges are not useful to identify the actual fault plane. The near-field tsunami waveforms at the tide gauge station just above the fault (about 60 km away from the epicenter) provide useful information for the identification of the actual fault plane of the earthquake, if the tsunami waveforms are sampled at least every 2 min. Current sampling rates at these stations (6 min and 15 min) are too coarse to make use of the near-field tsunamis for earthquake source study.

The dip of the shallow nodal plane is 17° (CMT) to 23° (MTRF), which is a typical underthrust dip angle found in other subduction zone. On the other hand, the T axis dips about 65° which is steeper than the apparent slab dip, about 40°. While this tectonic argument offers some preference for an underthrusting interpretation, it is not a proof of the actual fault plane. To resolve this question in the future, different data or qualitative tectonic arguments must be used. It is an important question to answer.

Acknowledgements

We thank J. Johnson for helpful comments. The tide gauge and ocean bottom pressure gauge records in Japan were provided by Yoshinobu Tsuji at Earthquake Research Institute, Tokyo University and Masami Okada at Meteorological Research Institute in Japan. The tide gauge record at Agana, Guam was provided by Isao Matsuoka at USGS Water Resource Division. The tide gauge record at Apra Harbor, Guam was provided by Frank González at NOAA's Pacific Marine Environmental Laboratory. This work was supported by NSF (EAR 9117800).

REFERENCES

CHRISTENSEN, D. H., and RUFF, L. J. (1985), *Seismic Coupling and Outer Rise Earthquakes*, J. Geophys. Res. *75*, 1637–1656.

DZIEWONSKI, A. M., CHOU, T. A., and WOODHOUSE, J. H. (1981), *Determination of Earthquake Source Parameters from Waveform Data for Studies of Global and Regional Seismicity*, J. Geophys. Res. *86*, 2825–2853.

FUKUSHIMA, T., SUETSUGU, D., NAKANISHI, I., and YAMADA, I. (1989), *Moment Tensor Inversion for Near Earthquakes Using Long-period Digital Seismographs*, J. Phys. Earth *37*, 1–29.

HWANG, L.-S., BUTLER, H. L., and DIVOKY, D. J. (1972), *Tsunami Model: Generation and Open-sea Characteristics*, Bull. Seismol. Soc. Am. *62*, 1579–1596.

KOWALIK, Z., and MURTY, T. (1984), *Computation of Tsunami Amplitudes Resulting from a Predicted Major Earthquake in the Shumagin Seismic Gap*, Geophys. Res. Lett. *11*, 1243–1246.

OKADA, M., *Ocean bottom pressure gauge for tsunami warning system in Japan*. In *Proc. 2nd UJNR Tsunami Workshop* (eds. BRENNAN, A. M., and LANDER, J. F.) (National Geophysical Data Center, Boulder, Colorado, USA 1991) pp. 219–227.

OKADA, Y. (1985), *Surface Deformation Due to Shear and Tensile Faults in a Half-space*, Bull. Seismol. Soc. Am. *75*, 1135–1154.

RUFF, L. J., and MILLER, A. D. (1994), *Rupture Process of Large Earthquakes in the Northern Mexico Subduction Zone*, Pure Appl. Geophys. *142*, 101–172.

SATAKE, K. (1995), *Linear and Non-linear Computations of the 1992 Nicaragua Earthquake Tsunamis*, Pure Appl. Geophys., this issue.

SATAKE, K., and KANAMORI, H. (1991), *Abnormal Tsunamis Caused by the June 13, 1984, Torishima, Japan, Earthquake*, J. Geophys. Res. *96*, 19933–19939.

SATAKE, K., YOSHIDA, Y., and ABE, K. (1992), *Tsunami from the Mariana Earthquake of April 5, 1990: Its Abnormal Propagation and Implications to Tsunami Potential from Outer-rise Earthquakes*, Geophys. Res. Lett. *19*, 301–304.

SHINDELE, F., RAYMOND, D., GAUCHER, E., and OKAL, E. A. (1995), *Analysis and Automatic Processing in Near Field of Eight 1992–1994 Tsunamigenic Earthquakes: Improvements Towards Real-time Tsunami Warning*, Pure Appl. Geophys., this issue.

SIGRIST, D. J. (1995), *Tsunamis in Guam*, Tsunami Newsletter *27*, 20.

UYEDA, S., and KANAMORI, H. (1979), *Back-arc Opening and the Mode of Subduction*, J. Geophys. Res. *84*, 1049–1061.

YOSHIDA, Y., SATAKE, K., and ABE, K. (1992), *The Large Normal-faulting Mariana Earthquake of April 15, 1990 in Uncoupled Subduction Zone*, Geophys. Res. Lett. *19*, 297–300.

(Received August 29, 1994, revised March 6, 1995, accepted March 16, 1995)

PAGEOPH, Vol. 144, Nos. 3/4 (1995)

0033-4553/95/040839-16$1.50 + 0.20/0
© 1995 Birkhäuser Verlag, Basel

Field Survey of the East Java Earthquake and Tsunami of June 3, 1994

YOSHINOBU TSUJI,[1] FUMIHIKO IMAMURA,[2] HIDEO MATSUTOMI,[3]
COSTAS E. SYNOLAKIS,[4] PUSPITO T. NANANG,[5] JUMADI,[6] SATOSHI HARADA,[7]
SE SUB HAN,[8] KEN'ICHI ARAI,[1] and BENJAMIN COOK[9]

Abstract — A field survey of the June 3, 1994 East Java earthquake tsunami was conducted within three weeks, and the distributions of the seismic intensities, tsunami heights, and human and house damages were surveyed. The seismic intensities on the south coasts of Java and Bali Islands were small for an earthquake with magnitude M 7.6. The earthquake caused no land damage. About 40 minutes after the main shock, a huge tsunami attacked the coasts, several villages in East Java Province were damaged severely, and 223 persons perished. At Pancer Village about 70 percent of the houses were swept away and 121 persons were killed by the tsunami. The relationship between tsunami heights and distances from the source shows that the Hatori's tsunami magnitude was $m = 3$, which seems to be larger for the earthquake magnitude. But we should not consider this an extraordinary event because it was pointed out by HATORI (1994) that the magnitudes of tsunamis in the Indonesia-Philippine region generally exceed 1-2 grade larger than those of other regions.

Key words: 1994 East Java Tsunami, aftershock area, large tsunami with weak shaking, house and human damage due to the tsunami, relationship between earthquake and tsunami magnitudes.

1. Introduction

A large earthquake of magnitude M_w 7.6 (M_s 7.2) occurred off the southeast coast of Java Island, Indonesia at 01h 17m local time on June 3, 1994 (at 18h 17m GMT on June 2). The epicenter was at 10.5°S, 113.0°E (by NEIC, USGS) about 240 km from the nearest coast. The shock was felt on east Java Island and on Bali Island. Only ten to twenty percent of the inhabitants of the villages on the nearest

---

[1] Earthquake Research Institute, University of Tokyo, Japan.
[2] Asian Institute of Technology, Thailand.
[3] Faculty of Mining, Akita University, Japan.
[4] School of Engineering, University of Southern California, U.S.A.
[5] Bandung Institute of Technology, Indonesia.
[6] Denpasar Meteorological Observatory, Indonesia.
[7] Meteorological Research Institute, JMA, Tsukuba City, Japan.
[8] Korean Meteorological Administration, Seoul, Korea.
[9] Department of Civil Engineering, University of Washington, Seattle, U.S.A.

coasts were awakened by the ground shaking. No earthquake damage was reported on land. In contrast, about 50 minutes after the main shock, a sizable tsunami hit the coast, inflicting heavy damage on several coastal villages in the Banyuwangi and Jember regencies of East Java Province. A total of 223 persons lost their lives and 15 persons were missing, mainly on Java Island; at the village of Pancer, about 50 km southwest of Banyuwangi, 121 persons died and 70 percent of the houses were swept away.

We organized a team and surveyed the south coasts of Bali and Java Islands from June 19th to 24th. We conducted interviews at damaged villages, searched for traces of submergence of sea water, and measured heights of sea-water inundation with reference to sea-surface elevation at the time of the survey. Correction for the astronomical tide was applied later to obtain our estimates of tsunami heights referenced to mean sea level.

In this paper, we report the distributions of the seismic intensities, the distributions of tsunami heights, and human and building damages due to the tsunami. We will discuss the relationship between the hypocentral distance and tsunami height, and will estimate the tsunami magnitude.

2. Survey Schedule

We divided all members into three subteams. On June 20th, all subteams surveyed on the coast of Bali Island. On the morning of the 21st, all members entered Pancer, the village with the severest damage (Fig. 1). In the afternoon, those three subteams surveyed independently.

Sub-team A, Imamura's team, visited Rajekwesi Village, about 15 kilometers west of Pancer. The next day, June 22nd, they moved to Blitar City, about 300 kilometers west of Banyuwangi and conducted surveys on the coasts between Sempu Island and west to Perigi about 40 kilometers west of Blitar.

After finishing the survey of Pancer, subteam B, Matsutomi's team, moved to Lampon, about 10 kilometers east of Pancer. They surveyed the coasts towards the easternmost point of Java Island.

The third subteam, Tsuji's team, collected the statistical tables of damage, maps of distributions of swept-away houses at Pancer, Lampon and Rajekwesi at the local government office, and conducted field surveys at Bambangan, Ngliyep, and Popoh villages.

On our survey, students of the University of K. Petra, Surabaya, and two engineers from BPPT Jakarta assisted and joined our team.

3. Aftershocks

About 1,500 aftershocks were recorded within 4 days after the main shock by the seismic network of the Indonesian Meteorological and Geophysical Agency.

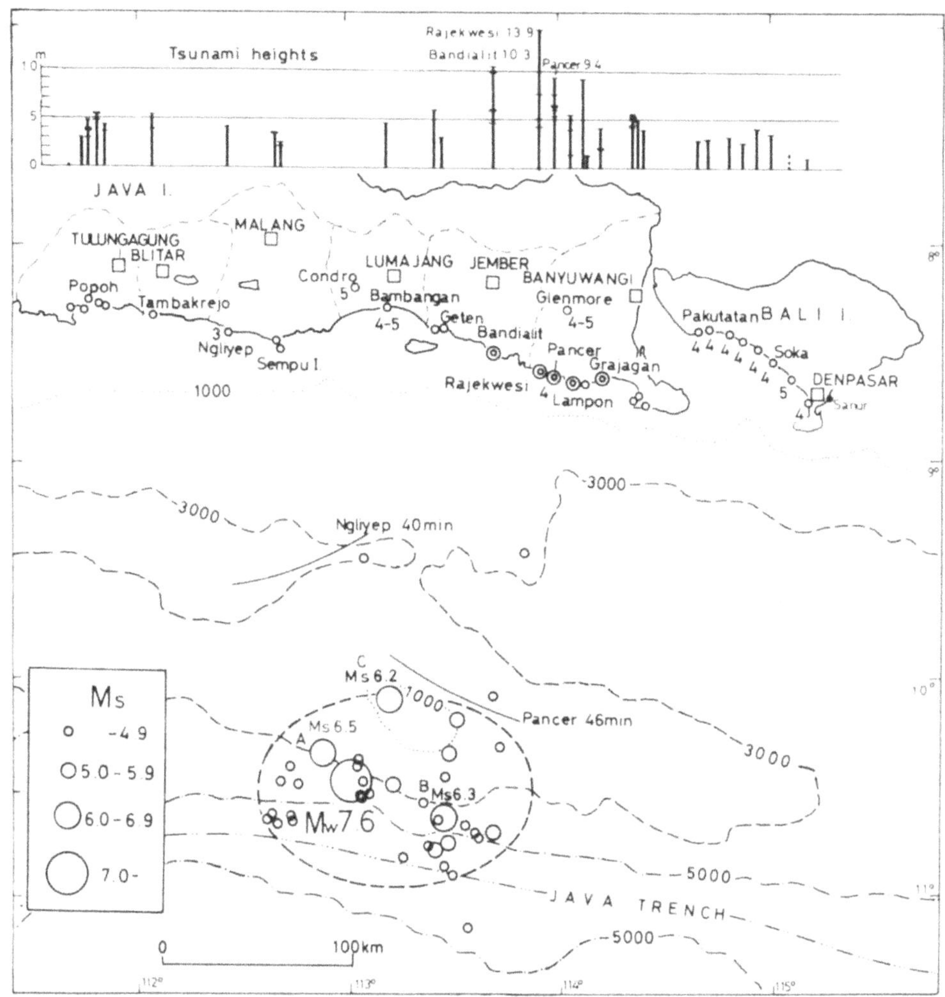

Figure 1

Distributions of tsunami heights, seismic intensity, and aftershocks of the 1994 East Java earthquake. Distribution of tsunami heights is denoted by solid bars in the upper graph. Small horizontal bars attached to each solid bar show the tsunami height at individual points where we measured at more than two points in the same village. Open circles on land illustrate the surveyed points, and attached numbers represent seismic intensity on the Modified Mercalli scale. Double circles show severely damaged villages due to the tsunami. Chain lines show regency boundaries, and squares show capitals of regencies. Circles in the sea area display the locations of the main shock and the aftershocks within 10 days after the main shock (by NEIC, USGS). Tsunami wavefronts estimated from the travel times to Pancer and Ngliyep are also shown. Sea bottom topography is expressed by contours and depth is shown in meter.

Figure 2 shows the daily change of the number of observed aftershocks. The daily number of aftershocks decreased rapidly five days after the main shock.

NEIC quick epicenter determination by USGS reported 34 aftershocks of the present event within ten days after the main shock (Fig. 1). The shape of the

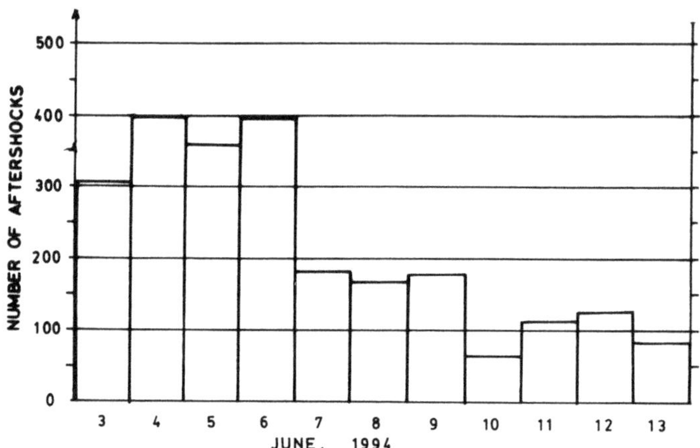

Figure 2
Daily change of aftershocks for the period from June 3 to June 13 observed by the Meteorological and
Geophysical Agency, Indonesia.

aftershock area is an ellipse with the longer diameter 120 km in an east-west direction and the shorter diameter 100 km in a north-south direction. The aftershock area is located near the axis of the Java trench. The northern boundary of the area is about 200 km distant from the southeast coast of Java Island.

Three big aftershocks occurred; event A (M_s 6.5) is at 21h 07min GMT on June 3 (04h 07min on June 4 local time), B (M_s 6.3) is at 00h 58min on June 4 (07h 58min), and C (M_s 6.2) is at 01h 45min on June 5 (08h 45min). It is probable that smaller tsunamis were also generated by these three aftershocks.

We also obtained eyewitness accounts of aftershocks at a few points. At Soka on Bali Island, 35 kilometers northwest of Denpasar, the inhabitants said that they felt an aftershock during the afternoon of the next day, and later another tsunami occurred. They testified that the height of the inundation limit of sea water was nearly the same as that of the main shock tsunami. We measured the height of both tsunamis as 3.7 meters above the mean sea level. We cannot identify the corresponding event in the table announced by NEIC.

On the coast of Bambangan (8°17′21.6″S, 113°06′30.3″E), 20 kilometers south of Lumajang City, we interviewed witnesses who related that at 7 o'clock on the morning of June 5th, two days after the main shock, another tsunami arrived. We measured the tsunami inundation heights for both the main and the aftershocks, on the basis of eyewitness accounts of the inhabitants, as 4.6 and 3.0 meters above the mean sea level, respectively. This aftershock tsunami was suggested to be generated by the abovementioned aftershock C.

4. Distribution of Seismic Intensity

As the native languages of the south coasts of the provinces of Bali and East Java are Javanese and Standard Indonesian, we interviewed the inhabitants with the assistance of Indonesian translators. We prepared the questionnaire sheets in the Indonesian language, asking the condition of the tsunami and the grade of seismic intensity on the Modified Mercalli scale. We asked Indonesian co-workers to judge the seismic intensity through natural conversation with the inhabitants in their native language.

As the main shock occurred at midnight, most people were asleep. About ten to twenty percent of the population at Pancer Village awoke, while the rest continued sleeping without noticing the shock. Seismic intensity is estimated as 4 on the Modified Mercalli scale there.

We estimated seismic intensity from interviews at 14 points (Fig. 1). It was clarified that seismic intensity did not exceed 5 at any point. There was no earthquake damage.

Through our interviews with the inhabitants we also noticed that most people on the coasts did not receive correct knowledge of tsunamis in order to be prepared for a tsunami attack after feeling a strong earthquake on the coast. But for the present case, even if they had the knowledge, we could not expect that people on the coast would take precautions against the tsunami, because they felt such a weak shaking which they experience several times every year.

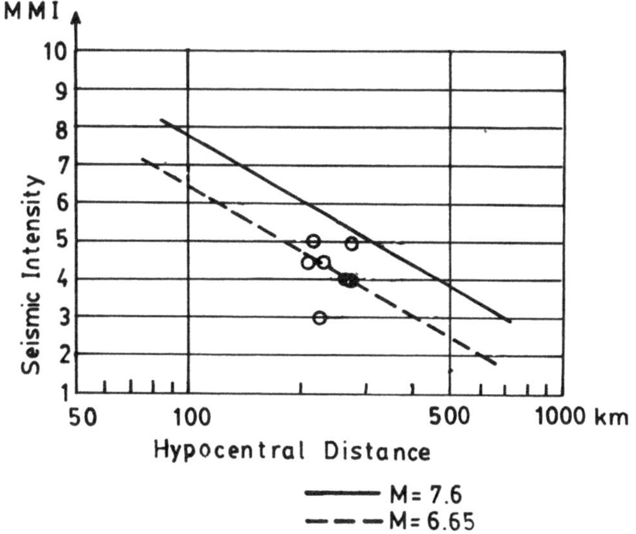

Figure 3

Relationship between the hypocentral distance and the seismic intensity. The solid line shows the relation given by the formula (1) for the case $M = 7.6$ and the dashed line is for $M = 6.65$.

The relationship between the seismic intensity and the hypocentral distance is formulated by ESTEVA *et al.* (1965) as follows

$$I = 8.16 + 1.45\,M - 2.46\ln r \tag{1}$$

where r(km) is the hypocentral distance, M is magnitude, and I is seismic intensity on the Modified Mercalli scale. For the present case we plotted the seismic intensities against the hypocentral distances as displayed in Figure 3. The solid line shows the expected intensity given by formula (1) for $M = 7.6$. The seismic intensity felt by the inhabitants at each point is evidently smaller than that expected by equation (1).

The formulae of the attenuation of Modified Mercalli intensity, with respect to the hypocentral distance, were also proposed by BRAZEE (1976) and ANDERSON (1978), mainly for earthquakes in the U.S.A. SATO (1948, 1955) and KAYANO (1990) also obtained empirical formulae from Japanese data for the relationship between seismic intensity and distance. These works illustrate that, strictly speaking, the attenuation of seismic intensity with respect to the distance varies in different localities and cannot be expressed by a universal formula. We suggest here that the present event was felt weaker in magnitude by the inhabitants.

5. Distribution of the Tsunami Heights

We found clear traces of tsunami submergence at many places on the coastal areas facing the source. We could easily distinguish the traces of inundation of sea water on walls of houses and on surface barks of trees in severely damaged villages (double circles in Fig. 1). We could measure the tsunami height using those traces. On the coasts far from the source, we measured the tsunami height mainly on the basis of eyewitness accounts of the inhabitants.

The south coasts of Java and Bali Islands are attacked by high swells from the subantarctic zone repeatedly every day, and the sea surface is disturbed usually. Thus, it is difficult to detect the sea-surface abnormality due to the tsunami with absolute height less than two meters.

A tide gauge station is located at Cilacap in Central Java, some 600 km west of the source, but we could not obtain information of the record there prior to this writing.

As we could not use bench marks on land, we measured the inundation height above mean-sea level at the time of the survey, and the astronomical tide component was compensated afterwards by the tide tables of Banyuwangi, Cilacap and Sanur Ports, which were computed on the data basis of 7 tidal elements (M2, S2, O1, K1, P1, N2, and K2) supplied by the Japan Ocean Data Center of the Hydrography Department, Maritime Safety Agency. Cilacap Port is on the south coast of Central Java about 600 kilometers west of Banyuwangi, and Sanur is on the

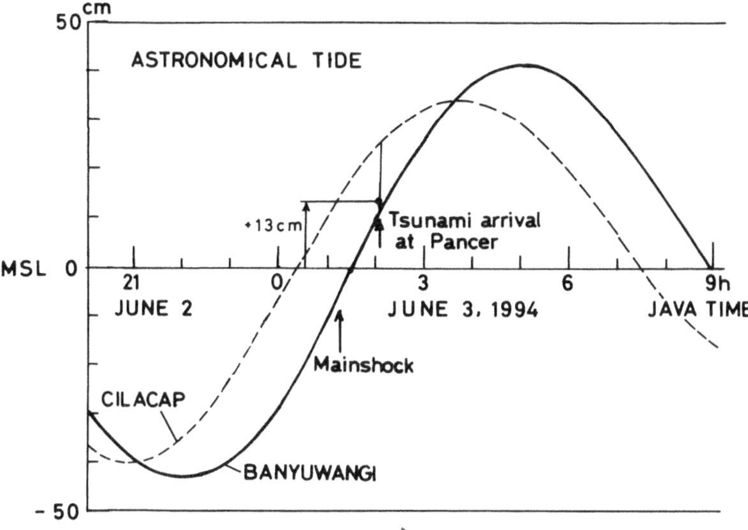

Figure 4

Computed astronomical tides at Banyuwangi and Cilacap ports for the day of the 1994 East Java earthquake tsunami.

southeast coast of Bali Island (Fig. 1). After observing that the tidal phase at Banyuwangi lags behind Cilacap by 60–90 minutes and that the amplitude of the former is 10–20 percent larger than that of the latter, we estimated the tidal level at survey time by interpolation for each point on Java. The same interpolation was made for points on Bali by using tidal data of Banyuwangi and Sanur. Hereafter we will denote tsunami height as the tide corrected value.

The astronomical tide change at the time of tsunami arrival is shown in Figure 4. Tsunami arrival time at Pancer was testified by the inhabitants at 02h 03min, that is 46 minutes after the main shock. At that time, astronomical tide was +13 cm above mean sea level and was in the rising phase.

The distribution of the inundation heights of the tsunami is displayed in the upper bar graph of Figure 1. We generally measured at more than two points in the severely damaged villages. Small horizontal bars attached with a fat vertical bar depict inundation height at an individually measured point in such a village. The highest inundation of 13.9 meters was measured at the east entrance road of the residential area of Rajekwesi village.

The length of the coast where tsunami height exceeded 4 meters is about 300 kilometers.

6. Tsunami Arrival Time

We obtained information of the tsunami arrival time in four villages. A person in Pancer Village continuously watched time and noticed the tsunami arrival at

02h 3m, that is 46 minutes after the main shock. We also obtained an eyewitness account of the tsunami arrival time at Ngliyep Village which the wave attacked shortly before 2 o'clock, from which we can estimate that tsunami arrival time was about 40 minutes.

Additionally, we obtained eyewitness accounts of the tsunami arrival time as 15–20 minutes, both at Bambangan and Popoh Villages. These were not checked by a clock, thus, we cannot expect accuracy from those witnesses.

We drew refraction diagrams inversely from Pancer and Ngliyep up to 46 and 40 minutes, respectively (Fig. 1). We can expect that the final progressive lines should touch the tsunami source region. The line from Pancer runs close to the boundary of the aftershock area. We can judge that the tsunami source area of the present event coincides with the aftershock area generally.

7. Damage due to the Tsunami

Most human and house damage was sustained in the territory of East Java Province. Table 1 lays out the statistics of damage by regions. The damage in the Banyuwangi Regency constitutes a large majority of the total damage. Killed and lost lives due to the tsunami totaled 223 and 15, respectively. The statistics of damage caused in the villages is reported by the rescue headquarters at Pancer in Table 2. The district office also announced the population and the total number of houses of the damaged three villages (Rajekwesi, Pancer, and Lampon), and we could calculate the mortality and the ratio of collapsed houses to the total (Table 3).

About seventy percent of the houses in Pancer were swept away by the tsunami. The rescue headquarters also made public the detailed maps of distributions of damaged houses in residential areas of those three villages (Figures 5a,b,c).

Table 1

Statistics of human and house damage by regencies in East Java Province, after the rescue headquarter at Pancer, up to June 20

Regency	Human Damage				House Collapsed			Damaged Ships
	Killed	Missing	Heavy	Slight	Totally	Partially	Slightly	
Tulungagung	2	—	20	—	62	59	—	84
Blitar	2	—	—	—	—	3	—	153
Malang	1	—	—	2	31	7	4	168
Jember	12	—	4	7	36	33	11	119
Banyuwangi	206	15	21	—	591	66	235	380
Total	223	15	45	9	720	168	250	904

Table 2

Statistics of human and house damage by villages in Banyuwangi Regency. After the table of the rescue headquarter at Pancer (number of injuries does not agree with Table 1)

Village	Killed	Missing	Human Damage Heavy	Injury Middle	Slight	House Collapsed	Population	Number of Houses	Reported Tsunami Height (m)
Lampon	39	1	4	29	30	112	645	171	12
Pancer	121	—	27	60	439	704	3,081	996	12
Rajekwesi	33	14	1	6	73	71	1,205	301	8
Purwosari	—	—	—	—	—	unknown	1,324	301	8
Grajagan	13	—	—	—	—	unknown	unknown	unknown	—
Kutorejo	—	—	—	—	3	unknown	2,688	647	8
Total	206	15	32	95	542	887	—	—	—

Table 3

Ratio of human and house damage to total population and number of houses by villages
**"House collapsed" contains both totally and partially collapsed houses*

Village	Killed + Missing A	Population B	Mortality A/B	House Collapsed* C	Number of Houses D	Ratio of Collapsed, C/D	Tsunami Height m
Lampon	40	645	6.2%	112	171	65.5%	5.4
Pancer	121	3,081	3.9	704	996	70.7	5.7–9.4
Rajekwesi	47	1,205	3.9	71	301	23.6	5.0–7.5

A) Rajekwesi

Figure 5a shows the schematic map of the residential area of Rajekwesi Village. We measured tsunami height at 7.5 m at point A, where a new vertical surface of sand step appeared on the seaside slope of coastal sand dunes due to the erosion of inundated sea water.

B) Pancer

Figure 5b shows the map of the residential area of Pancer, which is situated on a sand dune of a river mouth, and faces the open ocean. A row of palm trees is arranged in front of the village, and seems to valid for reducing the energy of the tsunami to some extent. The river runs behind the residential area. Because sea water also rose along the river, the inhabitants had difficulty finding the escape route to higher places. In Figure 5b we notice that houses were also washed away in some areas facing the river.

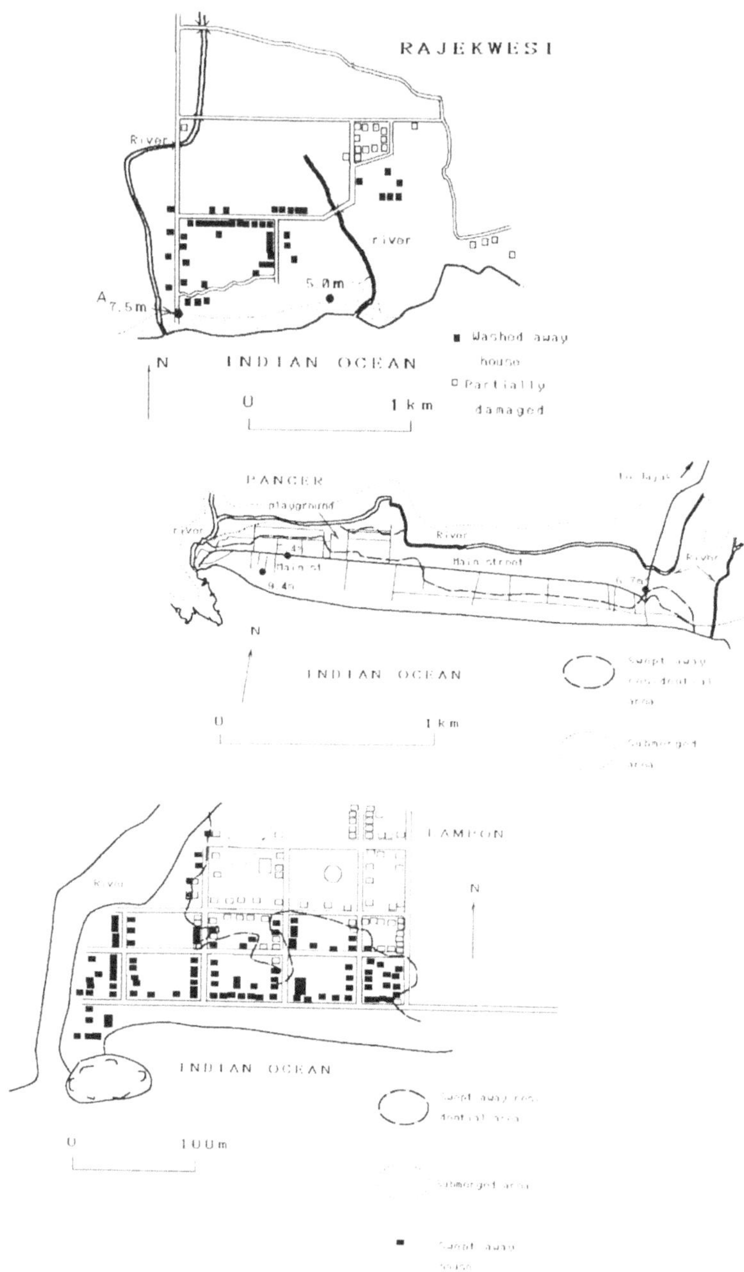

Figure 5a,b,c
Detailed maps of distributions of house damage of residential area of Rajekwesi, Pancer, and Lampon, respectively.

The surface height of the residential area is about 5 m, and the tsunami height was 9.4 m at the seaside part and 7.4 m at the main street in the central part. Thus, the water covered the surface with the thickness of 2–4 m. HATORI (1984) pointed out that when the thickness of inundating water exceeds 2 meters, wooden houses can be washed away. The main part of the residential area of Pancer seems to meet this condition. Actually, we saw nothing but foundations of houses on both sides of the main street of Pancer.

We observed that shore sand was carried into the residential area and formed a thin surface sediment layer. The shore bank was strongly eroded and the vertical surface of sand step with a height of 2–3 meters newly appeared.

The mortality of Pancer was only 3.9% in contrast to the ratio of collapsed houses which is about 70%. Considering that the time of tsunami arrival was midnight, and that tsunami height at the coast reached 9 meters or more, we should recognize that the mortality was fairly small in spite of the severe conditions.

C) Lampon

The residential area of the village of Lampon is also located on a sand dune at the mouth of a river (Fig. 5c). Sea water invaded the village, both from the ocean coast side and the river side. People living near the river mouth had difficulty escaping the tsunami due to its arrival from both sides. In the residential area near the mouth of a river, not only houses were swept away, but the ground itself was seriously eroded and even the foundations were lost. Only a few trees had been planted in front of the village, so sea water rushed into the residential area without impediment. The mortality of Lampon was 6.2 percent, and it seems to be influenced by the fact that there was a poor arrangement of trees in front of the residential area.

The height of the tsunami was measured as 5.4 m in the residential area, but we also measured the tsunami with a height of 9.1 m at a point on the seaside sand bank about 1 km east of the entrance to the village, where a new surface step with heights of 4–5 meters on the front side of the sand dune was formed by the erosion due to the tsunami.

8. Tsunami Magnitude

HATORI (1986) extended the definition of the tsunami magnitude m of Imamura-Iida's scale, which is defined with an interval of 0.5. He proposed two methods of estimating the tsunami magnitude. One is by using data of the averaged inundation heights H (unit; meter) and the epicentral distances D (km). He obtained an empirical formula

$$m = 2.7 \log H + 2.7 \log D - C. \tag{2}$$

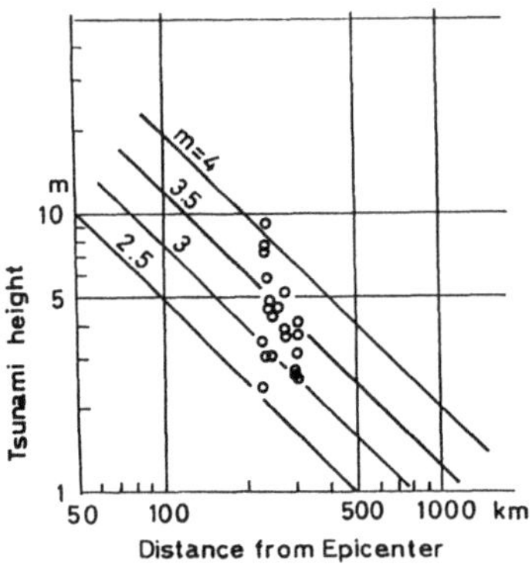

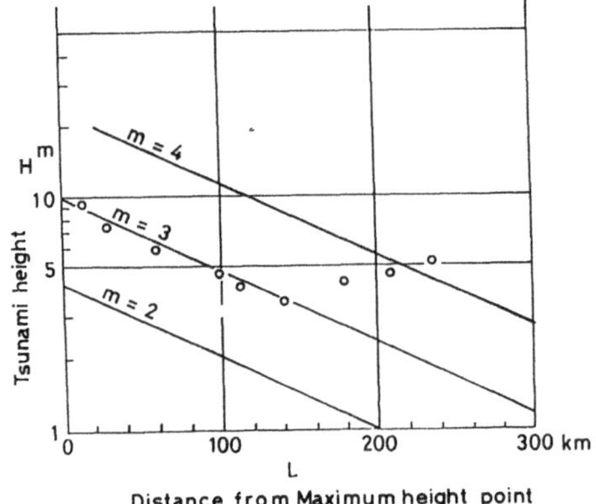

Figure 6(a,b)

(*Upper*) Attenuation of tsunami heights with epicentral distance for different tsunami magnitudes given by formula (2). Open circles show the measured data for the present event. (*Lower*) Relationship between tsunami heights H (unit: meter) and the distance L (unit: kilometer) from the location of the maximum height measured along the coast.

where C is constant and he assigned a value of 4.3. He confirmed that this formula provides good agreement with that defined by the amplitude of tide gauge records for many cases of small tsunamis. He also noted that for large tsunamis ($m > 1.5$), this formula produces about 0.5 larger value than that defined by tide gauge records. For the present case, tsunami magnitude evidently exceeds 1.5, therefore we selected the constant C as 4.8 instead of 4.3. In Figure 6a, we plotted the inundation height to the epicentral distance for each village (white circle). In cases of inundation heights measured at more than two points in the same village, we averaged them. Solid lines show tsunami magnitude defined by formula (2) with $C = 4.8$. Tsunami magnitude m of the present case can be estimated as 3 or 3.5.

The second method of estimating tsunami magnitude by Hatori is using the attenuation curve of inundation height H (in m) with the distance L (in km) from the location of the maximum height point. The empirical relationship between H and L with respect to m is expressed as,

$$m = 0.008\,L + 2.7\log H + 0.31. \tag{3}$$

Solid lines in Figure 6b show the relationship of (3), and white circles are plotted values of inundation heights H to L for the present event, from which we can estimate the tsunami magnitude of the present case as $m = 3$.

We should notice that the distance from the source region is not considered for the estimation of the m value in the second method.

Even though the definition of the m value by Hatori contains ambiguity, we can summarize that the m value for the present case is about 3, estimated by both methods.

WATANABE (1984) obtained a relationship between earthquake (M) and tsunami (m) magnitudes by using Japanese tsunami data in the modern ages. He obtained the relationship,

$$m = 2.30\,\dot{M} - 16.2 \tag{4}$$

by using the data of 60 events with removing four tsunami earthquake cases (dashed line in Fig. 7). KOYAMA and KOSUGA (1985) also proposed an empirical

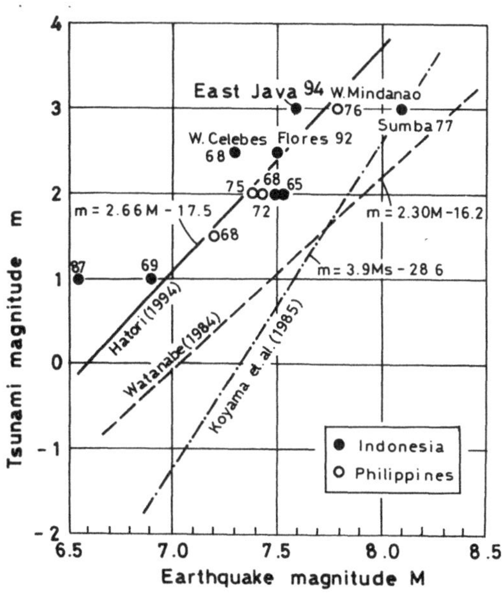

Figure 7

Relationship between the earthquake and tsunami magnitudes. Solid line shows the regressive line of Indonesia-Philippines Region by HATORI (1994). Dashed line shows the regressive line in the Japanese Islands by WATANABE (1984), and chain line shows that by KOYAMA and KOSUGA (1985) from 16 of the largest events in the world.

formula from 16 of the world's biggest events in addition to Japanese data,

$$m = 3.9\, M_s - 28.6, \tag{5}$$

which is expressed by the chain line in Figure 7.

Recently, HATORI (1994) pointed out that the magnitudes of the tsunamis in the sea regions of Indonesia and the Philippines exceed by one to two grades larger than those generated by the earthquakes of the same magnitude of the other regions, and gave the regression formula for those regions as

$$m = 2.66\, M - 17.5. \tag{6}$$

In Figure 7, the solid line shows the relationship (6). Circles represent tsunamis generated in the Philippine-Indonesia region and attached numbers designate years of occurrences. The plotted circle for the present event is also situated close to the line.

9. Discussion

The main shock was felt weaker on Java and Bali in contrast to its large magnitude 7.6. People could not imagine the attack of such a huge tsunami soon after the main shock, even if they comprehended tsunamis. Generally, earthquakes are frequently felt in the territory of Java and Bali Islands, and nobody considered that the present earthquake felt at midnight was an extraordinary one. The situation resembles that of the 1896 Meiji Sanriku Earthquake-Tsunami in Japan. Nobody was prepared for the gigantic tsunami, with height of 15 to 20 m which hit coastal villages thirty minutes after the earthquake and killed about 22,000 persons.

Our future task will be to clarify why the Indonesia-Philippine region has such characteristics that tsunamis one or two grades larger are generated by earthquakes of the same magnitude in comparison to other regions.

Acknowledgment

The authors wish to express their thanks to the staffs of the Meteorological and Geophysical Agency, Indonesia for their support of our field survey and for supplying us with the aftershock data.

We also thank Mr. Subandono, Mr. Presetya from BPPT, Mr. Z. L. Dupe from the Institute of Technology of Bandung, Ms. Maria Usman from the University of Southern California, and students from the University of K. Petra, Surabaya, for their assistance in the field survey. We also thank the Japan Ocean Data Center of the Hydrographic Department, Marine Safety Agency for offering data of tidal coefficients of Banyuwangi, Sanur, and Cilacap Ports. Dr. T. Hatori contributed useful suggestions in the estimation of tsunami magnitude.

Appendix

Table of the results of the measurement of tsunami heights

| Place Name | | Location | | | | Height |
Regency	Place	Lat.	S	Long	E	Abv. MSL (m)
	Kuta	8°	42.57'	115°	09.94"	1.0
	Tanah lot	8	37.19	115	05.15	<2.0
	Soka	8	31.59	114	59.59	3.7
Bali I.	Antor	8	31.65	114	59.72	4.1
	Surabratan	8	28.48	114	55.82	2.6
	Penggraguan	8	27.80	114	54.46	3.2
	Pakutatan	8	26.02	114	49.41	2.8
	Rambutsuiwi	8	24.30	114	46.03	2.7
	Trianḡḡul Asih*	8	39.41	114	21.64	4.9
	Trianḡḡul Asih*	8	39.42	114	21.64	3.6
	G-Land, Plenḡkunḡ	8	41.69	114	22.56	4.0
	Tg. Purwa 1	8	43.82	114	21.06	4.2
	Tg. Purwa 2	8	43.93	114	20.93	5.1
	Tg. Purwa 3	8	43.85	114	20.85	5.5
	Tg. Purwa 4	8	44.75	114	21.35	4.4
	Tg. Purwa 5	8	44.37	114	20.56	5.6
	Tg. Purwa 6	8	43.98	114	20.76	5.3
	Grajagan	8	35.78	114	13.40	2.5
	Grajagan West 1	8	36.0	114	13.5	2.3
	Grajagan West 2	8	36.49	114	13.60	4.1
Banyuwangi	Purwoasri	8	36.91	114	06.83	1.3
	Lampon East	8	36.0	114	05.8	9.3
	Lampon	8	36.93	114	05.19	5.4
	Lampon	8	36.93	114	05.19	1.3
	Lampon	8	36.93	114	05.19	3.8
	Pancer Center	8	35.35	114	00.26	6.7
	Pancer	8	35.4	114	00.3	7.5
	Pancer	8	35.4	114	00.3	9.4
	Pancer	8	35.35	114	00.28	6.7
	Pancer	8	35.16	114	00.47	5.7
	Pancer	8	35.36	114	00.50	6.3
	Rajekwesi	8	33.39	113	56.11	13.9
	Rajekwesi	8	33.40	113	56.11	4.2
	Rajekwesi	8	33.51	113	56.62	7.5
	Rajekwesi	8	33.32	113	56.69	5.0
	Bandialit	8	28.90	113	42.66	9.9
	Bandialit W	8	28.94	113	42.68	11.2
Jember	Bandialit E1	8	28.93	113	42.78	6.0
	Bandialit E2	8	28.97	113	42.75	5.9
	Bandialit E3	8	29.11	113	42.94	4.6
	Besini-Ngarpuḡer	8	22.82	113	27.93	5.9
	Geten	8	23.17	113	24.49	3.1
Lumajang	Bambangan	8	17.36	113	06.51	4.6

*Trianḡḡul Asih is in Kendalrejo Village

Appendix (Contd)

| Place Name | | Location | | | | Height |
Regency	Place	Lat. S		Long	E	Abv. MSL (m)
Malang	Sendanbru	8°	25.00'	112°	42.62'	3.6
	Sendanbru	8	25.81	112	41.08	3.4
	Sendanbru	8	26.00	112	40.92	3.6
	Sempu Is.	8	25.76	112	41.53	2.7
	Sempu Is.	8	26.00	112	41.43	2.1
	Ngliyep	8	21.	112	21.2	4.3
Tulungagung	Tambakrejo	8	18.	112	05.	5.4
	Tambakrejo	8	18.	112	05.	3.7
	Sine Gulf	8	16.	111	53.	4.2
	Sine Gulf	8	16.	111	53.	3.5
	Gerangan	8	15.43	111	50.38	4.6
	Gerangan	8	15.41	111	50.43	5.4
	Gerangan	8	15.42	111	50.39	5.5
	Brumburn	8	15.70	111	50.02	4.8
	Brumburn	8 ·	15.71	111	50.02	3.7
	Brumburn	8	15.72	111	50.04	3.8
	Popoh	8	15.58	111	48.43	2.9
	Popoh Port	8	15.78	111	48.26	3.9
	Sidem	8	15.34	111	48.00	3.1
	Prigi	8	17.23	111	43.43	

REFERENCES

ANDERSON, J. G. (1978), *On the Attentuation of Modified Mercalli Intensity with the Distance in the United States*, Bull. Seismol. Soc. Am. *68*, 1147–1179.

BRAZEE, R. J. (1976), *An Analysis of Earthquake Intensities with Respect to Attenuation, Magnitude, and Rate of Recurrence*, revised edition, NOAA, Technical Memorandum EDS, NGSDC-2, 53 pp.

ESTEVA, L., and ROSENBLUETH, E. (1964), *Spectra of Earthquakes at Moderate and Large Distances*, Bol. Soc. Mex. Ing. Sismica *2*, 1–18.

HATORI, T. (1973), *A Method for Determining Tsunami Magnitude,* IUGG, Tsunami Symp., 1971, Acad. Sci. USSR, Yuzhno-Sakhalinsk, "Tsunami" *32*, 86–96 (in Russian).

HATORI, T. (1984), *On the Damage to Houses due to Tsunamis*, Bull. Earthq. Res. Inst. *59*, 433–439 (in Japanese).

HATORI, T. (1986), *Classification of Tsunami Magnitude Scale*, Bull. Earthq. Res. Inst. *61*, 503–515 (in Japanese).

HATORI, T. (1994), *Tsunami Magnitudes in Taiwan, Philippines, and Indonesia*, Zisin 2 (47), 155–162 (in Japanese).

KAYANO, I. (1990), *Distribution of Various Effects and Damages Caused by Earthquakes and Seismic Intensities on the Basis of Questionnaire Surveys; A Newly Developed Group Survey Method*, Bull. Earthq. Res. Inst. *65*, 463–519 (in Japanese).

KOYAMA, J., and KOSUGA, M. (1985), *Tsunami Magnitude and Fault Parameters*, Zisin 2 (38), 610–613 (in Japanese).

SATO Y. (1948), *Relation between Seismic Intensity and Epicentral Distance (1)*, Bull. Earthq. Res. Inst. *26*, 91–93.

SATO Y. (1955), *Relation between Seismic Intensity and Epicentral Distance (2)*, Bull. Earthq. Res. Inst. *33*, 211–220.

WATANABE, H. (1984), *Statistical Studies on Tsunami Occurred in and near Japan Used Revised Tsunami Table of Watanabe (1983)*, Zisin 2 (37), 607–619 (in Japanese).

(Received September 29, 1994, revised April 12, 1995, accepted April 18, 1995)

PAGEOPH, Vol. 144, Nos. 3/4 (1995)

0033-4553/95/040855-20$1.50 + 0.20/0
© 1995 Birkhäuser Verlag, Basel

The 1994 Shikotan Earthquake Tsunamis

HARRY YEH,[1] VASILY TITOV,[2] VIACHESLAV GUSIAKOV,[3] EFIM PELINOVSKY,[4]
VASILY KHRAMUSHIN,[5] and VICTOR KAISTRENKO[5]

Abstract — The 1994 Shikotan earthquake was one of the greatest earthquakes in recent years with a magnitude of M_s 8.0. A tsunami survey was conducted by Russian and U.S. geophysicists from October 16–30, 1994, less than two weeks after the earthquake. The survey results and a numerical hindcast simulation are reported. Tsunami focusing effect at locations supposedly sheltered by the island chain is discussed. Based on the obtained data, tsunamis which attacked Shikotan Island are characterized as long waves (the order of 10–20 min wave. period) with a positive leading wave. Possible consequences of the positive leading wave form are discussed in relation to the observed minimal destruction of beach vegetation and relatively small transport of marine sediment onto the shore. The high-quality tide-gage record in Malokurilskaya Bay indicates the occurrence of a 53 cm subsidence at the site.

Key words: Earthquake, tsunamis, runup, South Kuril Islands, numerical simulation, tide gage, subsidence.

Introduction

On October 4, 1994, at 13:23 p.m. GMT, an earthquake of magnitude M_s 8.0 (based on Obninsk Seismic Center, Russia) struck the southern region of the Kuril Islands. On Shikotan Island, one of the South Kuril Islands, located closest to the earthquake epicenter, ground shaking was extremely intense: the intensity was reported to be between 9 and 10 on the Modified Mercalli Intensity Scale. An approximately 1.8 m tsunami runup in Nemuro, Japan, was reported approximately 90 minutes after the earthquake (SATAKE, 1994) and Pacific-wide tsunami warnings were issued, including the coastal regions of Hawaii and the west coasts of the United States and Canada. In the South Kuril Islands, 11 people were killed and 242 were injured. In Hokkaido, Japan, one person was killed and 140 were injured.

---

[1] Department of Civil Engineering, University of Washington, Seattle, Washington 98195, U.S.A.
[2] University of Southern California, Los Angeles, California, U.S.A.
[3] Computer Center, Novosibirsk, Russia.
[4] Institute Applied Physics, Nizhny Novgorod, Russia.
[5] Institute of Marine Geology and Geophysics, Yuzhno-Sakhalinsk, Russia.

None of the casualties were due directly to the tsunami in spite of its significant runup (approximately 10 m high runup on Shikotan Island). The total casualty toll was light for a magnitude M_s 8.0 earthquake. Perhaps this is because the earthquake occurred in the middle of the night (0:23 a.m. Oct. 5, local time) when people were sleeping in their houses; most of the residential houses were made of wood which tends to be invulnerable to shaking. On the other hand, inadequately reinforced masonry structures, commonly used as daytime working places, could not withstand the strong shaking and were destroyed. There was no fire caused by this earthquake; the climate was mild so that heating systems were not in use and the village-wide centralized and well-controlled heating and electric systems helped prevent igniting fire from domestic locations. More importantly, the South Kuril areas have experienced many large earthquakes and tsunamis (see Table 1), hence the people were well prepared for this natural disaster.

Remarkable effects of the earthquake to note were the formations of many large gaping fissures and landslides evidently due to very strong ground shaking. For example, near Malokurilsk, Shikotan, major fissure formations were observed on the grass covered hill: the main fissure is a maximum of 60 m wide and approximately 350 m long; one end of the fissure leads to the approximatley 80 m high shore precipice (see Fig. 1). Because of the strong ground shaking, the earthquake also generated significant secondary environmental impacts due especially to leakage of oil from many oil storage tanks.

Table 1

The tsunamigenic earthquakes occurred in the area within 42.0–46.0°N and 145.0° – 150.0°E. The data listed are limited to those with the tsunami intensity, I, greater than 0: the tsunami intensity I is based on SOLOVIEV *(1972). The data were compiled based on* SOLOVIEV *(1978) and* SOLOVIEV *et al. (1986).*

Year	Date	Latitude (N)	Longitude (E)	Depth (km)	Magnitude (M_s)	I	Max. tsunami runup height (m)
1843	4/25	44.70	149.70	40	8.2	2.5	4.50
1893	6/03	43.10	147.00	40	6.6	1.0	1.50
1894	3/22	42.50	145.10	40	7.9	2.0	4.00
1958	11/06	44.53	148.54	40	8.2	2.5	5.00
1958	11/12	44.39	148.70	40	7.4	0.0	1.00
1961	2/12	43.88	147.65	50	7.0	0.0	1.00
1963	10/13	44.80	149.50	47	8.1	2.5	5.00
1963	10/20	44.80	150.20	60	7.4	3.0	15.00
1969	8/11	43.58	147.82	40	8.2	2.0	5.00
1973	6/17	43.15	145.88	55	7.9	1.0	1.50
1973	6/24	43.38	146.52	57	7.4	0.0	1.20
1975	6/10	43.20	147.50	30	7.1	2.0	5.50
1978	3/23	43.70	149.30	40	7.6	1.5	0.17
1978	3/23	43.90	148.30	40	7.8	0.0	0.26
1978	3/24	43.90	149.10	30	7.9	0.0	0.65
1994	10/04	43.84	147.59	33	8.0	2.5	9.00

Figure 1

Major fissure formations near Malokurilsk, Shikotan. The fissure, maximum 60 m wide, 15 m deep, and approximately 350 m long, is formed parallel to the major landslide (approximately 80 m high) along a shoreline precipice. Note that the size of trees shown left of the fissure is approximately 8 m high and the white spots that appear on the lower hill next to the precipice are full-size cows.

From October 16 through 30, 1994, the tsunami reconnaissance survey was conducted by fifteen Russian scientists and two members from the United States. Results of the survey and geophysical aspects of this earthquake are reported in this paper.

Seismological Background and Earthquake Mechanism

This earthquake occurred in one of the most seismically active areas of the Kuril-Kamchatka region. There were at least 27 recorded tsunamigenic earthquakes in the area between 42.0–46.0°N and 145.0°–150.0°E. Significant ones with tsunami intensity (based on SOLOVIEV, 1972) greater than 0 are listed in Table 1.

The earliest and probably the largest event occurred on April 25, 1843 (magnitude $M_s = 8.2$). The exact position of epicenter for this event is unknown but severe seismic shaking and damage were reported in the large area between Urup Island in the north and Kushiro, Japan, in the south. The data on tsunami heights on the South Kurils are not available, but IIDA (1984) reported 4.5 m waves near Kushiro. The next large event in the area occurred on March 22, 1894 ($M_s = 7.9$) with a reported maximum tsunami runup of 4 m at Miyako, Japan. After this event there had been no large (with $M_s > 7.9$) earthquake until 64 years later on November 6, 1958 when an earthquake of $M_s = 8.2$ occurred near Iturup Island. This earthquake generated tsunamis and the maximum runup height of about 5 m was reported on the east coast of Iturup Island. In 1963, a couple of large earthquakes struck successively within a week at virtually the same location near Urup Island. Another large earthquake (on August 11, 1969; $M_s = 8.2$) struck the same region, about 1.5 degrees SW of the 1958 event and 3.5 degree SW of the 1963 events. The maximum runup height which reached 5 m was measured on the east coast of Shikotan Island. The aftershock areas of both the 1958 and 1969 events, based on the seismological observations, overlap by at least 50 km along the trench direction (HATORI, 1990).

The present Shikotan earthquake of October 4, 1994, occurred at the same location as the 1969 event. The aftershock clouds of both events almost coincided; the deviation in the cloud locations was only in the order of 20–30 km. Other recent and substantial earthquakes occurred at the same location. At the NE edge of the 1994 Shikotan earthquake source, and earthquake of $M_s = 7.1$ occurred on June 10, 1975. Despite its magnitude, tsunamis with 5.5 m height were measured at the nearest coast. The 1975 event is considered to be an example of a so-called tsunami earthquake (KANAMORI, 1972; FUKAO, 1979; KANAMORI and KIKUCHI, 1993). Furthermore, the SW edge of the 1994 Shikotan earthquake source is also overlapped with the source area of the June 17, 1973 earthquake ($M_s = 7.9$). Figure 2 shows the recent significant earthquakes which occurred in the South Kuril area. The numerous significant earthquakes in the same area in such a short time interval contradicts the framework of the seismic-gap theory which suggests that a large

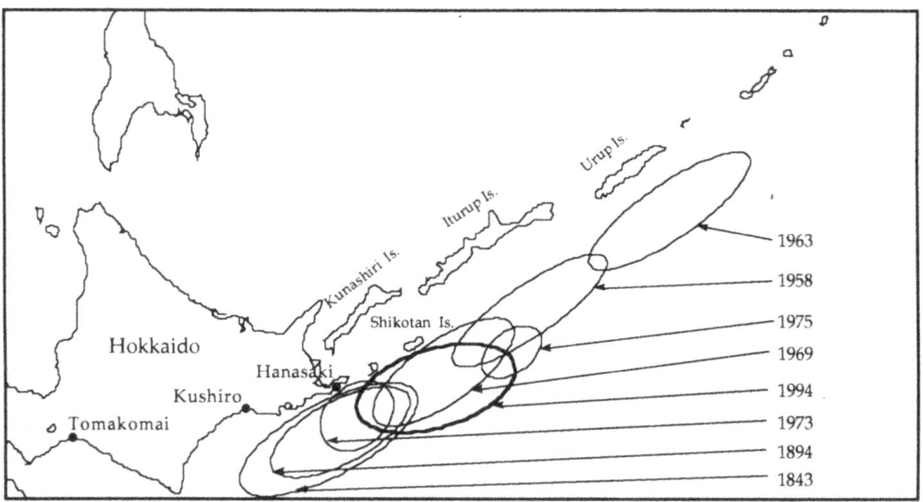

Figure 2
Distribution of source areas of some historical South Kuril tsunamis.

subduction earthquake tends to occur in a seismically quiet region adjacent to a seismically active neighborhood, for example, the Shumagin seismic gap along the Aleutian trench (JACOB, 1984).

The aftershock distribution shown in Figure 3 indicates that the seismic activities associated with the earthquake occurred at the continental slope. Harvard University's centroid moment tensor determination for the 1994 Shikotan earthquake was obtained about 6 hours after the main shock. The best double-couple solution consists of two planes, none of which can be considered to represent the low-angle thrust typical of the subduction region. The first plane dips at an angle 40° to the SW and strikes almost perpendicular to the trench axis. The second plane with strike direction 52°, which is roughly parallel to the trench axis, dips 77° to the SE so that its predominant source mechanism should be the reverse dip-slip fault with a considerable (slip angle 128°) strike component. This uncertainty with regard to source mechanism hopefully will be resolved later on the basis of detailed seismological investigation.

Note that this area is poorly covered by the regional seismic networks. To demonstrate the degree of accuracy in source parameter estimates, eight different main shock positions, determined by different seismological agencies, are listed in Table 2 and are presented in Figure 3. The widely scattered main shock locations exceed one degree in both latitude and longitude coordinates. Based on the seismological data available, it appears that the source mechanism of the 1994

Table 2

The main shock information determined by different agencies: the estimated locations of the main shock are also plotted on Figure 3

Latitude	Longitude	Time of of occurrence	Depth	Magnitude (M_s)	Agency
43.40 N	147.60 E	13:23		7.8	YS
43.68 N	147.63 E	13:22:57	33 km	8.0	OBN
43.22 N	147.40 E	13:23	30 km	8.1	JMA
43.39 N	147.08 E	13:23:05	46 km	8.1	HOKK
43.661 N	147.335E	13:22:58.1	33 km	8.1	NEIC
43.67	147.36 E	13:23:25	64 km	8.3	HARV
43.90 N	147.20 E	13:23	33 km	8.2	PTWC
43.80 N	148.60 E	13:23	33 km	8.1	ATWC

  YS: Yuzhno-Sakhalinsk Seismic Station, Russia
 OBN: Obninsk Seismic Center, Russia
 JMA: Japan Meterological Agency, Japan
HOKK: the Research Center for Earthquake Prediction at Hokkaido University, Japan
 NEIC: National Earthquake Information Center, USA
HARV: Harvard University, USA
PTWC: Pacific Tsunami Warning Center, USA
ATWC: Alaska Tsunami Warning Center, USA

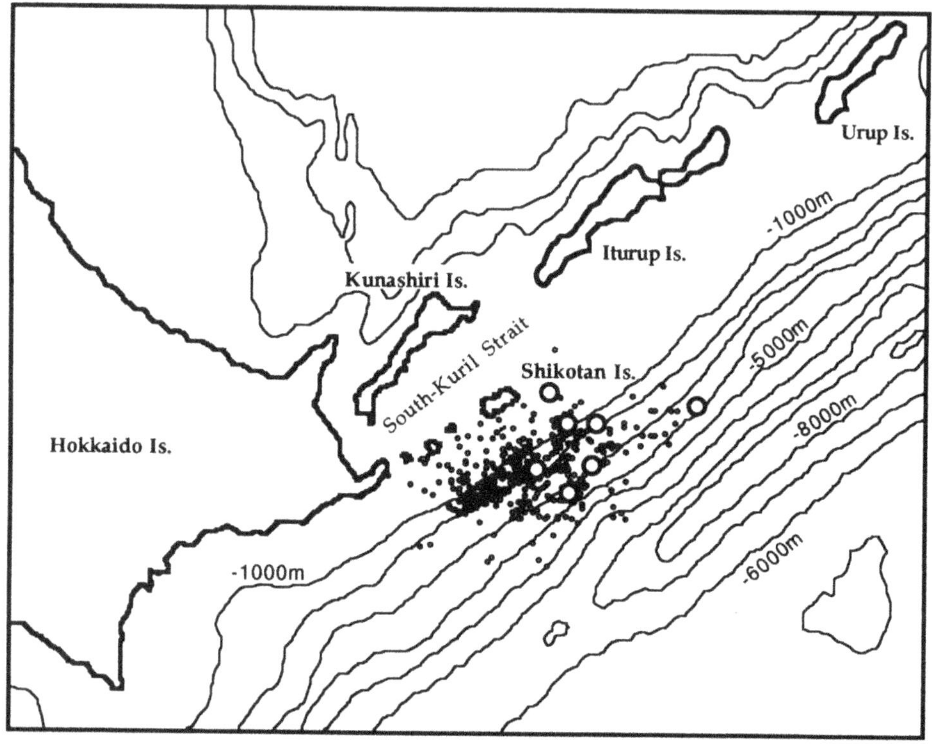

Figure 3

Bathymetry contours, and one-week-aftershock distribution, o, and several different estimates of epicenter locations, O, for the October 4, 1994 Shikotan earthquake. (The aftershock data are provided by the Research Center for Earthquake Prediction of Hokkaido University.)

Shikotan earthquake contained a complex nature in which several major planes ruptured during the event and several large crustal blocks displaced (possibly in different directions).

Tsunami Runup Distributions

A tsunami source condition was estimated based on the best-fit double-couple of the Harvard University centroid moment tensor solution and seismic moment 2.1×10^{28} dynes-cm. Estimating the source area to be 120×100 km based on the aftershock distribution (Fig. 3), with the consideration of an approximately 50 cm subsidence at Shikotan Island (as discussed later), an initial water-surface displacement was determined as shown in Figure 4. Based on this initial condition, the numerical simulation was carried out. The numerical model is a finite-difference scheme, based on the fully nonlinear shallow-water wave equation in the characteristic

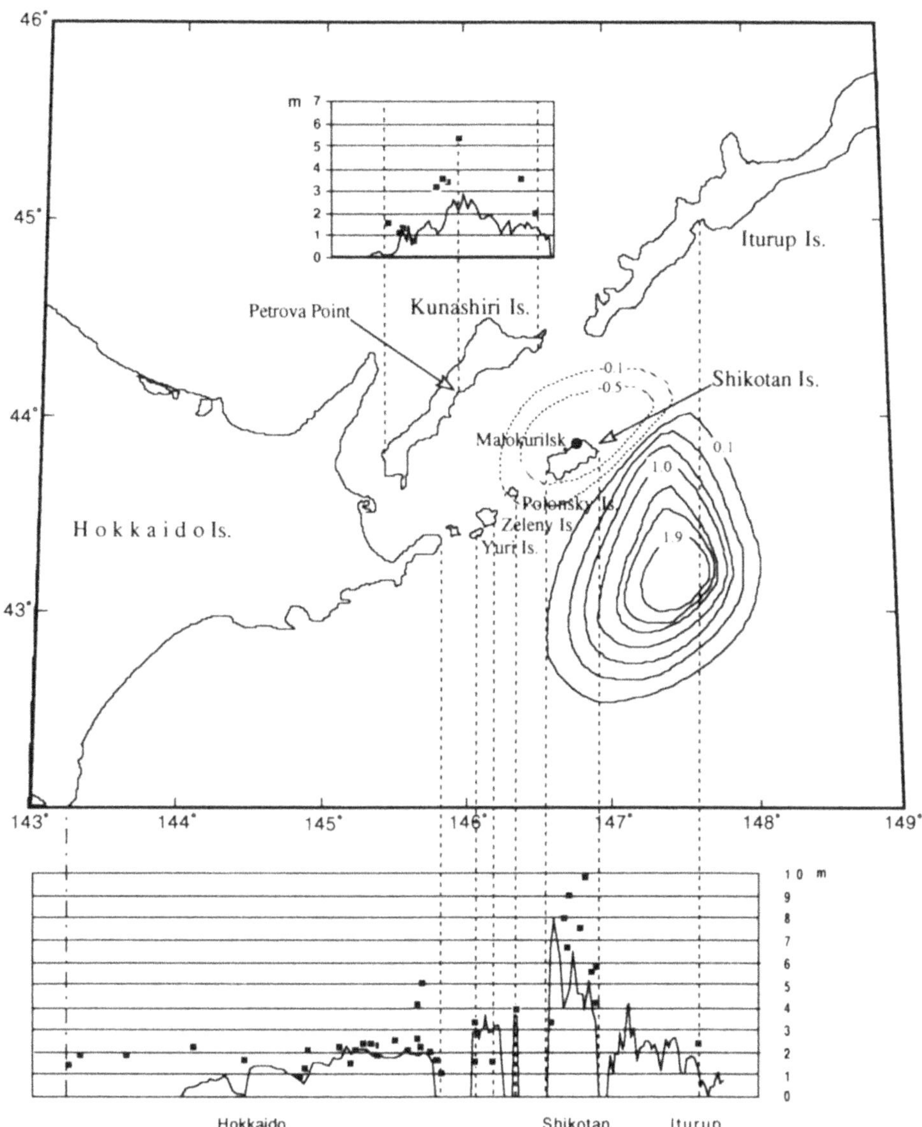

Figure 4

Initial sea-level displacement (in meters) used for the numerical model and tsunami runup heights from sea level at the time of the tsunami attack: ——, the numerical simulation model; ■, measured highest values of each location. All the runup values are those on the south-east side of the islands, i.e., facing to the tsunami source.

form (TITOV and SYNOLAKIS, 1993): the spatial grid sizes used are 4 km in deep water (depth greater than 300 m) and 0.9 km in shallow water (depth less than 300 m), and the computational time step used is 6 seconds. As pointed out earlier, the mechanism of this earthquake is complex and not fully understood at this point: it appears that this was not a simple subduction earthquake but, at least during the initial stage, the plate was buckled perpendicular to the subduction line. Consequently, it is difficult to determine an accurate estimate of the initial sea-bottom deformation; the results presented in Figure 4 (shown with our field data as discussed later) should be considered as initial attempts for the simulation. It is anticipated that more refinements and corrections for the simulations will be performed.

Tsunami runup height measurements were made in Shikotan, Iturup, Kunashiri and small islands between Shikotan and Hokkaido (e.g., Polonsky, Zeleny, and Yury Islands). Numerous measurements along the Hokkaido coast were carried out by the Tohoku University group in Japan (TAKAHASHI and SHUTO, 1994). The measurements were made using similar procedures as in other recent surveys (e.g., SATAKE *et al.*, 1993, in Nicaragua; Yeh *et al.*, 1993, in Flores Island; HOKKAIDO TSUNAMI SURVEY GROUP, 1993, in Okushiri Island and CHOI *et al.*, 1994, in Korea). No global positioning system (GPS) was used in our survey except on Shikotan Island. Considering that detailed maps and charts in the region were available, the lack of GPS caused no inconvenience nor did we lose any important information in our survey results. The compiled data are shown in Figure 4, together with the numerical simulation results. Note that the measured data were corrected to values of the vertical runup heights from sea level at the time of tsunami attack. In spite of the complexity of the rupture mechanism of this earthquake, the comparisons of the numerical predictions and measurements are in good agreement. Note that this degree of agreement between the measured and simulated runup values was not achieved in the cases of the 1992 Nicaragua, the 1992 Flores, the 1993 Okushiri, and the 1994 Java tsunamis. It is noted that the Nicaragua and Java tsunamis are considered to have been "tsunami" earthquakes (KANAMORI, 1972; FUKAO, 1979; KANAMORI and KIKUCHI, 1993), and the Flores and Okushiri tsunamis were caused by "backarc-thrust" earthquakes (PLAFKER and WARD, 1992), in which rupture patterns were very complex. Also there is considerable evidence for submarine landslides in Flores that might have enhanced the tsunami runup heights (PLAFKER, 1995).

The measured runup data shown in Figure 4 indicate that there is a locally high (5.3 m) runup location near Petrova Point, approximately 13 km north-east of Yuzhno-Kurilsk, on Kunashiri Island; the numerical simulation also exhibits a local maximum at the same location (see Fig. 4). Based on the assumed tsunami generation region, Kunashiri Island is located behind the Shikotan-Polonsky-Zeleny-Yury Islands chain, hence the tsunami propagation to Kunashiri Island must have been partially blocked by the island chain. As shown in Figure 5, the primary

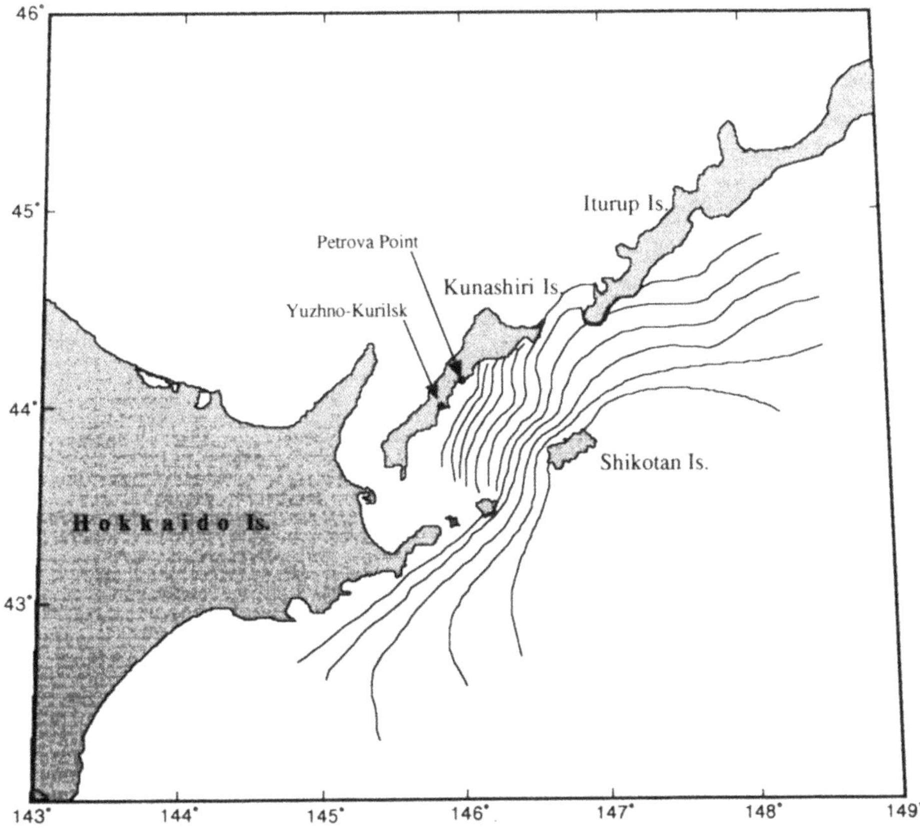

Figure 5

A sequence of the numerically simulated leading tsunami crests with the time intervals of 5 minutes. The tsunami focusing effect near Petrova Point is evident.

tsunami path to reach Kunashiri Island was through the mouth of the South Kuril Strait, between Shikotan and Iturup Islands (see Fig. 3 for the area of the South Kuril Strait). The water depth between Shikotan and Iturup Islands is deeper (approximately 200 m) than the depth (approximately 50 m) of the South-Kuril Strait, between Shikotan and Kunashiri Islands. The gap between Shikotan and Polonsky Islands (Fig. 4) is approximately 20 km wide and 40 m deep and some of the tsunami energy from the Pacific Ocean could have passed through. All other gaps are considered to be too shallow (less than 10 m deep) for substantial tsunami energy to penetrate into the South-Kuril Strait from the Pacific. Our explanation for the large runup height near Petrova Point on Kunashiri Island is as follows. The deeper channel along the mouth of the South-Kuril Strait caused tsunami refraction to guide the propagation westward into the South-Kuril Strait, i.e., directing the tsunami propagation to the middle of Kunashiri Island towards Petrova Point.

Meanwhile, small-but-sufficient tsunami energy which passed through the gap between Shikotan and Polonsky Islands reached Petrova Point; this tsunami met the previously mentioned tsunami that had traveled around the east side of Shikotan Island from the mouth of the South-Kuril Strait. The superposition of two waves caused higher runup near Petrova Point than those measured at other locations. Such effects are qualitatively verified in the numerical simulation model as demonstrated in Figure 5: similar effects were shown in the computer animation independently performed by TAKAHASHI and SHUTO (1994). As will be discussed later with Figure. 9, the water depth of the South-Kuril Strait is shallow (approximately 50 m) and the bottom slope is very mild (approximately 0.1 degree slope), hence submarine landslides are unlikely to occur near Petrova Point.

Runup Distribution Around Shikotan Island

On Shikotan Island, measurements were made at 85 different locations around the island. Note that approximatley 80% of the island's coastline is a series of nearly vertical cliffs and not readily accessible. Except in very few cases (one or two), measurements were made based on very conservative and strict tsunami runup marks, i.e., only marine-origin objects were considered to be appropriate for tsunami runup marks. The measured runup heights around the island are plotted in Figure 6. Note that the values presented in the figure are not all individual measurements but are the local maximum values. As a typical survey example, individual measurements made at Tserkovnaya Bay are presented in Figure 7.

Figures 6 and 7 show that the magnitudes of runup are fairly uniform: the average runup height is 6.1 m on the south side of the island with a standard deviation of 1.8 m, while on the north side of the island, the average runup height is 2.5 m with a standard deviation of 0.2 m. The higher runup heights on the south side of the island confirm that the primary tsunamis must have attacked from the south, and north side of the island was in the shadow zone for this tsunami event. The substantial difference in runup height also suggests that the order of magnitude of tsunami wavelength must have been comparable or smaller than the island dimension; if the tsunami wavelength had been much longer than the island dimension, then the runup heights would have been uniform all around the island, just like tides being uniform around the island. On the other hand, if the tsunami wavelengths were shorter than the length scale of local features (e.g., bays, coves, headlands, and small islands), then the runup distribution along the southern coast of the island would have been substantially affected by the local features: instead, we found the measured runup heights to be fairly uniform along the coast as shown in Figures 6 and 7, indicating that the tsunami wave length must have been long enough. The dimension of Shikotan Island is at the order of 15 km and the order of magnitude of the depth of the sea surrounding the island is approximately 50 m.

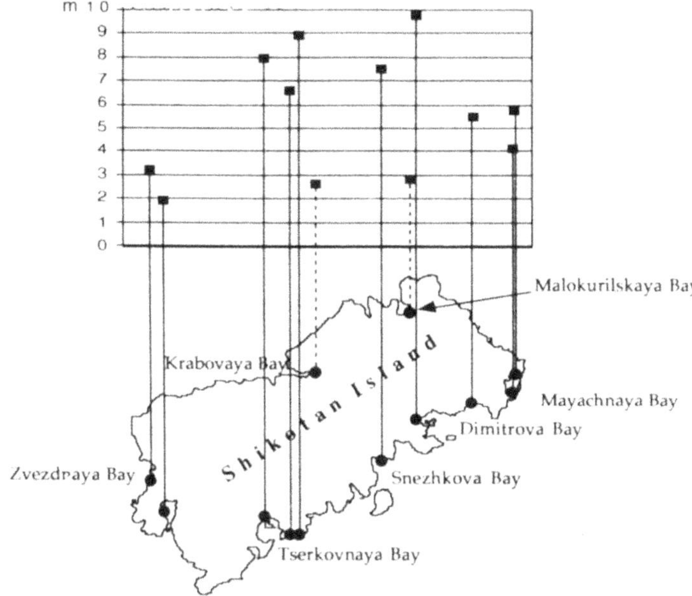

Figure 6

Tsunami runup heights around Shikotan Island. The values are from sea level at the time of the tsunami attack (we assume that the 53 cm island subsidence (see Fig. 11) occurred prior to the tsunami attack).

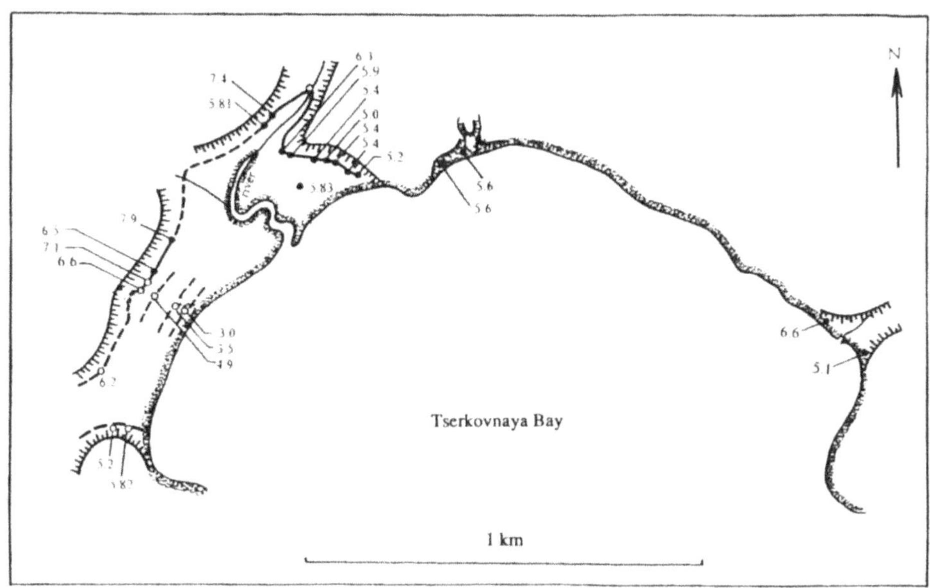

Figure 7

Individual runup measurements (vertical runup heights in meter from sea level at the time of the tsunami attack) at Tserkovnaya Bay.

If a half tsunami wave length is shorter than the island dimension and approximating the propagation speed by that of the shallow-water waves, i.e., a square root of the product of depth and gravity acceleration, then the tsunami wave period must have been less than 23 minutes. The dimensions of local features of the southern coast is at the order of 3 km in horizontal and 30 m in depth, thus the tsunami wave period must have exceeded 6 minutes. Therefore, we estimate the tsunami period to be between 6 and 23 minutes. This estimated range of tsunami period concurs with the numerical simulation result, approximately 20 minutes of the initial tsunami oscillation period. An additional explanation for the difference in runup height between the south and north sides of the island must be related to the shape of the island. As seen in Figure 6, Shikotan Island has a rectangular shape, which is not a shape favorable for waves to propagate around the shadow zone of the island. Most of the tsunami energy propagating along the east and west sides cannot be diffracted into the north side due to the sharp curvature of the shoreline at the north-east and north-west corners of the island.

At Tserkovnaya Bay, we found at least four different tsunami runup marks (3.0, 3.5, 4.9 and 7.1 m) along a transact perpendicular to the shoreline as shown in Figure 7. This means that there were at least four tsunamis attacked at the south coast of Shikotan. Another noteworthy observation from our survey in Shikotan was the fact that we could find no excessive sediment sheet formations nor noticeable transports of beach rocks on the tsunami runup areas. Furthermore, beach grass, bushes, and trees are still intact, although there are many clear signs of tsunami inundation. For example, one of the tsunami traces found at Tserkovnaya Bay is the marine-origin seaweed which clung to a tree branch as shown in Figure 8 which also shows the good vegetation conditions in the tsunami runup areas. It is well-known that tsunamis are capable of transporting substantial sediments onto the runup areas from offshore (ATWATER, 1987; BOURGEOIS *et al.*, 1988; DAWSON *et al.*, 1988; MINOURA and NAKAYA, 1991). Excessive sediment layers and transport of beach rocks and corals were found in the previous surveys of the 1992 Nicaragua tsunamis and the 1992 Flores tsunamis, as well as in the 1993 Okushiri tsunamis survey. Note that the runup heights, approximately 6 m high, measured on the south shore of Shikotan are comparable to the runup heights measured in Nicaragua and Flores, and the south-side Shikotan beaches are sandy beaches, at least near the shoreline. The lack of transported sediments and beach rocks, and the good vegetation conditions on the runup areas evidently suggest that the runup motion on Shikotan must be different from the other tsunami cases; the runup on Shikotan must have been a gradual process. This conjecture is consistent with the fact that the measured runup heights are fairly uniform regardless of the different local topography. Additional explanations could be related to the indication of the inital wave being positive. As discussed later, both tide-gage recordings in Malokurilskaya Bay, Shikotan, and Hanasaki Harbor, Hokkaido, show that the initial tsunami formation was a positive wave. In comparison to the case of an

Figure 8
A tsunami trace found at Tserkovnaya Bay, see the seaweed clinging in the shrub.

initial negative wave followed by a positive wave, the initial positive wave tends to cause smaller runup heights and less violent runup motions; these effects were first demonstrated by GOLUBTSOVA and MAZOVA (1989) and MAZOVA (1991) and the theory was refined by TADEPALLI and SYNOLAKIS (1994). Perhaps, those are the factors which minimized destruction of vegetation, scattering beach rocks, and the formation of sand layers.

Even though approximately 80% of the Shikotan coastal line was not accessible for the survey, regarding surveyed areas, no abnormally high runup region was found: the order of magnitudes of all the measured runup heights can be explained by the numerical simulations. As mentioned earlier, this was not the case for the past several tsunamis; for example, the maximum runup of 26 m measured in Riangkroko (in Flores, Indonesia) could not be explained until evidence of massive nearshore submarine landslides was discovered (PLAFKER, 1995). In spite of several massive landslides observed at many places on Shikotan Island, the occurrence of underwater landslides is very unlikely near Shikotan Island. Figures 9a, b and c display an evident contrast in bathymetry between Shikotan, Okushiri (Japan), and Flores (Indonesia) Islands. The bathymetry of both Okushiri and Flores Islands is very steep around the islands, while in spite of the very ragged terrain above the sea level, the sea bottom is uniform and shallow in the sea surrounding Shikotan Island. Hence, significant submarine landslides are not possible. Incidentally, the

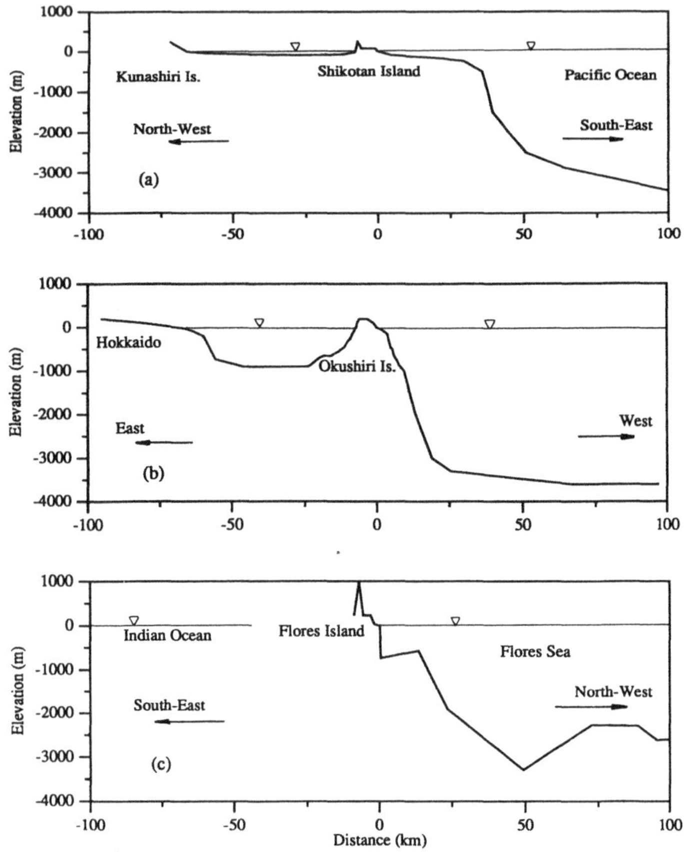

Figure 9

Comparison of sea bottom profiles: a) Shikotan, b) Okushiri Island, and c) Flores Island: the origin is set at the island shoreline.

shallow uniform sea bottom in the South-Kuril Strait (see Fig. 3) resulted in substantial sedimentary deposits with the maximum sediment thickness of more than 4,000 meters (SNEGOVSKOY *et al.*, 1987).

Tide Gage Data at Malokurilskaya Bay

This tsunami event left quality tide-gage records at Malokurilskaya Bay in Shikotan, Hanasaki Harbor in Hokkaido, Poronaisk in Sakhalin, and among others in Japan. The tide gage record at Hanasaki Harbor (see Fig. 2 for its location) shows that the first wave was a positive wave with the wave oscillation periods of 30 minutes (see Fig. 10a). Other than the tide gages, the ultrasonic wave gage placed in 18.5 m deep offshore of Tomakomai Port, Hokkaido (see Fig. 2 for

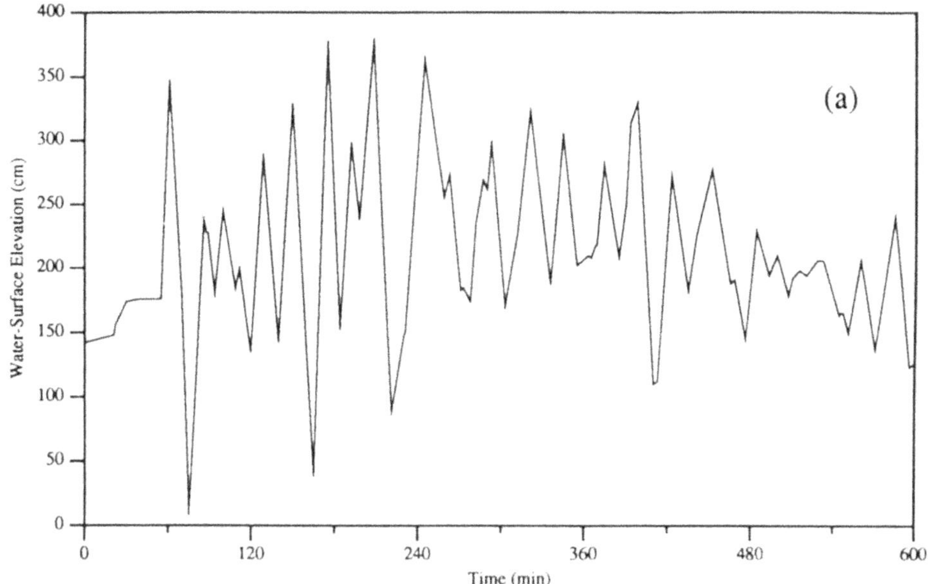

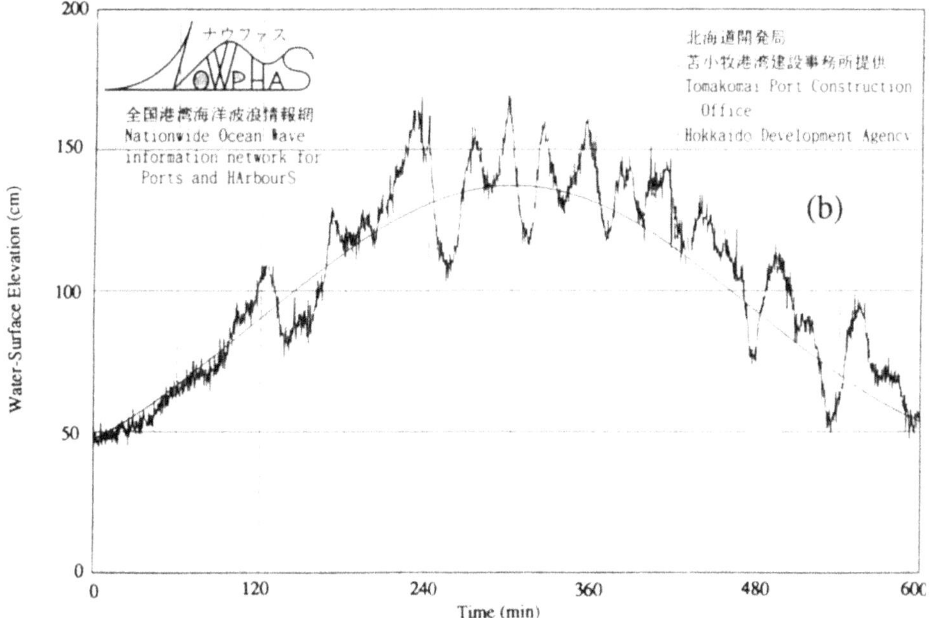

Figure 10

Tsunami measurements in Hokkaido, Japan. a) The tide-gage record in Hanasaki Harbor (the original analog data taken by the Japan Meteorological Agency were digitized and processed); b) Tsunami profile recorded by an ultrasonic wave gage (18.5 m deep) off Tomakomai (after NAGAI *et al.*, 1995). The time origin corresponds to 0:00 a.m. the South-Kuril local time and the earthquake occurred at 0:23 a.m. See Figure 2 for the locations.

its location), captured the tsunami record as shown in Figure 10b. This record was numerically low-pass-filtered because high wind-wave conditions with more than 2 m significant wave heights were present at the time of the measurements. The record reveals that the tsunami arrival time was approximately 1:35 a.m. Kuril local time (14:35 p.m. GMT, that is 1 hour after the earthquake). Just like the record at Hanasaki Harbor (Fig. 10a), the first wave at Tomakomai was a positive wave, and the tsunami periods were very long, in the order of 50 minutes (NAGAI *et al.*, 1995). In this paper, we will focus on the data taken in Malokurilskaya Bay, Shikotan, only 70 km away from the epicenter and closest to the tsunami generation area.

Figure 11a illustrates the tide gage record in the Malokurilskaya Bay (the original analog record was digitized and processed to eliminate the tracer-ink smudge presumably caused by the strong shake). The bay has nearly a circular shape (approximately 800 m in diameter) with the narrow entrance (approximately 350 m wide). The well maintained tide gage is located at the east shore of the bay with a well diameter of 65 cm. The Malokurilskaya Bay is known to have a distinct Helmholtz oscillation period of 18.5 minutes (GJUMAGALIEV and RABINOVICH, 1993), and Figure 11a evidently shows the contamination of the tsunami record due to this Helmholtz oscillation. According to the record shown in Figure 11a, the arrival time of the first tsunami peak was approximately 40 minutes after the main shock; it also took 40 minutes for the tsunami crest to arrive at Hanasaki Harbor (Fig. 10a). It is noted that our numerical model shows that the first-peak arrival time is 30 minutes after the earthquake. The data after removing astronomical tides are presented in Figure 11b which also displays the evident mean water-level shift of 53 cm after the earthquake. The data presented in Figure 11b are considered to be a rare quantitative recording of the land subsidence with the tide gage. According to Figure 11a, the initial rise of water level appears to take place immediately after the main shock and was a slow process: at Hanasaki (Fig. 10a), the initial rise was much faster than that shown in Figure 11a. This gradual rise of the water level could have been created by filling water in the subsided region from the surrounding area, rather than the main tsunamis generated off the Pacific Ocean.

Daily averaged tidal levels were computed from the tide gage records by GUSEVA *et al.* (1994). The data were further adjusted by eliminating the daily variations caused by astronomical tides, and the results are presented in Figure 11c. Note that the data presented in the figure is based on GMT, hence the value of the October 4 record represents the average of pre-earthquake sea level for 13.5 hours and the post-earthquake sea level for 10.5 hours. As shown in Figure 11b, the actual subsidence took place in much shorter duration (less than one hour). This evidently shows that the major subsidence of the island (53 cm) occurred at the main shock followed by the gradual subsidence of about 30 cm from Octover 9 to 13, 1994; note that there was a strong aftershock, M_s 7.4 on October 9. Subsequently, the daily averaged sea level appeared to decrease in the last two days of the

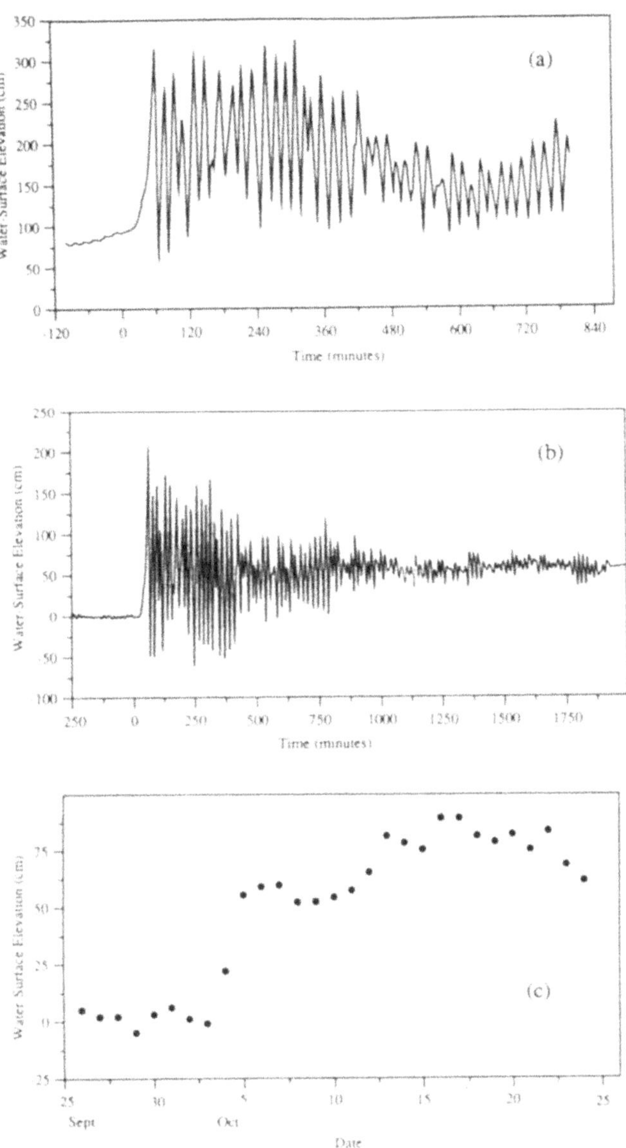

Figure 11

The tide-gage record in Malokurilskaya Bay: a) raw data after digitization (the tracer-ink smudge presumably caused by the strong shake, that appears on the original analog data, was eliminated in the digitization process): the time origin corresponds to 0:00 a.m. local time and the earthquake occurred at 0:23 a.m., b) after removal of the astronomical tides (the time origin is the same as that presented in a), and c) daily average data: the date is based on GMT.

data (October 23 and 24). Physical interpretations for this gradual increase and decrease of sea level are not clear at this time and no explanation is given here.

An eyewitness report in Malokurilskaya Bay indicates that the leading wave was a positive wave followed by approximately 2.5 m of negative wave, which is consistent with the tide gage data presented in Figure 11a. The resulting current was so swift that no ship could cross the bay entrance nor approach the shore for the first 2.5 hours, the current speed of approximately 6 m/sec was reported by an eyewitness. The bottom sounding showed that the water depth at the center of the bay changed from 5 m to 8 m, evidently due to tsunami scouring effects.

Summary

The 1994 Shikotan earthquake was undoubtedly one of the largest earthquakes in recent years. In this very seismically active region, in spite of uncertainty in the earthquake rupture mechanism, resulting tsunami effects were simulated reasonably with the numerical model based on the shallow-water wave theory.

Some noteworthy effects are that the tsunami runup was enhanced at the mid-south coast of Kunashiri Island: the location appears to be protected from the tsunami source because it is behind the Shikotan-Polonsky-Zeleny-Yury Islands chain. Tsunami refraction around the mouth of the South Kuril Strait and merging tsunamis which leaked from the gap between Shikotan and Polonsky Islands might be the expalantion for this large runup.

Based on the runup distribution around the island and the runup conditions, the leading tsunami form that attacked Shikotan Island appears to be a positive wave with a gradual water surface increase. Such leading wave characteristics are consistent with the tide gage data recorded in Malokurilskaya Bay. The positive and gradually rising leading wave might have resulted from the observed minimal destruction of beach vegetation and relatively small transport of marine sediment onto the shore. The tide gage recorded the subsidence of the gage site by 53 cm at the event of the earthquake.

In spite of the many significant landslides observed on Shikotan Island, any substantial underwater slumping was unlikely to occur because of the shallow and uniform sea bottom bathymetry. This is another reason why no abnormally high local tsunami runup was observed on Shikotan Island.

Acknowledgment

The analog tide gage data at Malokurilskaya Bay were digitized and processed by E. Novikova and the data at Hanasaki Harbor were digitized and processed by D. Akkeson. The survey team consisted of V. Djumagliev, S. Gusiakov, V.

Kaistrenko, A. Kharlamov, A. Klochkov, Y. Korolyov, A. Kruglyakov, E. Kulikov, V. Kurakin, B. Levin, E. Pelinovsky, A. Poplavsky, E. Shelting, V. Titov, H. Yeh, L. Zhukova and N. Zolotukhina. This international survey was arranged by A. Ivachenko. HY thanks the members of the Institute of Marine Geology and Geophysics, Sakhalin, and the Russian Army, Shikotan for their hospitality and friendship. The travel support for VT was made possible by a National Science Foundation grant funded to C. Synolakis. This survey was supported by the Russian Fund of Basic Research, the Presidium of the Far Eastern Division of the Russian Academy of Sciences and the US National Science Foundation.

REFERENCES

ATWATER, B. F. (1987), *Evidence for Great Holocene Earthquakes along the Outer Coast of Washington State*, Science *236*, 942–944.

BOURGEOIS, J., HANSEN, T. H., WIBERG, P., and KAUFFMAN, E. G. (1988), *A Tsunami Deposit at the Cretaceous-Tertiary Boundary in Texas*, Science *241*, 567–570.

CHOI, B. *et al.* (1994), *Tsunami Survey of East Coast of Korea due to the 1993 Southwest of the Hokkaido Earthquake*, J. Korean Soc. Coastal and Ocean Eng., *6*, 117–125.

DAWSON, A. G., LONG, D., and SMITH, D. E. (1988), *The Storegga Slides: Evidence from Eastern Scotland for a Possible Tsunami*, Marine Geology *82*, 271–276.

FUKAO, Y. (1979), *Tsunami Earthquakes and Subduction Process near Deep-sea Trenches*, J. Geoph. Res. *84*, 2303–2314.

GJUMAGALIEV, V. A., and RABINOVICH, A. B. (1993), *Long-wave Investigation at the Shelf and in the Bays of South Kuril Islands*, J. Korean Soc. Coastal and Ocean Eng. *5*, 318–328.

GOLUBSTOBA, T. C., and MAZOVA, R. Kh. (1989), *The runup of waves of sign-variable form*. In *Oscillation and Waves in Fluid Mechanics* (ed. Petrukhin, N. S.) (Gorky Polytechnical Institute Press, Gorky 1989) pp. 30–43 (Russian).

GUSEVA, T., GALAGANOV, O., YAKOVLEV, F., MISHIN, A., KOZHURIN, A., PHILIPOV, M., YANKUSH, A., and VASILENKO, N. (1994), *Results of the Geological Survey and Satellite Geodesic Measurements at Shikotan Island*, 18–30 October of 1994, Special Report 23–31 (Russian).

HATORI, T. (1990), *Behaviour of the Kuril Tsunamis at the Hokkado and Sakhalin Coast Facing the Okhotsk Sea*, Zisin *43*, 493–498.

HOKKAIDO TSUNAMI SURVEY GROUP (1993), *Tsunami Devastates Japanese Coastal Region*, EOS *74*, 417 and 432.

IIDA, K. (1984), *Catalog of Tsunamis in Japan and its Neighboring Countries*, Special Report, Aichi Institute of Technology, Japan, 52 pp.

JACOB, K. H. (1984), *Estimates of Long-term Probabilities for Future Great Earthquakes in the Aleutians*, Geophys. Res. Lett. *11*, 295–298.

KANAMORI, H. (1972), *Mechanism of Tsunami Earthquakes*, Phys. Earth Planet. Inter. *6*, 346–359.

KANAMORI, H., and KIKUCHI, M. (1993), *The 1992 Nicaragua Earthquake: A Slow Tsunami Earthquake Associated with Subducted Sediments*, Nature *361*, 714–716.

MAZOVA, R. Kh. (1991), *The runup description for monochromatic wave propagating from the deep water*. Presented at the International Workshop on Long Wave Runup, Catalina Island, Calif. (also In *Report on the International Workshop on Long-wave Runup*, by Liu, P., Synolakis, C., and Yeh, H., J. Fluid Mech. *229*, 675–688, 1991).

MINOURA, K., and NAKAYA, S. (1991), *Traces of Tsunami Preserved in Inter-tidal Lacustrine and Marsh Deposits: Some Examples from Northeast Japan*, J. Geology *99*, 265–287.

NAGAI, T., HASHIMOTO, N., SHIMIZU, K., and TAKAYAMA, T. (1995), *Tsunami profiles observed at the NOWPHAS offshore wave stations*, Second International Workshop on Wind and Earthquake Engineering for Offshore and Coastal Facilities, University of California, Berkeley, Calif.

PLAFKER, G., and WARD, S. N. (1992), *Backarc Thrust Faculting and Tectonic Uplift along the Caribbean Sea Coast During the April 22, 1991, Costa Rica Earthquake*, Tectonics *11*, 709.

PLAFKER, G. (1995), personal communication.

SATAKE, K. (1994), Communication through Tsunami Bulletin Board dated Oct 4/94.

SATAKE, K., BOURGEOIS, J., ABE, K., TSUJI, Y., IMAMURA, F., IIO, Y., KATAO, H., NOGUERA, E., and ESTRADA, F. (1993), *Tsunami Field Survey of the 1992 Nicaragua Earthquake*, EOS *74*, 145, 156–157.

SNEGOVSKOY, S. S., KRANSY, M. L., and SERGEYEV, K. F. *Thickness of sediment deposits*. In *Geology-Geophysics Atlas of the Kuril-Kamchatka Island System* (eds. Segeyev, K. F. and Kransy, M. L.) (Academy of Sciences of the USSR 1987).

SOLOVIEV, S. L., *Basic data on tsunamis on the Pacific coast of the USSR, 1737–1976*. In *Izuchenie tsunami v otkrytom okeane (Study of Tsunami in the Open Ocean)* (Moscow, Nauka Publishing House 1978) pp. 61–136 (in Russian).

SOLOVIEV, S. L., GO, Ch.N., and KIM, X. S. *Catalog of tsunamis in the Pacific Ocean, 1969–1982* (Moscow, Soviet Geophysical Committee 1986) 164 pp. (in Russian).

SOLOVIEV, S. L., *Earthquakes and tsunami recurrence in the Pacific ocean*. In *Volny Tsunami (Tsunami Waves)* (Proceeding of Sakh. Complex Res. Inst. 1972) 29, pp. 7–47 (in Russian).

TADEPALLI, S., and SYNOLAKIS, C. E. (1994), *The Run-up of N-waves on Sloping Beaches*, Proc. R. Soc. Lond. A. *445*, 99–112.

TAKAHASHI, T., and SHUTO, N. (1994), personal communication.

TITOV, V. V., and SYNOLAKIS, C. E. (1993), *A numerical study of wave runup of the September 1, 1992 Nicaraguan tsunami*, Proc. IEGG/IOC International Tsunami Symposium, Wakayama, Japan, 627–635.

YEH, H., IMAMURA, F., SYNOLAKIS, C., TSUJI, Y., LIU, P., and SHI, S. (1993), *The Flores Island Tsunamis*, EOS *74*, 369, 371–373.

Measured runup data are available upon request to HY.

(Received February 13, 1995, revised March 28, 1995, accepted April 8, 1995)

PAGEOPH, Vol. 144, Nos. 3/4 (1995)

Field Survey of the 1994 Mindoro Island, Philippines Tsunami

FUMIHIKO IMAMURA,[1] COSTAS E. SYNOLAKIS,[2] EDISON GICA,[1] VASILY TITOV,[2] EDDIE LISTANCO,[3] and HO JUN LEE[4]

Abstract—This is a report of the field survey of the November 15, 1994 Mindoro Island, Philippines, tsunami generated by an earthquake ($M = 7.0$) with a strike-slip motion. We will report runup heights from 54 locations on Luzon, Mindoro and other smaller islands in the Cape Verde passage between Mindoro and Luzon. Most of the damage was concentrated along the northern coast of Mindoro. Runup height distribution ranged 3–4 m at the most severely damaged areas and 2–4 in neighboring areas. The tsunami-affected area was limited to within 10 km of the epicenter. The largest recorded runup value of 7.3 m was measured on the southwestern coast of Baco Island while a runup of 6.1 m was detected on its northern coastline. The earthquake and tsunami killed 62 people, injured 248 and destroyed 800 houses. As observed in other recent tsunami disasters, most of the casualties were children. Nearly all eyewitnesses interviewed described the first wave as a leading-depression wave. Eyewitnesses reported that the main direction of tsunami propagation was SW in Subaang Bay, SE in Wawa and Calapan, NE on Baco Island and N on Verde Island, suggesting that the tsunami source area was in the southern Pass of Verde Island and that the wave propagated rapidly in all directions. The fault plane extended offshore to the N of Mindoro Island, with its rupture originating S of Verde Island and propagating almost directly south to the inland of Mindoro, thereby accounting for the relatively limited damage area observed on the N of Mindoro.

Key words: Earthquake, tsunami, runup, Mindoro Island, Philippines, lateral strike slip, field survey, N wave.

1. Introduction

An $M = 7.0$ ($M_w = 7.1$) earthquake struck at 3:17 a.m. Philippine time on November 15, 1994. Its epicenter (13.5N, 121.1E) was located in the southern pass of Verde Island passage along the Lubang fault (Fig. 1). According to the historical data shown in Figure 1, the last powerful earthquake to hit Mindoro occurred more

---

[1] School of Civil Engineering, AIT, G.P.O. Box 2754, Bangkok 10501, Thailand.
[2] Department of Civil Engineering, University of Southern California, Los Angeles, California 90089, U.S.A.
[3] Philippine Institute of Volcanology and Seismology, Manila, Philippines.
[4] Sung Kung Kwang University, Korea, and Disaster Control Research Center, Tohoku University, Japan.

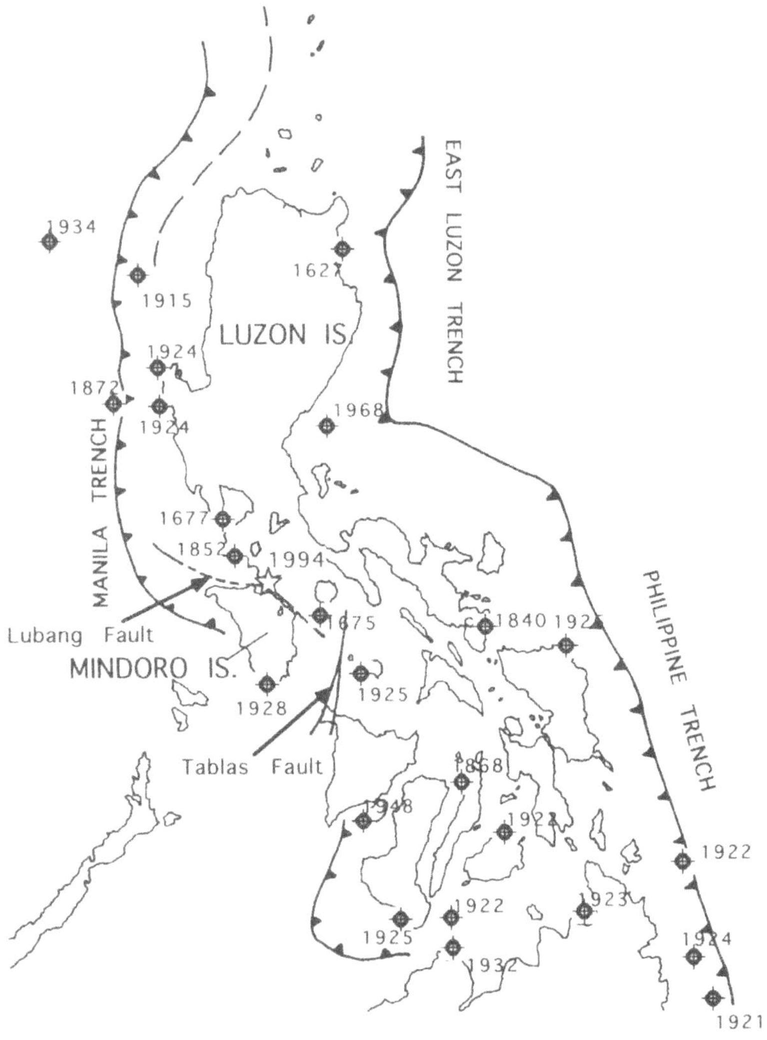

Figure 1
Tectonics and historical tsunami sources in the northern Philippines (after Uy and PUNSALAN, 1987),
and the epicenter of the 1994 Mindoro Island earthquake.

than a century ago on September 16, 1852. As of November 17, 1994 the
earthquake and tsunami killed 62 people with 9 still missing (most of the casualties
were children), injured over 248, and destroyed over 800 houses on Mindoro Island
in the Philippines. For reference, this was the most devastating earthquake to hit
the Philippines since 1990, when an $M = 7.7$ earthquake killed more than 1,600
people on Luzon Island. (The 1990 Luzon earthquake is not included in Figure 1
because no tsunami was generated.)

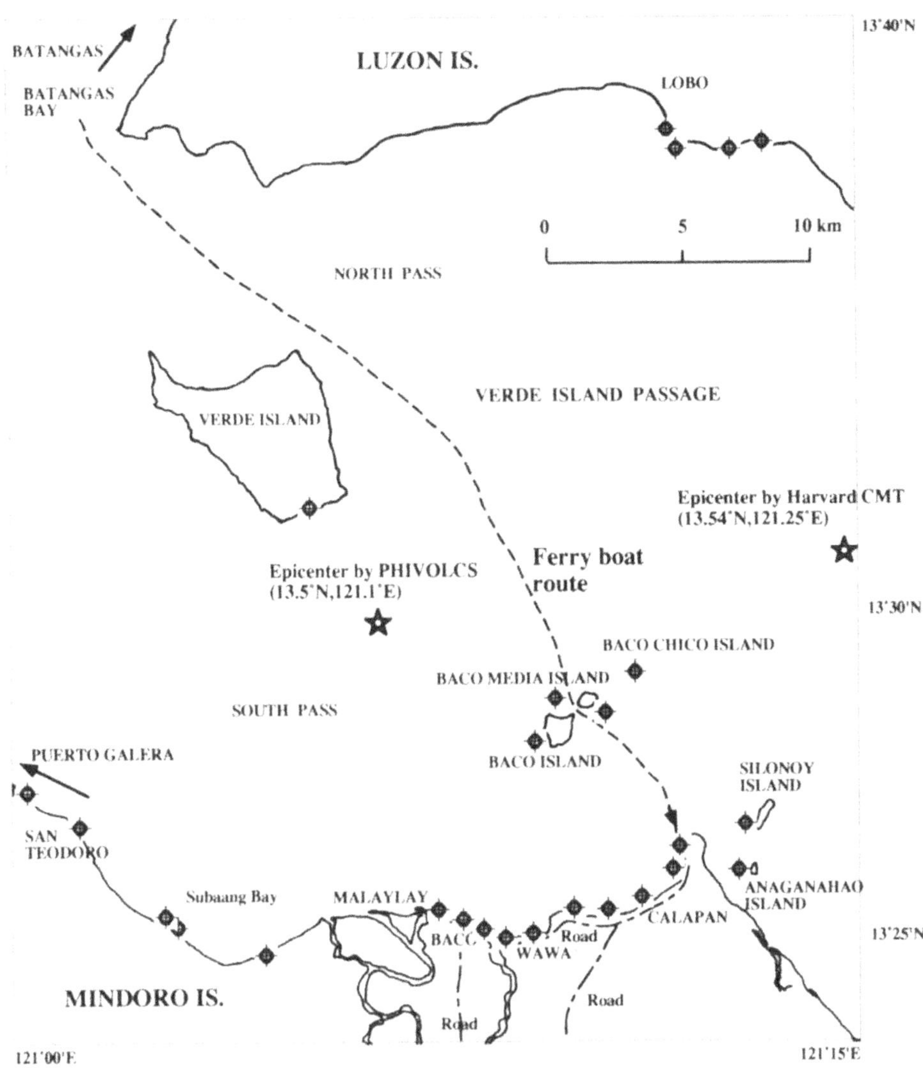

Figure 2

Epicenters estimated by PHIVOLCS and Harvard University and survey points. Dashed line shows the route of the ferry boat between Batangas and Calapan.

This field survey was conducted between November 26 to 29, 1994 on the northeastern coast of Mindoro Island, from Calapan Bay to Subaang Bay, on Verde, Baco, Silonay and Anaganahao Islands, and on the southeastern coast of Luzon near Lobo (Fig. 2). Our purpose was to gather tsunami and earthquake data on tsunami runup, crustal movements and tsunami deposition/erosion, in order to understand how the lateral strike-slip focal mechanism, as given by the Harvard

University CMT solution and posted on Internet, could have generated a 10–14 m wave as reported in Philippine newspapers. For example, assuming that the deformation area was 45 km × 10 km, the seismic moment was 5.4×10^{26} dyne-cm, rigidity of 3.0×10^{11} dyne/cm^2 and depth of 3 km, the maximum vertical deformation calculated by the MANSINHA and SMYLIE method (1971) is only 0.75 m, too small to generate a tsunami wave of more than 5 m.

Team members from Korea, Thailand, and the United States met on the morning of November 26, 1994 in Manila at PHIVOLCS (Philippine Institute of Volcanology and Seismology) headquarters for a briefing of the team's purpose and for a debriefing of PHIVOLCS data.

With the addition of PHIVOLCS senior scientist, E. Listanco, the survey started at mid-morning on November 27, with all team-members surveying Batangas Port, and later departing for Calapan Mindoro Island via ferry. On November 28, the survey team was divided into two groups. One group (F. Imamura, E. Gica, S. Kawashima, D. Esplanada) surveyed and conducted interviews on the small islands (Baco, Verde, Silonay and Anaganahao Islands) and Wawa, Charico, Balete and Pachuka on the northeastern coast of Mindoro (see Fig. 2). The other group (C. E. Synolakis, E. Listanco, V. Titov, H. J. Lee) surveyed by boat the towns of Baco, Malaylay, Wawa and San Teodoro along the northwestern coast of Mindoro, sailed the Baruyan River to Baruyan Lake, and then proceeded to Lobo and Lagadlarin on the southern coast of Luzon. The survey was completed by sunset November 29, making the data recovery of 10.8 measurements/group-day.

For the record, this is the sixth tsunami-expedition by an International Tsunami Survey Team (ITST) since the 1992 Nicaragua event (SATAKE *et al.*, 1993), Flores, 1992 Indonesia (Yeh *et al.*, 1993), Hokkaido SW, 1993 Japan (HOKKAIDO SURVEY GROUP, 1994) East Java, 1994 Indonesia (SYNOLAKIS *et al.*, 1995; TSUJI *et al.*, 1995) and 1994 Shikotan, Kuril Islands (YEH *et al.*, 1995).

2. Observations

1. Tsunami: Runup Heights

We recorded a total of 54 maximum runup measurements. The runup heights were corrected to account for tidal fluctuations, recorded every 15 minutes manually at the Batangas harbor pier (see Fig. 3). The runup height data are listed in Table 1 and the distribution is shown in Figure 4. The local coastal topography is quite complex with river inlets, shallow shoals and steep cliff sides and it would require high-resolution hydrodynamic calculations to reproduce these measurements from offshore wave heights. In Figure 5 the maximum runup height of 7.3 m was observed at Sitio Pino, on the southwestern coast of Baco Island (the lee side, facing Mindoro), while 6.1 m was measured on the northern coast of this island

TIDAL RECORD (BATANGAS PORT PROJECT)

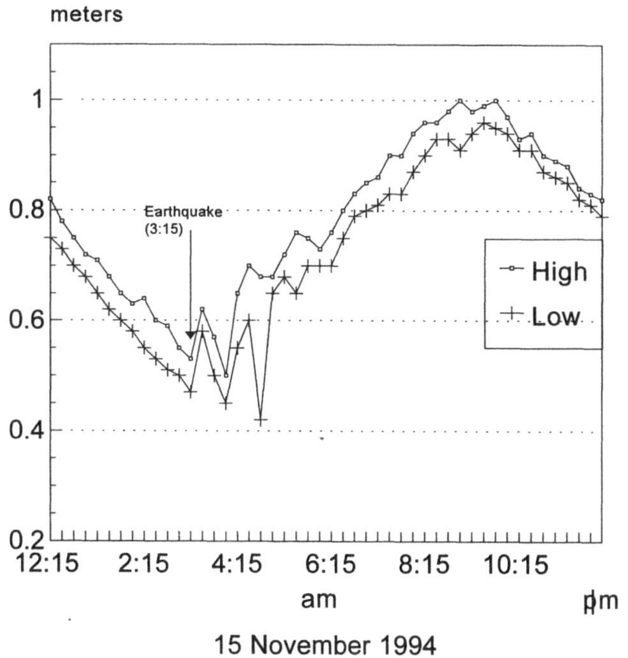

Figure 3
Tidal record at Batangas harbor.

(facing Luzon). Surprisingly, 1.6 m high runup was reported at Baco Media Island located very close to Baco Island. It is suggested that the high runup observed at Baco Island is due to the island trapping effect (YEH *et al.*, 1994; BRIGGS *et al.*, 1995) while the smaller runup recorded on the smaller Baco Media is due to the crest length/island diameter effect identified by BRIGGS *et al.* (1995).

Along the coast of northern Mindoro and within 10 km of the epicenter, the average tsunami heights were 3–4 m in the severely damaged areas and 2–3 m in the neighboring areas (see Figs. 6 and 7), which were smaller than those reported in the Philippine newspaper. The extensive damage associated with relatively small 3–4 m runup measurements is due to overland flow on these fairly shallow beaches sandwiched between river inlets and tidal flats. Also, south of Luzon Island near the resort town of Lobo, 2–4 m runup heights were measured; the damage here was considerably smaller than the damage at Mindoro locations with the same runup heights, because of steeper coastlines and no tidal inlets at Lobo, which could amplify a tsunami runup but reduce a tsunami current and force. Moreover, many houses in Lobo were placed in a more elevated area than those in Mindoro.

Table 1

Measured Tsunami: runup heights

No.	Place	Date Nov. 1994	Time (hh:mm)	Location (GPS)	Measured Ht (m)	Tidal Comp (m)	Corrected Ht (m)
1	Batangas Port	27	09:00	N13°45′14.60″ E121°02′39.36′	1.14	0.11	1.25
2	Calapan Port	27	13:00	N13°25′52.00″ E121°11′40.00″	4.02	0.06	3.96
3	Baco	27	16:20	N13°24′48.00″ E121°07′18.54″	1.03	0.43	1.46
4	Baco	27	16:39	N13°24′45.10″ E121°07′28.90″	3.29	0.45	3.74
5	Baco	27	16:39	N13°24′45.10″ E121°07′28.90″	3.12	0.45	3.57
6	Baco Is	28	09:18	N13°27′51.70″ E121°09′25.70″	$6.31 + 1 \sim 2$	0.27	$6.58 \pm 1 \sim 2$
7	Baco Is	28	09:18	N13°27′51.70″ E121°09′25.70″	6.63	0.27	6.90
8	Baco Is	28	09:18	N13°27′51.70″ E121°09′25.70″	7.15	0.27	7.29
9	Ibaba	28	10:00	N13°24′52.00″ E121°07′32.00″	1.80	0.17	1.97
10	Verde Is	28	10:48	N13°13′49.60″ E121°05′29.50″	2.74	0.07	2.81
11	Pampisan	28	11:00	N13°24′43.00″ E121°08′08.00″	2.0	0.05	2.05
12	Pampisan	28	11:20	N13°24′35.82″ E121°08′14.76″	2.15	0.02	2.17
13	Pampisan	28	11:40	N13°24′36.36″ E121°07′58.14″	3.00	−0.01	2.99
14	Verde Is	28	11:42	N13°33′00.03″ E121°05′40.30″	3.56	−0.01	3.55
15	Pampisan	28	12:15	N13°24′46.00″ E121°07′45.00″	4.10	−0.04	4.06
16	Pampisan	28	12:30	N13°24′44.16″ E121°07′42.84″	2.75	−0.04	2.71
17	Pampisan	28	12:35	30 m west from 15	4.30	−0.04	3.90
18	Pampisan	28	12:45	N13°24′47.70″ E121°07′37.08″	4.30	−0.04	3.90
19	Baco Media Is	28	13:38	N13°28′28.50″ E121°09′51.60″	1.63	0.01	1.64
20	Baco Is	28	14:00	N13°28′09.90″ E121°09′33.60″	6.10	0.04	6.14
21	Baco Is	28	14:00	N13°28′09.90″ E121°09′33.60″	3.50	0.04	3.54
22	Baco Chico Is	28	14:42	N13°29′03.90″ E121°10′50.70″	0.00	0.12	0.12
23	Malaylay	28	15:00	N13°25′07.56″ E121°05′39.66″	1.05	0.17	1.22
24	Malaylay	28	15:30	N13°24′27.00″ E121°04′16.00″	1.40	0.25	1.65

Table 1 (*Cont.*)

No.	Place	Date Nov. 1994	Time (hh:mm)	Location (GPS)	Measured Ht (m)	Tidal Comp (m)	Corrected Ht (m)
25	Wawa Is	28	15:50	N13°24'48.00" E121°02'48.90"	3.20	0.31	3.51
26	Wawa Is	28	15.55	N13°24'50.16" E121°02'52.44"	3.50	0.32	3.82
27	Calapan	28	16:23	N13°25'42.50" E121°11'40.90"	2.33	0.41	2.74
28	San Teodoro	28	16:35	N13°26'26.64 E121°01'04.32"	3.43	0.44	3.87
29	Calapan	28	16:56	N13°25'02.50" E121°11'01.50	2.64	0.50	3.14
30	Calapan	28	17:18	N13°24'53.60" E121°10'40.70"	1.86	0.55	2.41
31	Villaflor	28	17:20	N13°26'57.84" E121°00'17.16"	2.35	0.55	2.90
32	Wawa	28	17:20	N13°24'16.02" E121°08'26.30"	0.71	0.48	1.19
33	Wawa	29	08:29	N13°24'16.02" E121°08'26.30"	1.44	0.48	1.92
34	Wawa	29	08:56	N13°24'26.20" E121°08'26.70"	3.98	0.46	4.44
35	Wawa	29	08:56	N13°24'26.20" E121°08'26.70"	2.02	0.46	2.48
36	Wawa	29	09:19	N13°24'28.10" E121°08'34.80"	1.65	0.42	2.07
37	Wawa	29	09:19	N13°24'28.10" E121°08'34.80"	2.17	0.42	2.59
38	Wawa	29	09:28	N13°24'29.30" E121°08'33.70"	1.55	0.40	1.95
39	Wawa	29	09:40	N13°24'31.80" E121°08'39.40"	2.39	0.39	2.78
40	Wawa	29	09:50	N13°23'50.90" E121°08'27.30"	2.86	0.37	3.23
41	Baruyan	29	09:59	N13°24'22.62" E121°08'16.08"	2.40	0.37	2.77
42	Charico	29	10:39	N13°24'36.40" E121°09'07.60"	1.70	0.27	1.97
43	Charico	29	10:39	N13°24'36.40" E121°09'07.60"	2.04	0.27	2.31
44	Charico	29	10:39	N13°24'36.40" E121°09'07.60"	1.05	0.27	1.32
45	Baruyan River	29	10:48	N13°22'55.98" E121°08'29.58"	0.50	0.25	0.75
46	Charico	29	10:51	N13°24'36.90" E121°09'20.90"	2.11	0.25	2.36
47	Balete	29	11:08	N13°24'59.50" E121°09'36.20"	2.18	0.21	2.39
48	Balete	29	11:08	N13°24'59.50" E121°09'36.20"	1.90	0.21	2.11

Table 1 (*Cont.*)

No.	Place	Date Nov. 1994	Time (hh:mm)	Location (GPS)	Measured Ht (m)	Tidal Comp (m)	Corrected Ht (m)
49	Pachuka	29	11:27	N13°24′59.50″ E121°09′36.20″	2.23	0.21	2.40
50	Lobo	29	14:45	N13°38′10.20″ E121°11′40.56″	2.15	0.10	2.25
51	Lobo	29	15:00	N13°38′05.10″ E121°11′48.78″	2.55	0.13	2.68
52	Lagadlarin	29	15:35	N13°37′41.28″ E12°12′54.90″	3.15	0.22	3.37
53	Lagadlarin	29	15:50	N13°37′42.30″ E121°12′48.60″	3.37	0.26	3.63
54	Barangay Sawang	29	16:20	N13°37′39.36″ E121°13′32.70″	3.85	0.35	4.20

2. Direction of Tsunami Propagation and Other Wave-related Observations

We inferred from eyewitness accounts that the main directions of tsunami attacks were to the SW in Subaang Bay, to the SE in Wawa and Calapan, to the NE on Baco Island and to the N on Verde Island (Fig. 4), indicating that the relatively small tsunami source-area was located in the southern Pass of Verde Island Passage, and that the tsunami propagated radially from the source. The first waves observed at most locations were leading-depression waves. For example, at Sitio Pinto on Baco Island, where the largest runup measurement was recorded, eyewitnesses reported that the water first receded approximately 45 m offshore with a wave period of about 4 min. On the northern part of Mindoro Island the first depression wave was followed by a second elevation wave.

Fishermen working off Wawa, Mindoro at the time of the earthquake described the initial behavior of the tsunami. In the shallow region, after the groundshaking, there was vertical movement of the water, accompanied by white bubbles on the sea surface, indicating bore-like propagation. The oars of a small boat in this area hit sea bottom in attempting to row towards shore, implicating the leading depression wave nature of the tsunami. They reported that the tsunami attacked twice and that the first wave was the strongest. The waves were described as appearing like a "high vertical wall mixed with sand." A fisherman in deep water close to the epicenter reported a sudden downward movement of water level and white bubbles on the sea surface. Regarding leading depression waves (*LD* waves), TADEPALLI and SYNOLAKIS' (1994) calculations show that the runup of the *LD* wave is larger than the runup of the following elevation waves. This is one of the reasons that the destructive tsunami attacked the coast of Mindoro Island.

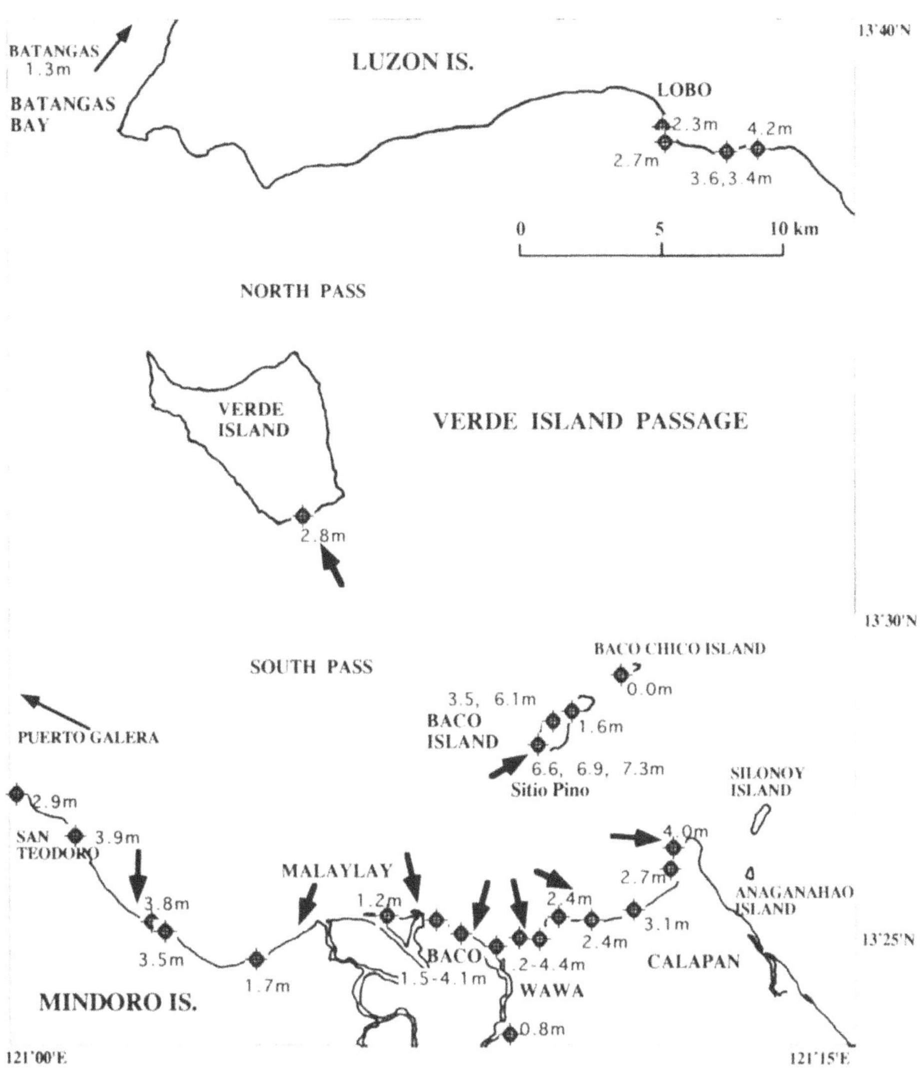

Figure 4

Measured runup heights and estimated direction of the first wave of the tsunami in southern Luzon, northern Mindoro, and other small islands. The runup heights are values at the time of the tsunami generation.

In Calapan, Mindoro (Fig. 6) eyewitnesses reported that the water first receded approximately more than 100 m, while they heard sounds of "drums." They described a "white-line" at sea, traveling from the western side of Calapan Point and striking the pier shown in Figure 6, indicating again bore-like LD waves.

At Sitio Pino, Baco Island, eyewitnesses reported that the water first receded for approximately 45 m from shore during the lateral groundshaking of around 10

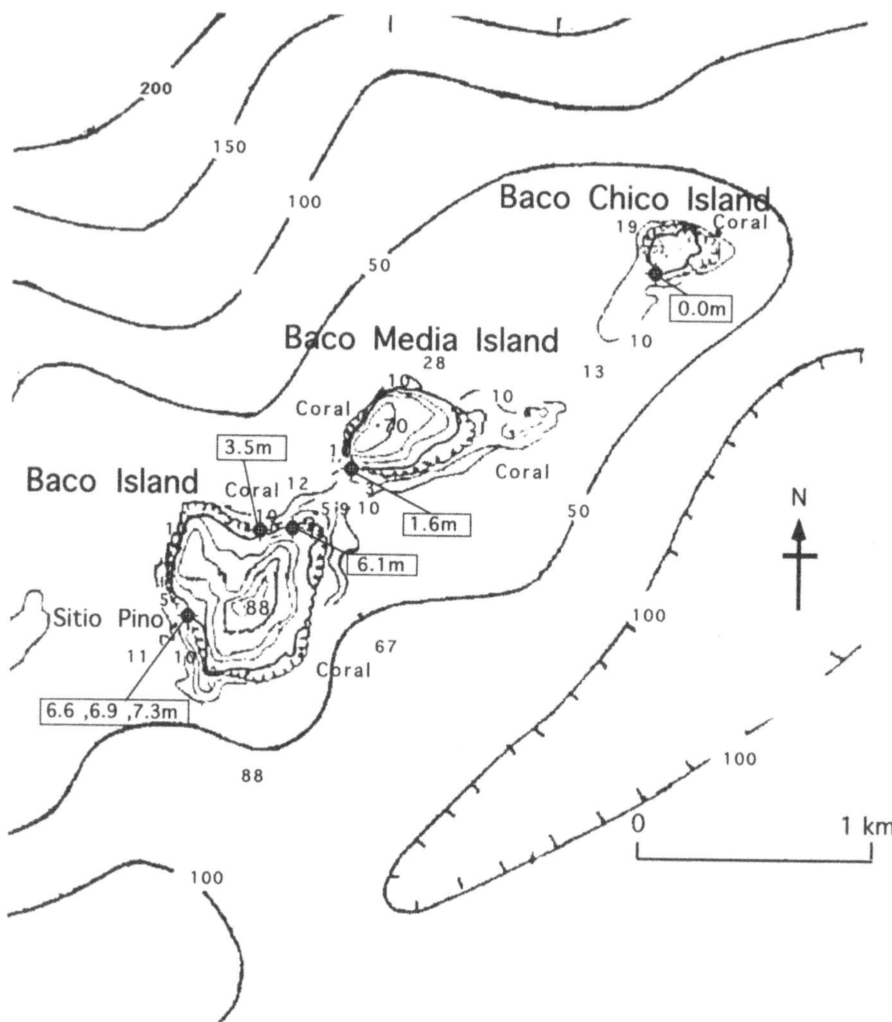

Figure 5
Measured runup heights in three Baco Islands and bathymetry in meters.

minutes, a time which in our opinion is exaggerated. The eyewitnesses heard sounds resembling a "landing jet-aircraft." Three waves were reported, with the second the highest, which is different from that in Wawa. On the north coast of Baco Island (Fig. 5) the tsunami attacked the lee side of the islands, no deaths occurred and only one house was damaged. Here, residents' accounts of the tsunami were similar to those reported at Sitio Pino, except that the first two waves were the strongest, followed by a smaller third wave.

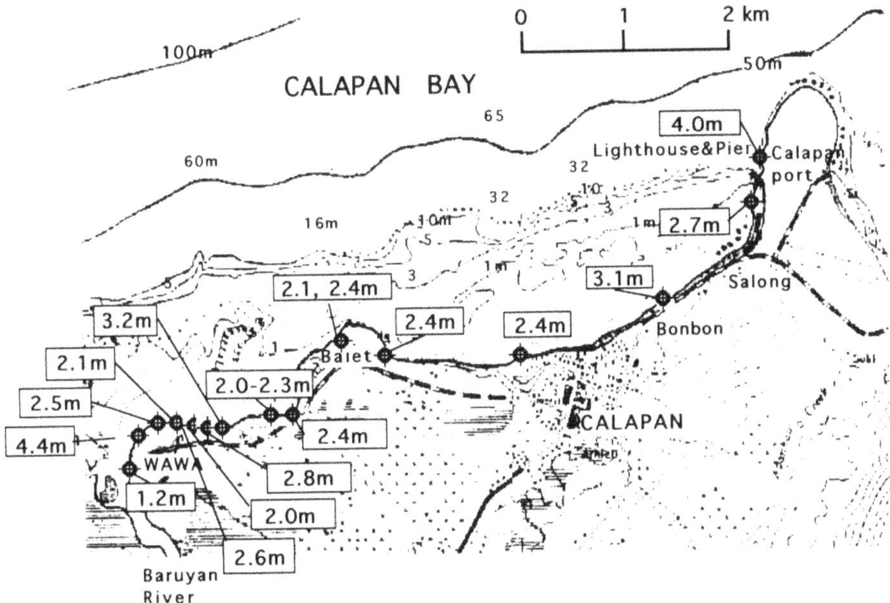

Figure 6
Measured runup heights in Wawa, Balete and Calapan, and bathymetry in meters.

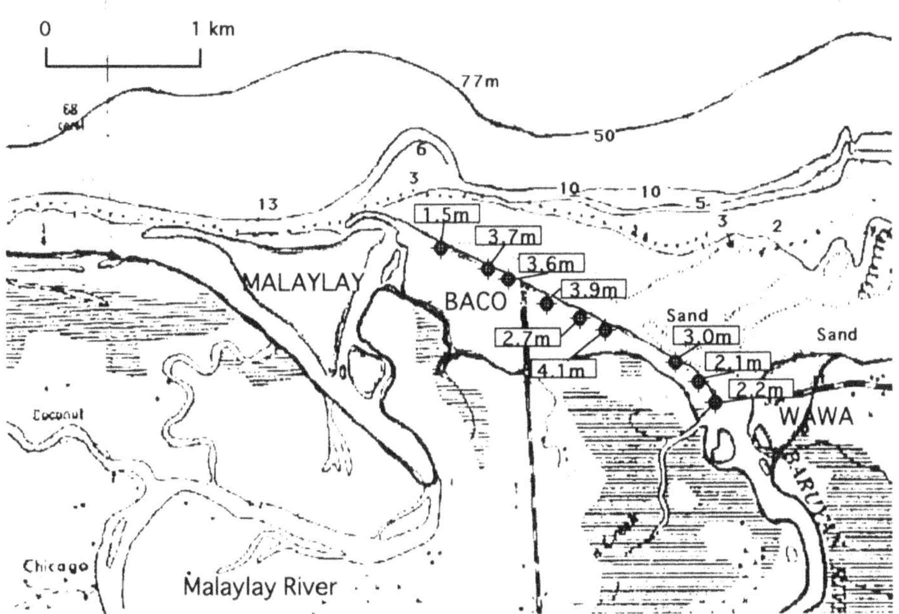

Figure 7
Measured runup heights in Malaylay, Baco and Wawa, and bathymetry in meters.

No casualties or damage was reported on Verde Island. The eyewitnesses recalled feeling the shaking almost simultaneously with the tsunami arrival accompanied by a "gong" style sound.

An unusual observation was made near the inlet of the Baruyan River near Wawa. Here, a 4000-ton, 7.2 MW power-generating barge moored at the inlet was swept 1.6 km upstream towards the interior of Mindoro Island. We speculate that this unusual motion resulted from a strong current set up in the river due to flow to ground cracks upstream rather than due to a tsunami wave caused by sea bottom deformation in the sea region. This is because the operation record clearly indicates that the barge was already swept and stopped the electricity supply before the first tsunami wave attacked the coast of Mindoro Island.

3. Ground Motions, Deposition and Landslides

At most areas of Mindoro and the smaller channel islands, the lateral ground motion was reported to be felt for a long duration, the longest being 10 minutes at Baco Island as mentioned in the previous section. On Mindoro Island, the ground

Photo 1
Closed river mouth due to the deposition of sea sand at Baco.

surface rupture was approximately 40 km, running north to south. The fault slip extended from the offshore epicenter onto the interior of Mindoro. 3.0 m and 2.0 m horizontal and vertical maximum displacements, respectively were measured by other PHIVOLCS team (personal communication, 1994). In spite of clear strike-slip motion, rather large vertical displacements were measured on the ground surface at Mindoro, indicating that similar displacements occurred at sea. Although the reason why such large vertical motions were caused by a strike-slip motion is still unclear, this information is needed to further elucidate the tsunami generation mechanism.

The river mouth in Baco village was completely obstructed by tsunami sand deposition. Photo 1 shows the white marine sediment presumably transferred and

Photo 2
Erosion of coast in Sitio Pino, Baco Islands.

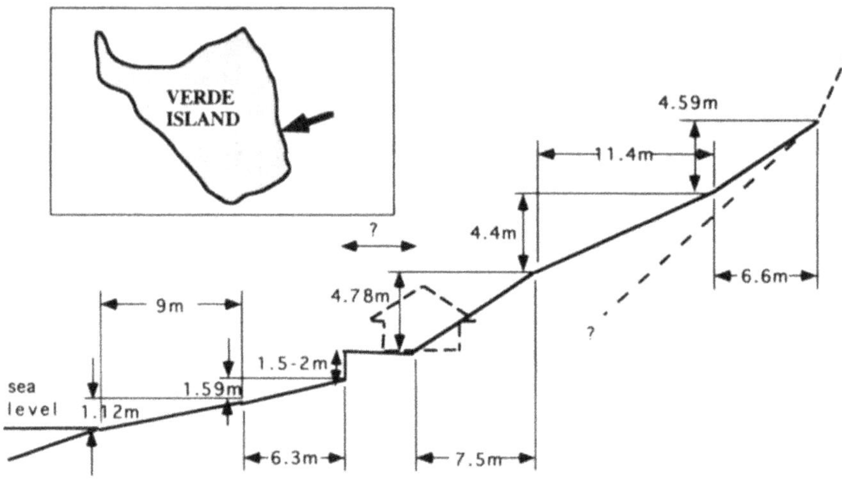

Figure 8
Sectional profile of landslide on Verde Island.

deposited on the black sand by the tsunami. Even though an argument could be made that the deposition could have been the result of wind waves, the direction of the bent-over vegetation associated with strong tsunami-induced currents suggests that this rather massive deposition was related to the tsunami.

By contrast, we observed 0.5–1.5 m erosion along the shoreline at Sitio Pino in Baco Island (Photo 2). We found arched trees towards the initial coastline, suggesting a rather large rundown induced current. This was consistent with our observations of uprooted coconut trees on the downstream side of other trees on a steep beach, suggesting large return currents.

Our sedimentation observations suggest that on steeper beaches the rundown-current moving sediment offshore is stronger than the runup current depositing sediment, while on gentler beaches the rundown current is weaker and there is deposition. This would be also related to the results of SYNOLAKIS (1987) and TADEPALLI and SYNOLAKIS (1994) that nonbreaking long waves on a gentle beach have higher minimum rundown velocities than breaking waves of the same shape on a steep beach, because the latter dissipate most of their energy during breaking near a shoreline or in a shallow region.

We observed more than 5 landslides along the steep cliffs on eastern Verde Island, responsible for at least two deaths. The sectional profile of one of the landslides as measured is shown in Figure 8. A dashed line shows the estimated ground slope before the landslide. The house was partially destroyed by landslide material, however, it did not reach the coast. These landslides did not displace any water masses, therefore generated no tsunamis.

4. Summary and Conclusions

We reported the field survey results of the November 15, 1995 earthquake and tsunami. The runup height distribution ranged from 0.8 to 7.3 m. We discovered that the tsunami was a leading depresssion wave in most locations surveyed. Both extensive sand deposition and erosion, which we have related to the coastal bathymetry and topography, were found. We recorded observations of rather large currents in the Baruyan river. Our findings indicate that vertical ground deformation itself cannot explain the relatively large tsunami inundation observed, suggesting that the horizontal motion of Mindoro Island during this strike-slip event must be included for hydrodynamic calculations to further evaluate exactly how this unusual tsunami was generated.

Acknowledgements

Thanks are expressed to Professor B. H. Choi of Sung Kung Kwan University, Korea, for providing tidal data and for his valuable comments, Director Dr. R. Punongbayan of PHIVOLCS for exchanging data on earthquakes and tsunamis and for valuable discussions, Mr. Seiichi Kawashima and Mr. Delfin Esplanada, Toyo Construction Co., Philippines for their assistance in the field survey, Mr. Yukihiko Ejiri, Japan International Cooperative Agency in Philippines, for his kind support for transportation, and Mr. Koji Nishigaki, NHK (Japanese Broadcasting Corp.), Manila, for allowing us to use their helicopter for the ground motion identification. Finally we would like to thank Dr. Cliff Astill of the Natural Hazards Mitigation program of the National Science Foundation of the United States for providing the funds to Professor Synolakis for conducting this survey.

REFERENCES

BRIGGS, M. J., SYNOLAKIS, C. E., HARKINS, G. S., and GREEN, D. R. (1995), *Laboratory Experiments of Tsunami Runup on a Circular Island*, Pure and Appl. Geophys., this volume.

HOKKAIDO SURVEY GROUP (1993), *Tsunami Devastates Japanese Coastal Region*, EOS Trans., Am. Geophys. Union *74*, 417 and 432.

MANSHINHA, L., and SMYLIE, D. E. (1971), *The Displacement Fields of Inclined Faults*, Bull. Seismol. Soc. Am. *61*, 1433–1440.

SATAKE, K., BOURGEOIS, J., ABE, KU, ABE, KA., TSUJI, Y., IMAMURA, F., IIO, Y., KATAO, H., NOGUERA, E., and ESTRADA, F. (1993), *Tsunami Field Survey of the 1992 Nicaragua Earthquake*, EOS, Trans. Am. Geophys. Union *74*, 156–157.

SYNOLAKIS, C. E., IMAMURA, F., TSUJI, T., MATSUTOMI, H., TINTI, S., COOK, B., CHANDRA, Y. P., and USHMAN, M. (1995), EOS, Trans. Am. Geophys. Union, in press.

SYNOLAKIS, C. E. (1987), *The Run-up of Solitary Waves*, J. Fluid Mech. *187*, 523–545.

TADEPALLI, S., and SYNOKALIS, C. E. (1994), *The Run-up of N-waves*, Proc. Roy. Soc. London, Series *A445*, 99–112.

TINTI, S., and VANNINI, C. (1995), *Tsunami Trapping near Circular Island*, Pure and Appl. Geophys., this volume.

TSUJI, Y., IMAMURA, F., MATSUTOMI, H., SYNOLAKIS, C. E., NANANG, P. T., JUMADI, HARADA, S., HAN, S. S., ARAI, K., and BENJAMIN, C. (1995), *Field Survey of the East Java Earthquake and Tsunami of June 3, 1994*, Pure and Appl. Geophys., this volume.

UY, E. A., and PUNSALAN, B. T. (1987), *Earthquake and Tsunami Prone Areas in the Philippines*, Geologic Hazards and Disaster Preparedness System, PHIVOLCS Report, 38–55.

YEH, H., IMAMURA, F., SYNOLAKIS, C. E., TSUJI, Y., LIU, P., SHI, S. (1993), *The Flores Island Tsunamis*, EOS, Trans. Am. Geophys. Union 74, 371–373.

YEH, H., LIU, P. L-F., BRIGGS, M., and SYNOLAKIS, C. E. (1994), *Propagation and Amplification of Tsunamis at Coastal Boundaries*, Nature 372, 353–355.

YEH, H., TITOV, V., GUSIAKOV, V., PELINOVSKY, E., KHRAMUSHIN, V., and KAISTRENKO, V. (1995), *The 1994 Shikotan Earthquake Tsunamis*, Pure and Appl. Geophys., this volume.

(Received January 31, 1995, revised May 8, 1995, accepted May 13, 1995)

PAGEOPH

Pageoph Topical Volumes

Aspects of Pacific Seismicity
Edited by
E.A. Okal
1991. 200 pages. Softcover
ISBN 3-7643-2589-5

Deep Earth Electrical Conductivity
Edited by
W.H. Campbell
1990. 96 pages. Softcover
ISBN 3-7643-2564-X

Earthquake Hydrology and Chemistry
Edited by
C.-Y. King
1985. 478 pages. Softcover
ISBN 3-7643-1743-4

Electrical Properties of the Earth's Mantle
Edited by
W.H. Campbell
1987. 312 pages. Softcover
ISBN 3-7643-1901-1

Experimental Techniques in Mineral and Rock Physics
The Schreiber Volume
Edited by
R.C. Liebermann and C.H. Sondergeld
1994. 457 pages. Softcover
ISBN 3-7643-5028-

Faulting, Friction, and Earthquake Mechanics
Edited by
C.J. Marone and M.L. Blanpied,

Part I:
1994. 399 pages. Softcover
ISBN 3-7643-5073-3
Part II:
1994. 516 pages. Softcover
ISBN 3-7643-5099-7

Fractals and Chaos in the Earth Sciences
Edited by
C.G. Sammis, M. Saito and G.C.P. King
1993. 188 pages. Softcover
ISBN 3-7643-2878-9

Fractals in Geophysics
Edited by
C.H. Scholz and B.B. Mandelbrot
1989. 313 pages. Softcover
ISBN 3-7643-2206-3

Induced Seismicity
Edited by
A. McGarr
1993. 460 pages. Softcover
ISBN 3-7643-2918-1

Intermediate-Term Earthquake Prediction
Edited by
W.D. Stuart and K. Aki
1988. 550 pages. Softcover
ISBN 3-7643-1978-X

Ionospheric Modeling
Edited by
J.N. Korenkov
1988. 376 pages. Softcover
ISBN 3-7643-1926-7

**Localization of Deformation
in Rocks and Metals**
Edited by
A. Ord, B.E. Hobbs and H.-B. Mühlhaus
1991. 158 pages. Softcover
ISBN 3-7643-2772-3

Middle Atmosphere
Edited by
A.R. Plumb and R.A. Vincent
1989. 472 pages. Softcover
ISBN 3-7643-2290-X

Oceanic Hydraulics
Edited by
R.L. Hughes
1990. 184 pages. Softcover
ISBN 3-7643-2498-8

**Physics of Fracturing
and Seismic Energy Release**
Edited by
J. Kozak and L. Waniek
1987. 376 pages. Softcover
ISBN 3-7643-1863-5

Quiet Daily Geomagnetic Fields
Edited by
W. H. Campbell
1989. 244 pages. Softcover
ISBN 3-7643-2338-8

**Scattering and Attenuation
of Seismic Waves**
Edited by
R.-S. Wu and K. Aki
Part I:
1988. 448 pages. Softcover
ISBN 3-7643-2254-3
Part II:
1989. 198 pages. Softcover
ISBN 3-7643-2341-8

Part III:
1990. 438 pages. Softcover
ISBN 3-7643-2342-6

Seismicity in Mines
Edited by
S.J. Gibowicz
1989. 404 pages. Softcover
ISBN 3-7643-2273-X

**Shallow Subduction Zones: Seismicity,
Mechanics and Seismic Potential**
Edited by
R. Dmowska and Eckström,
Part I:
1993. 220 pages. Softcover
ISBN 3-7643-2962-9
Part II:
1994. 220 pages. Softcover
ISBN 3-7643-2963-7

Subduction Zones
Edited by
L.J. Ruff and H. Kanamori
Part I:
1988. 352 pages. Softcover
ISBN 3-7643-1928-3
Part II:
1989. 282 pages. Softcover
ISBN 3-7643-2272-1

Tsunamis: 1992–1994
Their Generation, Dynamics, and Hazards
Edited by
K. Satake and F. Imamura,
1995. Approx. 520 pages. Softcover
ISBN 3-7643-5102-0

**Please order through your
bookseller or write to:**
Birkhäuser Verlag AG
P.O. Box 133
CH-4010 Basel / Switzerland
FAX. ++41 / 61 / 271 76 66
e-mail: 100010 2310@compuserve.com

**For orders originating
in the USA or Canada:**
Birkhäuser
333 Meadowlands Parkway
Secaucus, NJ 07096-2491 / USA

Birkhäuser

**Birkhäuser Verlag AG
Basel · Boston · Berlin**